Lecture Notes in Networks and Systems 906

The series "Lecture Notes in Networks and Systems" publishes the latest developments in Networks and Systems—quickly, informally and with high quality. Original research reported in proceedings and post-proceedings represents the core of LNNS.

Volumes published in LNNS embrace all aspects and subfields of, as well as new challenges in, Networks and Systems.

The series contains proceedings and edited volumes in systems and networks, spanning the areas of Cyber-Physical Systems, Autonomous Systems, Sensor Networks, Control Systems, Energy Systems, Automotive Systems, Biological Systems, Vehicular Networking and Connected Vehicles, Aerospace Systems, Automation, Manufacturing, Smart Grids, Nonlinear Systems, Power Systems, Robotics, Social Systems, Economic Systems and other. Of particular value to both the contributors and the readership are the short publication timeframe and the world-wide distribution and exposure which enable both a wide and rapid dissemination of research output.

The series covers the theory, applications, and perspectives on the state of the art and future developments relevant to systems and networks, decision making, control, complex processes and related areas, as embedded in the fields of interdisciplinary and applied sciences, engineering, computer science, physics, economics, social, and life sciences, as well as the paradigms and methodologies behind them.

Indexed by SCOPUS, INSPEC, WTI Frankfurt eG, zbMATH, SCImago.

All books published in the series are submitted for consideration in Web of Science.

For proposals from Asia please contact Aninda Bose (aninda.bose@springer.com).

Mohamed Ben Ahmed ·
Anouar Abdelhakim Boudhir · Rani El Meouche ·
İsmail Rakıp Karaş
Editors

Innovations in Smart Cities Applications Volume 7

The Proceedings of the 8th International Conference on Smart City Applications, Volume 1

Springer

Editors
Mohamed Ben Ahmed
Computer Science and Smart systems Laboratory, Faculty of Science and Techniques of Tangier
Abdelmalek Essaadi University
Tangier, Morocco

Anouar Abdelhakim Boudhir
Computer Science and Smart systems Laboratory, Faculty of Science and Techniques of Tangier
Abdelmalek Essaadi University
Tangier, Morocco

Rani El Meouche
École Spéciale des Travaux Publics
Paris, France

İsmail Rakıp Karaş
Computer Engineering Department
Karabük University
Karabük, Türkiye

ISSN 2367-3370 ISSN 2367-3389 (electronic)
Lecture Notes in Networks and Systems
ISBN 978-3-031-53823-0 ISBN 978-3-031-53824-7 (eBook)
https://doi.org/10.1007/978-3-031-53824-7

This Springer imprint is published by the registered company Springer Nature Switzerland AG
The registered company address is: Gewerbestrasse 11, 6330 Cham, Switzerland

Paper in this product is recyclable.

Preface

The content of this Conference Proceedings volume comprises the written version of the contributions presented at the 8th International Conference on Smart City Applications 2023.

This multidisciplinary event was co-organized by the ESTP in the partnership with Mediterranean Association of Sciences and Sustainable Development (Medi-ADD) sponsored by the digital twins' chair of construction and infrastructure at ESTP.

The contents of this volume delve into recent technological breakthroughs across diverse topics including geo-smart information systems, digital twins of construction and infrastructure, smart building and home automation, smart environment and smart agriculture, smart education and intelligent learning systems, information technologies and computer science, smart healthcare, etc.

The event has been a good opportunity for more than 110 participants coming from different countries around the world to present and discuss topics in their respective research areas.

In addition, four keynote speakers presented the latest achievements in their fields: Prof. Jason Underwood "Imagining a digital competency management ecosystem approach to transforming the productivity of people in the built environment", Prof. Isam Shahrour "Smart city: why, what, experience feedback and the future/challenges", Dr. Ihab Hijazi "Integrating system dynamics and digital twin for the circular urban environment", Prof. Mohammed Bouhorma "Challenges of cybersecurity in smart cities", Prof. Filip Biljecki "Advancing urban modelling with emerging geospatial datasets and AI technologies", Prof. Ismail Rakip Karas "Background of Smart Navigation".

We express our gratitude to all participants, members of the organizing and scientific committees, as well as session chairs, for their valuable contributions.

We also would like to acknowledge and thank the Springer Nature Switzerland AG staff for their support, guidance and for the edition of this book.

We hope to express our sincere thanks to Pr. Janusz Kacprzyk and Dr. Thomas Ditzinger for their kind support and help to promote the success of this book.

November 2023

Rani El Meouche
Mohamed Ben Ahmed
Anouar Abdelhakim Boudhir
Ismail Rakip

Committees

Conference Chair

Rani El Meouche	ESTP, Paris, France

Conference Co-chairs

Mohamed Ben Ahmed	FST, Tangier, UAE University, Morocco
Anouar Boudhir Abdelhakim	FST, Tangier, UAE University, Morocco
İsmail Rakıp Karaş	Karabuk University, Turkey

Conference Steering Committee

Rani El Meouche	ESTP, Paris, France
Rogério Dionisio	Polytechnic Institute Castelo Branco, Portugal
Domingos Santos	Polytechnic Institute Castelo Branco, Portugal
İsmail Rakıp Karaş	Karabuk University, Turkey
Alias Abdul Rahman	Universiti Teknologi Malaysia
Mohamed Wahbi	EHTP Casablanca, Morocco
Mohammed Bouhorma	FST, Tangier UAE University, Morocco
Chaker El Amrani	FST, Tangier UAE University, Morocco
Bernard Dousset	UPS, Toulouse, France
Rachid Saadane	EHTP Casablanca, Morocco
Ali Youness	FS, Tetouan, Morocco

Local Organizing Committee

Elham Farazdaghi	ESTP Paris, France
Mojtaba Eslahi	ESTP Paris, France
Muhammad Ali Sammuneh	ESTP Paris, France
Maryem Bouali	ESTP Paris, France
Mohamad Al Omari	ESTP Paris, France
Mohamad Ali	ESTP Paris, France
Zhiyu Zheng	ESTP Paris, France

Technical Programme Committee

Ali Jamali	Universiti Teknologi Malaysia
Ali Jamoos	Al-Quds University, Palestine
Alias Abdul Rahman	Universiti Teknologi Malaysia
Aliihsan Sekertekin	Cukurova University
Ana Paula Silva	Polytechnic Institute of Castelo Branco, Portugal
Ana Ferreira	Polytechnic Institute Castelo Branco, Portugal
Anabtawi Mahasen	Al-Quds University, Palestine
Anton Yudhana	Universitas Ahmad Dahlan, Indonesia
Arlindo Silva	Polytechnic Institute of Castelo Branco, Portugal
Arif Çağdaş Aydinoglu	Gebze Technical University, Türkiye
Arturs Aboltins	Technical University of Riga, Latvia
Assaghir Zainab	Lebanese University, Lebanon
Barış Kazar	Oracle, USA
Bataev Vladimir	Zaz Ventures, Switzerland
Behnam Atazadeh	University of Melbourne, Australia
Benabdelouahab Ikram	UAE, Morocco
Bessai-Mechmach Fatma Zohra	CERIST, Algeria
Beyza Yaman	Dublin City University, Ireland
Biswajeet Pradhan	University of Technology Sydney, Australia
Carlos Cambra	Universidad de Burgos, Spain
Damir Žarko	Zagreb University, Croatia
Darko Stefanovic	University of Novi Sad, Serbia
Domingos Santos	IPCB, Portugal, France
Edward Duncan	The University of Mines & Technology, Ghana
Eehab Hamzi Hijazi	An-Najah University, Palestine
Eftal Şehirli	Karabuk University, Türkiye
El Hebeary Mohamed Rashad	Cairo University, Egypt
EL Arbi Abdellaoui Allaoui	ENS, UMI, Morocco
Enrique Arias	Castilla-La Mancha University, Spain
Filip Biljecki	National University of Singapore
Francesc Anton Castro	Technical University of Denmark
Ghulam Ali Mallah	Shah Abdul Latif University, Pakistan
Gibet Tani Hicham	FP UAE University, Morocco
Habibullah Abbasi	University of Sindh, Pakistan
Ihab Hijazi	An-Najah National University and Technical University of Munich
Isam Shahrour	Lille University France
J. Amudhavel	VIT Bhopal University, Madhya Pradesh, India
Jaime Lloret Mauri	Polytechnic University of Valencia, Spain
José Javier Berrocal Olmeda	Universidad de Extremadura, Spain

Jus Kocijan	Nova Gorica University, Slovenia
Khoudeir Majdi	IUT, Poitiers university, France
Labib Arafeh	Al-Quds University, Palestine
Loncaric Sven	Zagreb University, Croatia
Lotfi Elaachak	FSTT, UAE, Morocco
Mademlis Christos	Aristotle University of Thessaloniki, Greece
Maria Joao Simões	Universidade da Beira Interior, Portugal
Mónica Costa	Polytechnic Institute of Castelo Branco, Portugal
Mohamed El Ghami	University of Bergen, Norway
Muhamad Uznir Ujang	Universiti Teknologi Malaysia
Mahboub Aziz	FSTT UAE University Morocco
Omer Muhammet Soysal	Southeastern Louisiana University, USA
Ouederni Meriem	INP-ENSEEIHT Toulouse, France
Rachmad Andri Atmoko	Universitas Brawijaya, Indonesia
R. S. Ajin	DEOC DDMA, Kerala, India
Rani El Meouche	Ecole Spéciale des Travaux Publics, France
Rui Campos	INESC TEC, Porto, Portugal
Rogério Dionisio	Polytechnic Institute Castelo Branco, Portugal
Sagahyroon Assim	American University of Sharjah, United Arab Emirates
Saied Pirasteh	University of Waterloo, Canada
Senthil Kumar	Hindustan College of Arts and Science, India
Sonja Ristic	University of Novi Sad, Serbia
Sonja Grgić	Zagreb University, Croatia
Sri Winiarti	Universitas Ahmad Dahlan, Indonesia
Suhaibah Azri	Universiti Teknologi Malaysia
Sunardi	Universitas Ahmad Dahlan, Indonesia
Xiaoguang Yue	International Engineering and Technology Institute, Hong Kong
Yasyn Elyusufi	FSTT, UAE, Morocco
Youness Dehbi	University of Bonn, Germany
ZAIRI Ismael Rizman	Universiti Teknologi MARA, Malaysia

Keynotes Speakers

Smart City: Why, What, Experience Feedback and the Future/Challenges

Isam Shahrour

Lille University, France

Prof. Isam was a graduate from the National School of Bridges and Roads (Ponts et Chaussées-Paris); he has been strongly involved in research, higher education and partnership with the socio-economic sector. During the period of 2007–2012, he acted as Vice President "Research and innovation" at the University Lille1. He is a distinguished professor at Lille University with about 35 years of intensive academic activity with strong involvement in the university management as well as in both socio-economic and international partnership. His research activity concerned successively: geotechnical and environmental engineering, sustainability and since 2011 Smart Cities and urban infrastructures. Associate Editor of Infrastructures Journal (MDPI).

Imagining a Digital Competency Management Ecosystem Approach to Transforming the Productivity of People in the Built Environment

Jason Underwood

University of Salford, UK

Prof. Jason Underwood is a Professor in Construction ICT & Digital Built Environments and Programme Director of the MSc. in Building Information Modelling (BIM) & Digital Built Environments within the School of Science, Engineering & Environment at the University of Salford. He holds a BEng (Hons) in Civil Engineering from Liverpool John Moores University, a Master's in Psychology from Liverpool Hope University and a PhD from the University of Salford. His doctoral thesis was on "Integrating Design and Construction to Improve Constructability through an Effective Usage of IT". He is a Chartered Member of both the Institution of Civil Engineering Surveyors (MCInstCES) and The British Psychological Society (CPsychol) and a Fellow of the Higher Education Academy (FHEA). He is actively engaged in the digital transformation of the UK construction industry. He is the present Chair of the UK BIM Academic Forum and Director of Construct IT For Business, an industry-led non-profit making collaborative membership-based network.

Challenges of Cybersecurity in Smart Cities

Mohammed Bouhorma

UAE University, Morocco

Prof. Bouhorma is an experienced academic who has more than 25 years of teaching and tutoring experience in the areas of information security, security protocols, AI, big data and digital forensics at Abdelmalek Essaadi University. He received his M.S. and Ph.D. degrees in Electronic and Telecommunications from INPT in France. He has held a Visiting Professor position at many Universities (France, Spain, Egypt and Saudi Arabia). His research interests include cyber-security, IoT, big data analytics, AI, smart cities technology and serious games. He is an editorial board member for over dozens of international journals and has published more than 100 research papers in journals and conferences.

Advancing Urban Modelling with Emerging Geospatial Datasets and AI Technologies

Filip Biljecki

National University

Prof. Filip is a geospatial data scientist at the National University of Singapore where he had established the NUS Urban Analytics Lab. His background is in geomatic engineering, and he was jointly appointed as Assistant Professor at the Department of Architecture (College of Design and Engineering) and the Department of Real Estate (NUS Business School). He hold a PhD degree (with highest honours, top 5%) in 3D GIS from the Delft University of Technology in the Netherlands, where he also did his MSc in Geomatics. In 2020, he has been awarded the Presidential Young Professorship by NUS.

Integrating System Dynamics in Digital Urban Twin

Ihab Hijazi An-Najah

National University and Technical University of Munich

Dr. Hijazi is an associate professor of Geographic Information Science at Urban Planning Engineering Department, An-Najah National University in Palestine. Also, he is a senior scientist at the chair of Geoinformatics at Technical Uni of Munich. He worked as a postdoc scholar at the chair of information architecture, ETH Zurich. He was a researcher at ESRI–the world leader in GIS and the Institute for Geoinformatics and Remote Sensing (IGF) at the University of Osnabrueck in Germany.

Background of Smart Navigation

Ismail Rakip Karas

Karabuk University, Turkey

Prof. Ismail Rakip Karas is a Professor of Computer Engineering Department and Head of 3D GeoInformatics Research Group at Karabuk University, Turkey. He received his BSc degree from Selcuk University, MSc degree from Gebze Institute of Technology and PhD degree from GIS and remote sensing programme of Yildiz Technical University, in 1997, 2001 and 2007, respectively, three of them from Geomatics Engineering Department. In 2002, he involved in a GIS project as a Graduate Student Intern at Forest Engineering Department, Oregon State University, USA. He has also carried out administrative duties such as Head of Computer Science Division of Department, Director of Safranbolu Vocational School of Karabuk University. Currently, he is the Dean of Safranbolu Fine Art and Design Faculty in the same university. He is the author of many international and Turkish publications and papers on various areas of Geoinformation Science.

Contents

Smart Cities

Smart Cities

Connections Between Smart City and Flood Management Against Extreme Weather Events

Fanny Josse[1](✉), Zhuyu Yang[1,2], and Bruno Barroca[1]

[1] Lab'urba, Université Gustave Eiffel, Champs-sur-Marne, France
fanny.normand.josse@gmail.com

[2] LATTS, UMR CNRS 8134 Université Gustave Eiffel/Ecole des Ponts ParisTech, Marne la Vallee, France

Abstract. Flooding is a highly dynamic phenomenon, and can occur due to several natural and anthropogenic causes, including flash floods, rising groundwater, gradual sea level rise, and coastal storm surges. With increased flood risk in urban areas, more and more studies suggest an integration of flood management and the concept of Smart City. This paper aims to discuss the connections between Smart Cities and flood management against extreme weather events potentially occurring in the future caused by the growing climate change. Flood prediction and warning are crucial for flood management, and their improvement could rely on the digital technological systems highlighted in many Smart City strategies. Moreover, the installation of technological equipment could benefit from flood maps that indicate high-risk flood zones in the short, medium, and long term.

Keywords: Smart city · Flood management · Flood prediction · Flood maps · Digital data

1 Introduction

Environmental changes, combined with the concentration of property and persons in urban territories foretell devastating events in the coming years. The consequences of the frequency and severity of climate extremes will lead directly to the possible occurrence of various recent heat waves and other damaging extreme weather. For example, increasing heat extremes develop into a key hazard due to their multifaceted damage, like a number of health risks, disturbances in urban activities (in the production and consumption of energy, water, supplies, etc.), and ecological imbalances. Of almost all natural hazards, floods are the natural hazard with the highest frequency and the widest geographical distribution worldwide, and the hazard that causes the most economic damage [1]. A flood is defined as the overflowing of the normal confines of a water body or the accumulation of water in areas normally without it. Floods can occur due to a variety of causes such as river floods, flash floods, heavy rainfall, storm surges in coastal areas, or failure of the sewerage system in urban areas [2].

Although flooding cannot be fully predicted, flood management can significantly reduce damage to people and infrastructure. This goal could be achieved by various

M. Ben Ahmed et al. (Eds.): SCA 2023, LNNS 906, pp. 3–10, 2024.
https://doi.org/10.1007/978-3-031-53824-7_1

practical measures, such as flood prevention, flood mapping, flood prediction, flood warning, etc. The main current challenges for disaster management can be summarized according to the four disaster phases, preparation, response, recovery, and mitigation [3], which, according to Josipovic and Viergutz [4], smart city solutions could address. More and more studies argue that an integration of flood management and the concept of “Smart cities” should be considered for sustainable urban development [5–8]. The concept of the smart city appeared in different forms before taking on its definitive name. A smart city could be defined as a place where traditional networks and services are made more efficient with the use of digital solutions for the benefit of its inhabitants and business [9].

As smart city strategies, apply mainly to big data applications, their main improvement to conventional flood management stems from the integration of different data streams to improve flood prediction [4]. However, little research has been conducted to discuss the connections between flood management and Smart cities with a perspective of inter-complementarity. This paper aims to figure out reinforcing measures for better urban management by presenting the connection between flood management and Smart city from different perspectives. These connections could be improved for better operationalisations of both two sectors, in particular for Smart City which is still at a theoretical stage and whose application to extreme events management from pure theory to reality is recently a significant challenge.

Section 2 will present the concept of Smart Cities from technological and data resource dimensions, based on which, it suggests the contribution of smart cities to flood prediction and warning. Section 3 shows how to identify urban zones that need smart technological systems through flood hazard maps with an example, a French district in coastal areas (Bocca district in Cannes). The variety of development and application of smart city features in coastal disaster management is recently required [8].

2 Smart Data Resources for Flood Management

The literature review on the origins of the smart city [10] observes that in 1987 the notion of "wired cities" made its appearance, that of “cyber cities” emerged in 1999 and the following year it was the turn of “digital cities”. From 2002 onwards, the "intelligent cities" approach became more widespread, and Hollands in 2008 [11] developed the term "smart cities". These different names refer to a digital city based on the exchange of information. Sustainable strategies for integrated urban flood management are considered crucial for smart city planning and development [6].

The smart city is a collection of data that is processed in real-time by ICT (Information Communication Technology), with applications providing a visual representation of this information base for greater visibility and are analysed to develop action plans based on the results obtained. Understanding the nature of the data is essential, as are the applications according to requirements. Based on a preview of the work on strategies to collect data and then implement them in smart cities that suffer from different types of flooding, a number of strategies are developed in different articles, some of which are evoked and others explored in more depth. This work studies the diversity of technical solutions that can be promoted.

Before analysing the strategies employed, a non-exhaustive study was carried out to understand what data and information were injected into the digital containers. Two types of data emerged from this work: dynamic data and static data. It is important to note that not all data is structured in the same way and comes from different sources. The sheer number of different types of data and the differences in their structure raises questions about their improtability and reliability.

Regulations, norms or standards are considered as static data to assess different types of conformity. This category also includes geospatial data such as Geographic Information Systems (GIS). The institutions share data relating to demography, the geography of the territory (relief, soil typology, location of infrastructure, etc.) and the geography of the area. Two- and three-dimensional cartographic models are used to support applications for visualising data sets. Static data can be updated in the event of modifications but does not require continuous monitoring. The Internet of Things (IoT) is defined by Sarker [12] as a connected network of heterogeneous components that detect, collect, transmit and analyse data for intelligent systems and services, introducing the notion of dynamic data.

City management uses information fromweather conditions and air and water pollution studies. Two articles, from Loftis et al., [13], and from Jon Derek et al., [14], present the monitoring of water levels using IoT sensors with an alert threshold that triggers emergency processes. As far as infrastructure is concerned, intelligent transport systems (ITS) technology presented in more than one study [15, 16] to assess transport occupancy rates. Other connected objects such as passenger cars can also be used as detection systems, as shown in the work in the article [15]. Ha et al. [17] propose a framework for using unmanned aerial vehicles (UAVs) to monitor urban roads, and this concept of camera surveillance is covered in the study of Gabbar et al. [18], which combines it with satellite scanning. Kurte et al., [19] present a semantic model called Dynamic Flood Ontology (DFO) that can be used to represent partially or fully flooded road segments, enabling targeted intervention. With regard to buildings, the guide of Kitchin et al., [20] suggests setting up building management systems, networks of sensors on equipment or intelligent appliances controlled by applications to control water and energy consumption in normal times, but also to respect protocols in the event of an emergency (power cuts, disconnection of urban systems, etc.).

Many businesses are offering different tools to contain, manage, process and visualise all this data. The work of Pettit et al. [21] suggests that the use of dashboards for the analysis of city data would allow consolidating of the information on a web page offering the public decision-makers a global view. For example the new Dublin district of the Docklands relies on this strategy to obtain reports and adjust their flood policies [22]. The information is also freely available to citizens, making it possible to set up intelligent digital spaces for discussion and consultation. Surveillance systems using cameras, drones and sensors are used to diagnose events and adjust intervention protocols. Communication systems are used by the authorities to communicate with residents and also to process their data in order to estimate specific needs in terms of functions. Event data is also collected for better anticipation and to adjust procedures. The smart city generates data that can be used to make this information more reliable.

3 Flood Maps: Basis for Smart Systems Planning

A smart city is made by a set of fundamental factors: technology (infrastructures of hardware and software), people (creativity, diversity, and education), and institutions (governance and policy) [23]. Smart City is also called Digital City [24], as recent smart urban plannings relate to digital content and services and incorporate pervasive computing [25]. Mondschein et al. [26] consider smart cities depending on large technological systems, which are immense, interconnected systems that consist of technological equipment. To build a smart city, advanced technological systems are required to provide citizens and enterprises with a powerful platform to connect city elements and let them interact effortlessly with each other and with their administration through electronic means. "Stable sturdy infrastructures, from optical fiber networks covering the city acting as a backbone to the installation of sensors, are the key for the development of intelligent solutions in cities" [27].

Evolving the concept of Smart City from pure theory for practice flood management requires the services of relevant technological systems (monitoring and warning systems). The location of technical equipment needs to be correctly identified to ensure the proper use of technological resources. For example, machines used to monitor railways inundation should be installed in vulnerable and high-risk areas, while public warning systems should be set up in schools and hospitals that are occupied by vulnerable people, or in industrial buildings that are important for economic development. Flood maps, involving various types like hazard maps, depth maps, event maps, damage maps, etc., can help recognise the more dangerous and sensitive areas. The installation of technology systems in these key areas will ensure that intelligent technology systems are used effectively. Plans for the installation of technical systems, such as water monitoring systems and alarm systems, can be added to or modified as the disaster area expands in the short, medium, and long term. The map of future flood hazards at different temporal scales is therefore significant to predict disasters and preparation.

The increasing disasters caused by climate change can be presented through a number of scenarios with different temporal phases. For instance, to analyse smart city strategies and their impact on different temporalities, Raven [28] suggests Arc 3.2 methodology [29], based on a prototype intervention aimed at mapping different scenarios (Current time, business as usual and Best Practices). This methodology implemented by UCCRN and NYIT since 2018 allows for bringing more concrete evidence to public and political decision-makers to argue in a spatiotemporal way [28]. The scenarios are allocated into three simulations:

- Current time (short term) that represents existing elements, considering current trends and events in the analysed environment;
- Business-as-usual (middle term) refers to a simulation of future scenarios that prolongs current trends without changing behaviour, which may even be accentuated if decisions are not taken.
- and Best practice (long term) based on the strict application of the recommendations of the various institutions.

This paper takes the Bocca district in Cannes as an example and explains the identification of critical zones appropriate for smart technological systems installation based

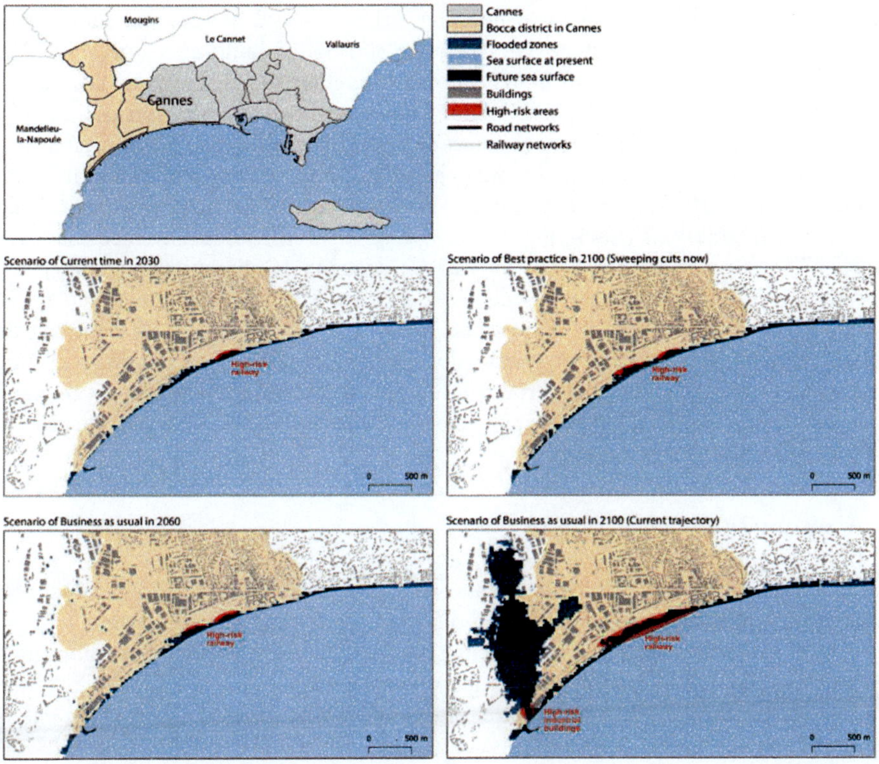

Fig. 1. Flooded zones in four scenarios and high-risk infrastructure in the Bocca disctrict (Cannes, Frence)

on flood maps in different scenarios. More and more publicly available flood maps exist online, for example, the web Géoportail [30] of the French government, the US federal agency NOAA's service [31] specifically for coastal areas, and Climate Central [32]. This paper decides to apply the map of "Land projected to be below annual flood level", produced by Climate Central and based on peer-reviewed science in leading journals. It has been chosen because the parameter of "Pollution Pathway or Sea Level Scenario" in this tool could be changed in corresponding to the requirement of the "Arc 3.2" scenarios. The "Current trajectory" option in this tool, signifies that Global emissions of heat-trapping pollution continue to rise, with annual emissions approximately doubling by the end of the century, which is suitable for the scenario "Biasness as usual" in "Arc 3.2". The "Sweeping cuts now" option, which means that Global emissions of heat-trapping pollution immediately begin to decline steeply, reaching net zero by mid-century, serves as the scenario of "Best practice".

Cannes, located on the French Riviera, is known for its association with the rich and famous, its luxury hotels and restaurants, and for several conferences. The Bocca district, to the west of Cannes, is now an industrial and touristic area following improvements to its economic and social structures. Cannes is a city facing the challenge of flood hazards, due to the flooding caused by river flooding, urban run-off, and marine submersion. Meanwhile, Cannes is a territory that has never considered building a smart city. This paper believes that the smart city concept could strengthen the city's flood management capacity in the future. According to Climate Central, the threat of sea level rise and coastal flooding to the Bocca district will rise and expand the flooding area and sea surface every year. Figure 1 shows the coastal flooding caused by sea level rise and annual floods in 2030, 2060, and 2100. The circumstances in 2030 and in 2060 could present as "Current time" and "Business as usual" in "Arc 3.2" respectively. Moreover, since the "Best Practice" scenario in 'Arc 3.2' is designed for sustainable climate management, two circumstances in 2100 illustrate respectively the scenario of "Business as usual" (Current trajectory) and the scenario of "Best practice" (Sweeping cuts now). In all scenarios, the railway networks in Cannes have flood risks, so that monitoring equipment for them is necessary. The flood risk for industrial buildings exists only in the scenario of "Business as usual" in 2100, and both monitoring and warning systems could be taken into account for better emergency management.

4 Conclusion

In the flooding risk management process, mitigation/prevention, and preparedness are before the events, whereas response and recovery are during and after events [3]. The connections between flood management and the concept of smart cities could contribute to the prevention, preparedness, and response phases. Gathering significant data, provided by smart monitoring systems, can effectively predict flood events. Then, intelligent communication and alarm systems could quickly and automatically transmit hazard information to managers and related populations. On the other hand, the planning for smart technological benefit from flood maps that clarify the urban areas more vulnerable. In identified high-risk areas, monitoring systems could be planned and set up to collect data to predict flood events and warning systems can be put in place for better warning and communication.

To go further in implementing the scenarios developed in this document, future work could explore the new emerging technology of the smart city, the digital twin [33]. This idea will enable to study this exact digital representation of the urban environment and how it will continuously enrich the prediction and simulation of natural hazards for an ever more resilient and dynamic smart city.

References

1. Flood hazard and risk assessment | UNDRR. https://www.undrr.org/publication/flood-hazard-and-risk-assessment
2. Summary for Policymakers — Global Warming of 1.5 °C. https://www.ipcc.ch/sr15/chapter/spm/
3. Alexander, D.E.: Principles of Emergency Planning and Management. Oxford University Press, Oxford (2002)
4. Josipovic, N., Viergutz, K.: Smart solutions for municipal flood management: overview of literature, trends, and applications in German Cities. Smart Cities. **6**, 944–964 (2023). https://doi.org/10.3390/smartcities6020046

5. Zhu, S., Li, D., Feng, H., Gu, T., Hewage, K., Sadiq, R.: Smart city and resilient city: differences and connections. Wiley Interdisciplinary Rev. Data Min. Knowl. Discov. **10** (2020). https://doi.org/10.1002/widm.1388
6. Yereseme, A.K., Surendra, H.J., Kuntoji, G.: Sustainable integrated urban flood management strategies for planning of smart cities: a review. Sustain. Water Resour. Manag. **8**, 85 (2022). https://doi.org/10.1007/s40899-022-00666-5
7. Hingmire, A.M., Bhaladhare, P.R.: A review on urban flood management techniques for the smart city and future research. In: Hemanth, J., Pelusi, D., Chen, J.I.-Z. (eds.) Intelligent Cyber Physical Systems and Internet of Things, pp. 303-317. Springer, Cham (2023). https://doi.org/10.1007/978-3-031-18497-0_23
8. Tonmoy, F.N., Hasan, S., Tomlinson, R.: Increasing Coastal Disaster Resilience Using Smart City Frameworks: Current State, Challenges, and Opportunities. Frontiers in Water 2 (2020)
9. Smart cities. https://commission.europa.eu/eu-regional-and-urban-development/topics/cities-and-urban-development/city-initiatives/smart-cities_en
10. Montes, J.: A Historical View of Smart Cities: Definitions, Features and Tipping Points (2020). https://papers.ssrn.com/abstract=3637617
11. Hollands, R.G.: Will the real smart city please stand up?: intelligent, progressive or entrepreneurial? City **12**, 303–320 (2008). https://doi.org/10.1080/13604810802479126
12. Anshu, Firduai, S.F.D.: Urban Resilience and Flash Floods: A Case Study of Chennai Metropolitan City. In: Making Cities Resilient. Urban Book Series. 127-140. Springer, Cham (2019). https://doi.org/10.1007/978-3-319-94932-1_10
13. Loftis, J.D., Katragadda, S., Rhee, S., Nguyen, C.: StormSense: a blueprint for coastal flood forecast ınformation & automated alert messaging systems. In: Présenté à Proceedings - 2018 3rd International Workshop on Science of Smart City Operations and Platforms Engineering, in Partnership with Global City Teams Challenge, SCOPE-GCTC 2018 (2018)
14. Loftis, Wang, H., Forrest, D., Rhee, S., Nguyen, C.: Emerging flood model validation frameworks for street-level inundation modeling with storm sense. In: Présenté à Proceedings - 2017 2nd International Workshop on Science of Smart City Operations and Platforms Engineering, in partnership with Global City Teams Challenge, SCOPE 2017 (2017)
15. Sanchez-Iborra, R., Bernal-Escobedo, L., Santa, J.: Eco-efficient mobility in smart city scenarios. Sustainability (Switzerland) **12**, 1–15 (2020). https://doi.org/10.3390/su12208443
16. Amari, H., Khoukhi, L., Belguith, L.H.: Prediction and detection model for hierarchical Software-Defined Vehicular Network. In: Présenté à Proceedings - Conference on Local Computer Networks, LCN (2022)
17. Ha, N.T., et al.: Leveraging deep learning model for emergency situations detection on urban road using ımages from CCTV cameras. In: Présenté à 8th International Conference on Engineering and Emerging Technologies, ICEET 2022 (2022)
18. Gabbar, A., et al.: HAIS: highways automated-ınspection system. Technologies **11** (2023). https://doi.org/10.3390/technologies11020051
19. Kurte, K., Potnis, A., Durbha, S.: Semantics-enabled spatio-temporal modeling of earth observation data: An application to flood monitoring. In: Présenté à Proceedings of the 2nd ACM SIGSPATIAL International Workshop on Advances in Resilient and Intelligent Cities, ARIC 2019 (2019)
20. Kitchin, R., Coletta, C., Evans, L., Heaphy, L.: Creating smart cities. In: Creating Smart Cities. p. 1-18 (2018)
21. Pettit, C., Lieske, S.N., Jamal, M.: CityDash: visualising a changing city using open data. In: Geertman, S., Allan, A., Pettit, C., et Stillwell, J. (éd.) Planning Support Science for Smarter Urban Futures. p. 337-353. Springer International Publishing, Cham (2017). Doi: https://doi.org/10.1007/978-3-319-57819-4_19
22. Sarker, I.H.: Smart city data science: owards data-driven smart cities with open research issues. Internet of Things. **19**, 100528 (2022). https://doi.org/10.1016/j.iot.2022.100528

23. Papa, R., Gargiulo, C., Galderisi, A.: Towards an urban planners' perspective on Smart City. TeMA J. Land Use Mob. Environ. **6**, 5–17 (2013). https://doi.org/10.6092/1970-9870/1536
24. Anthopoulos, L.: The ımplementation model of a digital city. The case study of the digital City of Trikala, Greece. J. E-Govern. **2**, 91 (2006)
25. Anthopoulos, L.G., Vakali, A.: Urban planning and smart cities: interrelations and reciprocities. In: The Future Internet: Future İnternet Assembly 2012: From Promises to Reality. pp. 178-189. Springer, Heidelberg (2012). https://doi.org/10.1007/978-3-642-30241-1_16
26. Mondschein, J., Clark-Ginsberg, A., Kuehn, A.: Smart cities as large technological systems: overcoming organizational challenges in smart cities through collective action. Sustain. Cities Soc. **67**, 102730 (2021). https://doi.org/10.1016/j.scs.2021.102730
27. Bakıcı, T., Almirall, E., Wareham, J.: A smart city initiative: the case of Barcelona. J. Knowl. Econ. **4**, 135–148 (2013). https://doi.org/10.1007/s13132-012-0084-9
28. Raven, J.: From climate science to design practice. Urban Des. J. (2019)
29. Raven, J., et al.: Urban planning and urban design. In: Rosenzweig, C., Solecki, W.D., Romero-Lankao, P., Mehrotra, S., Dhakal, S., et Ali Ibrahim, S. (eds.) Climate Change and Cities, pp. 139–172. Cambridge University Press, Cambridge (2018)
30. Géoportail. https://www.geoportail.gouv.fr/
31. Coastal Flood Exposure Mapper. https://coast.noaa.gov/floodexposure/#-10575352,4439107,5z
32. Sea level rise and coastal flood risk maps -- a global screening tool by Climate Central. https://coastal.climatecentral.org/
33. Riaz, K., McAfee, M., Gharbia, S.S.: Management of climate resilience: exploring the potential of digital twin technology, 3D city modelling, and early warning systems. Sensors. **23**, 2659 (2023). https://doi.org/10.3390/s23052659

Unleashing the Potential of Graph Database in Smart Asset Management: Enhancing Predictive Maintenance in Industry 4.0

Farah Ilyana Hairuddin(✉), Suhaibah Azri, and Uznir Ujang

3D GIS Research Lab, Faculty of Built Environment & Surveying, Universiti Teknologi Malaysia, Johor, Malaysia
farahilyana@graduate.utm.my, {suhaibah,mduznir}@utm.my

Abstract. The integration of predictive maintenance with Industry 4.0 technologies such as big data, the Internet of Things, and artificial intelligence has led to the possibility of the overwhelming high amounts of data together with the production of unstructured and structured data. Although varieties of analytics will be able to be conducted due to additional information obtained with the state-of-the-art technologies, the capability of managing these high amounts of data is concerning. An agile database is required to accommodate this dynamic digital environment that will help in sustaining the data as reliable data sources for future analytics of intelligent asset management. A graph database is known for its flexibility and scalability due to their schema-less and unfixed structures which makes it easier to add new data. Besides that, its unique structures that represent entities in the form of nodes and relationships in the form of edges have made it excels in dealing with complex join-style queries. This paper discussed the possibility of implementing a graph database as an agile database approach by storing asset information to provide reliable data sources for the predictive maintenance process that will help to realise smart asset management.

Keywords: Predictive Maintenance · Graph Database · Industry 4.0 · Smart Asset Management

1 Introduction

Predictive maintenance is a systematic approach that monitors assets' condition throughout their lifecycle to provide a prognosis once the possibility of faults is detected [1]. The prognosis depends on historical maintenance data such as maintenance logs and sensor readings. Due to its historic data-based features, it is capable to improve decision-making process by predicting trends, behaviour patterns, correlations between variables through data-driven models such as statistics, machine learning, and deep learning for possible failure downtime [2]. It is widely applied and discussed in many domains such as building management [3], facility management [4], high performance computing system [5], renewable energy sector [6] and manufacturing [2] which makes it necessary to have the knowledge on the domain that the predictive maintenance shall be applied, as its implementation might differ based on the discipline that it applies to.

M. Ben Ahmed et al. (Eds.): SCA 2023, LNNS 906, pp. 11–21, 2024.
https://doi.org/10.1007/978-3-031-53824-7_2

Current practise in predictive maintenance in this Industry 4.0 era is exposed to new challenges due to the rapid growth of data, data connectivity, and the emergence of new devices. Sensors are widely applied in monitoring assets. The information received from the monitoring sensors benefits predictive maintenance analytics, as detailed insights can be obtained to detect early warning of failure. Datasets for predictive maintenance can be received from internal and external sources, where vast pools of sensor data can be conducted internally, while environmental data such as temperature and humidity related to the degradation of the assets' failure can be used as external source [7]. These data acquisition technologies cause exponential growth of potential data to be supplemented in predictive maintenance process, which makes data sets for predictive maintenance to have both structured and unstructured data, which makes it a necessity to have flexible database that can accommodate with the growth of data sources and its various types. This makes the graph database the center of attention to cater to this digital growth scenario in asset maintenance.

In this paper, we discuss the possibility of implementing a graph database to facilitate the predictive maintenance process. The paper is organised into several sections. Section 1 includes the current state of predictive maintenance practises with a brief introduction to predictive maintenance and its application for immovable and movable assets. Section 2 discussed the implementation of graph database for predictive maintenance. Section 3 discussed the challenges and future direction. Lastly, Sect. 4 will conclude the paper.

1.1 Predictive Maintenance and Industry 4.0

There are four maturity stage in predictive maintenance implementation [7]. Starting with level 1: Visual inspections where physical inspections are conducted periodically with complete reliance on the inspector's expertise, level 2: Instrument Inspections where physical inspections are conducted periodically with reliance on the combination of inspector's expertise and instrument readings, level 3: Real-Time Conditions Monitoring where the assets are being continuously monitored in real time and alerts are given based on pre-established rules or level of criticality. Lastly, level 4: PdM 4.0 that continuously monitors the asset and sends alerts based on predictive modelling. The aspects that distinguish each of these maturing stages consist of inspection process, content form, performance measurement, IT, and organisation as shown in Fig. 1.

Despite the discussion on predictive maintenance 4.0 (PdM 4.0) having been active in recent years as the practise of predictive maintenance evolves with the technology development, only 11% of the survey taken from the total of 280 companies in Belgium, Germany and the Netherland that have applied predictive maintenance 4.0 while 2/3 of the survey practise predictive maintenance below level 3 with half of the survey respondents intend to implement PdM 4.0 in the future [7].

Faults and implications as essential knowledge for PdM in the domain of asset management [1]. For that, there are several approaches to classify failure in predictive maintenance, such as physical, knowledge-based, data-driven, and hybrid and the most commonly used prediction classification is data-driven [8]. In the current evolution of PdM solutions, the data-driven models applied are statistic-based and have pattern recognition capabilities, artificial intelligence, and machine learning algorithms.

Capability	1. Visual Inspections	2. Instument Inspections	3. Real Time Conditions Monitoring	4. PdM4.0
Processes	- periodic inspection (physical) - checklist - paper recording	- periodic inspection (physical) - instruments - digital recording	- continouos inspection (remote) - sensors - digital recording	- continuous inspection (remote) - sensors and other data - digital recording
Content	- paper based condition data - multiple inspection points	- digital condition data - single inspection points	- digital condition data - multiple inspection points	- digital condition data - multiple inspection points - digital environment data - digital maintenance history
Performance Measurement	- visual norm verification - paper based trend analyses - prediction by expert opinion	- automatic norm verification - digital trend analyses - prediction by expert opinion	- automatic norm verification - digital trend analyses - monitoring by CM software	- automatic norm verification - digital trend analyses - prediction by statistical software - advanced decision support
IT	- MS Excel/MS Access	- embedded instrument software	- condition monitoring software - condition database	- condition monitoring software - big data platform - wifi network - statistical software
Organisation	- experienced craftsmen	- trained inspectors	- reliability engineers	- reliability engineers - data scientists

Fig. 1. Predictive Maintenance Maturity Stages (Source: [7]).

1.2 Predictive Maintenance for Immovable and Movable Assets

Assets can be categorized as immovable and movable when differing in aspect of portability, physical characteristics, and legal considerations. In general, immovable assets are infrastructures that are fixed in a place, hardly removed, and require the involvement of legal expertise to dismantle since it consists of building services that are constructively attached to the system and moulded, while movable assets are portable as it is being used as supplements to infrastructures and can be in various types such as inventory, plant, machinery, vehicles, equipment, and any spare part for any equipment and furnishings [9].

The application of predictive maintenance has been actively discussed in the varieties of immovable and movable assets in many industries. In a study conducted by [10], schedule maintenance prediction was performed for a flexible manufacturing environment, focussing on factory assets such as machines and robotics. The predictive maintenance process also considers the relative positions of the assets and the time displacement to move the repair equipment to the machine location. Another example is in the maintenance of railway infrastructure where a study by [11] conducted a prediction for the maintenance of railway switches in the aspect of maintenance needs, type of maintenance, and maintenance trigger. The study uses several data sources, such as the location of the railway switch, the year of installation, condition, and other attributes. Lastly, [12] applied a machine-learning-based anomaly detection model in the building services sector to predict failures in heating, ventilation, and air conditioning (HVAC) systems, showcasing its potential for smart environment applications.

It can be seen that location is an important information in predictive maintenance for most of the assets in varies industries and predictive maintenance is applicable in both movable and immovable assets.

1.3 Data Challenges in Predictive Maintenance

Data are the key in generating insights to assist in decision-making in asset maintenance. In line with Industry 4.0, data collection technologies for asset predictive maintenance have also evolved with the implementation of embedded sensors, RFID tags, and processors built into the asset [13]. An integral part of this evolution involves the utilization of modern data acquisition techniques such as drones, which can efficiently capture asset-related data from varying perspectives.

For instance, employing drones enables the acquisition of both two-dimensional and three-dimensional data, enriching the asset information pool. However, the raw data collected via drones necessitates comprehensive processing to extract meaningful insights. This is where different software packages come into play. These packages, tailored for image processing and point cloud analysis, are instrumental in transforming the drone-captured data into a coherent format. By leveraging the capabilities of software like photogrammetry software for image stitching and GIS software for spatial analysis, the collected data is refined, structured, and aligned for optimal output [14].

Incorporating drone-based data acquisition, along with judicious processing utilizing specialized software packages, empowers the asset predictive maintenance process with richer, multidimensional insights. This comprehensive approach not only aligns with the advancements of Industry 4.0 but also paves the way for more accurate and effective decision-making in asset management.

This has caused a vast amount of data from varying sources that impact accessibility, quality, and integrating the information. Obtaining information from the equipment also requires the on-site experts' knowledge to understand the cause of the effect [15]. In addition to that, expert judgment and historical data also come in handy when assesing the criticality of failure modes [16]. The presence of different data sources taken by manual data entry and sensory data calls for needs of data management that are capable of relating both information types. The unionship of structured, semi-structured and unstructured data would be capable of assisting in identifying fault patterns.

Next, the overwhelming data raise the needs for Internet of Things development, which makes information of the assets can be stream wirelessly [13]. This offers flexibility in assesing the condition as more information can be obtained, which rises varieties of predictive maintenance data analytics associated with artificial intelligence, deep learning, and machine learning due to the incapacity of traditional data analytics to sort out useful and quality data from the vast amount of data collected with varies data type. For example, a research conducted by [17] implemented deep learning structure into Remaining Useful Life (RUL) modelling to run the process involving multisource data consists of sensor data and GIS data.. Both improvising to the data analytics help increase the prediction accuracy as more necessary information can be used in the analysis.

In the context of predictive asset management, the effectiveness of data analysis becomes crucial in identifying areas of infrastructure that warrant improvement. This approach also applies to strategic asset planning and anticipatory analytics for ensuring the safety and security of assets. Nevertheless, irrespective of the degree of technological advancement applied to predictive asset management, data doesn't inherently structure itself within a database. To meet these goals, the establishment of a sophisticated database structure is essential to yield informative data outputs [18]. Traditional database composed of tabular structures with predefine data type and any modification would effect the whole structure [19]. The rigid and schema-based approach of the traditional database would make dealing the diversity of unstructured, semi-structured and structured data a challenge, which has make NoSQL a popular choice due to its dynamic schemas and scalable architecture that it makes suitable to deploy for voluminous and high velocity data environment [20].

2 Graph Database as Agile Solution: Implementing Graph Database for Predictive Maintenance

The agility of a database can be demonstrated by its capability to evolve based on the data growth without the need to pre-defined again the whole structure. The discussion by [21] on this matter advocates that agility means that the database can be evolved and not necessarily be developed "once and for all". Since asset maintenance involves varieties of information and requires collaboration of information to have a thorough understanding of the effect of failure, there are presence of continuous data growth in asset maintenance which makes its database adaptable to changes.

2.1 Storing Asset Information Using Graph Database for Predictive Maintenance

Conceptually, the asset locations will be represented in the form of nodes with each asset defined its relationship to the asset that attached to it in the form of edges. Then, the properties of each asset are defined to the nodes. Additional nodes are established to the asset that contains maintenance logs, as shown in Fig. 2.

In predictive maintenance, focused has been limited research specifically on storing asset information. However, several studies have explored the storage of built-environment infrastructure in graph databases, as exemplified in the next subsection. Although it is applied for different applications, its demonstration would give further visualisation of graph database potential capabilities to store asset information.

2.2 Graph Database in Managing Built Environment Infrastructure

The basic component of the Predictive Maintenance 4.0 system architecture should consist of data acquisition, storage, analysis, and visualisation of the results to ensure that the system implemented is operational and reliable [22]. Since data availability is the most important component in predictive maintenance, its quality should also be considered, as the accuracy of failure prediction is dependent on the relevance, sufficiency, and quality of the data sources [22].

To accommodate in providing relevant data source, the graph data structure is capable of modelling temporal versioning such as the study conducted by [23] that uses graph data structure as shown in Fig. 3 to model temporal versioning of slots and openings for installation of building service systems such as piping networks, electrical networks, HVAC and communication system in Neo4j graph database.

Upon implementing it in a graph database, the user can query up-to-date information. It calls for graph solutions necessity because having the information for changes of

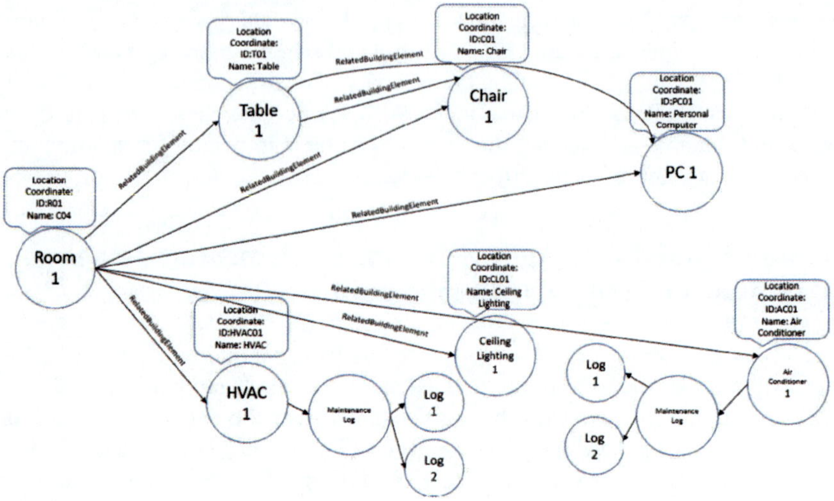

Fig. 2. Conceptual representation of asset information in graph database.

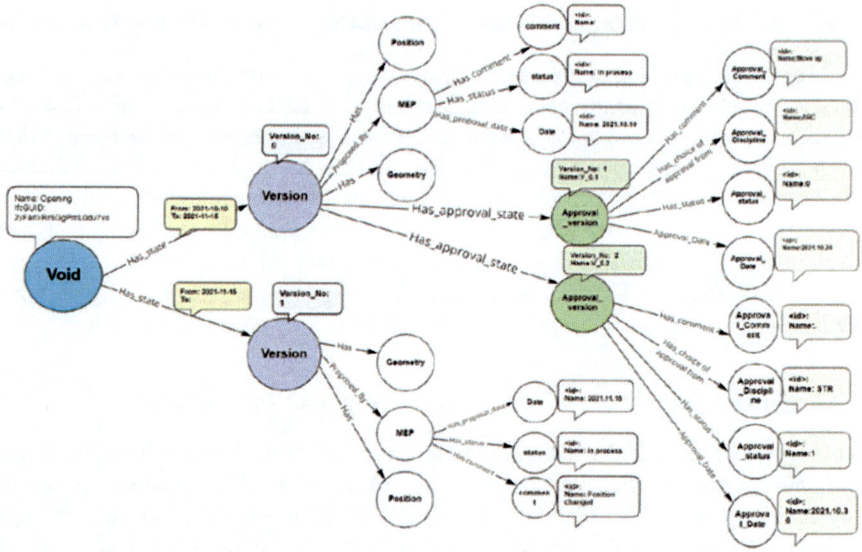

Fig. 3. A graph model of the slot and opening process (Source: [23]).

positions, geometry, and approval of the slots and opening helps in decision-making in building design application. In the context of predictive maintenance, this will help to provide predictive modeling with up-to-date information.

Next, in aspect of quality, graph data structure has the capability of querying information from other formats by converting it into graph data structure such as demonstrated by [24] that construct IFC-graph in Neo4j graph database to improve the query of building components. This data model has been proven to convert IFC data structure into the form of graph data structure which has benefitted the user to access and query building information for further information instead of using IFC parser, which has its limitations. For example, the user may query building components in the aspect of its adjacency and connectivity as shown in Fig. 4 and This capability helps in data analytics such as indoor pathfinding.

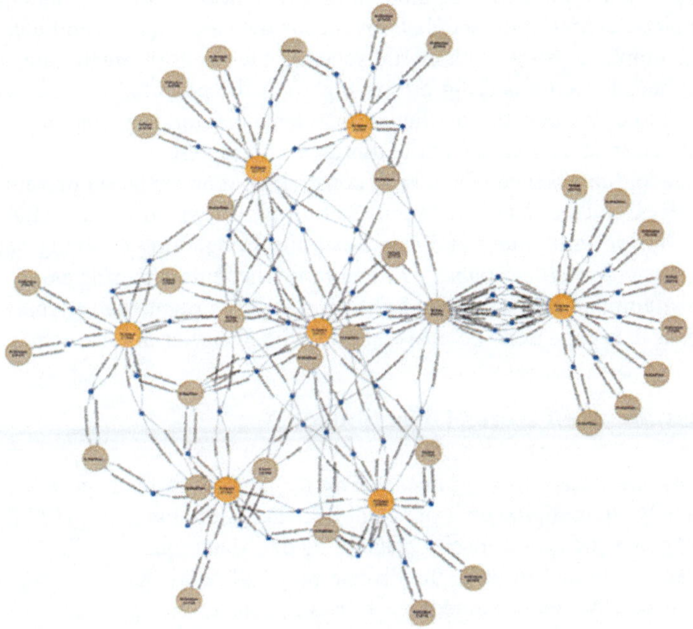

Fig. 4. Query result of elements adjacency of house model (Source: [24]).

This capability will enable more data sources with different formats to be supplemented into predictive maintenance analytics, which will increase the insight of the analysis. This makes graph database suitable to be applied in an environment involving collaboration and coordination of multidisciplinary processes which exists in asset management discipline.

2.3 Integrating Graph Database with Predictive Maintenance Workflows

In the context of predictive maintenance workflow, limited attention and research have been devoted to the implementation of graph databases. However, recent studies have emerged that focus on the application of graph data structures within this domain, such as conducted by [25] that developed a dynamic graph neural network structure to learn sensor spatial relation this helps improve sensory data acquisition where usually they would ignore the spatial relations between sensors. This further improves Remaining Useful Life (RUL) which is one of the predictive maintenance models.

Besides that, [8] conducted a comprehensive survey on the application of graph-based approach such as Graph Neural Network, Knowledge Graph, Bayesian Network and Graph Theoretic Model through the stages of predictive maintenance, which consists of anomaly detection, diagnosis, prognosis and maintenance decision-making.

The integration of graph data structure into predictive maintenance workflows is actively conducted for a data-driven approach. Integration led to positive results as the current predictive maintenance modelling can be improvised with new capabilities such as semantic causal inference, heterogeneous data association, and data visualisation through graphical form. Although displaying the result in graph database has not been mentioned, these pioneering efforts highlight the potential of integrating graph database technologies into the predictive maintenance workflow, paving the way for further exploration and advancement in this area of research.

To conclude, the absence of integrity constraints, representation of values through nodes and edges, and the absence of a pre-established schema in a graph database enable adaptability, foster agility, and facilitate flexibility in handling evolving data requirements [19]. The graph agile capability further brings benefits in storing asset information that involves temporal versioning and also in predictive maintenance analytics where heterogeneous data capable to be analyse as explained earlier in this section.

3 Discussion and Future Directions

The flexibility and scalability of graph database displayed significant potential in accommodating the dynamic digital environment of Predictive Maintenance (PdM) due to the ease in adding new data and capacity in handling data that requires intricate relationship which makes it potential to unite the diverse range of data sources presence in PdM. Previous research that incorporated graph data structures into predictive maintenance workflows has demonstrated improvements in predictive analytics, due to the capability of graph databases to analyse heterogeneous data.

Although the above matter displayed the effectivity and efficiency of graph database for PdM, they are not without challenges, particularly in the industry 4.0 era, where a vast amount of data is generated in varieties of structures and formats. To overcome these challenges, continuous research efforts are necessary to be conducted by testing different data structure and format as input into graph database. This will open up many potential research in aspects of database mapping from one format into graph data structure, data integration between different sources in graph database, and managing big data analytics in graph database.

Furthermore, it is worth noting that application of graph data structure in the context of data-driven approach in predictive maintenance modelling has received considerable attention which makes resources regarding implementation of graph structure in the perspective of database management for predictive maintenance quite limited. Therefore, further research focused on database-orientated approaches for predictive maintenance is suggested to address this gap.

In addition, to enhance the capability of graph data structure for predictive asset maintenance, the integration of advanced spatial analysis techniques and graph structures presents a comprehensive framework for optimized decision-making. [26] introduces the concept of space filling curves, which can be applied to organize asset locations as nodes within the graph, optimizing spatial queries. Building upon this, [27] emphasize the integration of higher-dimensional spatial attributes into the graph nodes, enriching the understanding of assets' contextual characteristics. The findings from [28] further enhance this structure, highlighting the significance of reflecting spatial adjacency analysis within the graph's edge connections. This collective approach allows for a holistic representation, where assets are nodes with relationships represented by edges. Additionally, nodes capturing maintenance logs further enrich the structure. Ultimately, this integration not only streamlines predictive maintenance planning but also ensures a more nuanced consideration of asset interactions and spatial dependencies, thus empowering asset managers with a robust framework for informed decision-making.

4 Conclusion

The drawback of corrective maintenance in providing an estimate of the time of failure of the components' failure causes the evolution of maintenance to predictive maintenance. It is seen as a proactive approach, as it helps increase the longetivity of the infrastructure by identifying potential problems by assessing the infrastructure's historical data of the infrastructure, such as its maintenance history and history of failures. The challenges that predictive maintenance shall face will be due to the growth of data and complexity of its interaction. The graph database is seen to have the capability for complex situations due to its unique data structure. In addition to that, to support predictive maintenance 4.0 and realise smart asset management, previous research that demonstrated the development graph data structure for temporal versioning and querying information from different data formats seems capable of accommodating the supply of relevant, sufficient and good quality data sources for predictive maintenance analytics.

Acknowledgements. This work was supported by the Ministry of Higher Education through Fundamental Research Grant Scheme (FRGS/1/2022/WAB07/UTM/02/3).

References

1. Aremu, O.O., Palau, A.S., Parlikad, A.K., Hyland-Wood, D., McAree, P.R.: Structuring data for intelligent predictive maintenance in asset management **51**(11), 514–519 (2018)
2. Zonta, T., da Costa, C.A., da Rosa Righi, R., de Lima, M.J., da Trindade, E.S., Li, G.P.: Predictive maintenance in the Industry 4.0: a systematic literature review. Comput. Ind. Eng. **150**, 106889 (2020)
3. Hu, Y., Cheng, X., Wang, S., Chen, J., Zhao, T., Dai, E.: Times series forecasting for urban building energy consumption based on graph convolutional network. Appl. Energy **2022**(307), 118231 (2021)
4. Sanzana, M.R., Maul, T., Wong, J.Y., Abdulrazic, M.O.M., Yip, C.C.: Application of deep learning in facility management and maintenance for heating, ventilation, and air conditioning. Autom. Constr. **141**(February), 104445 (2022)

5. Lima, A.L. da C.D., Aranha, V.M., Carvalho, C.J. de L., Nascimento, E.G.S.: Smart predictive maintenance for high-performance computing systems: a literature review. J. Supercomput. **77**(11), 13494–513 ()
6. Stadtman, F., et al.: Digital twins in wind energy: emerging technologies and industry-informed future directions 51–56 (2023)
7. Haarman, M., Mulders, M., Vassiliadis, C.: Predictive Maintenance 4.0: Predict the Unpredictable (2017)
8. Xia, L., Zheng, P., Li, X., Gao, R.X., Wang, L.: Toward cognitive predictive maintenance: a survey of graph-based approaches. J. Manuf. Syst. **64**, 107–120 (2022)
9. Nasir A.A.I.M., Azri, S., Ujang, U.: Asset management in Malaysia: current status. IOP Conf. Ser. Earth Environ. Sci. **1064**(1), 012035 (2022)
10. Graça, P., Camarinha-Matos, L.M.: Evaluating and influencing the performance of a collaborative business ecosystem – a simulation study. In: Camarinha-Matos, L.M., Afsarmanesh, H., Ortiz, A. (eds.) PRO-VE 2020. IAICT, vol. 598, pp. 3–18. Springer, Cham (2020). https://doi.org/10.1007/978-3-030-62412-5_1
11. Allah Bukhsh, Z., Saeed, A., Stipanovic, I., Doree, A.G.: Predictive maintenance using tree-based classification techniques: a case of railway switches. Transp. Res. Part C Emerg. Technol. **101**, 35–54 (2019)
12. Katona, A., Panfilov, P.: Building predictive maintenance framework for smart environment application systems. In: Annual DAAAM Proceedings of the International DAAAM Symposium, vol. 29, no. 1, pp. 0460–70 (2018)
13. Kwon, D., Hodkiewicz, M.R., Fan, J., Shibutani, T., Pecht, M.G.: IoT-based prognostics and systems health management for industrial applications. IEEE Access. **4**, 3659–3670 (2016)
14. Hinge, L., Gundorph, J., Ujang, U., Azri, S., Anton, F., Rahman, A.A.: Comparative analysis of 3D photogrammetry modeling software packages for drones survey. Int. Arch. Photogramm. Remote Sens. Spat. Inf. Sci. - ISPRS Arch. **42**(4/W12), 95–100 (2019)
15. Åkerman, M., Lundgren, C., Bärring, M., Folkesson, M., Berggren, V., Stahre, J., et al.: Challenges building a data value chain to enable data-driven decisions: a predictive maintenance case in 5G-enabled manufacturing. Procedia Manuf. **17**, 411–418 (2018)
16. Kauschke, S., Janssen, F., Schweizer, I.: On the challenges of real world data in predictive maintenance scenarios: a railway application. CEUR Workshop Proc. **1458**, 121–132 (2015)
17. Chen, C., Liu, Y., Sun, X., Di, C.-G., Titmus, S.: An integrated deep learning-based approach for automobile maintenance prediction with GIS data. Reliab. Eng. Syst. Saf. **216**, 107919 (2021)
18. Azri, S., Ujang, U., Abdul Rahman, A.: Dendrogram clustering for 3D data analytics in smart city. Int. Arch. Photogramm. Remote Sens. Spat. Inf. Sci. - ISPRS Arch. **42**(4/W9):247–53 (2018)
19. Benhissen, R., Bentayeb, F., Boussaid, O.: GAMM: graph-based agile multidimensional model. In: CEUR Workshop Proceedings, vol. 3369, pp. 23–32 (2023)
20. Zaki, A.K.: Nosql databases: new millennium database for big data, big users, cloud computing and its security challenges. Int. J. Res. Eng. Technol. **03**(15), 403–409 (2014)
21. Morien, R.: Agile development of the database: a focal entity prototyping approach. In: Proceedings of the - Agil Conference 2005, pp. 103–10 (2005)
22. Righetto, S.B., Izumida Martins, M.A., Carvalho, E.G., Hattori, L.T., De Francisci, S.: Predictive maintenance 4.0 applied in electrical power systems. In: 2021 IEEE Power Energy Social Innovation Smart Grid Technology Conference on ISGT 2021, pp. 2–6 (2021)
23. Surendran Sanila, G.: An object and history-based approval method for MEP Slot and Opening planning in openBIM projects using a Database-driven workflow (2022)
24. Zhu, J., Wu, P., Lei, X.: IFC-graph for facilitating building information access and query. Autom. Constr. **148**, 104778 (2023)

25. Zhang, Y., Li, Y., Wei, X., Jia, L.: Adaptive spatio-temporal graph convolutional neural network for remaining useful life estimation. In: Proceedings of the International Jt Conference on Neural Networks (2020)
26. Uznir, U., Anton, F., Suhaibah, A., Rahman, A.A., Mioc, D.: Improving 3D spatial queries search: newfangled technique of space filling curves in 3D city modeling. In: ISPRS Ann. Photogramm. Remote Sens. Spat. Inf. Sci., **2**(2W1), 319–27 (2013)
27. Suhaibah, A., Uznir, U., Rahman, A.A., Anton, F., Mioc, D.: 3D geomarketing segmentation: a higher spatial dimension planning perspective. Int. Arch. Photogramm Remote Sens. Spat. Inf. Sci. - ISPRS Arch. **42**(4W1), 1–7 (2016)
28. Salleh, S., Ujang, U., Azri, S., Choon, T.L.: Spatial adjacency analysis of Citygml buildings via 3d topological data structure. Int. Arch. Photogram. Remote Sens. Spat. Inf. Sci. - ISPRS Arch. **42**(4/W16), 573–9 (2019)

Deep Learning or Traditional Methods for Sentiment Analysis: A Review

Bellar Oumaima(✉), Baina Amine, and Bellafkih Mostafa

STRS Lab, National Institute of Posts and Telecommunications (INPT), Rabat, Morocco
oumaimabellar@gmail.com, baina@inpt.ac.ma, mbellafkih@yahoo.ma

Abstract. Sentiment analysis's main goal is to extract the context from the text. The digital world of today offers us a variety of raw data formats, including blogs, Twitter, and Facebook. In order to perform analysis on this raw data, researchers must transform it into useful information. Numerous researchers used both deep learning and traditional machine learning techniques to determine the text's polarity. In order to understand the work done, we reviewed both approaches in this paper. The best methods for classifying the text will be selected by the researchers with the aid of this paper. We select a few of the best articles and evaluate them critically based on various factors. The purpose of this study is to explore the different machine learning and deep learning techniques to identify its importance as well as to raise an interest for this research area.

Keywords: Sentiment Analysis · Opinion Mining · Machine Learning · Deep Learning · Natural Language Processing

1 Introduction

In order to better understand the text's sentiments, the researchers are able to explore new avenues with the assistance of opinion mining and natural language processing. Sentiment analysis extracts the text's context. In many tasks, including product forecasts, movie recommendations, and many others, machine learning is used to classify and analyze the emotions in the text. Humans use their interpretation of the emotional intent of the words to determine whether a passage of text is positive or negative or marked by another complex emotion, like surprise or disgust. Text mining software can be used to programmatically approach the text's emotional content.

By utilizing machine learning techniques, a great deal of work has already been done in the area of sentiment analysis. Opinion mining is the process of classifying unstructured data and text into positive, negative, and neutral categories. Millions of users have recently flocked to microblogging sites like Facebook and Twitter, according to [7], giving them an open forum to express their ideas and opinions. As described by [8] and [9], sentiments are typically categorized as either positive or negative in binary form. There are a number of machine learning techniques for classifying sentiment, such as Maximum Entropy (MaxEnt), which is a model based on features rather than independent assumptions, as explained by [10–12].

M. Ben Ahmed et al. (Eds.): SCA 2023, LNNS 906, pp. 22–33, 2024.
https://doi.org/10.1007/978-3-031-53824-7_3

SailAil Sentiment Analyzer (SASA) is another machine learning based sentiment classification algorithm as described by [2]. Multilayer Perceptron (MLP) is a robust and nonlinear neural network model developed using a machine learning approach by [15]. Another well-known, simple-to-use, and effective machine learning algorithm is naive bayes, which was first put forth by Thomas Bayes, as stated by [15]. One of the most well-known supervised machine learning-based classification algorithms is the Support Vector Machine (SVM) [16]. There are numerous studies and researches on machine learning-based sentiment analysis using various algorithms and tools, but SVM outperformed many other methods in terms of accuracy and efficiency because the results were comparatively higher, according to [17, 18].

According to (Feldman 2013), the analysis of sentiment is a technique where the dataset consists of emotions, actions, or assessments that take into account a person's thought process. According to Feldman (2013), polarity classification can be carried out at different levels, including the document, sentence, and aspect or feature levels. Any classification level that best fits a researcher's model may be used. Traditional machine learning and deep learning approaches are two of the most widely used approaches for sentiment analysis (Medhat et al. 2014). Other approaches include Lexicon-based (Taboada et al. 2011), hybrid (Prabowo and Thelwall 2009), and machine-learning (Sebastiani 2002).

Conventional neural networks have also been successfully used in a variety of IR applications, for example (Shen et al. 2014a; Shen et al. 2014b). Deep learning has emerged as the most sophisticated machine learning tool with state-of-the-art results in many application domains, including natural language processing, speech recognition, and image classification (Goodfellow et al. 2016). Recently, sentiment analysis using deep learning has also become commonplace. We won't define terms like a neural network, SVM, Random Forest, etc. because they are well-known, similar to other review papers. The remainder of the essay is divided into a conclusion, a literature review, and references. Since the literature review is the primary component of this paper, we primarily concentrate on it.

2 Literature Review

The development of NLP has drawn attention to sentiment analysis. The sentiment has been divided into negative, positive, and neutral categories using a variety of techniques. Accuracy, however, is not proven to be affected. Deep learning techniques have gained popularity recently among researchers. To give an overview of the developments made in this area, we will compare the conventional methods with the deep learning methods.

3 Machine Learning Tools

3.1 Naïve Bayes (NB)

Naive Bayes Classifier, first proposed by Thomas Bayes, is simple to implement and performs computing efficiently in comparison to other machine learning algorithms. It is a supervised classifier that determines whether a piece of data is more likely to be positive or negative. [23] described various difficulties that the researchers encountered and the need for additional study to resolve them.

Naive Bayes is the most effective and efficient inductive learning algorithm for machine learning and data mining. It is based on the Bayes Theorem and makes the assumption that predictors are independent. Its competitive classification performance is surprisingly rarely true for real-world applications. Simply put, the Naive Bayes classifier makes the assumption that the existence of other features has no bearing on the predefined properties. Based on the Bayes theorem, the Nave Bayes Model specifies the relationship between the probabilities p of two events, c and Z, and is represented as P(c) and P(Z), as well as the conditional probability of event c that is conditioned by event Z and vice versa, which is represented as

P(c|Z) and P(Z|c). The Baye's Formula would be as follows:

$$P(c|Z) = \frac{P(c)P(Z|c)}{P(z)} \quad (1)$$

A typical way to represent an example Z is as a tuple of attribute values, where it is the value of the attribute Ti. Assume that C is a classification variable and that c represents C. Let's take a look at two classes: the positive class (represented by +) and the negative class (represented by −). [24] assessed the accuracy by training the Nave Gauss algorithm with 5000 sentences and obtained an accuracy of 0.789, with n = 2. Naive Bayes classification methods for sentiment analysis have the major benefits of being simple to understand and producing accurate results. The algorithm's dependence on the independence of the attributes is a weakness because it may not always hold true.

3.2 Multinomial Naïve Bayes

The Multinomial Nave Bayes classifier, which is frequently used for document level sentiment classification and generally produces good output and performance, is also based on the Bayes Theorem as previously mentioned. As it plays simply for updating the counts needed to estimate the conditional and algorithmic probabilities, this algorithm can be applied and used for the data stream in a trivial manner.

The classification algorithm treats a given document as a collection of words with each word belonging to a different class, with P(w|c) representing the likelihood that the given words w will be found in that class. Then, using the provided training data, an estimate is made by simply computing the relative frequency for all the words in the set of training documents.

$$P(c|d) = \frac{p(c) \prod w \in d \; P(w|c) n_{wd}}{P(d)} \quad (2)$$

The normalization factor P(d) is used, and to get around the zero-frequency issue, the Laplace correction is applied to each conditional and algorithmic probability involved, initializing all the count values to one instead of zero.

3.3 Support Vector Machine (SVM)

According to [16], Support Vector Machine (SVM) is a supervised learning model that outperforms Naive Bayes and Max Ent in the traditional text categorization process.

SVM was first proposed by [25]. It basically identifies the best boundaries to distinguish between positive and negative training samples, and it is widely used because of its superior performance to other methods in most machine learning models, as mentioned by [26] and [27]. However, there are some complexities, as discussed by [28] and [29], that require additional research to be resolved.

SVM can be enhanced with a variety of extensions, increasing its effectiveness and adaptability to practical needs. One of the SVM extensions called Soft Margin Classifier classifies the majority of the data while ignoring any outliers and noisy data because the data may occasionally be linearly visible for multi-dimensional problems and can be separated linearly. Another extension of SVM originally proposed by [7] is Non-Linear Classifier. In No-Linear SVM, the kernel is used to maximize the margin hyper planes. SVM and its extensions are typically used for binary class tasks, but they can also be used for multi-class problems. According to [15], a multi-class SVM extension is available with labels made for objects selected from a limited number of different elements.

In order to improve the parameters of a general SVM, SVMPSO, as described by [30], uses PSO, which is based on Swarm Intelligence Optimization Technique. In this study, SVMPSO outperformed SVM in terms of accuracy and productivity.

Using the right kernel, SVM can handle linear separation on high-dimensional nonlinear input data. There are numerous kernel functions available, including the polynomial kernel of degree, the Gaussian Radial Basic Kernel, and the sigmoid kernel.

According to [31], the Gaussian Radial Basic Kernel Function (RBF) is noticeably superior because it has the soft margin constant C and the kernel hyperparameter (gamma) for the kernel.

According to [32], LIBSVM is a well-known library for support vector machines (SVM), which are used in machine learning operations like classification, regression, and other learning methodologies. It was first proposed in (2001). A training dataset is typically used by LIBSVM to create a model, which is then used to predict data from a testing dataset.

3.4 Maximum Entropy (MaxEnt)

Maximum Entropy (MaxEnt) models are feature-based and do not make any independence assumptions, according to [10–12]. With no feature duplication, we can add new features to MaxEnt by using bigrams and phrases.

The intriguing concept behind Maximum Entropy models is that, according to [19], one should favor the most uniform models that satisfy a particular constraint. It is possible to estimate any probability distribution using these feature-based models. Similar to a two-class situation, logistic regression can be used to find a distribution over classes. Belonging to the teacher class in a four-way text classification where we know that 40% of the documents contain the word "teacher".

It makes sense that a document with the word "teacher" in it would have a 40% chance of being in the teacher class and a 20% chance in the other three classes. When the word "teacher" is absent from the document, we can estimate the uniform class distribution at 25% for each student. Entropy is maximized in this model. In this case, it is simple to calculate the model, but when there are numerous constraints, rigorous techniques are needed to identify the best solution.

MaxEnt does not make the independent assumption for its features, whereas Naive Bayes does. There won't be any feature overlap when we add new bigrams and phrases to MaxEnt.

The model can be seen in the following equation:

$$\mathrm{P_{ME}}(c|d, \lambda) = \frac{\exp[\Sigma_i \lambda_i f_i(c,d)]}{\Sigma_{c'} \exp[\Sigma_i \lambda_i f_i(c,d)]} \tag{3}$$

Class, tweet, and weight vector are all represented by the letters c and d, respectively, in the formula above. Each fi (c, d) stands for a feature. The longer wait time suggests that the feature is a reliable predictor for the given class. Through the use of weight vectors, the importance of a feature in classification is determined. For MaxEnt classification, the authors used Stanford classifier and conjugate gradient ascent for weight training. A feature is a strong pointer if the weight is higher. Naive Bayes (NB) performs well in practice on a variety of problems, but Maxent is better at handling overlapping features, according to [19].

3.5 Stochastic Gradient Descet (SGD)

According to [13], Stochastic Gradient Descent (SGD) can adapt changes over time and offers an effective way to learn some classifiers, even if they are based on the same no-differentiable loss function (hinge loss) used in SVM. They experimented with a vanilla SGD implementation with a fixed learning rate, and hinge loss was optimized using an L2 penalty, which is frequently used to study support vector machines. The formula for document classification that was developed is as follows.

$$\frac{\lambda}{2}\|w\|^2 \sum [1 - (y\ xw + b)]_+, \tag{4}$$

Here, the class labels y is assumed to be (+1, −1) while the optimized loss function x, w represents the weight vector, b represents the bias, and represents the regularization parameters.

This methodology's performance was compared to the Pegasos method, which was developed by [20] and does not call for the explicit specification of a learning rate, but no performance gains were found when using the latter discussed algorithm. On the other hand, with the shifting twitter data streams, the algorithm's capacity to calculate the explicit learning rate seemed to be a crucial task. The learning rate for the per-example updates to the classifier's parameters was kept at 0.1 and the researchers used = 0.0001 in their experiments.

3.6 Multilayer Perceptron (MLP)

According to [15], the Multilayer Perceptron (MLP) is a reliable and non-linear neural network model. It operates as a universal function approximator and has at least one hidden layer and numerous non-linear units, making it effective for learning any relationship between input variable sets. The data flow in a multilayer perceptron (MLP) is unidirectional, just like when data flows from the input layer to the output layer. Every node in the input layer of the neural network used by the multilayer perceptron (MLP) begins as a predictor variable. Neurons (input nodes) in the forward-flowing layer and the following layer (known as the hidden layer) are connected to one another.

The output layer's structure is similar to that of the hidden layer in that the hidden layer neurons are connected to each other and so on.

i. If the prediction is binary, the output layer only contains one neuron.
ii. If the prediction is non-binary, N neurons make up the output layer.

When the neurons are patterned in this way, an efficient information transfer from the input to the output layer. There is an input layer and an output layer, as shown in the image below. Output layer similar to a single layer perceptron, but the same algorithm also includes a hidden layer network in parallel (Fig. 1).

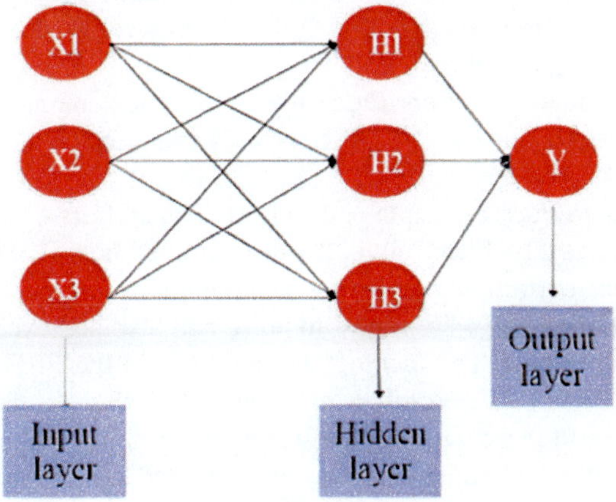

Fig. 1. MLP architecture.

The forward phase of a multi-layer perceptron, which has two phases, is where activation moves from the input layer to the output layer. The errors between the real and operational values and the requested nominal values are replicated in the second phase but in the opposite direction. MLP is a well-known algorithm because of its use as a universal function approximator. It has at least one hidden layer with multiple non-linear objects that can learn almost all functions or relationships within a given input and output variable set thanks to its "back propagation" feature. MLP does not enforce any constraints on the input data and does not begin with any specific assumptions. It can still evaluate the data even if there are noises or other irregularities.

3.7 Random Forest

Random Forest is an ensemble learning technique that focuses on enhancing and storing classification trees, according to [14]. The format of the tree predictors is such that every tree is dependent on independently patterned values of random vectors, and every tree is distributed evenly throughout the forest. According to its definition, random forest is a classifier made up of tree-structured classifiers with the structure h(x,Ok), k = 1,…, where Ok is a set of independently identically distributed random vectors, and each tree casts a single vote for the class that is the most well-known and popular at input x.

In the past few years, random forests have been successfully used to solve a variety of challenging problems in genetic epidemiology and microbiology. Among other established techniques, random forest quickly rose to prominence as a key data analysis technique.

The authors claim that classifying things is a difficult job. They create a classifier on the basis of ensembles. Robust classification procedure. This classifier foresees how the entire set of predictions will be randomly divided. In comparison to the original classifier, the proposed method dramatically improves prediction by combining the effects of multiple classifiers. This strategy is intended especially for high-dimensional data sets where a classifier is desired. This classifier goes by the name CERP. Classification trees (C-T CERP) and logistic regression (LR-T CERP) trees were used as the base classifiers for this new ensemble-based classifier.

In comparison to other classifiers, the results of this one demonstrated high accuracy. The findings demonstrate that the C-T CERP is less dependent on the threshold selection than the LR-CERP.

The fuzzy Random Forest is explained in [22]. The authors of this study presented a fuzzy decision tree as a FRF ensemble. It was examined how fuzzy trees and random forest were combined for training purposes. The projected ensembles benefited from poor data management in that they were robust to noise and also had excellent classification range with small ensembles. The incomplete datasets and the FRF ensembles' results show great promise. The FRF ensembles perform well with datasets that contain fuzzy values. When applied to these datasets, the weighted combination method outperforms the non-weighted method. Compared to the non-weighted method usually applied when comparing the FRF ensembles at random.

4 Deep Learning Tools

After carefully reading the aforementioned papers, we discovered that the majority of researchers only used a small number of datasets to train and test their models. Additionally, they did not take into account the neutral data points. All of the papers use supervised machine learning; none of them have ever used unsupervised learning methods like the Knn algorithm. In terms of accuracy, conventional machine learning algorithms fell short of expectations. All of the papers use the same feature extraction method, preferring to use the uni-gram approach while taking into account approaches like BOW, Word2vec, OneHotshot encoding, and TF-IDF. We will now turn to deep learning techniques, we divided the deep learning models into CNN's, Word Embedding, LSTM (Long Short-term memory), Recurrent Neural network, and DBN's.

4.1 CNN's

The most widely used CNN model for categorizing sentence-level sentiment is Kim's (2014) work. The author tested CNN using pre-trained word2vec as the foundation.

The experimental findings demonstrate that pre-trained vectors can effectively use deep learning as a feature extractor for NLP tasks. Zhang and Wallace proposed a one-layer CNNN architecture for sentence classification as a result of their observations (Zhang and Wallace 2015) (Fig. 2).

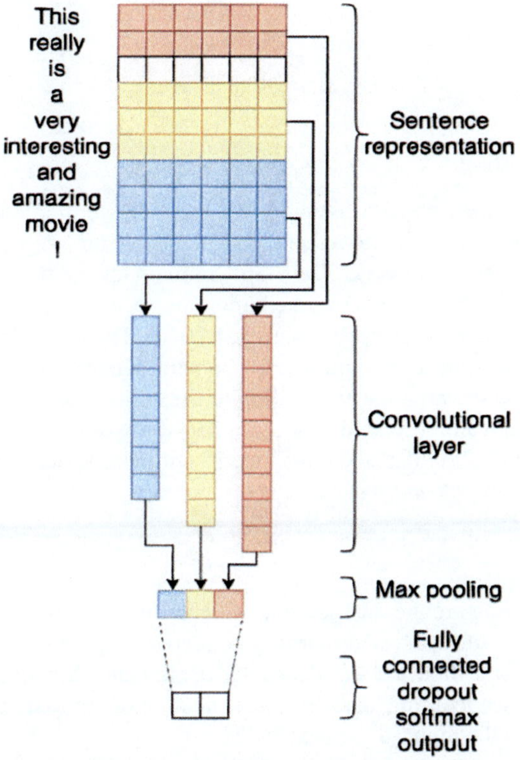

Fig. 2. CNN architecture for sentiment classification for an example.

4.2 Recurrent Neural Networks (RNN)

RNN takes into account the handling of the components in a series' time factor. RNN efficiency depends on both the current input and the output calculated from the network's previously hidden state (Fig. 3).

4.3 LSTM

Long short-term memory in a regular RNN can control the vanishing gradient and capture long-term dependencies.

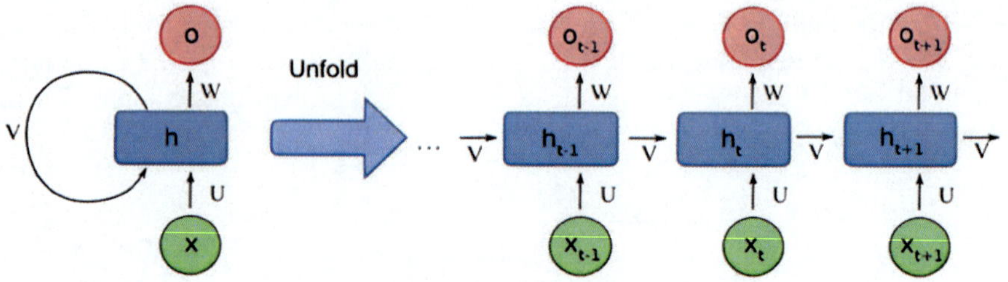

Fig. 3. RNN architecture.

4.4 Word Embedding

Word2Vec is one of the well-liked methods for learning word embedding (Joulin et al. 2016). Before incorporating it into a deep learning algorithm, they process a text using an already-existing neural network. The Skip Gram model and the Bagofwords Typical model (CBOW) can both be used for embedding. The vector encoding of a word (GloVe) is also produced by GlobalVectors (Faruqui et al. 2016). The Glove model has the benefit of being easily trained on more data due to the parallelization of the implementation. However, rather than learning the embedding of the entire word, char2vec c (Sun et al. 2019) learns the embedding related to a character of a word from the other side. 2018 (Xu et al.) used sentiment intensity scores from sentiment lexicons to suggest a model for learning sentiment embeddings.

4.5 DBN's

A type of deep learning architecture known as Deep Belief Networks (DBNs) combines neural networks and unsupervised learning concepts. They are made up of layers of unsupervised, one-at-a-time trained Restricted Boltzmann Machines (RBMs). When performing supervised learning tasks like classification or regression, the final output of one RBM is used as the input to the next RBM.

On ten sentiment datasets, Jin (Naja and Mohamed 2017) implemented DBNs with the delta rule for sentiment classification. The Delta rule employs gradient descent in a single layer neural network to adjust the weights. In (Ruangkanokmas et al. 2016), Ruangkanokmas et al. used DBNs with feature selection (DBNFS) to distinguish between sentences. In Yuan et al. (Chen et al. 2018), Zhang et al. (Li et al. 2019), Jiang et al. (Hailong et al. 2014), and Song et al. (Yoo et al. 2018), attention-based network analysis is used to analyze emotions. Additionally, capsule networks are gaining popularity in natural language processing for a variety of text classification tasks (Ke et al. 2019; Yang et al. 2019; Kim and Jeong 2019) (Fig. 4).

While deep learning techniques have shown promising results in sentimental research, there are some limitations. Deep learning networks require a large amount of labeled data for training in order to produce better results. In contrast to conventional machine learning or lexical approaches, where we know, what features are selected to forecast a particular feeling, it is challenging to ascertain the true reason for the neural network to forecast a specific sentiment in a body of text by pointing at weights in multiple elements. Many researchers find it challenging to understand how neural networks

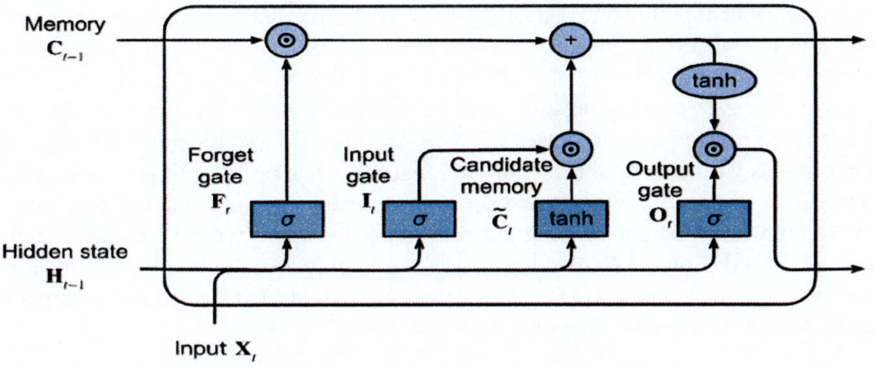

Fig. 4. Architecture of LSTM.

work as a "black box" and a prediction mechanism. Selecting ideal circumstances is a difficult task as well. Due to the numerous parameters, deep learning methods have been resource-intensive.

5 Discussion and Possible Directions

Several papers and researches on deep learning and machine learning-based tools and approaches for sentiment analysis and categorization are consulted in this article. We examine well-known algorithms and methods that are frequently applied to sentiment analysis. A comparison of accuracy across several datasets is provided, and this comparison can be utilized as a guide in additional research.

The majority of the suggested sentiment analysis method is static. As a result, dynamic sentiment analysis and tracking is one of the main challenges facing the use of deep learning algorithms for sentiment analysis. For instance, the information, vocabulary, and user base on the social network Twitter can change at any time, making dynamic analysis difficult. Dealing with language structure, which includes slang, is challenging, on the other hand. Additionally, complex analysis and specialized processing are needed for heterogeneous information (such as short or long phrases) [48].

Subsequently, well known algorithms and techniques are examined, generally used for classification of feelings. A comparison is provided in terms of precision on different datasets and can be used as a reference for future research. We advise researchers to adopt cutting-edge deep learning approaches to resolve these issues, such as Bert's algorithm for learning context representation [8], deep reinforcement learning (DRL), and extremely effective generative adversarial networks (GANs).

6 Conclusion and Future Works

Sentiment classification at the sentence level is the foundation of the study. It is helpful for determining how people feel about any subject or occasion. It can be used in a variety of domains, including economics, business development, and social phenomenon research to forecast socioeconomic occurrences like stock market predictions. Academic research and industrial applications have skyrocketed as a result of its extensive use in potential implementation. Deep learning algorithms for sentiment analysis have attracted a lot of academic interest recently, leading to the development of a wide range of effective methods for various tasks. We started off by going over the development of sentiment analysis and its various levels. Both conventional and machine learning methodologies have been discussed, with a focus on deep learning techniques and their applications.

As future work, we intend to examine additional algorithms typically used for sentiment analysis and compare them to the metrics we employed in this study. Additionally, we intend to offer an architecture to enhance the outcomes of each algorithm for compare their mechanism and known the performance of them.

References

1. Gautam, G., Yadav, D.: Sentiment analysis of Twitter data using machine learning approaches and semantic analysis. In: 2014 Seventh International Conference on Contemporary Computing (IC3), pp. 437–442, IEEE. https://doi.org/10.1109/IC3.2014.6897213
2. Hasan, A., Moin, S., Karim, A., Shamshirband, S.: Machine learning-based sentiment analysis for Twitter accounts. Math. Comput. Appl. **23**(1), 11 (2018). https://doi.org/10.3390/mca23010011
3. Wawre, S.V., Deshmukh, S.N.: Sentiment classification using machine learning techniques. Int. J. Sci. Res. **5**(4), 819–821 (2016). https://doi.org/10.21275/v5i4.NOV162724
4. Joulin, A., Grave, E., Bojanowski, P., Mikolov, T.: Bag of tricks for efficient text classification. arXiv preprint arXiv:1607.01759 (2016). https://doi.org/10.18653/v1/E17-2068
5. Naja, M.M.F., Mohamed, M.I.I.: Analysis of Systematic Data Mining Approaches for Achieving Competitive Advantage by Monitoring social media (2017)
6. Ramadhani, A.M., Goo, H.S.: Twitter sentiment analysis using deep learning methods. In: 2017 7th International annual engineering seminar (InAES), pp. 1–4. IEEE. https://doi.org/10.1109/INAES.2017.8068556
7. Severyn, A., Moschitti, A.: Twitter sentiment analysis with deep convolutional neural networks. In: Proceedings of the 38th International ACM SIGIR Conference on Research and Development in Information Retrieval, pp. 959–962 (2015). https://doi.org/10.1145/2766462.2767830
8. Faruqui, M., Tsvetkov, Y., Rastogi, P., Dyer, C.: Problems with an evaluation of word embeddings using word similarity tasks. arXiv preprint arXiv:1605.02276 (2016). https://doi.org/10.18653/v1/W16-2506
9. Pang, B., Lee, L., Vaithyanathan, S.: Thumbs up: sentiment classification using machine learning techniques. In: Proceedings of the 2014 Conference on Empirical Methods in Natural Language Processing, pp. 79–86 (2002)
10. Mouthami, K., Devi, K.N., Bhaskaran, V.M.: Sentiment analysis and classification based on textual reviews. In: 2013 International Conference on Information Communication and Embedded Systems, pp. 271–276 (2013)
11. Nigam, K., Lafferty, J., McCallum, A.: Using maximum entropy for text classification. In: IJCAI-99 Workshop on Machine Learning for Information Filtering, vol. 1, pp. 61–67 (1999)
12. Singh, P.K., Shahid Husain, M.: Methodological study of opinion mining and sentiment analysis techniques. Int. J. Soft Comput. 5(1), 11–21 (2014)
13. Argamon, S., Whitelaw, C., Chase, P., Hota, S.R., Garg, N., Levitan, S.: Stylistic text classification using functional lexical features. J. Am. Soc. Inform. Sci. Technol. **58**(6), 802–822 (2007). https://doi.org/10.1002/asi.20553

14. Taşkin Kaya, G., Ersoy, O.K., Kamaşak, M.E.: Support vector selection and adaptation for remote sensing classification. IEEE Trans. Geosci. Remote Sens. **49**(6) PART 1, 2071–2079 (2011)
15. Vaghela, V.B., Jadav, B.M., Scholar, M.E.: Analysis of various sentiment classification techniques. Int. J. Comput. Appl. **140**(3), 975–8887 (2016)
16. Cortes, C., Vapnik, V.: Support vector machine. Mach. Learn. 1303–1308 (1995)
17. Pang, B., Lee, L.: Opinion mining and sentiment analysis. Found. Trends Inf. Retr. **2**(1–2), 1–135 (2008)
18. Dang, Y., Zhang, Y., Chen, H.: A lexicon-enhanced method for sentiment classification: an experiment on online product reviews. IEEE Intell. Syst. **25**(4), 46–53 (2009). https://doi.org/10.1109/MIS.2009.105
19. Samad, D., Gani, G.A.: Analyzing and predicting spear-phishing using machine learning methods. Multidiszciplináris Tudományok, **10**(4), 262–273 (2020). https://doi.org/10.35925/j.multi.2020.4.30
20. Vohra, M.S., Teraiya, J.B.: A comparative study of sentiment analysis techniques. J. Inform. Knowl. Res. Comput. Eng. ISSN 0975–6760
21. Feldman, R.: Techniques and applications for sentiment analysis. Commun. ACM **56**(4), 82–89 (2013). https://doi.org/10.1145/2436256.2436274
22. Li, X., Bing, L., Zhang, W., Lam, W.: Exploiting BERT for end-to-end aspect-based sentiment analysis. arXiv preprint arXiv:1910.00883 (2019). https://doi.org/10.18653/v1/D19-5505
23. Ahmad, M., Aftab, S., Muhammad, S.S., Waheed, U.: Tools and techniques for lexicon driven sentiment analysis: A review. Int. J. Multidiscip. Sci. Eng. **8**(1), 17–23 (2017)
24. Wang, H., Can, D., Kazemzadeh, A., Bar, F., Narayanan, S.: A system for real-time twitter sentiment analysis of 2012 U.S. presidential election cycle. Proceedings of the 50th Annual Meeting of the Association for Computational Linguistics, no. July, pp. 115–120 (2012)
25. Breiman, L.: Random forests. Mach. Learn. **45**(1), 5–32 (2001)
26. Nirmala, D.K., Jayanthi, P.: Sentiment classification using Address for Correspondence. Int. J. Adv. Eng. Technol. 1–3 (2016)
27. Khairnar, J., Kinikar, M.: Machine learning algorithms for opinion mining and sentiment classification. Int. J. Sci. Res. Publ. **3**(6), 1–6 (2013)
28. Xu, H., Liu, B., Shu, L., Yu, P.S.: Double embeddings and CNN-based sequence labeling for aspect extraction. arXiv preprint arXiv:1805.04601 (2018). https://doi.org/10.18653/v1/P18-2094
29. Amrani, Y.A., Lazaar, M., Kadiri, K.E.E.: Random forest and support vector machine based hybrid approach to sentiment analysis. Procedia Comput. Sci. **127**, 511–520 (2018). https://doi.org/10.1016/j.procs.2018.01.150
30. Ahmad, M., Aftab, S., Muhammad, S.S.: Machine learning techniques for sentiment analysis: a review. Int. J. Multidiscipl. Sci. Eng. **8**(3), 27 (2017)
31. Bellar, O., Baina, A., Bellafkih, M.: Application of machine learning to sentiment analysis. In: Hassanien, A.E., et al. (eds.) The 3rd International Conference on Artificial Intelligence and Computer Vision (AICV 2023). https://doi.org/10.1007/978-3-031-27762-7_13

Knowledge Infrastructure Data Wizard (KIDW): A Cooperative Approach for Data Management and Knowledge Dissemination

Ammar Aljer(✉), Mohammed Itair, Mostafa Akil, and Isam Sharour

Civil and Geo-Environmental Laboratory (LGCgE), Lille University, Lille, France
Ammar.aljer@univ-lille.fr

Abstract. The advent of the digital revolution has witnessed the proliferation of data generation, from the use of machines in computing to their use in storing data and then to their use in the automatic generation of data, as in the case of the IoT (Internet of Things). Consequently, the volume of data has exponentially increased, necessitating enhanced storage capabilities and processing power for knowledge extraction. This paper highlights the significance of extracting information and knowledge from the early stages of data generation. It introduces key concepts and demonstrates their implementation in the "Knowledge Infrastructure Data Wizard" (KIDW) tool, which utilizes a dynamic tree topology. The tool facilitates human-machine interaction through user-friendly interfaces, enabling effective project management by enhancing cooperative data management. The article showcases the practical application of the developed tool within a European Interreg project, where it facilitated data and knowledge sharing, as well as fostering cooperation and participant management.

Keywords: Data Wizard · Knowledge Transfer · Information Oriented Management · Knowledge Dissemination · Common Data Environment · Project Management

1 Introduction

In today's data-driven world, the volume, variety, and velocity of data generated worldwide are enormous. According to a study conducted by the University of California, San Diego, the average individual generates approximately 1.7 megabytes of data per second [1]. This generated data per person is roughly equivalent to 140 million books per person per year. Furthermore, produced data is continuously expanding. According to the International Data Corporation (IDC), the global data sphere tends to grow from 64.2 zettabytes in 2020 to 180 zettabytes by 2025, with a compound annual growth rate (CAGR) of 23% [2].

Despite the abundance of data, effectively managing and utilizing it remains challenging. Increasing data needs storing and processing for extracting information and consequently needs more energy which exacerbates the greenhouse effect; data centers are responsible for 2% of the global CO2 emission [3].

M. Ben Ahmed et al. (Eds.): SCA 2023, LNNS 906, pp. 34–43, 2024.
https://doi.org/10.1007/978-3-031-53824-7_4

Conceptualizing this massive influx of data using the DIKW pyramid, which is a model that represents the relationships between Data, Information, Knowledge, and Wisdom [4], can be shown as a massive expansion in the pyramid's base. This vast and complex data cannot be efficiently employed without outstanding instruments.

This paper adopts a modified version of the DIKW pyramid to illustrate two key aspects. First, the base of the pyramid represents the immense volume of data, which is not directly generated by humans but rather by utilizing machine-generated data. Second, the knowledge or wisdom positioned at the pinnacle of the pyramid should possess a global scope, requiring suitable formatting and worldwide accessibility to facilitate dissemination (Fig. 1).

Fig. 1. Revising DIKW toward Global Knowledge

The effective dissemination of information is essential, and its accessibility should align with its significance. The quality and structure of information significantly impact its availability. This paper explores the implementation of this principle within the Knowledge Infrastructure Data Wizard (KIDW).

To leverage data derived from experiential knowledge building, it is crucial to convert the data into information and place it within the appropriate context of human knowledge. This entails presenting the information to users through suitable tools that facilitate comprehension and utilization.

The exponential growth of data witnessed in the past decade can be attributed to the utilization of machines for data generation. Although machines operate under the control and management of human algorithms and decisions to accomplish specific objectives, it is ultimately the programmers or users who possess an understanding of "why" and "what" the machines produce. They possess the expertise to extract valuable information and knowledge from the generated data. However, a prevalent mistake nowadays is relying solely on machines for answers, overlooking the fact that machines function based on human instructions.

On the other hand, the utility of this information for humanity depends on its availability, ease of access, and integration within the appropriate educational and cognitive framework, ensuring its accessibility to the widest possible audience of beneficiaries.

The KIDW was a part of the Polder2C (P2C) European Interreg project. Fourteen main partners and fifty observers cooperated to realize Polder2C over 42 months. The University of Lille (U-Lille) was one of these partners, and the development of the KIDW was one of its tasks. It is a key component of the "Knowledge Infrastructure" work package within the P2C project, which aims to transform experimental data and observations into accessible knowledge for professionals, students, administrators, and communities [5]. The KIDW acts as a didactical tool for the dissemination of knowledge, a tool to access the P2C's database, an approach for managing data in multi-activity projects, and a tool for project development, assessment, and monitoring.

The remainder of this paper is structured as follows: Sect. 2 provides a brief review of the main challenges facing data management in multi-activity projects, highlighting the gaps and limitations that the KIDW aims to address. Section 3 describes the methodology and architecture of the KIDW, detailing its main features and functionalities. Section 4 introduces the main components of the data wizard. Section 5 presents the testing phase and the feedback. Finally, Sect. 6 concludes the paper by summarizing the main contributions of the KIDW, the key insights obtained from this study, and highlights future works.

2 Data Management Challenges in Multi-Activity Projects

In multi-activity projects, managing data can be an arduous task due to the complexity and interrelatedness of the various activities involved [6, 7]. The challenge is to collect, process, and analyze the massive amounts of data generated by different activities and to derive meaningful insights that can support decision-making [8].

One of the primary challenges in multi-activity projects revolves around the integration of data from diverse sources and the standardization of this data to enable seamless collaboration [9, 10]. Stakeholders involved in the project often utilize different data formats, tools, and systems, which pose significant obstacles in consolidating the data and ensuring consistency throughout the project.

Ensuring data security and privacy presents another noteworthy challenge in multi-activity projects. With multiple stakeholders accessing and sharing data, the risk of data breaches and unauthorized access is heightened [11, 12]. It becomes crucial to maintain data privacy, particularly when dealing with sensitive or confidential information related to infrastructure, government, or citizens [13].

Mass-producing data is a common issue in multi-activity projects, where various stakeholders generate and collect vast amounts of data throughout the project lifecycle. However, not all of this data turns into useful information that can inform decision-making and improve project outcomes. A significant portion of the data may be irrelevant or redundant, leading to information overload and decreased data quality [14, 15]. This issue can have a profound impact on the quality of multi-activity projects outcomes, as stakeholders may struggle to extract meaningful insights from the data and make informed decisions [16].

The KIDW was developed to address such challenges by providing an efficient approach for managing information sharing and data accessing multi-activity projects. The tool streamlines data collection and processing by providing a centralized platform for data management, enabling partners to access and contribute to the project's databases easily. By using the KIDW, partners can reduce the time and effort required to access, process, and analyze data, allowing them to focus on deriving insights and knowledge from the data.

Moreover, the KIDW provides a didactic approach to disseminating knowledge, making it accessible to partners with varying levels of expertise. Thus, it ensures that everyone involved or intended to be involved in the project can understand and contribute to the project's objectives and outcomes, promoting transparency and collaboration.

3 Methodology

A prerequisite for the success of any endeavor to advance knowledge is the incorporation of the initiative's outcomes into the practices of a larger community. The Data Wizard concept was introduced primarily to consolidate the obtained knowledge of the P2C Living Lab through a variety of activities. As one of the main partners in P2C, U-Lille was responsible for developing this concept within the context of P2C and its activities.

U-Lille adopted a spiral model for the software development of the Data Wizard platform, consisting of multiple software components. Each component, including the platform itself, undergoes a defined process of requirement specification, design, implementation, testing, deployment, and integration with existing components. The development process follows an iterative approach, incorporating user feedback to drive continuous improvement (Fig. 2).

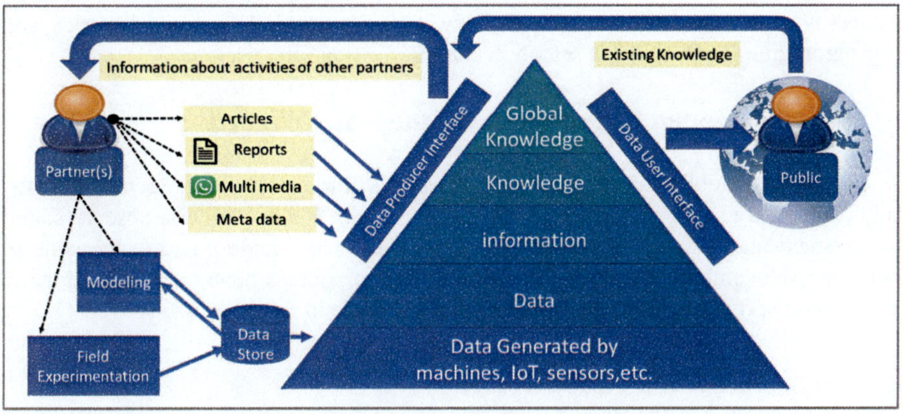

Fig. 2. Project partners cooperation in Knowledge Creation & Dissemination via data wizard

U-Lille collaborated with other P2Cs to establish the project's data workflow. Discussions with partners, incorporating questionnaires and focus group meetings, helped define the Data Wizard based on diverse needs, interests, and perspectives. The resulting specification was presented in an official meeting, followed by the design, implementation, deployment, and testing of the software component. The continuous improvement led to the sequential lifecycle of different components, ensuring a dynamic development process.

The goal was to achieve the following objectives:

1. Involving Polder2Cs partners in meta-data, information, and knowledge generation: The wizard team collaborated with partners and analyzed data production conditions and goals. Interfaces were developed to encourage partners to provide meaningful descriptions of their data in the early stages and gradually enhance these descriptions with relevant knowledge. Existing resources, such as videos, texts, reports, etc., were utilized to link data with corresponding information and knowledge.
2. Tree topology for representation: To handle system complexity, a tree topology was employed to structure the information and the resources, where more abstract components are at the root and detailed ones are at the leaves. This structure facilitated cooperation among partners and managers at different hierarchical levels. It represented existing and new knowledge, project management structure, etc.
3. Virtual linking of data, metadata, and knowledge: Metadata describing datasets were kept separated from the datasets themselves in an abstract layer that has the necessary links to access the data when necessary, allowing for easier communication, project management, and access. This approach bypassed the constraints of storage, access rights, and security related to data.
4. Gathering and structuring related existing knowledge: The wizard team developed a methodology to select relevant references related to Polder2Cs topics and presented them alongside project outputs.
5. User-friendly interfaces: The roles of different partners in data production and the presentation of data, information, and knowledge to the public were analyzed. Interfaces were designed and implemented, leveraging available IT options, libraries, and programming languages for seamless communication with users.

4 Main Developed Components of the Data Wizard

Based on the established methodology, several components have been systematically developed and implemented in collaboration with P2C's partners, observers, and management team. Notably, Polder2C's Wizard has been made openly accessible to users since November 2021 [17]. Many of these components have been created using open-source software, allowing for their potential reuse in future projects.

4.1 The Interactive Knowledge Tree (IKT)

This part presents a database of existing knowledge and practices about flood protection infrastructures with a focus on the design, management, challenges, and needs for new knowledge. This database is based on links to reports, websites, videos, and images. To improve usability, the contents of the database are classified according to topics, pedagogic level, media type, etc. (Fig. 3)

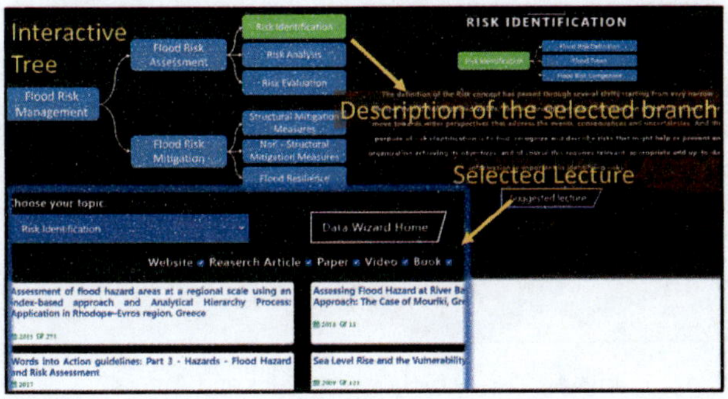

Fig. 3. The Interactive Knowledge Tree and Existing Knowledge Library

4.2 The Typical Experimentations

The presented experiments on erosion and overflow adopt a digital twin approach, where the primary object (levee) and its associated components are represented with their key 3D features. Interactive graphs are utilized to visualize the resulting data, while vlogs and textual information provide explanations regarding metadata, goals, pre-and post-conditions, and outcomes of the experiments.

4.3 The Activity in the Project

The core of the wizard lies in the concept and representation of activities. An activity is any meaningful action realized by a partner in the context of the project, big or small, functional or operational, planned to be done or already executed. Examples of activities are exercises, modeling, vlogs, measurements, work packages, etc. The wizard enables any partner to gradually create his activities and link them together or with the activities of other partners. Depending on the insight of the activity creator, many activities can be grouped into a single, more complex one. An activity may be created using merely a name and a brief description; it is supposed that the activity creator would gradually fill the maximum possible of the activity fields to describe the project: temporal and space references, project management references such as the related work package, links to the produced data, meta-data, realization pre-and post-conditions, outcomes, produced data and knowledge, etc. (Fig. 4)

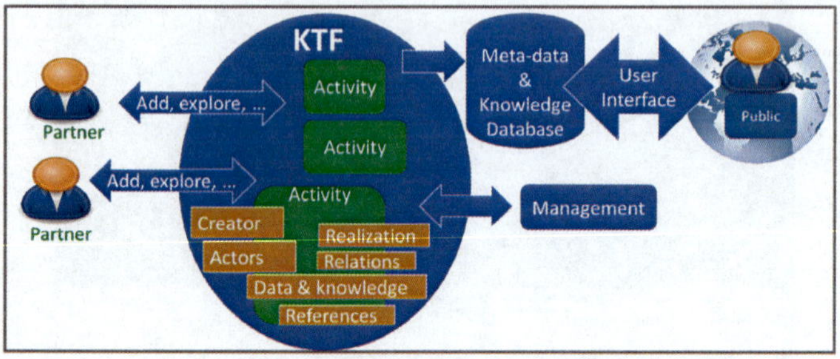

Fig. 4. The Activity's Attributes and Actors

4.4 The Relations and Dependency Graph

This is a dynamic visual tool that provides a comprehensive project overview, utilizing nodes to represent activities and arrows to depict their interrelationships. Clicking on a node yields a detailed description of the corresponding activity. Arrows are color-coded to denote specific types of oriented relationships. Users have the flexibility to select activity groups and relation types for exploration. Notably, the "makes part of" relationship type plays a crucial role, enabling users to uncover the structural hierarchy of the project (Fig. 5).

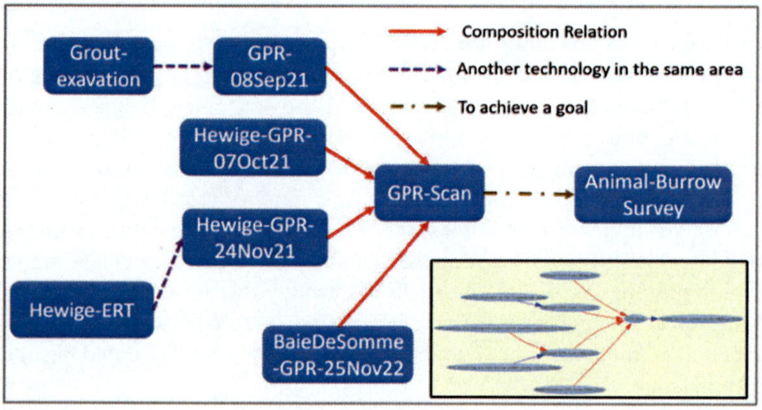

Fig. 5. The Relations and Dependency Graph

4.5 The Knowledge Transfer Form (KTF)

An intermediate tool developed to facilitate communication between P2Cs' partners, the management team, and the Data Wizard team, this tool helps Polder2Cs partners leave meaningful traces (or meta-data) about their activities and the produced data, information, and knowledge. Starting in the early stages of data production and continuing throughout the project, this gradually and cooperatively builds the structure of the project and provides an overview of the entire project. This facilitates structured access to the data and knowledge produced by these activities.

4.6 The Storybook

As part of the knowledge dissemination and user-friendly interface, a storybook has been developed to provide an engaging and accessible means of showcasing the activities associated with the levee project. The storybook adopts a storytelling approach, presenting a high-level overview with simplified details and gradually delving into more intricate knowledge nodes. Through this interactive and visually appealing interface, users can easily navigate and explore the project's activities, fostering a deeper understanding and appreciation of its significance (Fig. 6).

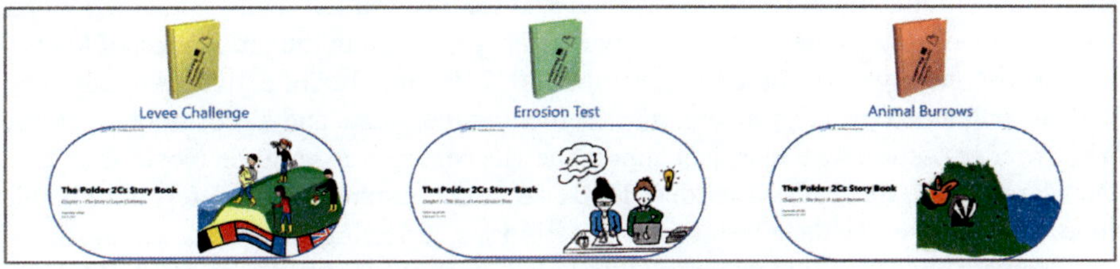

Fig. 6. The Levee Storybook's Chapters

5 Put into Operation and Feedbacks

In November 2021, the Data Wizard launched with three key components: "existing knowledge," "typical experimentation," and "an experimental site." The wizard had 12 datasets and a database with fewer than 100 rows. Partners and the public could view information but not add or modify it (Table 1).

The concept of the Knowledge Transfer Form (KTF) was proposed and adopted in November 2021. After analyzing partner and manager needs, a prototype was developed and improved until April 2022, when the platform launched. Partners could describe their activities using about 50 fields for each one, they could link them to other partners' activities, and structure their activities in the KTF. Over time, partners used the platform to report their progress and learn from other partners. However, sporadic project management intervention was needed to persuade partners to create or update activities. Partners and the Data Wizard team collaborated to improve the wizard's structure, information, and references. The table gives quantitative measures of KTF key field usage on three occasions.

Table 1. Statistics on the evolution of the use of the various fields of the KTFs during nine months: 2-Number of Users 3-Number of Activities. 4-Objectives. 5-With brief description. 6-With full description. 7-With pre-condition. 8-Description of the state during the activity. 9- with post conditions. 10-with start date. 11-With final conclusion. 12-with Data. 13-with knowledge

Date	2 Us	3 Ac	4 Ob	5 BD	6 F D	7 Pr C	8 D C	9 Po C	10 S D	11 FC	12 D	13 K
June 2022	8	49	23	22	13	15	11	9	20	9	14	1
Nov. 2022	21	106	49	60	23	22	24	19	54	21	33	15
March 2023	15	116	87	86	51	41	43	37	74	46	78	33

During the period from April to June 2022, over 49 activities were established, characterized by limited meta-data and minimal knowledge products. Out of these activities, only 14 had available data, and merely one activity resulted in the production of knowledge. Subsequently, from June to November 2022, the number of activities nearly doubled, accompanied by a significant enhancement in meta-data and a substantial increase in knowledge output. As the project approached its conclusion, between November 2022 and March 2023, there was a modest 10% rise in the number of activities, but notable improvements were made in the activity descriptions, reflecting the project's final stages.

The main challenge in developing KIDS was the ability to examine potential knowledge in a wide range of activities of various partners, develop appropriate components to obtain it, and then present it in appropriate form(s) to other partners and observers. The involvement of the partners since the early steps of the developing of a a tool, that they had not yet seen its benefits, was crucial. In a future project, it would be useful to use the KIDS from the beginning of partners' activities. This may need a previous step to analyze the specific needs of each partner and for training.

6 Conclusion and Future Works

This work emphasizes the significance of correlating information dissemination with its level in the DIKW Pyramid and highlights the principles that facilitate effective data documentation and knowledge extraction from the early stages of data generation in a project. A dedicated tool, the Data Wizard, has been developed and successfully implemented in a European Interreg project. Practical application of the tool has also revealed its added value in project management. However, it has become evident that the mere presence of a knowledge transfer tool is insufficient to encourage partners to document their knowledge and information. Collaborating with the project management team is crucial to promoting partner engagement and ensuring proper utilization of the tool. Additionally, the availability of the wizard team to provide explanations and assist in selecting the appropriate information format for dissemination is essential.

One of our objectives is to open access to the software components for the public. Moving forward, future work will involve extending the usability of the Data Wizard beyond a single project. By introducing an additional level in the conceptual tree, the tool can integrate results from different projects simultaneously. Currently, the KTF, EKT, and storybook structures are separate, with partners managing activities in the KTF while the EKT and storybook require direct involvement from the wizard team for communication with the partners. To streamline the process and enhance efficiency, these structures can be integrated with specific stereotypes and interfaces, allowing partners to directly contribute to the contents of the EKT and the storybook in a controlled manner. This integration not only simplifies the task of the wizard team but also enables the direct linking of stories and ENT items to specific activities.

References

1. Bohn, R.E., Short, J.E.: How Much Information? 2009 Report on American Consumers (2009)
2. Reinsel, D., Gantz, J., Rydning, J.: The Digitization of the World from Edge to Core (2018)
3. Khosravi, A., Buyya, R.: Energy and Carbon Footprint-Aware Management of Geo-Distributed Cloud Data Centers: A Taxonomy, State of the Art, and Future Directions. In: Sustainable Development: Concepts, Methodologies, Tools, and Applications, pp. 1456–1475. IGI Global (2018). https://doi.org/10.4018/978-1-5225-3817-2.ch064
4. van, M.H.J.: Revising the DIKW pyramid and the real relationship between data, information, knowledge, and wisdom. Law Technol. Hum. **2**, 69–80 (2020). https://doi.org/10.3316/agispt.20210112042035
5. Living Lab Hedwige-Prosperpolder | Polder2C. https://polder2cs.eu/. Accessed 14 May 2023
6. https://www.facebook.com/ClicDataBI: Guide to Transform Data into Actionable Insights. https://www.clicdata.com/blog/transform-data-into-actionable-insights/. Accessed 14 May 2023
7. Marle, F., Jaber, H., Pointurier, C.: Organizing project actors for collective decision-making about interdependent risks. Complexity **2019**, e8059372 (2019). https://doi.org/10.1155/2019/8059372
8. Lehtinen, J., Peltokorpi, A., Artto, K.: Megaprojects as organizational platforms and technology platforms for value creation. Int. J. Project Manage. **37**, 43–58 (2019). https://doi.org/10.1016/j.ijproman.2018.10.001
9. Surani, I.: Challenges of Integrating Heterogeneous Data Sources. https://www.dataversity.net/challenges-of-integrating-heterogeneous-data-sources/. Accessed 14 May 2023
10. Sibenik, G., Kovacic, I.: Assessment of model-based data exchange between architectural design and structural analysis. J. Build. Eng. **32**, 101589 (2020). https://doi.org/10.1016/j.jobe.2020.101589
11. Bairu, G.: Forum Guide to Data Governance (2020)
12. Seh, A.H., et al.: Healthcare data breaches: insights and implications. Healthcare (Basel) **8**, 133 (2020). https://doi.org/10.3390/healthcare8020133
13. Lee, W.W., Zankl, W., Chang, H.: An Ethical Approach to Data Privacy Protection. https://www.isaca.org/resources/isaca-journal/issues/2016/volume-6/an-ethical-approach-to-data-privacy-protection. Accessed 14 March 2023
14. Schneider-Kamp, A.: The potential of AI in care optimization: insights from the user-driven co-development of a care integration system. Inquiry **58**, 00469580211017992 (2021). https://doi.org/10.1177/00469580211017992
15. Beck, S.F., Abualdenien, J., Hijazi, I.H., Borrmann, A., Kolbe, T.H.: Analyzing contextual linking of heterogeneous information models from the domains BIM and UIM. ISPRS Int. J. Geo Inf. **10**, 807 (2021). https://doi.org/10.3390/ijgi10120807
16. Hemp, P.: Death by Information Overload (2009). https://hbr.org/2009/09/death-by-information-overload
17. Data Wizard. https://www.smartdata2030.com/. Accessed 14 March 2023

YOLOv5 Model-Based Real-Time Recyclable Waste Detection and Classification System

Leena Ardini Abdul Rahim[1], Nor Afirdaus Zainal Abidin[1], Raihah Aminuddin[1](✉), Khyrina Airin Fariza Abu Samah[1], Asma Zubaida Mohamed Ibrahim[2], Syarifah Diyanah Yusoh[3], and Siti Diana Nabilah Mohd Nasir[4]

[1] College of Computing, Informatics and Mathematics, MARA University of Technology Malacca Branch (Jasin Campus), Malacca, Malaysia
raihah1@uitm.edu.my
[2] City University, Selangor, Malaysia
[3] Multimedia University, Malacca, Malaysia
[4] University of Hull, Hull, UK

Abstract. Emerging nations, driven by population growth and rapid urbanization, generate significant waste. Inadequate waste management systems prevail in many countries, including Malaysia, due to a lack of understanding and insufficient infrastructure. Despite poor waste management, there needs to be an automated classification system, leading to time-consuming manual recycling processes. The project aims to develop a real-time waste identification and classification system. The project's objectives are: 1) design a prototype using a web application and a real-time video platform to detect and categorize recyclable waste; 2) develop the prototype utilizing the YOLOv5 model; and 3) test the model's accuracy. In the real-time video environment, the system can identify the type of waste and the corresponding recycle bin colors for proper disposal. The model achieved an accuracy rate of 86.25% in identifying and detecting the waste.

Keywords: YOLOv5 · recycle · deep learning · image processing

1 Introduction

According to [1], the continued population growth and resource shortage have led to a problem with the disposal of raw materials, which has severely impacted the environment. The waste is disposed of in landfills or water, and if the current trends continue, the world may become a massive landfill with harmful environmental effects. Prioritizing trash sorting is essential to improving material recycling and reducing the danger of contamination. As stated by [2], recycling is essential for protecting the environment. [3] also emphasizes the need to set up effective recycling practices to avoid contamination, which slows down recycling and creates environmental risks.

M. Ben Ahmed et al. (Eds.): SCA 2023, LNNS 906, pp. 44–54, 2024.
https://doi.org/10.1007/978-3-031-53824-7_5

Recyclable materials such as paper, plastic, metal, and glass can be disposed of in recycling bins. The recyclable bin illustrated these materials with four distinct colors: green, blue, yellow, and red, based on Hering's theory of primary colors [4]. This color scheme facilitates accurate waste categorization, preventing contamination, [5] and [6].

The advancement of technology, particularly in image processing and artificial intelligence, has transformed waste into valuable resources and energy. [7] indicate that machines, computer vision, and artificial intelligence can detect and sort recyclable materials effectively. Real-time image processing is essential for image analysis and information extraction.

This project proposes a real-time trash detection and classification system to assist users in identifying and classifying waste types, thereby reducing the risks associated with improper disposal. The system can accurately identify the type of waste used in real-time videos by utilizing image processing techniques. The project emphasizes the significance of waste sorting and recycling in addressing environmental challenges such as air and water pollution through appropriate waste management practices.

1.1 Project Scope

The project focuses on classifying four common types of recyclable waste: paper, plastic, glass, and metal. [4] conducted a project with 313 Taiwanese respondents and 105 participants from various East Asia and the Pacific (EAP) nations. Based on EAP's preferences, the proposed system aims to categorize garbage in Malaysia using the colors green for paper, blue for glass, yellow for plastic, and red for metal, as can be seen in Fig. 1.

Fig. 1. The four common types of recyclable waste and recycle bin colors are metal (red), plastic (yellow), paper (green), and glass (blue) [8].

2 Related Works

Recent advancements in waste classification have shown promising results. One notable study by [9] introduced a system that can do real-time detection with a classification algorithm for recycled materials. The system uses Convolutional Neural Networks (CNNs) and the You Only Look Once Version 4 (YOLOv4) technique to identify common waste materials on moving belts in waste management facilities. With an impressive mean average precision (mAP@0.50) score of 92.43%, the algorithm accurately detects and classifies materials like plastic, paper, and aluminum. Another related work by [2] presents a system with cloud-based capabilities for in-house recycling and waste classification. The dataset used in their study included 2527 photos divided into six categories: cardboard, glass, metal, paper, plastic, and trash. Their CNN-based system achieves an accuracy of 93.4% and assists in automating waste classification for citizens at home. The authors employed image processing, advanced computer vision techniques, classifiers like K-nearest neighbor (KNN), and neural network algorithms for waste level detection, collection, and management. [10] developed a project on trash classification for recyclability status, utilizing Support Vector Machines (SVM) and CNN techniques. The project aimed to categorize images of garbage or trash into specific classes, including glass, paper, metal, plastic, cardboard, and trash. To accomplish this, they collected a dataset of approximately 400 to 500 images for each class and a training and testing dataset with a ratio of 70:30. The test accuracy showed a result of 63% with a training error of 30%. These results were obtained using the SVM technique, which outperformed the CNN model. On the other hand, the CNN model achieved a test accuracy of only 22%. While their results were modest, the study highlights the challenges and complexities of training complex models. Overall, the study by [9] stands out as the most similar to the proposed system, achieving high accuracy in waste detection with a score of 92.43% using CNNs and YOLOv4 for the classification.

3 System Design

The project involves several steps, and Fig. 2 illustrates the research design of the system. Firstly, the dataset is collected from TrashNet and consists of 2527 images categorized into six waste types. The dataset is a JPEG file. However, the focus is narrowed to four waste types: plastic, paper, glass, and metal. The four types of waste are also divided into four recycle bin colors: red for glass, yellow for metal, green for paper, and blue for plastic. The dataset is split into 80% for training (1590 images) and 20% for testing (397 images). In the pre-processing step, bounding box labelling is performed using an online tool called MakeSense AI. The training stage utilizes the You Only Look Once (YOLO) algorithm, specifically YOLO version 5, known for its speed and lightweight nature. The training process involves using the training and testing datasets to generate a weight file for later use. YOLOv5, with its lightweight model and equivalent accuracy to YOLOv4, consists of three main components: Cross Stage Partial Networks (CSPDarknet) for feature extraction, Path Aggregation Network (PANet) for feature fusion, and the YOLO Layer for detection results [11]. Images captured using a webcam and 397 images from the dataset were tested with the YOLO v5 model after choosing the best weighted that can be used in the model. The final step involves classifying waste categories and corresponding recycle bin colors using Convolutional Neural Networks (CNN). CNN, as part of the YOLO method, enables instant object recognition. The waste is classified based on both CNN and YOLO algorithms, ensuring accurate identification of the waste type and recycle bin color.

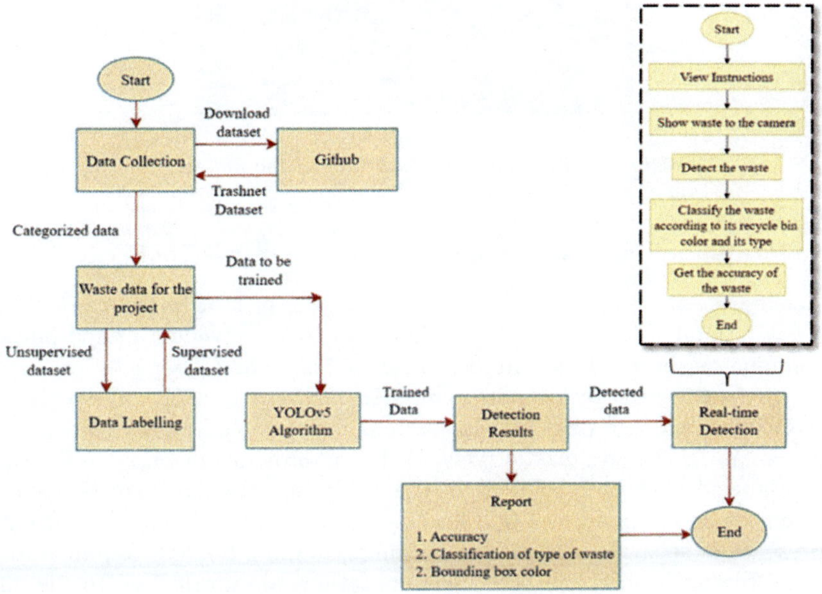

Fig. 2. The research design of the system.

In the real-time environment of the user's front-end interface, as shown in Fig. 3. The user will go to the detection page to show the trash in front of the camera, and the system will detect the waste and classify it based on its bin type. The system will also display the accuracy score of the machine learning during real-time detection.

3.1 Dataset

The data used in this project was downloaded from the TrashNet Dataset, and can be accessed at [12]. The images were captured using a white poster board background and natural and room lighting with Apple iPhone 7 Plus, 5S, and SE devices. Then, the images were resized to 512×384 pixels. Then the 2,527 images are divided into six categories, with the following distribution: 501 glass, 594 paper, 403 cardboard, 482 plastic, 410 metal, and 137 trash. However, this project focuses on the categories of glass, paper, plastic, and metal.

Fig. 3. The example of interface of the system.

3.2 Data Labelling Using the Make Sense Online Tool

The makesense.ai tool [13] is used for the purpose of labelling data in YOLO, VOC XML, VGG, JSON, and CSV. The tool is a convenient online platform that eliminates the need for complex installations. The labeling technique surrounds the waste with rectangular bounding boxes of specific colors indicating the associated waste types. The labels and boxes include glass (blue), paper (green), metal (red), and plastic (yellow). In the provided example in Fig. 4, the glass is labeled with blue colored. In the final step, the data will be exported to a file that can be read in YOLO format. The YOLO labeling format entails the presence of separate text files containing bounding box (BBox) annotations for each object within the image. The annotations are normalized to the image size, ranging from 0 to 1, with each element denoting object-class-ID, X center, Y center, Box width, and Box height. An example of the text can be seen in Fig. 5.

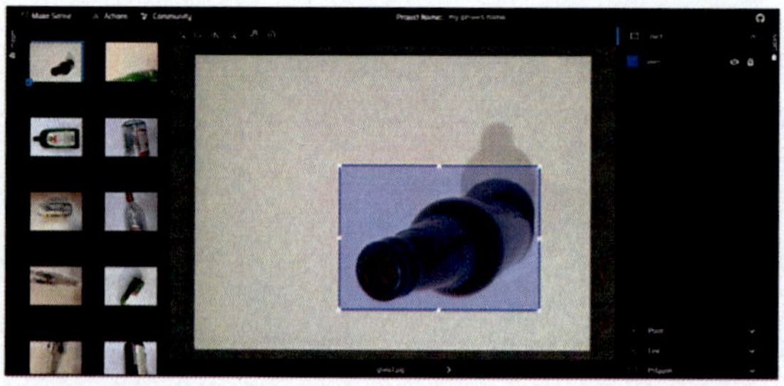

Fig. 4. The example of the image's bounding box from the dataset.

3.3 YOLOv5 Algorithms

YOLO (You Only Look Once) is an object detection algorithm that uses convolutional neural networks (CNN) as the backbone of its architecture, which can help in the detection of real-time images and videos [14]. The YOLO algorithm

0 0.635787 0.645516 0.482234 0.390863

< object-class-ID> <X center> <Y center> <Box width> <Box height>

Fig. 5. The annotations of the bounding box.

is an update of the traditional method that performs tasks for region proposal and classification in a single neural network. YOLO's advantage is its ability to detect multiple partially overlapped objects in a single frame or real-time video by dividing the image into a grid. Each grid will assign the image a bounding box and class label. Each bounding box predicts the object's coordinates and class probabilities based on the object that needs to be detected. However, one limitation of the grid-based approach is that the algorithm may need help accurately detecting small objects and objects in low-contrast backgrounds. There are various versions of YOLO, such as YOLOv2, YOLOv3, and the latest version, YOLOv8. Over the years, the YOLO algorithms have been improved based on their accuracy of detection and speed. However, in this study, YOLOv5 will be used for detecting and classifying recyclable waste types.

The YOLOv5 algorithm is available in the "YOLOv5 Ultralytics" repository [15]. YOLOv5 offers five different models sizes: YOLOv5n (nano), YOLOv5s (small), YOLOv5m (medium), YOLOv5l (large), and YOLOv5x (extra large). These models vary in terms of depth and width values, impacting the model's capacity and performance. However, for this project, the YOLOv5 algorithm has been trained using three different models: YOLO-v5s, YOLOv5m, and YOLOv5l. The parameter for the image size is set to a default value of 640 pixels [16]. The "batch" parameter is a hyperparameter indicating the number of training samples processed before updating the model's internal parameters. In this project, the batch size of 3 is used. 100 epochs or 100 iterations have been used as parameter controls in the training dataset. The speed tests of the three models have been conducted on Google CoLab Pro notebook.

The YOLO algorithm typically involves the following stages: input and output, backbone layers, detection head layers, post-process and non-max suppression (prediction result) [17,18] and [19]. The pseudocode of the YOLO is shown in Fig. 6.

In this project, the confidence threshold is set to 0.6 to improve the system's accuracy. The IoU threshold of 0.45 is also set as a benchmark for measuring the model's prediction accuracy. In order to ensure consistency, all images are standardized to 640 pixels in size. Afterwards, transformation removes any dimensional elements from the image array shape. The images are then arranged in the Blue, Green, and Red (BGR) sequence compatible with OpenCV. Finally, these BGR images are converted into RGB format to visualize the output. The Intel(R) Core(TM) i5-8250U CPU with 1.60 GHz and 8.00 GB of RAM is used for the training and testing phases.

```
Input: Image 640, batch 3, epoch 100, YOLOv5 model
parameters, weight
Output: Prediction results

//Model receives pre-processed input image.

i. Backbone layers: Extracting the high-level
   features of objects and their patterns

i. Detection head layers: Prediction of bounding
   box coordinates, confidences level and classes

i. Post-process:
   a. Decode the predictions from detection head
      layers into actual coordinates.
   b. Use a confidence value to rule out
      detections with low confidence scores.

ii.Non-Max Suppression:
   a. Eliminate duplicate detections.
   b. Prioritizes the selection of the most
      confident detection among overlapping
      boxes.
   c. Produce final detection result.
```

Fig. 6. Pseudocode of basic YOLO model.

4 Result and Discussion

The training and testing datasets are divided in an 80:20 ratio, with 1590 samples in the training set and 397 samples in the testing set. The evaluation metric is used to evaluate the precision of detection performed by the machine learning model. The evaluation metric includes the mAP (mean Average Precision) metric and accuracy testing to evaluate the YOLOv5 model's performance.

4.1 Mean Average Precision Metric

mAP is a measurement used in object detection models like YOLO to assess the model's performance. It relies on several metrics such as IOU (Intersection over Union), Precision, Recall, Precision-Recall Curve, and Average Precision (AP) to determine the overall accuracy of the model [16].

Average precision (AP) is used to calculate the mAP measure. The mAP uses the precision and recall of a classifier to get the AP score. Precision measures the "false positive rate", or the ratio of correct object detections to all predicted detections, in this context. When a model's precision score approaches 1.0, there is a high likelihood that the classifier's positive detections are correct predictions. On the other hand, Recall measured the "false negative rate", or the proportion of accurate item detections to all objects in the dataset. A recall score close to 1.0 suggests that the model can identify almost all the dataset objects.

During the mAP testing, a dataset comprising 1590 images from the training set was utilized. The objective was to compare the accuracy score of three different YOLOv5 model sizes, as shown in Table 1: YOLOv5s, YOLOv5m, and YOLOv5l. The waste detection model demonstrated impressive performance, achieving high mAP50-95 scores. Specifically, YOLOv5s scored 0.904, YOLOv5m scored 0.924, and YOLOv5l obtained the highest score of 0.932. The mAP50-95 metric calculates the average mAP across a range of IoU thresholds (0.5–0.95) to assess overall detection accuracy.

The formula can be seen in Eq. 1, 2 and 3.

In contrast to the previous investigation done by [20] with YOLOv4, which had a detection accuracy of 0.64%, the outcome score shows that YOLOv5 is more accurate when it comes to identifying and classifying waste categories.

$$Precision\ rate = \frac{True\ Positive}{True\ Positive + False\ Positive} \tag{1}$$

$$Recall\ rate = \frac{True\ Positive}{True\ Positive + True\ Negative} \tag{2}$$

$$mAP = \frac{1}{n} * sum(AP) \tag{3}$$

Table 1. Comparison scores of mAP50 and mAP50-95 Metric for YOLOv5s, YOLOv5m, and YOLOv5l architecture

Architecture	Total Images	Precision	Recall	mAP50	mAP50-95
YOLOv5s	1590	0.992	0.992	0.995	0.904
YOLOv5m	1590	0.995	0.993	0.995	0.924
YOLOv5l	1590	0.995	0.994	0.995	0.932

4.2 Accuracy Testing

The accuracy test aims to test if the model can detect the object correctly. This project's accuracy testing evaluates four waste types: glass, paper, metal, and plastic. The accuracy score in percentage is measured by dividing the total number of correct predictions by the total number of test images; the formula is shown in Eq. 4.

$$Accuracy\ score\ in\ \% = \frac{Total\ number\ of\ correct\ predictions}{Total\ number\ of\ test\ images} \tag{4}$$

Table 2 presents the accuracy testing results based on the YOLOv5 model. Eighty images (20 for each category) are used for this accuracy test, drawn from the testing dataset of 397 images. However, the object detection model has failed to accurately classify some of the test images of waste objects. Notably, there was frequent misclassification of glass waste as plastic waste and plastic waste as paper waste.

Out of the total 80 tested images, 69 were predicted correctly based on the accuracy results. Among the four categories, plastic waste had the lowest accuracy in terms of class prediction. Consequently, the overall accuracy percentage of the detection model was 86.25%.

Table 2. Accuracy score in percentage for each waste type

Waste Type	Total number of test images	Total number of correct predictions	Percentage Accuracy (%)
Glass	20	16	80
Paper	20	20	100
Metal	20	18	90
Plastic	20	15	75
Total	80	69	86.25

5 System Limitation

This project has identified three limitations. Firstly, the dataset used in the project lacks variations as it only includes waste on a white background. Consequently, the system's ability to detect waste on different backgrounds is limited since it is designed to identify waste specifically on a white background. Secondly, waste detection accuracy depends on the web-cam's stability. If the camera is stationary and positioned on a table, it can detect waste with high precision. However, any movement or instability may affect its performance. Lastly, the research lacked multiple object detections. The dataset primarily consisted of images with one waste item per image, and the system can detect only one object at a time.

6 Recommendation

Waste detection has a number of potential areas for improvement. First, besides the YOLO approach, it would be advantageous to include additional deep-learning models. The Single Shot Multi-box Detector is one such model, and contrasting the outcomes of multiple models can assist in choosing the best detection strategy. Additionally, various data-gathering techniques are advised because they can produce broader results and cover various recyclable waste from various backgrounds. Additionally, as waste products are not restricted to particular features, enabling multiple detections in a single video would be helpful. It is advised to increase and improve the system's capabilities in terms of usability. This might involve adding extra features, such as login access and benefits for successful waste recognition.

7 Conclusion

In conclusion, the first objective of designing a prototype that utilizes a web application and real-time video platform to detect and classify recyclable waste has been accomplished. Furthermore, using bounding boxes, the prototype can also sort the trash based on the color of the recycle bin. The second objective of developing a prototype that employs the YOLOv5 algorithm to detect and classify recyclable waste has also been achieved. The third objective, verifying and evaluating the model's accuracy, has also been met.

Acknowledgement. Grateful thanks to all authors for their valuable contributions towards funding the research. The authors thank Universiti Teknologi MARA Malacca Branch, Center of Vision and Algorithms Analytics (C-VAA), and Information Technology for Organisations (ITFO) Research Groups for their support throughout this research.

References

1. Zhang, L., Xu, M., Chen, H., Li, Y., Chen, S.: Globalization, green economy and environmental challenges: state of the art review for practical implications. Front. Environ. Sci. **10**, 870271 (2022). https://doi.org/10.3389/fenvs.2022.870271
2. Baras, N., Ziouzios, D., Dasygenis, M., Tsanaktsidis, C.: A cloud based smart recycling bin for in-house waste classification. In: 2020 International Conference on Electrical, Communication, and Computer Engineering (ICECCE) (2020). https://doi.org/10.1109/MOCAST49295.2020.9200283
3. Cho, R.: Recycling in US. is broken. How do we fix it? (2020). https://news.climate.columbia.edu/2020/03/13/fix-recycling-america. Accessed 13 Mar
4. Chang, E.: Conceptual compatibility of recycle bin color: from a cross-cultural perspective. Color. Res. Appl. **45**(3), 558–566 (2020). https://doi.org/10.1002/col.22479
5. Kalatzi, I.K., Nikellis, A.E., Menegaki, A.N., Tsagarakis, K.P.: The preferred bin color for recycling plastic bottles: evidence from a student's sample. Progress Industrial Ecol. Int. J. **9**(3), 256–268 (2015). https://doi.org/10.1504/PIE.2015.073429
6. Montazeri, S., Gonzalez, R., Yoon, C., Papalambros, P.Y.: Color, cognition, and recycling: how the design of everyday objects prompt behavior change. In: DS 70: Proceedings of DESIGN 2012, the 12th International Design Conference, Dubrovnik, Croatia, pp. 1363–1368 (2012)
7. Fang, B., et al.: Artificial intelligence for waste management in smart cities: a review. Environ. Chem. Lett. 1–31 (2023). https://doi.org/10.1007/s10311-023-01604-3
8. DeawSS.: Plastic trash four colorful recycling bins stock vecto. Shutterstock. (2023) https://www.shutterstock.com
9. Ziouzios, D., Baras, N., Balafas, V., Dasygenis, M., Stimoniaris, A.: Intelligent and real-time detection and classification algorithm for recycled materials using convolutional neural networks. Recycling **7**, 9 (2022). https://doi.org/10.3390/recycling7010009
10. Yang, M., Thung, G.: Classification of trash for recyclability status. CS229 project report (2016)
11. Tamin, O., et al.: On-shore plastic waste detection with YOLOv5 and RGB-near-infrared fusion: a state-of-the-art solution for accurate and efficient environmental monitoring. Big Data Cogn. Comput **7**(2), 103 (2023). https://doi.org/10.3390/bdcc7020103
12. TrashNet Dataset. https://github.com/garythung/trashnet
13. MakeSense Ai Online Tool. https://www.makesense.ai/

14. Ahmad, T., Ma, Y., Yahya, M., Ahmad, B., Nazir, S., Haq, A.U.: Object detection through modified YOLO neural network. Sci. Program. 1–10 (2020). https://doi.org/10.1155/2020/8403262
15. Kaur, R. and Singh, S.: A comprehensive review of object detection with deep learning. Digit. Signal Process. 103812 (2023). https://doi.org/10.1016/j.dsp.2022.103812
16. Glučina, M., And'elić, N., Lorencin, I., Car, Z.: Detection and classification of printed circuit boards using YOLO algorithm. Electronics **12**(3), 667 (2023). https://doi.org/10.3390/electronics12030667
17. Huang, D., Yan, X., Hao, X., Dai, J., Wang, X.: Low SNR Multi-Emitter Signal Sorting and Recognition Method Based on Low-Order Cyclic Statistics CWD Time-Frequency Images and the YOLOv5 Deep Learning Model. Sensors **22**(20), 7783 (2022)
18. Pham, T.N., Nguyen, V.H., Huh, J.H.: Integration of improved YOLOv5 for face mask detector and auto-labeling to generate dataset for fighting against COVID-19. J. Supercomput. **79**(8), 8966–8992 (2023)
19. Terven, J., Cordova-Esparza, D.: A comprehensive review of YOLO: from YOLOv1 to YOLOv8 and beyond. arXiv preprint arXiv:2304.00501 (2023)
20. Chen, Q.: Garbage classification detection based on improved YOLOv4. J. Comput. Commun. **8**(12), 285 (2020)

Reviewing the Effect of Indoor Living Walls on Air Quality, Energy Consumption in Different Climates

Atina Ghunaim(✉) and Young Ki Kim

United Arab Emirates University, Abu Dhabi, Al Ayn, UAE
201735457@uaeu.ac.ae

Abstract. Living Walls and greenery systems importance is increasing as a sustainable building design elements to enhance environmental building's impact. Numerous amounts of studies have demonstrated that outdoor greening, such as trees and landscaped areas can reduce a building's energy consumption indirectly, as well as studies that have been conducted on the impact of indoor vegetation on energy use and air quality. Indoor vegetation and planting decrease the carbon dioxide (CO_2) concentration through absorbing carbon dioxide (CO_2) and releasing oxygen (O_2) where energy expenditures get reduced. Individuals spend around 90% of their time indoors which threatens their health and wellbeing due to poor indoor air quality (IAQ). Therefore, this paper recognizes the impacts of indoor living walls on air quality and energy consumption in various countries. The results of previous studies include pre and post results of analysis and simulation that proves the reduction of concentration, total volatile organic compounds (TVOC), PM10, temperature, and an increase in relative humidity that reduces energy consumption, and boosted air quality using indoor living walls.

Keywords: Indoor air quality · Energy consumption · Living Wall · Temperature · Air Temperature

1 Introduction

The demand for energy use and poor air quality increase due to climate change. Lands with lots of artificial structures tend to have a striking increase in surface temperature which concludes an increase in the use of more energy and air conditioning [1, 2]. According to a synthesis report made in 2014, temperature will eventually start increasing in the future that will conclude longer heat waves [3]. Unfortunately, global warming and climate change are resulting in serious health effects and higher energy demand [4, 5].

Poor air quality has many factors and unhealthy substances such as volatile organic compounds, fine dust, nitrogen oxide, sulfur oxide, ozone, air born microorganisms [6]. This can have major negative impacts on human health and wellbeing as an observation was conducted on the Sick Building Syndrome (SBS). SBS is a syndrome of respiratory issues, allergies, and irritations that is found to be one of the poorest health factors

M. Ben Ahmed et al. (Eds.): SCA 2023, LNNS 906, pp. 55–66, 2024.
https://doi.org/10.1007/978-3-031-53824-7_6

because of substandard air quality [7]. Studies have proven that the implementation of indoor vegetation in a closed work office environment positively changed individuals' perception of the room, improved occupants' thermal comfort and enhanced IAQ [8]. Vertical greening is previously addressed in two categories, either "living walls" or "green facades". Green facades are usually implemented as exterior applications while indoor vertical greening is addressed as "living walls". Living walls can have numerous benefits if they are placed indoors such as air purification, biofiltration and stabilized VOCs [9]. Living walls are complex structured system of vertical [10]. In order to define living walls, living walls are recent approaches that have vertical substrates on the face of the wall. Researchers named this as "wall based" greening that allows the plant to grow to tied back wall. Different systems such as water irrigations and fertigation are implemented to allow the water and nutrients to pass using specific controls and monitoring systems depending on different applications, living walls could come as or "modular living wall" or "continuous living wall" showing Fig. 1 below.

Consequently, this paper aims to review and answer the impact of indoor planting on the indoor air quality (IAQ), building's energy consumption and individual's wellbeing in countries with different climates by examining previous studies published on living walls and indoor planting respecting energy consumption and IAQ aspects. The objectives of this research would be (1) to quantify how indoor planting can affect air quality and energy consumption, (2) to review how living walls improve air quality and energy consumption in closed spaces and (3) to identify concerns needs to be addressed for future living walls research.

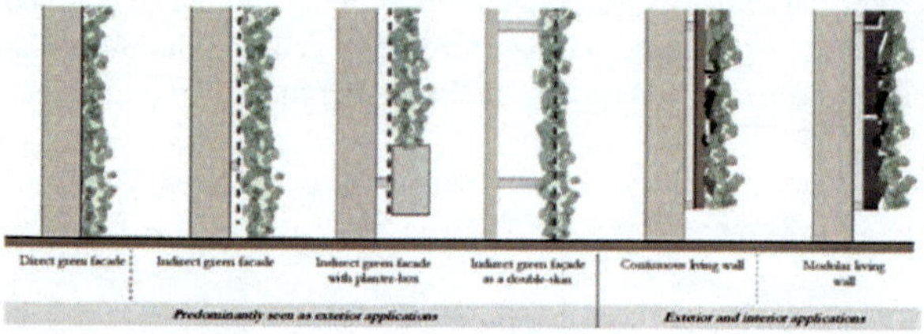

Fig. 1. Vertical Greening Categories, Kanchan, Koen, Living walls in indoor environments, 2019 [11].

2 Methodology

This paper considered the review of experimental studies and reviewed journals from 2000 to 2023 obtained through keyword searching and database as shown in Fig. 2 below. The database obtained was Google Scholar, ScienceDirect and ResearchGate. The keyword searching included "living walls", "greenery systems", "indoor vegetation", and their variants "energy consumption", "indoor air quality", and "outdoor greenery". These were distilled to put into consideration "51" of papers based on keywords and number of citations that addressed the effect of indoor living walls and planting on air quality and energy consumption being an experimental, simulation or a case study based paper as shown in Fig. 3 below. However, additional studies and literature review was added from building science and psychology to be obtained. In addition to the interviews and surveys obtained in different case studies (Table 1).

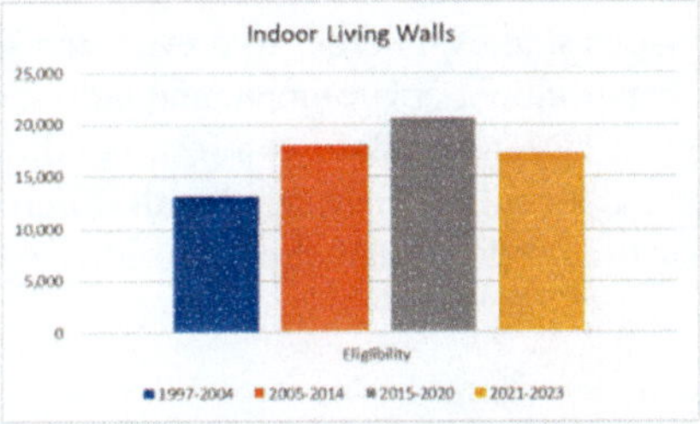

Fig. 2. Case Studies Eligibility

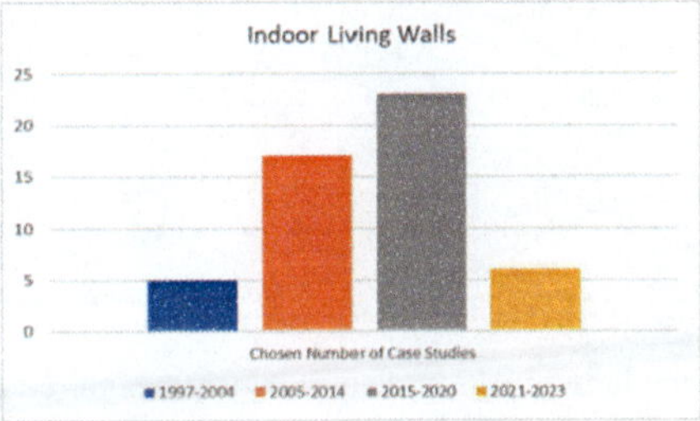

Fig. 3. Chosen Number of Case Studies

3 Findings and Discussion

3.1 Climate Change, Pandemic (Covid-19) and Energy Consumption

All the climate change, polluting the environment, and COVID-19 crises might have tremendous negative attributes on mental and physical health of an individual [12–16]. As it is known that most people spend 90% of their time indoors, with the pandemic of COVID-19, individuals spent all their time in their homes. IAQ is still one of the main attributes to Sick Building Syndrome (SBS) and occupant's comfort, satisfaction, and productivity. In fact, indoor air pollution has been shown to affect health conditions. Figure 2 shows the environmental factors influence indoors, outdoors and on occupants' health. Poor IAQ can come from lack of ventilation, air filtering and air circulation in space. Heating, ventilation and air conditioning (HVAC) systems can result in poor air quality and infections in the building. After the COVID-19 pandemic, prioritizing different approaches to enhance indoor air quality (IAQ) was important to focus on. However, these strategies should be focused on from policy changes that include urban planning, architectural design sectors, and public health sections. Therefore, air quality can be enhanced in several strategies such as designing different ventilation systems, air filtration and source of poor air quality control [17–21]. There have been changes in energy usage in buildings since the global pandemic began.

An increase in energy consumption was confirmed in residential buildings due to the increased time spent inside the buildings and to make the indoor environment comfortable. As known from previous studies, in order to prevent the spread of viruses such as COVID-19, to provide a better indoor environment and indoor air quality to occupants, enhancing the air circulation systems, indoor comfort, buildings' heating, ventilation, air conditioning (HVAC) and air filtration systems with different approaches and considerations are needed such as indoor plantation and air circulation system [22–25] (Fig. 4).

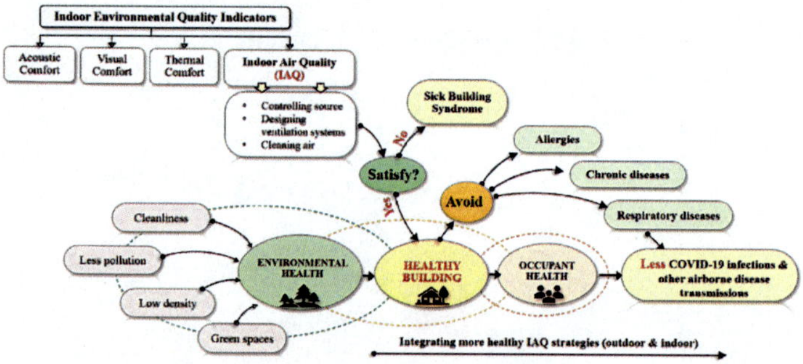

Fig. 4. Relationship between individual's and environment health (Naglaa A. & Ehab M. 2021) [26].

3.2 Indoor Air Quality

Indoor Air Quality is the measurement of indoor air pollutants that come from different sources. These indoor air pollutants are transmitted from fabric of buildings and many other activities such as total volatile organic compounds (TVOC) and PM10.Sources can be defined as being associated with the occupants of the buildings, substances and emissions from building materials. Sick Building Syndrome (SBS) is also a phenomenon that can happen in buildings where occupants in the buildings have health complaints of fatigue, headaches, respiratory issues etc. [27]. The importance of outdoor sources can be the contributor that is happening indoors that result in pollutants found in indoor air especially buildings surrounded by streets, industrial areas, and heavy traffic. Poor IAQ resulted in 65,000 to 150,000 deaths per year in the United States [28]. It has been found that most respiratory illnesses and lung cancer occur due to poor air quality that had been avoided. The sources that affect indoor pollution can come from radon, ammonia, combustion by-products, and VOC's [29]. Volatile Organic compounds (VOC's) are usually known to be the most frequent indoor pollutants [30]. These studies show the role of indoor environmental parameters such as temperature, relative humidity (RH) concentration, total volatile organic compounds (TVOC) and PM10. Many components of the environment can affect the indoor air quality such as concentration, toxic gases and compounds, temperature, and relative humidity. The health of climate, temperature and humidity have been studied for many years. The concept of how heat influences individuals is complex, although temperature is fixed as an individual parameter, heat is a combination of temperature, radiation, and humidity. More hours of radiation and sunlight with lower humidity has been found to increase chronic obstructive pulmonary disease (COPD) [31]. Temperature and relative humidity can have huge effects on human health and comfort. A study conducted in Trinidad by Ivey [32] found that asthma could increase in relation to high relative humidity in warm, wet climates from May to December [32].

3.3 Living Walls

Living walls are systems made and identified by the different plant types growing either indoor or outdoor. The life of the building could be enhanced if finishes and other actions were taken on the envelope because the envelope is the part of the building that faces the extreme weather [33]. Actions taken on the envelope could help achieve less CO_2 emissions and further reduce energy consumption [34]. In order to be able to achieve lower temperature and thus reduce energy consumption, a living wall (LW) could be one of the strategies. Temperature that is being reduced is a potential consideration of LW's for energy demand reduction 20–23% [35]. LW's provide insulation by vegetation and substrate and act as a shading material to the envelope. It also provides evaporative cooling by evapotranspiration and wind barrier effect [36, 37]. Fur thermore, green walls, living walls or vertical greenery systems are a beneficial solution to help enhance indoor air qualities and environments and reduce energy consumption [38, 39]. Living walls could reduce indoor pollutants and contaminants such as CO_2 concentration, VOCs, and PM [40].

3.4 Indoor Living Walls

Indoor living walls are getting popular in the sustainability sector and built environment. In terms of indoor environment, living walls allow to increase visual comfort and reduce poor IAQ [41, 42]. Although HVAC systems and air-conditioning is known to cool and enhance IAQ, poor air could result from low amount of natural ventilation, high CO_2 concentration and other air circulation from HVAC systems [43]. In fact, indoor planting and living walls could enhance indoor thermal comfort by sustaining the air temperature and the relative humidity of the space itself in the process of transpiration. Research was conducted by Smith and Pitt to study the advantages of green walls and indoor vegetation for enhancing the IAQ. It was proven to have an increase of relative humidity in the space due to the indoor plants implementation and another research evaluated and studied indoor plants effect on thermal comfort of office work up to four months. Results have shown that adding indoor vegetation had a positive effect on thermal comfort in occupants [44]. Also, Fernandez Canero et al. have shown that having indoor green walls for four months could benefit in decreasing the indoor air temperature by 4 °C–6 °C [45] (Fig. 5).

Fig. 5. Indoor Living Wall, (Source: Tudiwer., & Korjenic., 2017; Energy and Buildings, 146, 73–86) [46].

3.5 Living Walls in Various Climates

LW's in different countries with various climates could implement different results according to the type of plants used in specific location, the type of methods used, and the limitations of the study. As shown in Table 2 below, Ying Shao [47], China conducted experimental research on LW's impact on CO_2 concentration and energy use, the results show that the CO_2 absorption from the plants reached up to 9.2 times higher and the energy use got reduced by 12.7% to 58.4%. The limitation of the study is not having the trial of other air purification systems. Liping Wang and Michael J. [48], conducted research in Los Angeles California (CA), a medium office building that proved indoor living walls (indoor planting), can cool down the building energy consumption by 25.1% with an experiment on the floor area ratio (LFAR) and the orientations that equal to 15 for the south perimeter zone, 0.5 LFAR of east perimeter zone with an energy reduction of 14.4%, 0.3 LFAR at the north perimeter zone with 0.3% savings. Lastly, west perimeter zone with 0.5 LFAR, and an energy saving of 14.5% using a machine learning algorithm. Also, evapotranspiration (ET) model was conducted on experimental data with the machine learning algorithm. The (ET) process will allow the sensible and latent loads to appear from the indoor vegetated module; following the use of a software called (Energy Plus) that had plug in called python feature. The missing gap in this study is measuring only one type of plant "epipremnum aureum" for ET performance, while many other types can be measured. Biofiltration is one of the benefits of using vegetated walls inside a building. Therefore, there are specific methods implied using active living walls (ALW) or indoor vertical planting in the University of Seville, Spain that showed results of air temperature getting reduced by almost 0.8 °C and 4.8 °C through implementing indoor vegetation at the university while having various distances from the ALW. Results prove that the cooling process was better when the room was warmer using ALW. ALW was passing through the vegetated wall to take advantage of their evaporative cooling by reducing ventilation requirements, measurements of the air temperature, and humidity with the use of sensors model HOBO Pro Temp-HR U23-00. This study had few research gaps such as studying the plant types, materials and ALW size deeply and how it affects the performance of air quality and energy [49]. Moreover, research conducted by Zuzana Poórova et al. [50], in Kosice TUKE campus, a city in Slovakia, using green wall set in a classroom with an oriented strand board (OSB) that relates to plastic drainage. Different types of plants were implemented on the OSB with its own filtration layer connected to the plastic drainage pipe. This study used subjective matter type of questions that had different results according to humidity, air temperature and wellbeing. Results show that air humidity is 28.7% without green wall and 30.1% with green wall. This study demonstrates that indoor vegetation is beneficial for indoor buildings. However, different distances and sizes of the OSB could be recommended in this study [50]. Swedish schools have poor indoor environment, and many people worry about the climate change due to the infectious diseases resulting in severe health conditions. Studies have been proving that indoor living walls with active systems in mid to normal climates have been reducing the use of energy when implemented indoors. Itai Danielski et al. [51], have conducted research in Ostersund, Sweden, educational facility on green plants using experimental methods in two classrooms with the same radiation and window orientation to measure the effects of indoor planting on the indoor environment and wellbeing. Results of this research have demonstrated positive effects on wellbeing, health, 10% less CO_2, stable temperatures, less worry and better air quality as teachers found there was an improvement in air quality.

According to the methods, there were 4 prototypes of indoor plants implemented, waterfall, green working space, plant pillars, and plants picture with different materials. In addition, 310 green plants were installed, and specific sensors were applied. These sensors were used to collect the humidity and air temperature level. Measurements such as final energy model, subjective matter of questions about the well-being and qualitative interviews were obtained. Some of the limitations found in the study include measuring pre and post installments effects which should have been implemented to have more accurate results. Moreover, identifying the number of the plants to investigate whether the number of plants have bigger effects on air quality [51]. The conducted research in Slovakia was green walls implemented in classrooms and monitored to result in having better air quality, raised humidity level, and positive effects on wellbeing [50, 51]. In order to reduce the effects of temperature and relative humidity on human health, the previous studies and findings' show the positive impact of living walls and indoor planting on energy consumption, IAQ and therefore, the improvement in human health wellbeing.

Table 1. Indoor Living Walls Case Studies

Indoor Living Walls				
Years	1997–2004	2005–2014	2015–2020	2021–2023
Eligibility	13,000	17,900	20,600	17,200
Chosen Number of Case Studies	5	17	23	6
Total of Chosen Case Studies	51			

Table 2. Types of Living Walls in Different Countries

Source	Location	Methods	Outcomes	Limitation
[47]	China	Experimental and	CO_2 absorption up to 9.2 times reduced by 25.7%–34.3%	other air purification equipment
		stimulation	energy use 12.7%–58.4%	
[48]	Los Angles CA	ET model was made using a machine algorithm	The cooling savings: 25.1% for the south perimeter zone, 4.4% the east perimeter zone –	One type of plant was measured
			3% for the north perimeter zone and 14.5% for the west zone	
[49]	Spain	ALW or active living wall passing through the vegetated wall	Air temperature got reduced by almost 0.8 and 4.8 celsiuis	different sizes of the ALW plant types, materials and ALW size,
[50]	Slovakia	Green wall set in a classroom	–without green wall, humidity is 28.7%	different distances of (OSB) and sizes
		an oriented strand boards	–With green wall, humidity is 30.1%	
		Different types of plants	–Better air quality	
		Subjective type of questions		
[51]	Ostersund, Sweden	Two classrooms 4 prototypes of indoor plants questionnaire specific sensors	Positive effects on wellbeing health and worry. Improvement in air quality	number of plants Measuring pre and post installments

4 Conclusion

Previous studies present evidence into the implementation of living walls and greenery systems could have a positive impact on indoor air quality, energy consumption and individuals' health and wellbeing such as a cooling influence during summer seasons and improved performance of individuals during hard times. Previous studies prove the results are in a better state of performance in drier and arid climates.

The above review needs to be reassessed on how different types of plants used on living walls and greenery systems could affect the temperature and relative humidity in relation to the energy expenditure. Given the limited evidence could still be a potential and have a beneficial reduction of cooling loads in summer where various countries annual temperatures differ as seen in Table 3 below. The highest impacts on energy from studies reviewed is between 12.7% and 58.4%, CO_2 concentration got reduced by 9.2 times while (RH) and temperature was 1.4% higher with living wall and 4.8 °C less. This observation is applicable to other building sciences, psychological impacts on productivity and comfort. This review study explores the necessity for future studies to put into consideration and integration plant science to acoustic evidence for the implementation of indoor living walls.

Table 3. Reviewed Studies Impact on Energy and IAQ

Highest Impacts from Indoor Living Wall				
Case Studies	Energy Use	Indoor Air Quality		
	Cooling Load	CO_2 Concentration	Relative Humidity	Temperature
[47]	energy use 12.7%-58.4%	9.2 times reduced		
[49]				reduced by almost 0.8 °C and 4.8 °C
[50]			1.4% higher	

Acknowledgements. This work required no funding, the author is thankful for the support received from the United Arab Emirates University (UAEU) and the professors that were very helpful into enhancing the quality of this review study; Dr. Young Ki Kim.

References

1. Sims, A.W., Robinson, C.E., Smart, C.C., O'Carroll, D.M.: Mechanisms controlling green roof peak flow rate attenuation. J. Hydrol. **577**, 123972 (2019)
2. Santamouris, M.: On the energy impact of urban heat island and global warming on buildings. Energy Build. **82**, 100–113 (2014)
3. IPCC, A. (2014). IPCC Fifth Assessment Report—Synthesis Report
4. Liu, F., Yan, L., Meng, X., Zhang, C.: A review on indoor green plants employed to improve indoor environment. J. Build. Eng. **53**, 104542 (2022)
5. Paull, N.J., Irga, P.J., Torpy, F.R.: Active green wall plant health tolerance to diesel smoke exposure. Environ. Pollut. **240**, 448–456 (2018)

6. Stamatelopoulou, A., Asimakopoulos, D.N., Maggos, T.: Effects of PM, TVOCs and comfort parameters on indoor air quality of residences with young children. Build. Environ. **150**, 233–244 (2019)
7. Jafari, M.J., et al.: Association of sick building syndrome with indoor air parameters. Tanaffos **14**(1), 55 (2015)
8. Qin, J., Sun, C., Zhou, X., Leng, H., Lian, Z.: The effect of indoor plants on human comfort. Indoor Built Environ. **23**, 709–723 (2014)
9. Soreanu, G., Dixon, M., Darlington, A.: Botanical biofiltration of indoor gaseous pollutants–A mini-review. Chem. Eng. J. **229**, 585–594 (2013)
10. Kontoleon, K.J., Eumorfopoulou, E.A.: The effect of the orientation and propor- tion of a plant-covered wall layer on the thermal performance of a building zone. Build. Environ. **45**, 1287–1303 (2010)
11. Gunawardena, K., Steemers, K.: Living walls in indoor environments. Build. Environ. **148**, 478–487 (2019)
12. Berry, H.L., Bowen, K., Kjellstrom, T.: Climate change and mental health: a causal pathways framework. Int. J. Public Health **55**, 123–132 (2010)
13. MacDonald, J.P., et al.: Protective factors for mental health and well-being in a changing climate: Perspectives from Inuit youth in Nunatsiavut, Labrador. Soc. Sci. Med. **141**, 133–141 (2015)
14. Buoli, M., et al.: Is there a link between air pollution and mental disorders? Environ. Int. **118**, 154–168 (2018)
15. Attademo, L., Bernardini, F.: Air pollution as risk factor for mental disorders: in search for a possible link with Alzheimer's disease and schizophrenia. J. Alzheimer's Disease **76**(3), 825–830 (2020)
16. Ma, Y., et al.: Effects of temperature variation and humidity on the mortality of COVID-19 in Wuhan. Sci. Total Envron. **724**, 138226 (2020)
17. Brittain, O., Wood, H., Kumar, P.: Prioritising Indoor Air Quality in Building Design Can Mitigate Future Airborne Viral Outbreaks, Cities and Health. Accessed 25 June (2020)
18. Cui, Y., et al.: Air pollution and case fatality of SARS in the People's Re-public of China: an ecologic study. Environ. Health **2** (2003)
19. Guo, H., Murray, F., Lee, S.: The development of low volatile organic compound emission house—a case study. Build. Environ. **38**, 1413–1422 (2003)
20. Li, H., Xu, X.-L., Dai, D.-W., Huang, Z.-Y., Ma, Z., Guan, Y.-J.: Air Pollution and temperature are associated with increased COVID-19 incidence: a time series study. Int. J. Infect. Dis. **97**, 278–282 (2020)
21. Kumar, P., Morawska, L.: Could fighting airborne trans-mission be the next line of defence against COVID-19 spread? City Environ. Interact. **4**, 100033 (2019)
22. Chan, I., Liu, A.: Effects of neighborhood building density, height, greenspace, and cleanliness on indoor environment and health of building occupants. Build. Environ. **145**, 213–222 (2018)
23. Van den Berg, A., Maas, J., Verheij, R., Groenewegen, P.: Green space as a buffer between stressful life events and health. Soc. Sci. Med. **70**, 1203–1210 (2010)
24. Wang, Z., Zhang, J.: Characterization and performance evaluation of a full-scale activated carbon-based dynamic botanical air filtration system for improving indoor air quality. Build. Environ. **46**, 758–768 (2011)
25. World Health Organization (WHO): WHO Guidelines for Indoor Air Quality: Selected Pollutants. the WHO European Centre for Environment Health (2010). Accessed 25 June 2020
26. Megahed, N.A., Ghoneim, E.M.: Indoor Air Quality: Re-thinking rules of building design strategies in post-pandemic architecture. Environ. Res. **193**, 110471 (2021)
27. Horvath, E.P.: Building-related illness and sick building syndrome: from the specific to the vague. Clevel. Clin. J. Med. **64**, 303–309 (1997)

28. Fisk, W.J., Black, D., Brunner, G.: Changing ventilation rates in U.S. offices: implications for health, work performance, energy, and associated economics. Build Environment (2012)
29. Sidhes waran, M.A., Destaillats, H., Sullivan, D.P., Cohn, S., Fisk, W.: Energy efficient indoor VOC air cleaning with activated carbon fiber (ACF) filters. Build. Environ. **47**, 357–367(2012)
30. Wolkoff, P., Nielsen, G.D.: Organic compounds in indoor air—their relevance for perceived indoor air quality? Atmos. Environ. **35**(26), 4407–4417 (2001)
31. Tseng, C.M., et al.: The effect of cold temperature on increased ex-acerbation of chronic obstructive pulmonary disease: a nationwide study. PloS one **8**(3), e57066 (2013)
32. Ivey, M.A., Simeon, D.T., Monteil, M.A.: Climatic variables are associated with seasonal acute asthma admissions to accident and emergency room facilities in Trinidad, West Indies. Clin. Exper. Allergy **33**(11), 1526–1530 (2003)
33. Widiastuti, R., Zaini, J., Caesarendra, W.: Field measurement on the model of green facade systems and its effect to building indoor thermal comfort. Measurement **166**, 108212 (2020)
34. Koroxenidis, E., Theodosiou, T.: Comparative environmental and economic evaluation of green roofs under Mediterranean climate conditions–Extensive green roofs a potentially preferable solution. J. Clean. Prod. **311**, 127563 (2021)
35. Dabaieh, M., Serageldin, A.A.: Earth air heat exchanger, Trombe wall and green wall for passive heating and cooling in premium passive refugee house in Sweden. Energy Convers. Manage. **209**, 112555 (2020)
36. Djedjig, R., Bozonnet, E., Belarbi, R.: Analysis of thermal effects of vegetated envelopes: Integration of a validated model in a building energy simulation program. Energy Build. **86**, 93–103 (2015)
37. Kenaï, M.A., Libessart, L., Lassue, S., Defer, D.: Impact of plants occultation on energy balance: Experimental study. Energy Build. **162**, 208–218 (2018)
38. Irga, P.J., Paull, N.J., Abdo, P., Torpy, F.R.: An assessment of the atmospheric particle removal efficiency of an in-room botanical biofilter system. Build. Environ. **115**, 281–290 (2017)
39. Pearlmutter, D., et al.: The circular economy with nature-based solutions in the built urban environment: green building materials, systems and sites. Blue-Green Syst. **2**, 46–72 (2020)
40. GubbC, B.T., Griffiths, A., et al.: Can house plants improve indoor air quality by removing CO2 and increasing relative humidity? Air Qual. Atmos. Health **11**, 1191–1201 (2018)
41. Hui, S.C.M., Ma, T.C.: Analysis of environmental performance of indoor living walls using em- bodied energy and carbon. Int. J. Low-Carbon Technol. **12**, 67–74 (2017)
42. Irga, P.J., Torpy, F.R., Burchett, M.D.: Can hydroculture be used to enhance the performance of indoor plants for the removal of air pollutants? Atmos. Environ. **77**, 267–271 (2013)
43. Pérez-Urrestarazu, L., Fernández-Cañero, R., Franco-Salas, A., Egea, G.: Vertical Greening systems and Sustainable cities. J. Urban Technol. **22**, 65–85 (2015)
44. Mangone, G., Kurvers, S.R., Luscuere, P.G.: Constructing thermal comfort: investigating the effect of vegetation on indoor thermal comfort through a four-season thermal comfort quasi-experiment. Build. Environ. **81**, 410–426 (2014)
45. Fernández-Cañero, R., Urrestarazu, L.P., Franco Salas, A.: Assessment of the cooling potential of an indoor living wall using different substrates in a warm climate. Indoor Built Environ. **21**(5), 642–650(2012)
46. Tudiwer, D., Korjenic, A.: The effect of an indoor living wall system on humidity, mould spores and CO2-concentration. Energy Build. **146**, 73–86 (2017)
47. Shao, Y., et al.: The Effect of Vertical Farming on Indoor Carbon Dioxide Concentrations and Fresh Air Energy Consumption in Office (2021)
48. Wang, L., Witte, M.J.: Integrating building energy simulation with a machine learning algorithm for evaluating indoor living walls' impacts on cooling energy use in commercial buildings. Energy Build. **272**, 112322 (2022)
49. Pérez-Urrestarazu, L., Fernández-Cañero, R., Franco, A., Egea, G.: Influence of an active living wall on indoor temperature and humidity conditions. Ecol. Eng. **90**, 120–124 (2016)

50. Poórová, Z., Turcovská, A., Kapalo, P., Vranayová, Z.: The effect of green walls on humidity, air temperature, Co2 and well-being of people. Environ. Sci. Proc. **2**(1), 56 (2020)
51. Danielski, I., Svensson, Å., Weimer, K., Lorentzen, L., Warne, M.: Effects of green plants on the indoor environment and wellbeing in classroooms: a case study in a Swedish School. Sustainability **14**(7), 3777 (2022)

Acoustic Emission and Machine Learning for Smart Monitoring of Cable Damages in Bridges

Abdou Dia[1(✉)], Lamine Dieng[2], and Laurent Gaillet[2]

[1] ESTP, Ecole Spéciale des Travaux Publics, 28 Av. du Président Wilson, 94230 Cachan, France
adia@estp.fr

[2] Université Gustave Eiffel, Allée des ponts et chaussées - CS 5004, 44344 Bouguenais, France

Abstract. This article discusses the use of acoustic emission and machine learning tools for smart monitoring of cable damages in bridges. The need for discovering and measuring the degradation of metallic bridges cables comes out as a must for users' safety, transportation facility and economic evidence. And the application of acoustic emission for this purpose is emphasized since 1969. However, using this approach to diagnose civil engineering cables is scarce and previous research on friction in the industry and on damaged cables has shown that acoustic emission (AE) signals depend on materials, surface condition, pressure, and relative velocity. Therefore, separating the different sources of acoustic emissions recorded during cable monitoring to identify the signals caused by wire breakages remains a significant challenge. The study focuses on investigating intrafilamentary friction by leveraging the differences in acoustic signatures between cables with broken wires and intact cables. To achieve that, the article suggests an original experimental approach combined to machine learning algorithms to isolate AE sources recorded during data collection. After the experimental setup, parametric analysis, clustering, and classification techniques have been employed to separate different AE sources. The proposed approach could enable tracking the evolution of friction from broken wires over time and recognizing wire break signals even without a reference state for the cable.

Keywords: Acoustic emission · Machine Learning · cables · damages · monitoring

1 Introduction

The availability and proper functioning of infrastructure such as bridges is crucial for the safety of daily users, the continuous operation of transportation routes and economic activity in cities [1, 2]. The normal functioning of these bridges also implies a structurally sound condition of their load-bearing components, such as metallic cables. These cables are subjected to cyclic mechanical, chemical, and thermal loads that lead to damages resulting from pathologies, mainly corrosion and fretting fatigue [3, 4]. Though, different studies have pointed out the need to make "transition from a culture of urgency

M. Ben Ahmed et al. (Eds.): SCA 2023, LNNS 906, pp. 67–76, 2024.
https://doi.org/10.1007/978-3-031-53824-7_7

to a heritage management approach for bridges by creating tools to enhance knowledge and monitoring, and by investing more in preventive actions through the establishment of a multi-year work program" [5, 6]. Thus, the integration of Structural Health Monitoring (SHM) and nondestructive techniques (NDT) into the management of structures is essential for achieving a comprehensive characterization of damages and ensuring the long-term safety and functionality of the bridges' cables. In response to these needs, several NDT techniques such as ultrasonic methods, electromagnetic field methods, acoustic emission, vibration-based methods, among others, have been developed [7].

Acoustic emission (AE) is defined as the generation of transient elastic waves resulting from the release of energy within a material or from other processes [8, 9]. It is commonly used as a non-destructive testing method to characterize and monitor damage mechanisms in materials and structures. This article focuses on the usage of this AE technique for cable damages characterization due to its capability to detect low-energy sliding-friction mechanisms between surfaces in different states [10–12], even in inaccessible or hard-to-reach parts such as cable anchorages [13, 14]. However, this high sensitivity of acoustic emission can be considered its main drawback as it can lead to the recording of numerous acoustic signals from various sources during testing campaigns. Though, regarding cables used in bridges, AE sources include surrounding noise, corrosion, friction between intact wires, friction between intact and broken wires, and wire breakages [15]. Near the anchorages, acoustic emission can also originate from the filler materials used, such as resin [16].

By analyzing the characteristics of AE signals recorded by sensors, such as the signal amplitude, duration, energy, rise time, etc., the challenge is to identify the origin of each signal to isolate the friction between intact and broken wires of cables. To achieve this, we propose adopting an experimental approach tailored to this situation. In addition to correlation analysis between AE parameters, Machine Learning algorithms are used to separate different AE signals and identify the acoustic signatures of signals resulting from friction between intact and broken wires.

2 Experimental Setup

The principle of AE acquisition is schematically depicted in the Fig. 1.

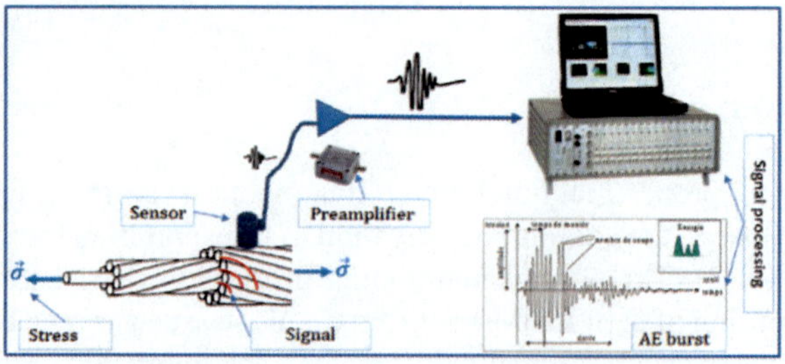

Fig. 1. Principle of AE acquisition system

In the process of implementing an experimental protocol, it was deemed necessary to start with the simplest case and apply the chosen methods to it before moving towards a more complex scenario. As a result, the initial tests were conducted on a T15.7 one-layer cable (see Fig. 2) and subsequently on a bi-layer cable (see Fig. 3).

Fig. 2. T15.7 one-layer cable

Fig. 3. Bi-layer cable

The geometric and mechanical characteristics of these two cables are provided in Table 1.

Table 1. Geometric and mechanical characteristics of the used cables

Cable	Length (mm)	Young's modulus E (MPa)	Density (g/cm^3)	Breaking strength (kN)
T15.7	6980	200	7.8	265
Bi-layer	6230	200	7.8	513

To assess the detectability of acoustic emissions due to friction between wires and identify the configurations that are most likely to facilitate this detection, campaign tests were deployed for the simplest T15.7 one-layer cable, with three broken wires out of the seven. The tests were conducted using the eccentric of the loading bench, applying harmonic-type loadings with different amplitude-frequency couples. During these tests, several parameters were evaluated, including the type of sensor used, the sensor-to-defect distance, the AE detection threshold, the amplitude-frequency couples of solicitations with the so-called eccentric, and the temporal parameters: PDT (Peak Definition Time), HDT (Hit Definition Time), and HLT (Hit Lockout Time).

We performed the tests with an eight-channel DISP system equipped with the AEWin software commercialized by Physical Acoustics. HBM's X60 adhesive was used for sensor-cable coupling. Prior to sensor-cable coupling, the sensor is equipped with a brass plate glued to the sensor's ceramic using cyanoacrylate adhesive. To ensure the proper functioning of the sensors attached to the cable, "Hsu Nielsen" tests, also known as pencil lead break tests [17], were conducted.

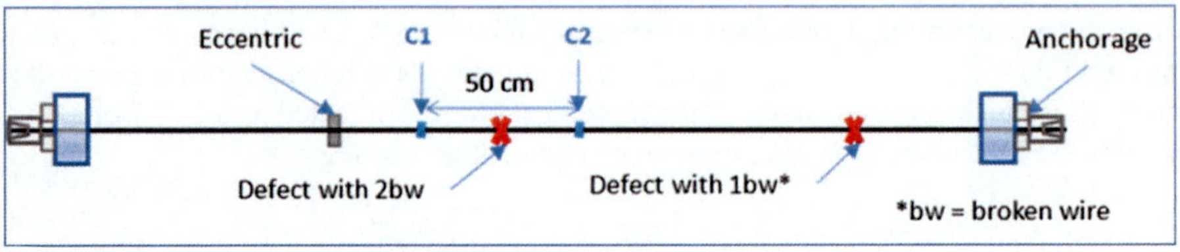

Fig. 4. Sensor configuration on T15.7

One of these configurations, where the two sensors (C1 and C2 of the same type) are positioned at 25 cm each from the defect of two broken wires, is depicted at Fig. 4.

For simplification, the following nomenclatures are adopted for these tests:

- Tests where the sensors are located 25 cm on each side of the defect with 2 broken wires (see Fig. 4) are referenced as: 25 cm2Fc
- Tests where the sensors are located 75 cm on each side of the defect with 2 broken wires are referenced as: 75 cm2Fc
- Tests where the sensors are located 25 cm on each side of the defect with 1 broken wire are referenced as: 25 cm1Fc

A summary, in terms of the number of acoustic hits (signals) from all these configurations tests conducted on this T15.7, is provided in Fig. 5. Therefore, we can observe that the closer the sensors are to the source (the defect), the higher the number of recorded acoustic hits. Similarly, a more severe defect (more broken wires) results in more AE hits.

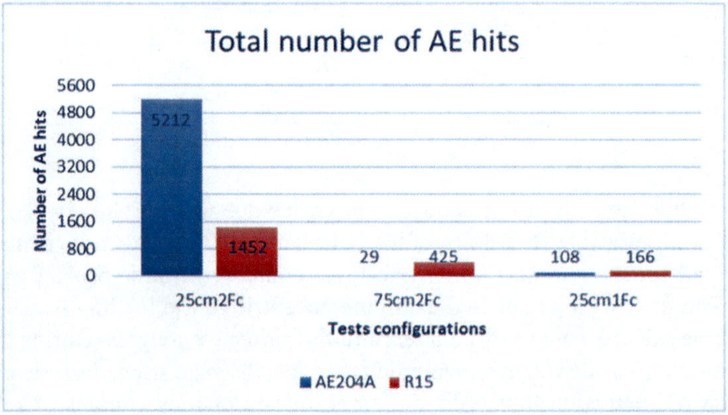

Fig. 5. Number of hits as a function of the sensor, the defect-sensor distance, and the number of broken wires

The quantitative analysis of all the test data allowed us to establish the test parameters for the bi-layer cable where sources separation will be done:

- Sensor type: AE204A. This sensor has a resonance frequency of 375 kHz, a bandwidth ranging from 170 to 700 kHz, a diameter of 8 mm, and a height of 18 mm.

- The detection threshold, which was 31 dB in the tests, will be set at 32 dB, 2 dB above the background noise level.
- The PDT, HDT and HLT parameters are respectively set to 300 μs, 600 μs, and 1000 μs.
- The excitation frequencies with the eccentric are fixed at 2 Hz, 3 Hz, and 4 Hz for amplitudes deviation ranging from 5 mm to 20 mm with a 5 mm increment.

For the bi-layer cables, three scenarios are tested sequentially and in the following order: the first scenario where the cable is free from any wire breakage, the second scenario with wire breakages in the central portion, and the third scenario with wire breakages at the anchorages. These different cases are shown in Fig. 6 indicating the sections of the cable where the wire breaks and the sensors are located.

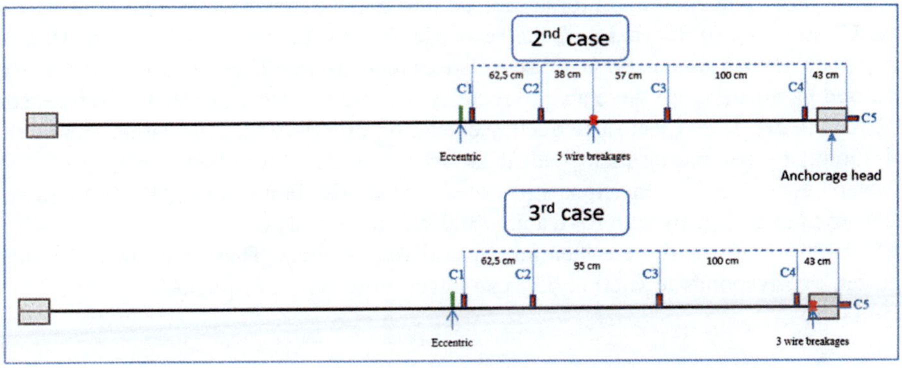

Fig. 6. Bi-layer cables with wire breakages and sensors positions

This arrangement of sensors will help in separating the sources that may generate AE during the tests. The AE signals attributed to the eccentric noise will mostly be detected by sensor C1 positioned 10 cm from the cable-eccentric coupling. As for the AE signals resulting from the cracking of the anchorage resin due to excitation, they will primarily be recorded by sensors C5 and C4 positioned outside the anchorage on the resin and 20 cm from the anchorage outlet, respectively. In the presence of wire breaks, we can expect that the signals resulting from the friction between broken wires and intact wires will be mainly recorded by sensors C2 and C3 positioned at 38 cm and 57 cm, respectively, on either side of the break zone.

For the data analysis, the AE tests were initially conducted on the cable without any defects. Then, after each wire break in the central section, up to five wire breaks, the same tests as those conducted on the intact cable were repeated. For each frequency-amplitude excitation pair, the data were recorded for 10 min to ensure a sufficient acoustic activity for conducting a statistical analysis.

3 Results and Discussion

The data analysis is performed using Jupiter Notebook and the Scikit-Learn library. We first analyzed the data collected from the safe cable to characterize the signature of the acoustic signals resulting from friction between intact wires. This characterization involves separating this source of AE from other sources such as cracking of the resin in the anchorage, eccentric noise, and any potential interference signals.

3.1 Parametric Analysis

Based on correlation analysis and existing literature [18], four parameters have been selected for the parametric analysis of the safe cable data: amplitude (Amp), rise time (Rise), peak frequency (PFrq), and centroid frequency (CFrq). For the safe cable, at all excitation frequencies and for amplitude deviations below 15 mm, only sensors C1 and C2 (see Fig. 6), located near the eccentric, recorded data. However, from 15 mm amplitude deviation and above, all sensors recorded signals. Then, based on the above selected parameters, we were able to identify the three clusters of signals from sensor C2 for the 2Hz15mm test (a frequency excitation of 2Hz with a deviation amplitude of 15mm) by comparing them with those from the 4Hz10mm test (see Fig. 7). The cluster 1 corresponds to the signals recorded for the 4Hz10mm test and therefore to the eccentric noise. Signals with PFrq above 300 kHz (cluster 3) were only recorded during the 2Hz15mm test and were therefore considered as noise. Therefore, the remaining cluster 2 corresponds to friction between intact wires with PFrq values of 130 kHz to 250 kHz.

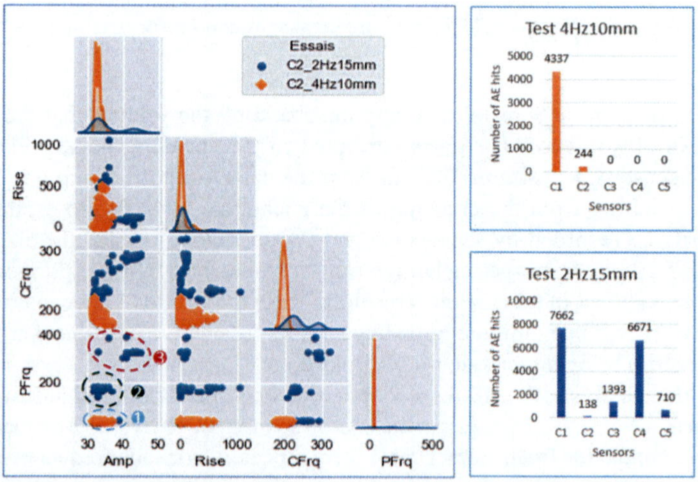

Fig. 7. Distribution diagrams of AE parameters of safe cable from sensor C2 (left) and total AE hits from 4Hz10mm and 2Hz15mm tests.

However, an overlap between PFrq intervals of AE signals from eccentric and friction sources was observed in data recorded with sensors C1 and C4. To improve signals separation, simultaneous examination of signals evolution with the others AE parameters is then needed. Such a complex task requires advanced statistical tools. Therefore, supervised Machine Learning analysis is employed to address this challenge.

3.2 Unsupervised and Supervised Machine Learning Analyses

For this Machine Learning analysis, we will proceed in three steps. First, we will perform dimensionality reduction using Principal Component Analysis (PCA). Then, we will conduct an unsupervised analysis and finally a supervised analysis.

To obtain more representative data of the different phenomena generating acoustic activity, we will consider all tests conducted with an eccentric deviation amplitude of 15 mm and 20 mm. The parasitic data, indicated by the 2Hz15mm test (0.13% of the total data) with PFrq above 300 kHz, will be filtered out beforehand. The preliminary quantitative analysis showed that micro-cracks in the resin with PFrq of 0 kHz disappear after a certain test time. Due to the transient nature of these signals, they will also be filtered out. Thus, after filtering out these indicated signals, the total number of signals for each sensor in these tests with 15 mm and 20 mm excitation amplitudes is shown in the Fig. 8.

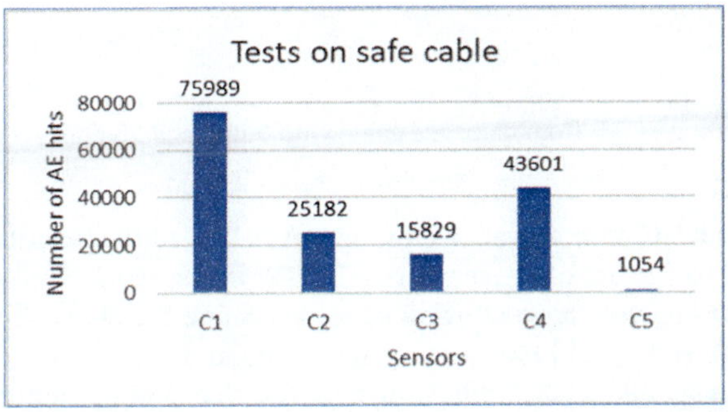

Fig. 8. Number of AE signals on the safe cable with deviation amplitudes of 15 mm and 20 mm.

After dimensionality reduction using PCA [19], unsupervised analysis (clustering) of the data is performed using the K-Means [20] and Gaussian Mixture Model (GMM) [21] algorithms with scikit-learn [22]. Only the results obtained with the GMM algorithm are presented. To ensure the validity of the clustering performed by this algorithm, the data from each cluster are compared to the grouped data from the tests with 5 mm and 10 mm excitation amplitudes, where only signals from the eccentricity are observed. For this purpose, the Gaussian Naive Bayes (GNB) algorithm is used through a supervised analysis (classification). The result of this classification is shown in the Fig. 9.

This result shows that the signals recorded by sensors C1 and C2 are classified as originating from the eccentric at 82% (Exc), while the remaining 18% are classified as friction between safe wires (FriSW).

The same procedure as for the intact cable was applied to the cable with wires breakages in the mid-span (see case 2 of Fig. 6). The classification of AE signals for the

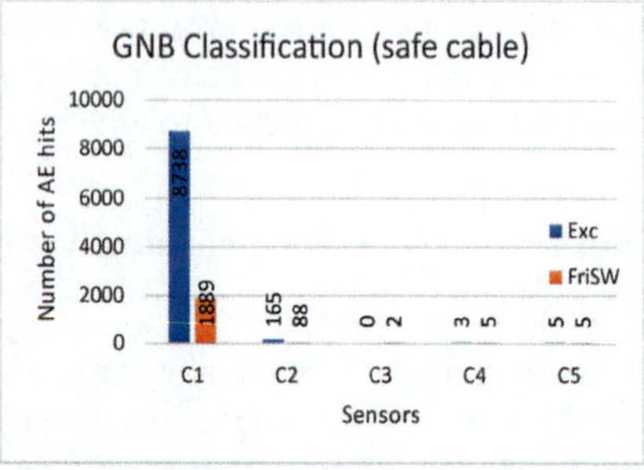

Fig. 9. Classification of AE from safe cable with 5 mm and 10 mm deviation amplitudes

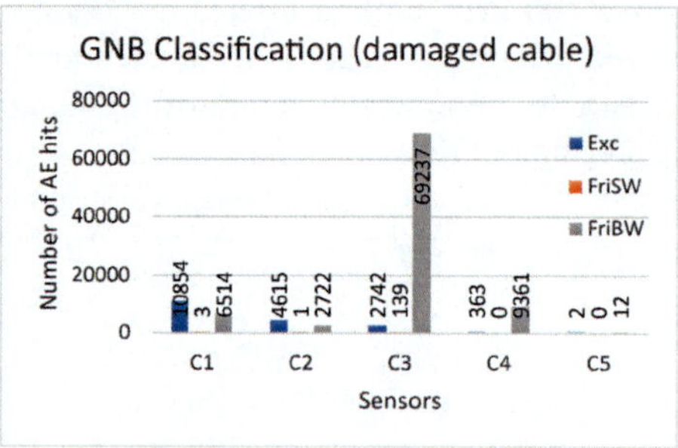

Fig. 10. Classification of AE from damaged cable with 5 mm and 10 mm deviation amplitudes.

tests with 5 mm and 10 mm deviations on this case allowed the identification of a new cluster attributed to friction of broken wires (FriBW in Fig. 10).

These various analyses have allowed us to determine the characteristics of signals generated by each source and to create a labeled signal database, where each signal is assigned to a corresponding label indicating its source (eccentric, friction between intact wires, or friction between cut wires).

From this database, we were able to directly identify, through supervised analysis, the signals attributed to friction between the broken wires in the cable anchorage (case 3 in Fig. 6). Specifically, 79.07% of the signals recorded by sensor C4 (see Fig. 11), located at the anchor outlet, were identified as friction between the broken wires. We note that 11% of the signals from sensor C4 are classified as friction between intact wires (FriSW), and the 9% attributed to eccentric noise (Exc) may be due to imperfections in the classification process.

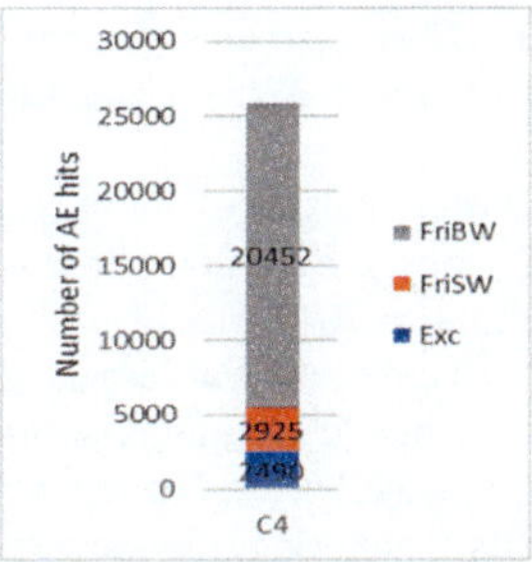

Fig. 11. AE signals from sensor C4 of broken cable in anchorage

4 Conclusion

The tests were conducted on two types of cables: a single-layer T15.7 cable used to determine the parameters for AE tests, and a bi-layer cable. The bi-layer cable was tested under three conditions to characterize different sources of AE signals: safe state, wire breaks in the central section, and wire breaks in the anchorages. Parametric analysis and Machine Learning techniques were employed to identify the signal characteristics of different AE sources, such as friction of intact or broken wires, resin micro-cracks, and noise from the eccentric machine used to perform the tests. By strategically positioning AE sensors and creating a database of labeled signals through unsupervised and supervised analyses, signals resulting from wire breaks in the anchorages of these bi-layer cables were successfully detected and isolated.

References

1. Albouy, D., Labourie, L., Billon, J., Lucas, V.: Surveillance et entretien courant des ouvrages d'art routiers (2011)
2. Commission d'Inspection Ministérielle: Commune de Genes Autoroute A10 - effondrement du viaduc de Polcevera (2018)
3. Gaillet, L.: Les câbles dans le génie civil : De l'importance de comprendre et connaitre leurs états de dégradation, de les protéger. IFSTTAR (2013). https://doi.org/10.3829/erlpc.oa72-fr.
4. Dieng, L., Périer, V., Gaillet, L., Tessier, C.: Mécanismes de dégradation et moyens de protection des câbles du génie civil. Méc. Ind. **10**(1), 33–42 (2009). https://doi.org/10.1051/meca/2009030
5. Watson, S.C., Stafford, D.: Cables in trouble. Civ. Eng. **58**(4), 38 (1988)
6. Maurey, H., Chaize, P., Dagbert, M.: Sécurité des ponts : éviter un drame (2019)
7. Boller, C.: Structural health monitoring—an introduction and definitions. In: Boller, C., Chang, F.-K., Fujino, Y. (eds.) Encyclopedia of Structural Health Monitoring. Wiley (2009). https://doi.org/10.1002/9780470061626.shm204
8. AFNOR: Essais non destructifs. terminologies. partie 9 : Termes utilisés en contrôle par émission acoustique (2017)
9. A. E1316–17a: Standard Terminology for Nondestructive Examinations. West Conshohocken, PA (2017)
10. Skåre, T., Krantz, F.: Wear and frictional behaviour of high strength steel in stamping monitored by acoustic emission technique. Wear **255**(7–12), 1471–1479 (2003). https://doi.org/10.1016/S0043-1648(03)00197-2

11. Sung, K.Y., Kim, I.S., Yoon, Y.K.: Characteristics of acoustic emission during stress corrosion cracking of inconel 600 alloy. Scr. Mater. **37**(8), 1255–1262 (1997). https://doi.org/10.1016/S1359-6462(97)00222-4
12. Li, D., Yang, W., Zhang, W.: Cluster analysis of stress corrosion mechanisms for steel wires used in bridge cables through acoustic emission particle swarm optimization. Ultrasonics (2017). https://doi.org/10.1016/j.ultras.2017.01.012
13. Kretz, T., Brevet, P., Crémona, C., Godart, B., Paillusseau, P.: Haute surveillance et évaluation de l'aptitude au service du pont suspendu d'Aquitaine. In: LCPC, pp. 4581–13 (2006)
14. Zejli, H., Laksimi, A., Tessier, C., Gaillet, L., Benmedakhene, S.: Detection of the broken wires in the cables' hidden parts (anchorings) by acoustic emission. Adv. Mater. Res. **13**, 345–350 (2006)
15. Lau, K., Lasa, I.: Corrosion of prestress and post-tension reinforced-concrete bridges. In: Corrosion of Steel in Concrete Structures, pp. 37–57. Elsevier (2016). https://doi.org/10.1016/B978-1-78242-381-2.00003-1
16. Casey, N.F., Laura, P.A.A.: A review of the acoustic-emission monitoring of wire rope. Ocean Eng. **24**(10), 935–947 (1997). https://doi.org/10.1016/S0029-8018(96)00052-2
17. Hsu, N., Breckenridge, F.R.: Characterization and calibration of acoustic emission sensors. Mater. Evalution **39**, 60–68 (1981)
18. Zejli, H., Gaillet, L., Laksimi, A., Benmedakhene, S.: Detection of the presence of broken wires in cables by acoustic emission inspection. J. Bridge Eng. (2012). https://doi.org/10.1061/(ASCE)BE.1943-5592.0000404.
19. Wold, S., Esbensen, K., Geladi, P.: Principal component analysis. Chemom. Intell. Lab. Syst. **2**, 37–52 (1987)
20. Likas, A., Vlassis, N., Verbeek, J.J.: The global k-means clustering algorithm. Pattern Recognit **36**(2), 451–461 (2003). https://doi.org/10.1016/S0031-3203(02)00060-2
21. Reynolds, D.: Gaussian Mixture Models * (2023)
22. Pedregosa, F., et al.: Scikit-learn: machine learning in Python. J. Mach. Learn. Res. **12**, 2825–2830 (2011)

Seismic Waves Shielding Using Spherical Matryoshka-Like Metamaterials

Brahim Lemkalli[1(✉)], Sébastien Guenneau[2], Youssef El Badri[3], Muamer Kadic[4], Hicham Mangach[3], Abdellah Mir[1], and Younes Achaoui[1]

[1] Laboratory for the Study of Advanced Materials and Applications, Department of Physics, Moulay Ismail University, B.P. 11201 Zitoune, Meknes, Morocco
brahim.lemkalli@edu.ac.ma

[2] UMI 2004 Abraham de Moivre-CNRS, Imperial College London, SW7 2AZ London, UK

[3] Laboratory of Optics, Information Processing, Mechanics, Energetics and Electronics, Department of Physics, Moulay Ismail University, B.P. 11201 Zitoune, Meknes, Morocco

[4] Institut FEMTO-ST, UMR 6174, CNRS, Université de Bourgogne Franche-Comté, 25000 Besançon, France

Abstract. Among the most destructive natural mechanical events on the planet, seismic waves cause substantial damage and degradation to infrastructure throughout the world, posing a threat to humankind. The development of seismic metamaterials opens up a new frontier for shielding buildings and infrastructure against earthquakes. Furthermore, vibrations from seismic waves propagating at the surface of the earth are mostly due to Rayleigh waves, which have low frequencies, typically below 10 Hz. Within this critical range of frequencies, we report a novel architecture optimized for shielding against seismic waves, which is denoted as a spherical Matryoshka-like seismic metamaterial. We explore the response of this system using numerical analysis based on the finite element method. The band diagram in the irreducible Brillouin zone reveals, most notably, omnidirectional stop bands. Its frequency response analysis was carried out to better explore this metamaterial's ability to attenuate seismic waves. The findings of this study open up new pathways for designing and optimizing seismic metamaterials. Thus, offering new avenues for improving earthquake shielding.

Keywords: Seismic metamaterials · Spherical Matryoshka · Seismic shielding

1 Introduction

Seismic activity, more than any other natural hazard, has the greatest potential to bring about catastrophic devastation, life casualties, and significant economic loss. Deploying seismic metamaterials into smart city designs provides a new

M. Ben Ahmed et al. (Eds.): SCA 2023, LNNS 906, pp. 77–85, 2024.
https://doi.org/10.1007/978-3-031-53824-7_8

frontier of research to enhance the resilience of smart buildings and transportation, thereby reducing the disastrous repercussions of earthquakes [1]. There are different types of seismic waves that propagate with different wavelengths and polarization states. The bulk waves that propagate inside the soil and rocks are divided into two categories: the primary or compression waves, and the secondary or shear waves. Then come the surface waves that are the slowest and propagate on the surface of the ground [2]. The main causes of ground displacement and surface rupture that occur during an earthquake are vibrations from seismic waves propagating on the surface [3]. It is worth noting that these surface waves have low frequencies and large amplitudes that decrease with the depth of the ground, which explains their damaging nature [4]. Generally, the movements of earthquakes observed at the surface of the ground are mostly Rayleigh waves with frequencies below 10 Hz [5]. In order to develop seismic protection in this range of frequencies, researchers have integrated composite materials known as seismic metamaterials into civil engineering applications. These systems are based on two phenomena: Bragg's scattering and local resonance to deflect the trajectory of waves by reflection or radiation into the bulk [6].

Brulé et *al.* were the first to make use of metamaterials in civil engineering. They designed an array of empty vertical inclusions drilled into the ground to attenuate seismic waves [4]. Colombi et *al.* outlined the role of forests in reducing the propagation of Rayleigh waves, giving rise to a new generation of seismic metamaterials known as forest metamaterials [7]. Achaoui et *al.* theoretically created complete stop bands by designing resonators based on cubic unit cells made of concrete containing an iron sphere connected by rubber or iron ligaments [8]. Besides, Miniaci et *al.* proposed a large-scale metamaterial to attenuate both bulk and surface waves using mechanical and phononic crystal concepts [9]. Recently, Achaoui et *al.* experimented with anchoring columns to the underlying bedrock to create zero-frequency bandgaps [10]. Although the main drawback of their work lies in the fact that such a configuration is not easy to implement. More recently, Zeng et *al.* have proposed a metamaterial formed by an inertial amplification process to achieve the prospect of broadband low-frequency seismic waves attenuation [11]. Elsewhere, Liu et *al.* developed a topology optimization system for porous seismic metamaterials to optimize a series of structures that exhibited low forbidden frequency ranges starting at 1.6 Hz [12]. Developing defense structures to block earthquakes from reaching structures is therefore of considerable significance, especially for seismic waves with frequencies less than 10 Hz, which coincides with the inherent resonance frequency of many artificial structures [13–15].

In this study, we propose mitigating low-frequency seismic wave propagation using a novel design dubbed spherical Matryoshka-like metamaterial, which has significant attenuation characteristics, as each cell in the lattice is designed to present complete bandgaps in the low-frequency range. We explore the dynamic response of this metamaterial using numerical analysis based on the finite element method to solve the harmonic and eigenfrequencies equations of seismic wave propagation. We determined the dispersion curves of the Matryoshka cell and identified the forbidden frequency ranges. In addition, we calculated the Rayleigh wave transmission through a 2D model of 10×10 cells to further highlight the attenuation capabilities of the spherical Matryoshka-like metamaterial.

2 Model and Methods

The unit cell we designed is inspired by the Matryoshka dolls, in which many dolls are arranged in descending order of size, the smallest being inside. The Matryoshka-like seismic metamaterial (Fig. 1a) is composed of $N_x \times N_y \times N_z$ unit cells. The unit cell is comprised of a centered full steel sphere with R_1 radius that is encircled by a rubber shell with $R_2 - R_1$ thickness encapsulated by a shell made of steel with R_3 radius, and a rubber sphere with R_4 radius, as depicted in Fig. 1b.

Table 1. The geometrical parameters in meters of the unit cell of the spherical Matryoshka-like metamaterial.

a	R_1	R_2	R_3	R_4
2.5	0.5	0.7	1.15	1.26

The geometrical parameters of the Matryoshka cell are given in Table 1, taking into account the value of the periodicity order found in the literature [16]. The materials chosen for the study are assumed to be linearly elastic, and their parameters are given in Table 2.

Table 2. The material parameters used in the simulation.

Materials	Density (kg/m^3)	Poisson's ratio	Young's modulus (Pa)
Rubber	1300	0.47	1.02×10^5
Steel	7850	0.28	205×10^9
Soil	1800	0.3	2×10^7

Generally, the seismic wave propagation equation is written as follows [17]:

$$[C]\frac{\partial^2 u}{\partial r} = \rho\frac{\partial^2 u}{\partial t^2}, \tag{1}$$

u is the displacement; r is the coordinate system; ρ is the density; $[C]$ is the elastic stiffness matrix, which is related to Young's modulus and Poisson's ratio. In the harmonic analysis, the Eq. 1 is rewritten as:

$$[C]\frac{\partial^2 u}{\partial r} = -\rho w^2 u, \tag{2}$$

w is the frequency. We employed the Floquet-Bloch conditions on the six borders of the Matryoshka cell, and the boundary conditions are therefore written as follows:

$$u(a + r) = u(r)e^{ika}. \tag{3}$$

From the Eq. 2, we deduce the following eigenfrequency equation:

$$(K(k) - M(k)w^2)u = 0, \tag{4}$$

$M(k)$ and $K(k)$ represent the mass and stiffness matrices, respectively, which depend on the reduced wavenumber k. For a given value of k, there is an eigenfrequency from which the dispersion relation can be derived.

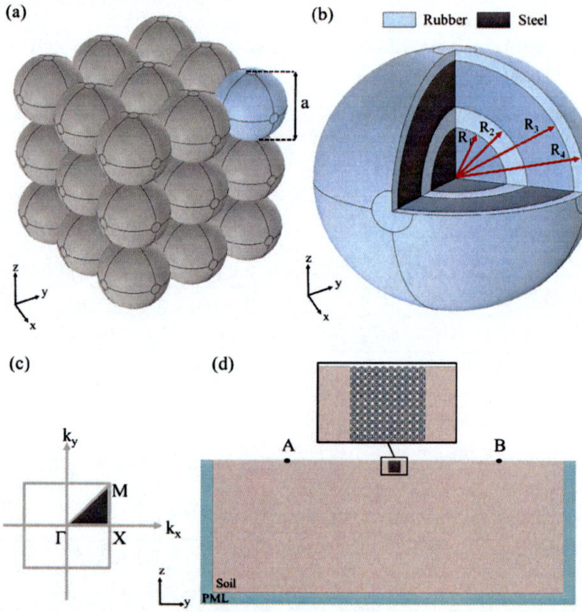

Fig. 1. Geometrical illustrations. (a) Metamaterial composed of $N_x = N_y = N_z = 3$ Matryoshka-like cells. (b) The spherical Matryoshka unit cell used to calculate the dispersion curves. (c) The first irreducible Brillouin zone: $\Gamma = (k_x = 0, k_y = 0, k_z = 0)$, $X = (k_x = \pi/a, k_y = 0, k_z = 0)$, and $M = (k_x = \pi/a, k_y = \pi/a, k_z = 0)$. (d) The 2D FEM model used to evaluate transmission.

To delineate the elastodynamic behavior of the suggested metamaterial, we first determined the eigenfrequencies in the first irreducible Brillouin zone (Fig. 1c). The FEM is employed to solve the eigenfrequencies equation (Eq. 4) of seismic wave propagation by applying the Floquet-Bloch periodicity conditions (Eq. 3) on the Matryoshka cell along the x- and y-directions. Two scenarios were specifically investigated: the first is for a single unit cell in the z-direction ($N_z = 1$) and the second is for an infinite number of cells ($N_z = \infty$), applying a condition of periodicity $k_z = 0$ along z-direction. Afterwards, we identify the obtained omnidirectional band gaps, within which seismic waves cannot propagate. The corresponding transmission spectrum for 10×10 cells in the xy-plane with the absorbing regions is calculated based on Eq. 2, as depicted in Fig. 1d. A harmonic excitation is applied at point A and the output response is recorded at point B. The frequency is swept from 0.01 to 11 Hz.

3 Results and Discussion

The findings from the dispersion diagram demonstrate the presence of multiple forbidden bands that prevent seismic waves of specific frequencies from propagating. According to the dispersion diagram, the first forbidden band started immediately after the frequency $f_1 = 0.64$ Hz and ended at $f_2 = 3.12$ Hz. The second forbidden band is located between $f_3 = 3.15$ Hz and $f_4 = 5.12$ Hz, while the third energy band is located between $f_5 = 5.15$ Hz and $f_6 = 10.5$ Hz. The most notable feature is that the array with finite unit cells along the z-direction maintains the same fundamental aspect as the infinite, as evidenced by the matching position of the forbidden energy bands for both cases. The degeneracy lift observed between the transverse and bending modes can be explained by the preclusion of periodicity along the z-direction for the first scenario (Fig. 2a). Otherwise, the two modes are degenerated for the infinite array (Fig. 2b).

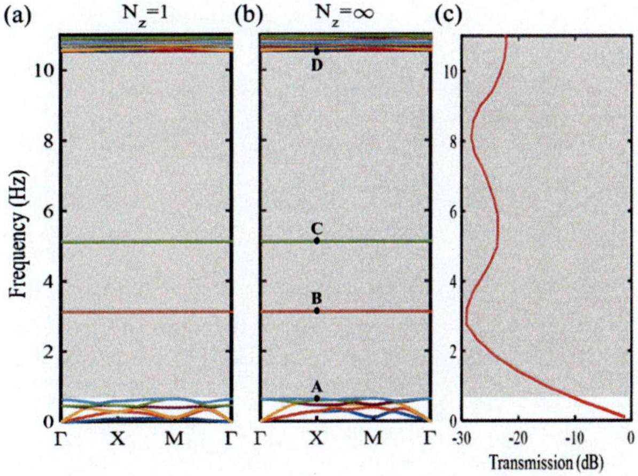

Fig. 2. The band structure of the spherical Matryoshka cell for the first irreducible Brillouin zone in the xy-plane. (a) A single cell along z-direction. (b) With periodicity conditions along z-direction. The gray areas represent the band gaps. The points (A, B, C, and D) indicate the different modes along the x-direction, as shown in Fig. 3. (c) Rayleigh wave transmission spectrum for a periodic array of 10×10 Matryoshka cells.

Furthermore, Fig. 3 depicts screenshots of the total displacement for four cases A, B, C, and D. Where B (Fig. 3b) and C (3c) correspond to the localization of energy inside the central sphere made of steel, which is consistent with the first and second flat modes with near-zero phase and group velocities depicted in the dispersion diagram. Meanwhile, A (Fig. 3a) and D (3d) are located at the limits of the first and last band gaps, respectively, where the energy is partially concentrated at the top border of our spherical Matryoshka-like metamaterial, enabling some energy to propagate within the structure. In summery, the proposed metamaterial has a significant stop band at low frequencies, ranging from 0.5 to 11 Hz. According to the eigenfrequency analysis, any earthquake with a frequency in this range and any polarization within the xy-plane cannot pass through our structure. A frequency response study was also carried out to better explore the ability of the proposed metamaterial to attenuate seismic waves with a finite number of cells, as shown in Fig. 2c.

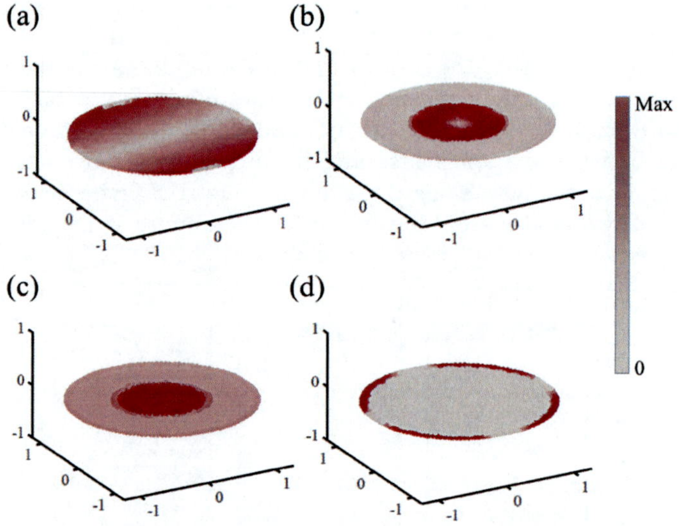

Fig. 3. Screenshots of eigenmodes at X point in the first irreducible Brillouin zone. (a) A matches the frequency of $f = 0.643$ Hz. (b) B matches the frequency of $f = 3.127$ Hz. (c) C matches the frequency of $f = 5.122$ Hz. (d) D matches the frequency of $f = 10.534$ Hz.

Figure 2c depicts the transmission spectrum of seismic waves with excitation frequencies from 0.1 Hz to 11 Hz, which corresponds to the range of the world's most devastating earthquakes, through 10×10 Matryoshka cells array. The proposed array exhibits significant Rayleigh wave attenuation, which is a product of the presence of the forbidden bands independently of the direction of propagation. It should be noted that the spherical symmetry of the proposed metamaterial is the main factor in obtaining omnidirectional stop bands. Figure 4 depicts the total displacement distribution maps for four different frequency values.

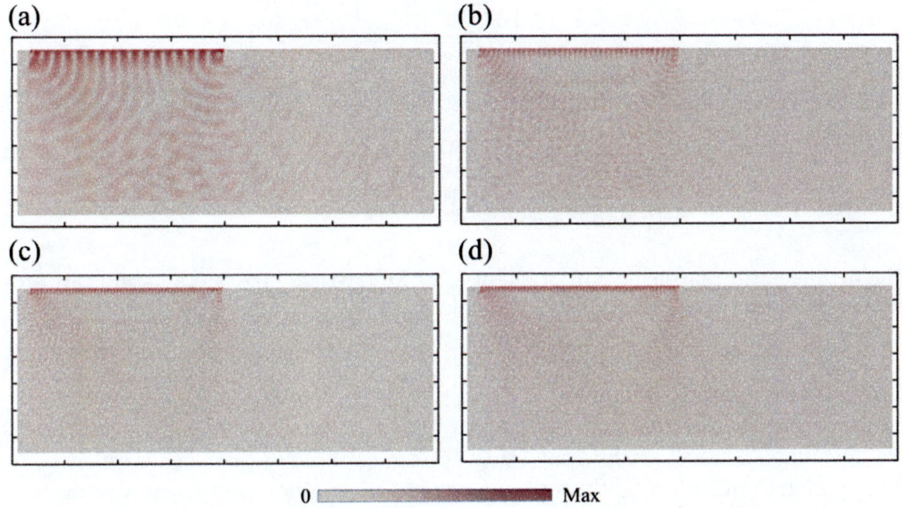

Fig. 4. The total displacement distribution for the harmonic responses. (a) For a frequency of $f = 1.43$ Hz. (b) For a frequency of $f = 3.2$ Hz. (c) For a frequency of $f = 5$ Hz. (d) For a frequency of $f = 8.10$ Hz.

We opted to compare our findings to the literature in the frequency range of 1 Hz and 10 Hz since it contains the most harmful surface waves. Zeng et *al.* used cylindrical Matryoshka structures with the same periodicity as our case. Their metamaterial opened three low-frequency band gaps, the first being between 5 Hz and 7 Hz, the second between frequencies 7 Hz to 10.5 Hz, and the third 11 Hz to 13 Hz [16]. Our structure opened complete bandgaps in the frequencies from 0.5 Hz to 11 Hz, as delineated by both the eigenfrequency and transmission analysis. As a consequence, the suggested spherical Matryoshka-like metamaterial is a promising candidate for shielding against ultra-low-frequency seismic waves, beyond the typical limits of the values reported.

This study provides novel prospects for the design and application of spherical Matryoshka-like metamaterials for handling seismic waves and protecting infrastructure. The construction of a smart city by surrounding the buildings with spherical Matryoshka-Like metamaterials will be resist to the destructive seismic, as shown in Fig. 5.

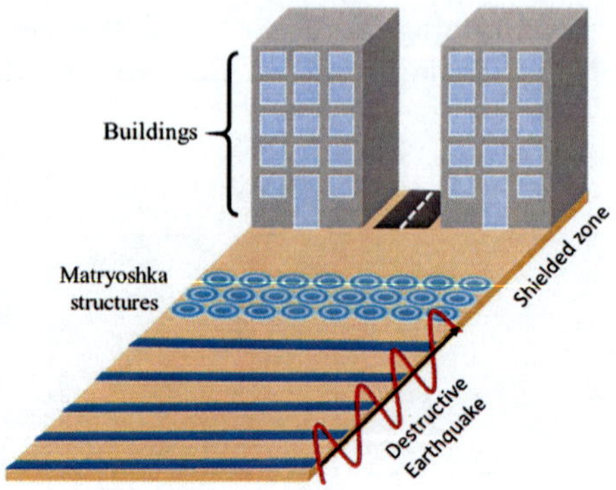

Fig. 5. Shematic representation of the role of Matryoshka-like Metamaterial in protecting buildings from disastrous Earthquakes.

4 Conclusion

We developed a novel seismic shielding system capable of protecting sensitive or strategic structures using concepts of structural engineering, namely, spherical Matryoshka-like metamaterials. We have numerically investigated the presence of omnidirectional band gaps on the dispersion. We analyzed the attenuation of Rayleigh waves in a bidirectional array consisting of 10×10 unit cells using a transmission simulation. We revealed that surface waves are deflected and transformed into bulk waves. We demonstrated that the metamaterial attenuated surface waves with frequencies located between 0.5 Hz and 10 Hz, which could be a solution to protect structures from ultra-low-frequency seismic waves.

References

1. Palermo, A., Krödel, S., Marzani, A., Daraio, C.: Engineered metabarrier as shield from seismic surface waves. Sci. Rep. **6**, 39356 (2016). https://doi.org/10.1038/srep39356
2. Mu, D., Shu, H., Zhao, L., An, S.: A review of research on seismic metamaterials. Adv. Eng. Mater. **22**, 1901148 (2020). https://doi.org/10.1002/adem.201901148
3. Kim, S.H., Das, M.P.: Seismic waveguide of metamaterials. Mod. Phys. Lett. B **26**, 1250105 (2012). https://doi.org/10.1142/S0217984912501059
4. Brûlé, S., Javelaud, E.H., Enoch, S., Guenneau, S.: Experiments on seismic metamaterials: molding surface waves. Phys. Rev. Lett. **112**, 133901 (2014). https://doi.org/10.1103/PhysRevLett.112.133901
5. Brûlé, S., Enoch, S., Guenneau, S.: Emergence of seismic metamaterials: current state and future perspectives. Phys. Lett. A. **384**, 126034 (2020). https://doi.org/10.1016/j.physleta.2019.126034
6. Kadic, M., Bückmann, T., Schittny, R., Wegener, M.: Metamaterials beyond electromagnetism. Rep. Prog. Phys. **76**, 126501 (2013). https://doi.org/10.1088/0034-4885/76/12/126501

7. Colombi, A., Roux, P., Guenneau, S., Gueguen, P., Craster, R.V.: Forests as a natural seismic metamaterial: rayleigh wave bandgaps induced by local resonances. Sci. Rep. **6**, 19238 (2016). https://doi.org/10.1038/srep19238
8. Achaoui, Y., Ungureanu, B., Enoch, S., Brûlé, S., Guenneau, S.: Seismic waves damping with arrays of inertial resonators. Extrem. Mech. Lett. **8**, 30–37 (2016). https://doi.org/10.1016/j.eml.2016.02.004
9. Miniaci, M., Krushynska, A., Bosia, F., Pugno, N.M.: Large scale mechanical metamaterials as seismic shields. New J. Phys. **18**, 083041 (2016). https://doi.org/10.1088/1367-2630/18/8/083041
10. Achaoui, Y., Antonakakis, T., Brûlé, S., Craster, R.V., Enoch, S., Guenneau, S.: Clamped seismic metamaterials: ultralow frequency stop bands. New J. Phys. **19**, 063022 (2017). https://doi.org/10.1088/13672630/aa6e21
11. Zeng, Y., et al.: Seismic metamaterials: generating low-frequency bandgaps induced by inertial amplification. Int. J. Mech. Sci. **221**, 107224 (2022). https://doi.org/10.1016/j.ijmecsci.2022.107224
12. Liu, Z., Dong, H.W., Yu, G.L., Cheng, L.: Achieving ultra-broadband and ultra-low-frequency surface wave bandgaps in seismic metamaterials through topology optimization. Compos. Struct. **295**, 115863 (2022). https://doi.org/10.1016/j.compstruct.2022.115863
13. Krasnok, A., et al.: Hierarchical large-scale elastic metamaterials for passive seismic wave mitigation. EPJ Appl. Metamaterials. **8**, 14 (2021). https://qmro.qmul.ac.uk/xmlui/handle/123456789/74964 https://qmro.qmul.ac.uk/xmlui/handle/123456789/74964
14. Wu, X., Wen, Z., Jin, Y., Rabczuk, T., Zhuang, X., Djafari-Rouhani, B.: Broadband rayleigh wave attenuation by gradient metamaterials. Int. J. Mech. Sci. **205**, 106592 (2021). https://doi.org/10.1016/j.ijmecsci.2021.106592
15. Zeng, Y., et al.: Low-frequency broadband seismic metamaterial using I-shaped pillars in a half-space. J. Appl. Phys. **123**, 214901 (2018). https://doi.org/10.1063/1.5021299
16. Zeng, Y., et al.: A Matryoshka-like seismic metamaterial with wide bandgap characteristics. Int. J. Solids Struct. **185–186**, 334–341 (2020). https://doi.org/10.1016/j.ijsolstr.2019.08.032
17. Zhang, K., Luo, J., Hong, F., Deng, Z.: Seismic metamaterials with cross-like and square steel sections for low-frequency wide band gaps. Eng. Struct. **232**, 111870 (2021). https://doi.org/10.1016/j.engstruct.2021.111870

Investigating the Spatial Suitability of the Location of Urban Services Using Space Syntax Theory

Saleh Qanazi[1](✉), Ihab H. Hijazi[1], Isam Shahrour[2], and Rani El Meouche[3]

[1] Department of Urban Planning Engineering, An-Najah National University, Nablus, Palestine
saleh.qanaze@najah.edu

[2] Le Laboratoire Génie Civil et géo-Environnement, Université de Lille 1 Sciences et Technologies, 59650 Villeneuve d'Ascq, France

[3] Institut de Recherche en Constructibilité, Université Paris-Est, ESTP, F-94230 Cachan, France

Abstract. This research paper examines the efficacy of the Space Syntax theory in analyzing the spatial suitability of the location of urban services within a Palestinian urban community, specifically in Hajjah. The study emphasizes the importance of incorporating specific Space Syntax parameters, such as integration, connectivity, and choice, with spatial configuration analysis to understand the performance of urban services in a given area. The study utilized axial map analysis to generate a comprehensive set of maps and graphs, providing an in-depth understanding of the spatial distribution and performance of different urban services in the study area. The analysis revealed that while some facilities, such as the health clinic and park, are located in favorable locations, others, such as the gas stations and some schools, require better placement. These findings can guide policymakers and urban planners in designing more efficient and accessible urban environments that cater to the community's diverse needs. The study underscores the potential of the Space Syntax theory to inform more equitable and efficient planning and design in urban areas.

Keywords: space syntax theory · spatial analysis · urban planning

1 Background

Space syntax theory is a human-focused approach that examines the relationship between the spatial configuration of urban environments and the way people use and experience that space [1]. This theory allows urban planners and policymakers to analyze the physical structure of a city and understand how it influences movement, social interaction, and the provision of urban services [2]. The spatial structure of a city can significantly impact the quality of life of its residents [3].

Space syntax theory provides several ways to represent the spatial structure of urban environments [4]. The most common representation is the axial map, which shows the primary paths of movement through the city [5]. Axial lines are a significant aspect of space syntax theory, representing the longest continuous lines of movement through a

M. Ben Ahmed et al. (Eds.): SCA 2023, LNNS 906, pp. 86–98, 2024.
https://doi.org/10.1007/978-3-031-53824-7_9

city, such as major streets or boulevards [5]. These lines form the backbone of a city's spatial structure, and their connectivity is critical to the functioning of urban systems [6]. Other representations include convex spaces which are areas of the city that are visible from a single point [7]. Isovists are spaces that can be seen from a particular vantage point, taking into account occlusion by walls and other obstacles [1, 8].

Connectivity, integration, and choice are central to space syntax theory [9, 10]. Connectivity refers to the degree to which streets and other urban elements are connected to one another, and how easy it is to move between them [6]. Integration refers to the level of access and connectivity that different parts of a city have to one another [11]. Choice refers to the number of different paths that people can take through a city and the extent to which they can navigate the urban environment freely [12].

Recent research has highlighted the impact of space syntax theory on the provision of urban services. Gündüz and Arslan (2020) found that areas with higher street connectivity had greater accessibility to urban amenities, such as schools and healthcare facilities. This research suggests that a well-connected street network can improve the provision of essential urban services [13]. Yildirim and Turgut (2020) found that areas with a high degree of integration in the street network had a lower incidence of crime. The authors argue that the spatial configuration of urban environments can have a significant impact on the safety and security of urban services [14]. Chen and Lu (2021) found that areas with a more connected street network had a greater number of ride-hailing pickups and shorter travel times. This research suggests that the spatial structure of a city can have a significant impact on the performance of transportation services beyond public transportation [15].

Xing and Guo (2022) proposed a new method for analyzing urban space using space syntax and geographic information systems. The study conducts a visual and quantitative analysis of spatial information data, including the integration of urban road networks, building height, architectural style, and more. The method was tested in the old city of Wuxi, where the regression model analysis revealed a positive correlation between the integration of the area and the presence of certain building characteristics [16]. This new analysis method of urban space is significant in exploring urban characteristics, proposing urban strategies, and addressing urban problems [17].

This paper utilizes space syntax theory to analyze the spatial suitability of the location of urban services based on their spatial distribution. By categorizing services into local public services, regional public services, and economic services, the paper highlights the specific criteria and values that should be considered when applying space syntax theory, such as integration, connectivity, and choice. The findings of this analysis can inform policymakers and urban planners in designing more efficient and accessible urban environments that meet the diverse needs of their communities, as they provide a framework for creating more livable and accessible cities [18, 19].

2 Methods

2.1 Study Area

The chosen case study for this paper is Hajjah, a Palestinian urban community located in the eastern part of the Qalqiliya governorate (see Fig. 1). Hajjah is considered a local center for the surrounding communities in the province, which is reflected in the types and levels of services provided to its residents and nearby villages. This makes it an ideal case for applying the Space Syntax theory to analyze the spatial distribution and performance of different types of urban services in the area. The Space Syntax approach will provide a comprehensive understanding of how urban services are distributed within the area and how the accessibility to these services varies based on their spatial relationships to the axial lines. By using this approach, we can identify the strengths and weaknesses of the current spatial distribution of urban services, which can assist policymakers in developing better planning strategies for service provision in urban areas.

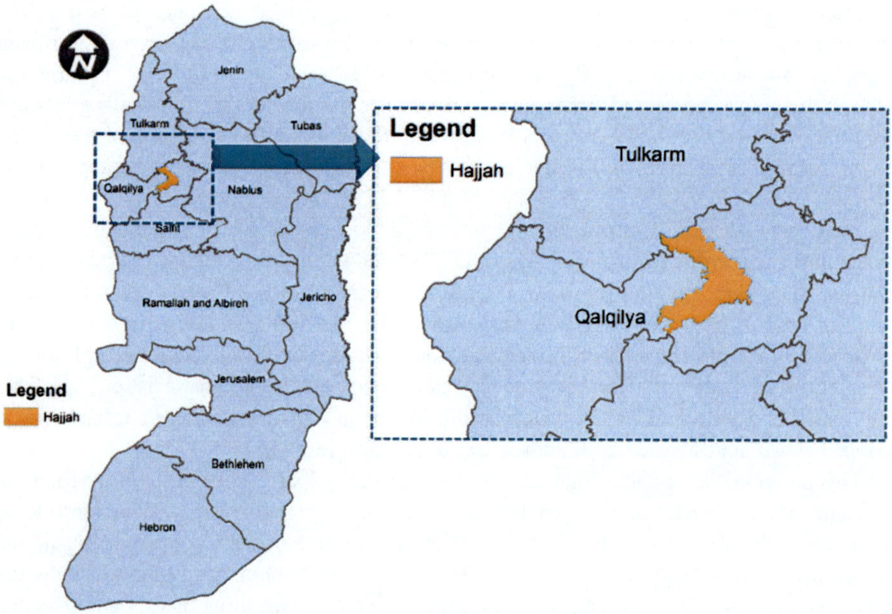

Fig. 1. Study area

2.2 Methodology

This study utilized a comprehensive methodology to analyze the distribution and performance of urban services in the Hajjah study area. There are various methods to represent urban spaces, including axial lines, convex spaces, and isovists. Among these, axial lines are the most widely used representation, showing the longest continuous lines of movement in a city. In our research, we chose to use axial maps since they offer an effective means of analyzing the connectivity and accessibility of urban networks. While other methods such as convex spaces and isovists may be more suitable for analyzing open spaces, axial maps remain the preferred choice for urban analysis due to their ability to highlight the importance of street networks. So, the first step was to prepare the data, and roads were converted into axial lines using DEPTHMAP software.

Axial lines in space syntax theory can be analyzed through different parameters, including integration, choice, and connectivity. Integration measures the degree to which a street segment lies on the shortest path between all other segments in the network, indicating its overall importance for movement. Choice measures the degree to which a street segment allows for different possible routes, indicating its potential for diversifying movement. Connectivity measures the degree to which a street segment is connected to other segments in the network, indicating its potential for accessibility. The local values of these parameters can be used to inform decisions about the placement of urban services, such as locating a school on a high-connectivity street. So, then the values of integration, choice, and connectivity were calculated for all the urban networks using Space Syntax theory. These values were used to create an index for each type of service, which presented the ideal locations based on the available values to maximize performance. The current location of each service was then compared to the ideal value to assess their suitability and performance.

The results of this analysis were presented spatially on maps and in graphs. The results provided insights into the application of Space Syntax theory to different urban contexts and its potential to inform more efficient and equitable planning and design (see Fig. 2). Overall, the methodology employed in this study offers a valuable contribution to the field of urban planning and design by combining theoretical and empirical approaches to analyze the spatial structure and performance of urban services.

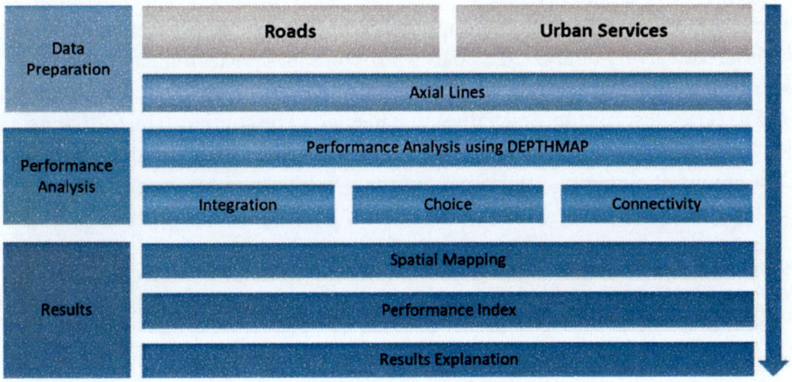

Fig. 2. Research methodology

Data Preparation

The data used in this study was collected from the Palestinian Ministry of local government geospatial portal (GeoMOLG), which is the official source for geospatial data in Palestine. The data consisted of two primary datasets: the road network and the location and types of urban services. The data was collected and digitized by specialized surveyors, ensuring its accuracy and reliability. The necessary data was prepared using ArcGIS software (version 10.7) by cutting the streets within the administrative boundaries of the study areas. The data was then transferred to the AutoCAD program to generate the axial lines. To ensure accuracy and consistency, the axial lines were double-checked for errors and omissions. The metadata, which includes information on the data collection methods, accuracy, and reliability, was reviewed and verified by the data provider to ensure its accuracy. The resulting data was then entered into the DEPTHMAP program, which converted the roads into axial lines and assigned values for different Space Syntax theory concepts, including integration, connectivity, and choice. The resulting data was used to analyze the spatial structure and performance of different urban services in the study areas. The data preparation stage was critical for ensuring the accuracy and reliability of the subsequent analyses. By using multiple software tools and verifying the data at each step, this study aimed to minimize errors and ensure the validity of the results.

Performance Analysis

In the analysis phase, the axial lines of the road network were analyzed based on three main values: choice, integration, and connectivity. Each type of urban service was then spatially located to the nearest axial line, and its corresponding values were analyzed. For instance, educational services were assigned a local value as they serve the residents of the community, and were analyzed based on the value of connectivity.

In the case of Hajjah, it is a local center with regional services that serve the residents of the community and surrounding villages. Therefore, some services, such as the fire station service, have a global value. On the other hand, other services in Hajjah, such as certain schools, serve only the local population, and thus have a local value.

To comprehensively analyze the spatial structure and performance of different types of urban services in the study area, a detailed description was prepared for each service, including its type, description, and values that needed to be studied. The following Table 1 provides an overview of the various services in Hajjah, along with their corresponding values for analysis:

Table 1. Services type and description in Hajjah, and their corresponding values.

Service type	Service description	Value type	Values needed
Commercial services	Serving the residents of the village as well as those passing through the village	Global	Choice Integration
Health Clinic	Serve residents of the village and neighboring villages	Global	Integration
Schools	Serves residents in various parts of the village and some nearby villages	Global & Local	Integration Connectivity
Mosques	Serve the population in different parts of the village	Local	Connectivity
Filling station	Serve those who pass through the village	Global	Choice Integration
Fire station	Serve the village and all the neighboring villages	Global	Choice Integration
Park	Serve all the inhabitants of the village	Global	Integration

The spatial analysis based on the values of choice, integration, and connectivity for each service, aims to provide a comprehensive understanding of the performance of different types of urban services in the study areas.

Each value represents unique characteristics of urban services, which are derived from mathematical and statistical formulas. These formulas have been developed to provide a solid foundation for analyzing urban services and understanding their features. The values include the following [1]:

Integration: Integration refers to the degree to which a location in the urban network is connected to other locations. The value of integration can be calculated using the following formula:

$$I(x) = \sum (d(x, y))^\wedge(-\alpha) \tag{1}$$

where I(x) is the integration value of location x, d(x,y) is the shortest topological distance between x and y, and α is a parameter that determines the rate of decay of the influence of distance on integration.

Choice: Choice refers to the degree to which a location is reached from other locations through a sequence of minimal turns. The value of choice can be calculated using the following formula:

$$C(x) = \sum (n(x, y))^\wedge(-\beta) \tag{2}$$

where C(x) is the choice value of location x, n(x,y) is the number of minimal turns required to reach y from x, and β is a parameter that determines the rate of decay of the influence of turns on choice.

Connectivity: Connectivity refers to the degree to which a location is directly connected to other locations. The value of connectivity can be calculated using the following formula:

$$K(x) = \sum k(x, y) \tag{3}$$

where K(x) is the connectivity value of location x, and k(x,y) is a binary variable that equals 1 if there is a direct link between x and y, and 0 otherwise.

3 Results and Discussion

3.1 Spatial Mapping

Mapping the calculated values of integration, choice, and connectivity in Space Syntax is a common practice used to visually represent the urban network's spatial properties. These maps can help identify the relationships between different parts of the urban environment, and how the services are distributed throughout the area. By displaying the performance of each service on these maps, it is possible to assess the effectiveness of the current location of each service and how it relates to the values of integration, choice, and connectivity. These maps also provide a comprehensive view of the urban environment, highlighting areas that may be underserved and areas that are well-connected and integrated with the rest of the network.

By utilizing the aforementioned formulas and conducting a spatial analysis through the use of DEPTHMAP software, we have generated maps (see Fig. 3) that represent each type of value for the community of Hajjah:

The maps demonstrate a clear trend in the distribution of integration, choice, and connectivity values throughout the town. The highest concentration of integration values is observed in the central region and its surrounding areas, with values decreasing gradually as we move away from the center. Moreover, the main roads of the town exhibit higher choice values, indicating that commercial services and other facilities located along these roads have greater demand and are more easily accessible to a larger population. Additionally, the connectivity values are highest in the connections of the branches near the main roads, highlighting the importance of the strategic placement of public facilities to ensure easy accessibility and connectivity.

(a)

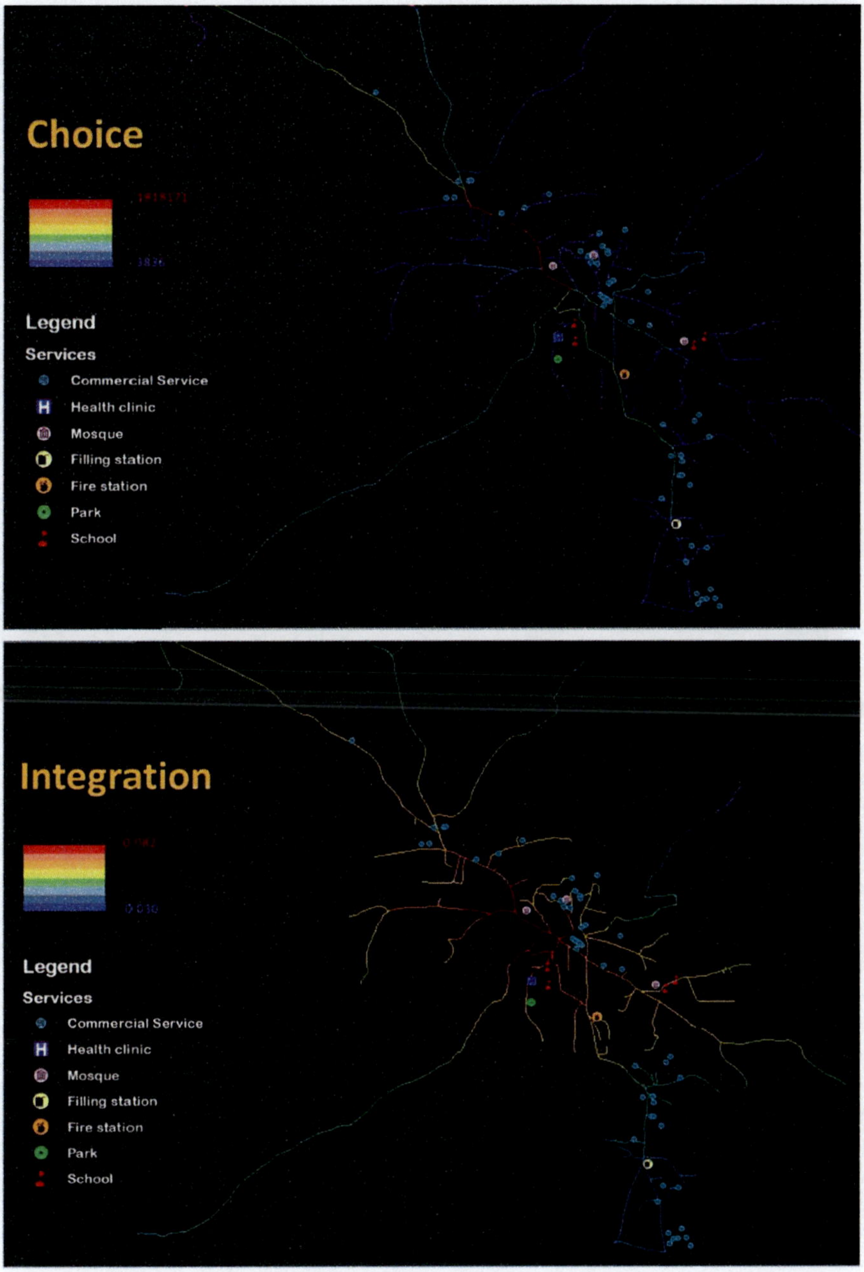

Fig. 3. **a.** Spatial configuration of urban services in Hajjah based on the values of integration, and choice. **b.** Spatial configuration of urban services in Hajjah based on the value of connectivity.

(b)

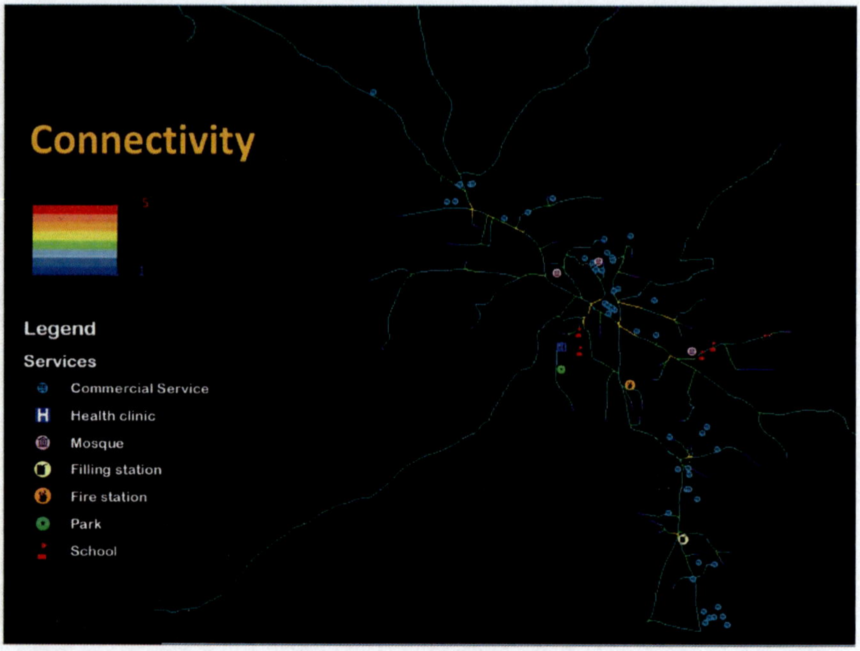

Fig. 3. (*continued*)

3.2 Performance Index

To evaluate the performance of various urban services, an index was developed based on integration, choice, and connectivity values. This index represents the ideal values for each service type, indicating locations on the map that offer the best potential for optimal performance. The index was calculated based on the maximum and minimum values, with integration values ranging from 0.030 to 0.082 and choice values ranging from 3836 to 1818171. The connectivity values ranged from 1 to 5. It's important to note that the ideal value would be less than the maximum based on the availability of the best location. For instance, commercial services are ideally located in areas with the highest integration values, while schools should be in areas with high connectivity values. Next, the current value of each service was compared to the ideal value using the index. The average value for each type of service was calculated and compared to its corresponding ideal value. The results were presented spatially on maps to provide a visual representation of the performance of each type of service in relation to its ideal value.

The following graphs (see Fig. 4) illustrate the average values for different services in comparison with their corresponding ideal values. The comparison allows for a clear understanding of the extent to which the current location of each service meets its potential for optimal performance. The maps and graphs serve as important tools for assessing the suitability and performance of the current location of each type of service in the study area and can inform future decision-making regarding urban planning and service provision.

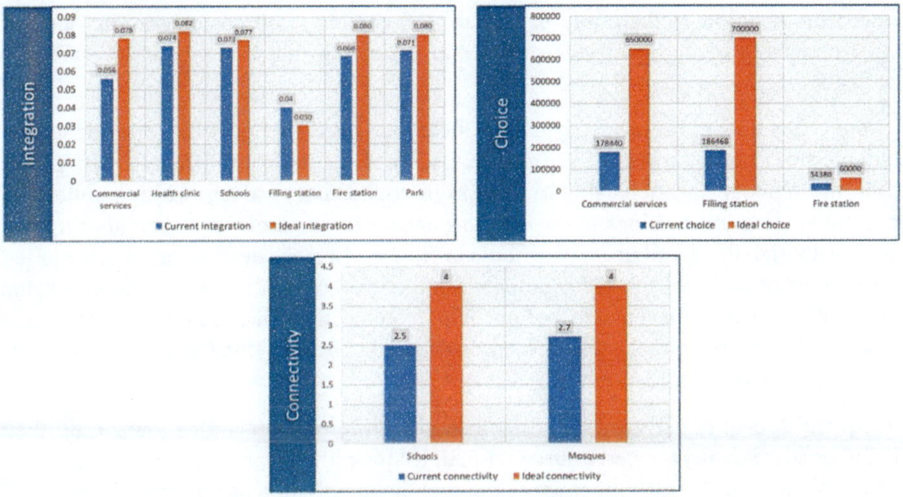

Fig. 4. Comparison between the average values of integration, choice, and connectivity and their corresponding ideal values.

3.3 Results Explanation

Based on the rigorous analysis of the spatial distribution and performance of different types of urban services in Hajjah village using Space Syntax theory, the study has generated a comprehensive set of maps and graphs that provide an in-depth understanding of the existing patterns of service distribution and their suitability to meet the functional requirements of the local population. By comparing the measured values of integration, choice, and connectivity of each service type with their ideal values, the study has revealed a significant mismatch between the current distribution of services and the optimal locations that would maximize their performance. These findings have important implications for urban planning and policy-making in Hajjah and similar Palestinian communities, as they highlight the need for more systematic and evidence-based approaches to service delivery and spatial development. In the following sections, we will present the main results of our analysis, discuss their implications, and suggest some recommendations for future research and action.

Commercial Services: Commercial services are vital for meeting the needs of both village residents and passersby. To ensure easy accessibility and high integration and choice values, commercial services should be strategically located. The distribution of commercial services in Hajjah has been evaluated based on calculated values, and while the integration index of 0.056 is close to the ideal value of 0.078, the choice value is suboptimal. Therefore, there is room for improvement in the distribution of commercial services, particularly along the main roads where the choice values are highest.

Health Clinic: The health clinic is a critical facility that serves not only the residents of the town but also the surrounding areas. As such, its location should offer high control values and easy accessibility from all parts of the town. According to the calculated values, the location of the clinic in Hajjah is favorable, with an integration value of 0.074 compared to the ideal value of 0.082, indicating that it is easily accessible to residents.

Schools: Schools in Hajjah cater to the educational needs of both village residents and students from nearby towns. To effectively serve students from different parts of the town, the schools' locations should have high connectivity values, while maintaining acceptable integration values to enable easy access for students from neighboring villages. Although the distribution of schools in Hajjah village is deemed acceptable based on the integration index of 0.73 (compared to the ideal of 0.77), the connectivity value is suboptimal due to uneven distribution. With a connectivity index of 2.5 (compared to the ideal of 4), some schools need to be redistributed to better serve the needs of students in different areas.

Mosques: Mosques play a crucial role in serving different parts of the town; thus, they should be strategically located in areas with high connectivity values. Currently, there are three mosques located near the town center, but the distribution can be further improved to ensure accessibility for all residents. According to the analysis, while the connectivity index of 2.7 suggests that the distribution of mosques in the town is relatively good, there is still room for improvement, as the ideal value is 4. Therefore, it may be beneficial to redistribute the mosques to better serve the needs of the entire town.

Filling Station: The ideal location for a filling station is where vehicles pass frequently to ensure easy accessibility while avoiding areas with high integration values to prevent pollution and disturbance to residents. However, based on the analysis, the current location of the filling station in Hajjah is less than ideal. The integration index of 0.04 is higher than the recommended ideal value of 0.03. Moreover, the low choice index of less than 190,000 falls short of the desired ideal value of 700,000. Therefore, a review of the filling station's location is necessary to ensure it is situated in a suitable area that meets the necessary criteria for accessibility and minimizes negative impacts on the community.

Fire Station: The fire station's location is crucial in ensuring timely responses during emergencies. To achieve this, it should have high integration values and easy access to all areas of the town. However, it should also be situated away from areas with high choice values to prevent traffic congestion during emergencies. Based on the analysis, the current location of the fire station in Hajjah is excellent. The calculated integration value of 0.068 is very close to the ideal value of 0.08. Additionally, the low choice value of less than 50,000 is consistent with the ideal value. Therefore, the current location of the fire station in Hajjah meets the necessary criteria for a suitable location and does not require relocation.

Park: As a common public facility that serves all residents of the town, the park should be easily accessible to everyone. Currently, there is only one park located near the heart of Hajjah. Upon analysis, the location of this park is deemed excellent, with an integration index of 0.071, which is very close to the ideal value of 0.08. This indicates that the park is conveniently located for all residents to access and enjoy.

In summary, the analysis shows that the distribution of services in Hajjah is generally acceptable, while some facilities, such as the health clinic and the park, are found to be situated in favorable locations, others, such as the filling station and some schools, require improvement in their locations. The results can be used to inform decision-making and urban planning for the village and similar communities.

4 Conclusion

The application of space syntax theory to analyze the distribution of services in the town of Hajjah has highlighted the importance of understanding the social logic of human interaction and its impact on different elements and human projects. It has also emphasized the significance of achieving acceptance and interaction with the community to ensure the success of these projects.

The overall analysis of the distribution of services in Hajjah and the evaluation of various public facilities in Hajjah, including commercial services, health clinics, schools, mosques, filling stations, fire stations, and parks, shows that it is generally good. The locations of these facilities are assessed based on integration, choice, and connectivity values. Although some facilities such as the health clinic and the park are in favorable locations, others like the filling station and some schools require better locations. The paper emphasizes the importance of strategic location to ensure easy accessibility and convenience for all residents.

The results of this analysis can inform future planning and development in the town of Hajjah and other similar communities. By considering the social logic of human interaction and using tools such as space syntax theory, planners and decision-makers can better understand the needs and requirements of the community and ensure that the distribution of services is optimized for the benefit of all resident.

References

1. Hillier, B., Hanson, J.: The Social Logic of Space. Cambridge Univ. (1984)
2. Jiang, B., Claramunt, C., Klarqvist, B.: Integration of space syntax into GIS for modelling urban spaces. Int. J. Appl. Earth Obs. Geoinf. **2**(3–4), 161–171 (2000)
3. Bafna, S.: Space syntax: a brief introduction to its logic and analytical techniques. Environ. Behav. **35**(1), 17–29 (2003)
4. Jiang, B.: A space syntax approach to spatial cognition in urban environments. In: Position Paper for NSF-Funded Research Workshop Cognitive Models of Dynamic Phenomena and Their Representations, p. 59. Pittsburgh, PA, USA: University of Pitts-burgh, (1998)

5. Figueiredo, L., Amorim, L.: Continuity lines in the axial system. In: The Fifth Space Syntax International Symposium. Delft University of Technology, Delft, The Netherlands (2005)
6. Hillier, B.: Space is the Machine: A Configurational Theory of Architecture. Cambridge University Press (1996)
7. Hillier, B., Vaughan, L.: The city as one thing. Prog. Plan. **67**(3), 205–230 (2007)
8. Benedikt, M.L.: To take hold of space: isovists and isovist fields. Environ. Plann. B. Plann. Des. **6**(1), 47–65 (1979)
9. Pafka, E., Dovey, K., Aschwanden, G.D.: Limits of space syntax for urban design: Axiality, scale and sinuosity. Environ. Plan. B Urban Analytics City Sci. **47**(3), 508–522 (2020)
10. Penn, A., Turner, A.: Space syntax-based urban design research and pedagogy: experiences with students and practitioners. J. Urban Des. **6**(3), 269–293 (2001)
11. Read, S.: Space syntax and the Dutch city. Environ. Plann. B. Plann. Des. **26**(2), 251–264 (1999)
12. Kropf, K.: Space syntax methodology. Handbook of Research on Computational Science and Engineering: Theory and Practice, pp. 546–568. IGI Global (2013)
13. Gündüz, M., Arslan, S.: The impact of street connectivity on the accessibility of urban amenities: a space syntax analysis in the Konyaalti district of Antalya. Turkey. Cities **96**, 102461 (2020)
14. Yildirim, A.S., Turgut, B.: The impact of space syntax on urban safety. Saf. Sci. **121**, 295–302 (2020)
15. Chen, Y., Lu, X.: Investigating the relationship between spatial structure and ride-hailing service performance using space syntax theory. J. Transp. Geogr. **93**, 103051 (2021). https://doi.org/10.1016/j.jtrangeo.2021.103051
16. Xing, Z., Guo, W.: A new urban space analysis method based on space syntax and geographic information system using multisource data. ISPRS Int. J. Geo Inf. **11**(5), 297 (2022)
17. Qanazi, S., et al.: Covid-19 severity and urban factors: investigation and recommendations based on ma-chine learning techniques. Pal. Med. Pharm. J. **7**(2), 237–254 (2022)
18. Turner, A., Penn, A.: Encoding natural movement as an agent-based system: an investigation into human pedestrian behaviour in the built environment. Environ. Plann. B. Plann. Des. **29**(4), 473–490 (2002)
19. Qanazi, S., Zawawi, Z.: Stone-industry in Palestine: bridging the gap between environmental sustainability and economical value. Pap. Appl. Geogr. **8**(1), 12–34 (2022)

The Impact of Influencer Marketing Versus Paid Ads on Social Media: Moroccan Perspective

Kawtar Mouyassir(✉), Mohamed Hanine, and Hassan Ouahmane

National School of Applied Sciences, The Information Technology Laboratory, Chouaib Doukkali University, El Jadida, Morocco
kawtar096@gmail.com

Abstract. Social media has brought many changes to the way businesses are moving beyond their traditional ways of promoting their products utilizing new strategies like influencer marketing and paid advertising. Influencers are becoming increasingly important, and businesses are racing to hire the most famous people on social media to boost their brand image. However, businesses still use paid advertising as an efficient and effective method to spread information about a product or service and reach consumers. The objective of this study is to provide a comparative analysis of the impact and profitability of paid advertising vs influencer marketing. This study aims to reach a Moroccan national population by putting a cosmetic product from a well-known Moroccan brand to the test. The organization wants to learn more about its consumers' habits and determine which of the two marketing strategies outlined above is the most successful. According to the findings of this research we confirm that influencer marketing is the most successful marketing method for e-commerce companies, but there's no doubting that consistent results from sponsored advertisements redirect people to the website who click on the ADs and are thus more likely to make purchases.

Keywords: Influencer marketing · Paid ADs · Influencers · Social media · Instagram

1 Introduction

Online social networks have become key engines for spreading information in recent years. These social networks are online platforms used by millions of individuals and are extended over several computers and vast distances. People have formed relationships through sharing their interests, sentiments, and actions. Social media provide an interesting knowledge that is particularly valuable for businesses trying to use social media to reach consumers and support opinions analysis, information system development and marketing.

Companies may employ a variety of techniques to communicate with their target customers on social media, including brand pages, sponsored adverts, etc. But these companies have recently realized the impact of classic media's loss of power, prompting them to alter their communication strategy by engaging the services of social media

M. Ben Ahmed et al. (Eds.): SCA 2023, LNNS 906, pp. 99–109, 2024.
https://doi.org/10.1007/978-3-031-53824-7_10

influencers to promote their products [1]. Paid advertisements are another common approach to market products. These advertisements are representing the company's product, and which allow the user to be redirected from the platform to the brand's landing page. Paid adverts allow businesses to target consumers based on their location, age, gender, language, interests, and behaviors [2].

Social media influencers (SMIs) are online personalities with a large number of followers, on one or more social media platforms. These individuals may represent celebrities, artists, or well-known public figures on the media who have acquired an audience by posting content on social media [3]. Social media influencers are considered as people who have created a massive social network of other people who follow them. Additionally, these consumers that are labeled followers may trust social media influencers the same way they trust their friends. The reason why corporations are increasingly relying on influencers to promote their products to their followers and beyond, rather than traditional advertising approaches.

There are various platforms that allow information to be freely shared around the world, and Instagram is currently one of the most popular social media platforms that attracts a massive number of influencers [4]. Influencers on Instagram frequently amass large followings by sharing pictures, ambitious reels using hashtags, and even live videos, which has caused an increase in organizations' willingness to cede partial control of their posts to a community of social media influencers, leveraging their higher credibility.

2 Literature Review

The number of social networks users is rising exponentially, as the existing platform seeks to make information exchange more efficient through processes such as viral marketing, suggestions, and information dissemination techniques. Indeed, social user interactions may lead to considerable outcomes because they are impacted by paid advertising or influencers that encourage and persuade people to pick specific products or services.

Influencer marketing is a concept that relates to marketing through SMIs, and it has prompted researchers to learn more about their impact on customers [5]. Influencers are persons who use their power to encourage and influence others to adopt a certain way of life, known as a lifestyle. Influencer marketing is defined by academic definitions proposed over the last decade [6], as a strategy that allows influencers to be used as opinion leaders on social media, to develop an interactive relationship with consumers. As a result, businesses seek to go beyond traditional advertising methods. Influencers are more likely than others to promote new products and services linked to their values [7], providing an intimate preview at how these products and services fit into their lifestyle. Influencers utilize their popularity to make suggestions, and provide advice to their followers, which helps them develop expert authority.

According to recent studies, finding the right influencer is a challenging job. Indeed, experts show that Instagram influencers with a large number of followers are more desirable, since they are more popular [8]. Collaborating with influencers with many subscribers, is not always a criterion to follow, which is why it's best to go through several criteria to ensure that the influencer shares the same values as the company in order to preserve the company's identity and cohesion.

Paid advertisements distributed through social media platforms such as Facebook, Instagram, and YouTube, among others, may be an effective way to reach a huge number of potential customers [9]. Anonymity, privacy, ease, a large potential reach, cost-effectiveness, user-friendliness, and the capacity to target hard-to-reach audiences are all possible benefits of paid advertisements.

According to some research, Paid advertising, have a great potential to increase interaction between the consumer and the brand. Companies pay the platform of their choosing to run adverts, and in exchange, social media shows their postings to predefined users [10]. Furthermore, paid advertising has a beneficial impact on sales performance, allowing an increase in the number of buyers and sales. According to recent study, paid advertising on social media permit users to locate intriguing products, particularly during the Covid-19 period [11], allowing individuals to avoid the markets and get their daily necessities online from the comfort of their own homes.

3 Methodology

3.1 Background to the Study

The research presented in this article is based on empirical data from a survey aimed to compare the success rates sales of the company that makes part of this study via social media influencers vs the success rates resulting from paid advertising.

In the first step, the company did select the most appropriate influencer to promote its product, as influencers have evolved into the most effective communication channels in terms of marketing. The company relied on three essential aspects in order to succeed in its marketing. The first factor is consistency between the influencer chosen, the brand image, and the relevance of their community. Second, the company did ensure that the influencer shares the same brand's values, since his authenticity will be called into question, and reputation will be severely impacted. Third, the chosen influencer has credible digital communication channels and is in sync with the brand's target audience. The company chose the influencer carefully, because it is necessary to properly identify the person that the company wishes to engage on social media.

The company has selected an influencer to represent it as an ambassador, since the influencer has a total of 460,000 Instagram followers, which may be regarded one of the most important considerations for the company's decision. The chosen influence is a very active social media content producer, and his goal on Instagram is to come up with fresh innovative ideas and share them with his followers.

On the other hand, the company will also rely on paid ADs as a traditional technique of attracting customers to its brand, goods, and services. Their objective is to design advertising that creates a sensory experience for the viewer. Paid Instagram advertising looks to be a good way to reach a large-scale worldwide costumers. The advertising department seeks to create an emotional connection between the audience and the company through their advertisements.

3.2 Data Collection

This study started with the creation of a questionnaire based on 20 research questions that were customized to the gathering of primary data to develop empirical patterns for quantitative and qualitative comparisons.

The option of using a questionnaire to gather data allows a collection of a vast quantity of data for a very broad population. Furthermore, this decision will make statistical data collection and processing easier, allowing for a clearer picture of the customer's behavior as near to him as possible. Moreover, we had direct interaction with the organization to better understand its strategic goal and to discuss the considered principles that allowed them to take the decision of hiring the influencer.

3.3 Population of the Study

The survey was sent as a poster on the Instagram platform. People who already had an idea about the goods or who had seen an advertisement for this brand's items made up the target demographic.

This survey revealed a total of 1413 responses, of which we excluded those that represented missing or incomplete fields, leaving us with a total of 892 participants who completed the questionnaire. Respondents who did not properly identify the product and who had never seen advertisement of the company were then eliminated from the study. This resulted in 650 valid replies from people who had heard about the product before, either through an influencer or through advertisements.

4 Results

4.1 Participants

The number of participants with legitimate answers is a total of 650 respondents, after deleting all the answers not completed or not clearly identified. We notice on the Fig. 1 that most participants in the questionnaire 30% ($n = 195$) live in Casablanca, followed by Marrakech 22% ($n = 143$), Subsequently, 16% of respondents ($n = 104$) were based in Rabat, and for Tangier 14% ($n = 91$), Fes 10% ($n = 65$), and for people living in other cities we had 8% ($n = 52$) of participants.

Participation in the questionnaire was completely voluntary and accessible from the Instagram page of the company. The participants in our survey are composed of men and women, as shown on the Table 1, and we notice that the participation rate of women is 66% ($n = 416$), which makes it higher than the percentage of men 34% ($n = 234$). In addition, data from the questionnaire shows that 10% ($n = 65$) of respondents are under the age of 18, while participants between the age of 18 and 30 have the highest percentage 54% ($n = 351$) of participation. For people aged between 30 and 60 they had a percentage of 32% ($n = 208$), and finally people aged of 60 or more had the lowest participation percentage which is 4% ($n = 26$).

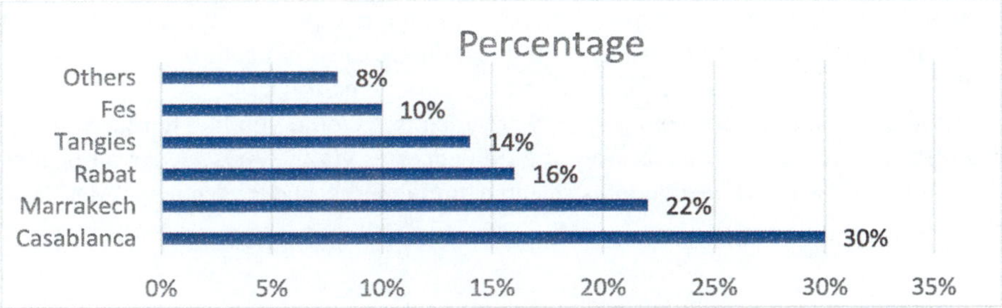

Fig. 1. Cities of residence of the participants.

Table 1. Demographic informations about participants.

Demographics	Percentage	Frequency
Age group		
- 18	10 %	65
18 - 30	54 %	351
30 - 60	32 %	208
60 +	4 %	26
Gender		
Female	66 %	429
Male	34 %	221

4.2 Social Media Platform

The Company chose Instagram as its marketing platform because it is one of the most recognizable marketing channels. Instagram has a 10 times higher active interaction rate between users and brands than Facebook and other platforms [12]. Instagram has turned planned shopping on its head by combining product discovery and spontaneous purchase into our natural behaviors, making it the most powerful social media network on our buying habits [13]. People believe that Instagram makes it easy for them to discover or find new products or services, and that influencers help them get closer to the product.

These statements are in accordance with the findings of our research, which revealed that 54% of individuals prefer to use Instagram to discover new brands. These statistics are based on responses to the following question that featured in our survey: "Which social network do you use the most?".

We then addressed another question: "How many hours a day do you use social media?", to better understand the frequency of use of these social media platforms, the responses to this question were impressive as we can observe on the Table 2, since 16% of people spend 2 h or less on social media per day, and the largest percentage 42% of people are spending between 2 to 4 h, while, 28% of respondents are spending 4 to 8 h on social media. Finally, 14% indicated that those people spend more than 8 h every day on social media.

The survey of this study proposes a very important question to the participants to understand how the product reaches the consumer via social media. This question is:

"How did you discover the company's products in the first place?", the answers to this question showed that 66% of those surveyed recognized the product through influencers. On the other hand, 34% saw it for the first time through paid ADs.

Another question arises from our survey which is: "Would you like to receive more information on brand's products from the influencer or via the ADs? ". As a result of that question 68% of the respondents selected "influencer", and 32% chose "ADs" as a response.

Table 2. Findings about Social media platform.

Social media platform	Percentage	Frequency
Time spent on social media:		
- 2h	16 %	104
2h - 4h	42 %	273
4h – 8h	28 %	182
8h +	14 %	91
Product discovered for the first time by:		
Influencer	66 %	429
Paid AD	34 %	221
Receiving more updates by:		
Influencer	68 %	442
Paid AD	32 %	247

4.3 Influencer Marketing

Influencer marketing is focused exclusively on the exploitation of popularity and reputation to increase word-of-mouth advertising by trustworthy people termed "Influencers" in the domain they regard to be experts [14]. The celebrity who is either a musician, an artist, or just a known person on social media can contribute at selling things on the particular platforms. Companies may take advantage of these strong relationships by asking influencers to promote their products to consumers who already accept the influencers 'opinion or ideas [15].

We asked the participants a few questions in our survey regarding the influencer who promoted the company's product and via whom they identified it. The first question is: "Do you follow the influencer?", as Table 3 indicates, 86% answered yes and 14% answered no. To gain a better understanding of customer behavior, a second question was asked: "Do you believe this influencer is the proper person to represent this company?". The findings of this question were expected, since 98% believe that the influencer has a significant impact on the image of this brand, while just 2% of people disagree.

Two more questions were offered to get as near as reasonable to the perspectives of the participants. "Do you trust the opinion of this influencer?" is the first question, and the results indicate that 88% of people trust the opinion of the influencer who marketed the product. Then, the second question is: "Do you think the influencer is credible?", and the answer rate is like the prior question, with 94% of individuals believing the influencer is credible.

Table 3. Findings about SMI.

Influencer Marketing	Percentage	Frequency
Influencer followers:		
Yes	86 %	559
No	14 %	91
The influencer is the right person to represent the brand:		
Yes	98 %	637
No	2 %	13
Trustworthiness of the influencer:		
Yes	88 %	572
No	12%	78
Credibility of the influencer:		
Yes	94 %	611
No	6 %	39

4.4 Paid ADS

With over one billion monthly active users, Instagram is one of the most popular social media networks on the planet. Since November 2013, traditional sponsored advertising has been available on Instagram, allowing businesses to advertise and post images and videos of their products [16]. As advertisers and marketers have started to understand the power of this network, product placement, advertisements, and sponsorship, they have taken advantage of the power of this platform to promote their products.

We asked participants additional questions concerning sponsored advertising in our questionnaire that did not involve influencers. The company has created advertising video that exclusively highlight its products, as well as influencer advertising, in order to compare the effectiveness of influencer and paid advertising. The number one question about ads is, "Do you like the ad video the company posted on Instagram?", 42% answered "Yes" (Table 4). The second question is "Do you think the advertisement effectively represent the product and give you an impression of it?", 38% answered "Yes" while the large percentage of answers which is 62% was "No".

Table 4. Findings about Paid ADs.

Paid ADs	Percentage	Frequency
Video appreciation:		
Yes	42 %	273
No	68 %	442
Video effectiveness:		
Yes	38 %	247
No	62 %	403

5 Discussion

Nowadays, companies enhance their brand image through social networks and employ a variety of approaches to present their products to consumers and promote their products according to several strategies. The most used strategies to promote products are paid advertising and influence marketing [17]. On the one hand, companies take advantage of the popularity of influencers to reach Internet users. Since people tend to follow these influencers via social networks, either to know the details of their private life, or for advice or criticism on topics such as products or services [18]. On the other hand, paid ADs allow advertisers to target certain demographics such as age, geography, interests, and behaviors of individuals who see their ads. Since these latter were initially shown to people that shared similar interests [19].

Our research is based on a comparison of the efficacy of paid advertisements versus influencers marketing on Instagram. To begin, we may determine that most people in this study's population are between the ages of 18 and 30. This finding aligns with recent research [20], findings that claim Instagram is heavily followed by young adults. The researchers state that this platform gives these users a sense of community.

The results that follow are interested in the platform used to spread the advertising established by the company. According to the findings, Instagram is the most popular social media network among young people, because most people and especially influencers are increasingly using this platform to communicate their lifestyles, activities, and the products they prefer.

People are also increasingly using social media and spending a substantial amount of time browsing web content and sites. Recent scientific research [21], has evaluated many elements such as pragmatics, attractiveness, communication, and usage expectations as they relate to social media addiction. This research showed that people have become attached to social networks and therefore spend a lot of time on these platforms. Our study came to the same results since 42% of participants spend 2 to 4 h each day on social media, which is a significant amount of time, while 28% of individuals spend 4 to 8 h each day on social media, which is a considerable amount of time, and we may conclude that it leads to addiction [22].

Based on the research findings of our study, we think that influencers play a critical role in expressing the company's image and reputation to social media users. Paid ADs are certainly important in promoting a product, but the results indicated that 66% of participants identified the product through the influencer. This influencer was selected according to a set of criteria in order to have a match with the value of the company. To sustain the company's identity and coherence, it will help the organization to build a more common social structure of connections and values. The chosen influencer has been recognized as having the ability to purposefully influence shopping behavior on Instagram.

In this research, we also measured trust in brand messages generated by the influencer through a set of questions. We began by measuring the percentage of followers, and results confirmed that 86% of participants follow this influencer on Instagram. This indicates that the influencer is very popular on social media, the reason why they were recruited to market the product by the brand. Moreover, we captured that the majority 98% of participants voted that this influencer is the ideal individual to represent this company. This affirms that followers trust the posts generated by the influencer and that he had a positive effect on brand awareness and purchase intentions. Our results suggest that, to a considerable extent, social media users regard influencers as quality information sources, cultivating their trust [23] or purchasing decisions based on the influencer's content's informative value rather than its entertainment appeal.

The Table 3 shows that 88% of respondents develop trust with influencers, which means that followers are more likely to consider the influencer's advice and suggestions. Indeed, the result demonstrates that 94% of respondents believe the influencer is trustworthy and efficient. The influencer received a high vote rate because the latter better explains the product to the consumer, bringing the product closer to the consumer than the Ads. Additionally, the influencer tests the product himself and provides feedback to the public on its quality and benefits, whereas the ADs only present the product.

Comparing the results between influencer and paid advertising, we notice that this comparison reveals several key differences. Influencers may offer a realistic example and promote the product to make it more authentic, whereas paid advertising promotes the brand in an unpractical way. Paid advertising, on the other hand, requires more time and effort to create high-quality videos to attract the attention of consumers [24], as well as it requires the involvement of specialist agencies.

Secondly, unlike paid advertising, influencer marketing may target the influencer's followers in addition to the business's own followers, resulting an increased brand visibility on social media [25]. Although influencer marketing may be a successful marketing strategy for e-commerce businesses, there is no disputing that sponsored adverts consistently lead users to the website who click on the ads and are therefore more willing to buy the product.

Moreover, we addressed the overall expenses determined between the costs of influencer recruiting and the costs of paid advertising with the company that was a part of our research [26]. The company confirmed to us that the cost of the influencer was higher than paid ads. Finally, there is no one-size-fits-all solution for organizations to manage their marketing budgets. Based on our findings, we suggest that businesses should utilize a combination of paid advertising and influencer marketing wherever possible.

6 Conclusion and Future Research

To summarize, influencer marketing and sponsored advertisements are two techniques that companies can use to promote their products. Both strategies have the potential to have a significant impact on consumers, and directly contribute to increase profitability of product sales. The results of the comparison study conducted in this research were used to determine which of the two ways is the most efficient. Paid advertising provides more precisely measurable outcomes that the enterprise can continually enhance, because it first presents the product to the customer and targets the consumer, allowing for continuous and predictable sales. Furthermore, paid adverts redirect visitors to the brand's Instagram page links, where they can easily subscribe to the page and receive all the updates. So, when it comes to establishing a product promotion campaign, paid adverts are the safest and most cost-effective solution.

Once the company has the money to pay influencers then can grow sales and orders, the ideal method at this point is influencer marketing, which is not regarded less expensive. However, this method brings the company a lot of money. Working with influencers may be incredibly rewarding, but finding the right individual requires a set of criteria to prevent a bad reputation, since influencer collaborations can help you grow your brand and connect with your target audience.

Current research identifies several significant aspects on which future research might be built to improve this work. For starters, the study was based on a small study population in Morocco, but it may be extended out to a bigger demographic scale to collect more client feedback. On the one hand, we derived these results using a single platform, but we hope to vary the type of platform in the future by comparing people's responsiveness rates, for example, between Instagram, Facebook, and YouTube. The study, on the other hand, is focused on a single company. But the following research suggests a comparison analysis of multiple companies in various industries, to gain more exact understanding of the investment returns of the two advertising strategies adopted. Additionally, it is considered that the company selected only one influencer, and maybe if we go through a series of interviews of many influencers before hiring, we can compare the profitability of male and female influencers.

References

1. Huang, J.P., Depari, G.S.: Paid advertisement on Facebook: an evaluation using a data mining approach. Int. J. Comput. Appl. **58**(13), 26–31 (2018)
2. Dinner, I.M., Heerde, H.J.V., Neslin, S.A.: Driving online and offline sales: the cross-channel effects of traditional, online display, and paid search advertising. J. Mark. Res. **51**(5), 527–545 (2014)
3. Kay, S., Mulcahy, R., Parkinson, J.: When less is more: the impact of macro and micro social media influencers' disclosure. J. Mark. Manag. **36**(3–4), 248–278 (2019)
4. Serrano, R., Martínez-García, A.: Personal style bloggers: the most popular visual composition principles and themes on Instagram. Observatorio (OBS*), vol. 10, no 2 (2016)
5. Crisafulli, B., Singh, T.Q.J.: Competence is power: how digital influencers impact buying decisions in B2B markets. Ind. Mark. Manag. **104**, 384–399 (2022)
6. Leung, X.Y., Sun, J., Asswailem, A.: Attractive females versus trustworthy males: explore gender effects in social media influencer marketing in Saudi restaurants. Int. J. Hospital. Manag. **103** (2022)
7. Lou, C., Yuan, S.: Influencer marketing: how message value and credibility affect consumer trust of branded content on social media. J. Interact. Advert. **19** (2017)
8. Gräve, J.F.: What KPIs Are Key? Evaluating Performance Metrics for Social Media Influencers (2019)
9. Zheng, X., Li, W., Wong, S.W., Lin, H.C.: Social media and E-cigarette use among US youth: longitudinal evidence on the role of online advertisement exposure and risk perception. Addictive Behav. J. **119** (2021)
10. Coleman, P.C., Hanson, P., Rens, T., Oyebode, O.: A rapid review of the evidence for children's TV and online advertisement restrictions to fight obesity. Prevent. Med. Rep. **26** (2022)
11. Eriksson, N., Stenius, M.: Online grocery shoppers due to the Covid-19 pandemic - an analysis of demographic and household characteristics. Procedia Comput. Sci. **196**, 93–100 (2022)
12. Laor, T.: My social network: group differences in frequency of use, active use, and interactive use on Facebook, Instagram and Twitter (2022)

13. Liao, S.H., Widow, R., Cheng, C.J.: Investigating Taiwan Instagram users' behaviors for social media and social commerce development. Entertain. Comput. **40** (2022)
14. Dwidienawati, D., Tjahjana, D., Abdinagoro, S.B., Gandasari, D., Munawaroh: Customer review or influencer endorsement: which one influences purchase intention more? Heliyon **6** (2020)
15. Farivar, S., Wang, F.: Effective influencer marketing: a social identity perspective. J. Retail. Consum. Serv. **67** (2022)
16. Staniewski, M., Awruk, K.: The influence of Instagram on mental well-being and purchasing decisions in a pandemic (2022)
17. Zhou, S., Blazquez, M., McCormick, H., Barnes, L.: How social media influencers' narrative strategies benefit cultivating influencer marketing: tackling issues of cultural barriers, commercialised content, and sponsorship disclosure. J. Bus. Res. **134** (2021)
18. Karagür, Z., Becker, J.M., Klein, K., Edeling, A.: How, why, and when disclosure type matters for influencer marketing. Int. J. Res. Mark. **39**(2), 313–335 (2022)
19. Aily, J., et al.: From clicks to consent: recruitment yield from social media advertisements and associated costs for a telehealth RCT. Osteoarthritis Cartilage **30** (2022)
20. Mayukh, N., Maznina, A., Manaf, A.: Relationship between Instagram usage, social comparison, and self-esteem among young adults during the Covid-19 pandemic. J. Pengajian Media Malaysia **24**(1), 93–108 (2022)
21. Mendini, M., Peter, P.C., Maione, S.: The potential positive effects of time spent on Instagram on consumers' gratitude, altruism, and willingness to donate (2022)
22. Davoud, N., Taghizadeh S.K., Rahman S.A.: Linking dark triad traits to instagram addiction: the mediating role of motives. Technol. Soc. **68** (2022)
23. Zolfaghari, B., Madjdi, F.: Building trusting multicultural organizations: rethinking the influence of culture on interpersonal trust development in the workplace. J. Int. Manag. **28** (2022)
24. SaadiAl-Subhi, A.: Metadiscourse in online advertising: exploring linguistic and visual metadiscourse in social media advertisements. J. Pragmat. **187**, 24–40 (2021)
25. Belanche, D., V.Casaló, L., Flavián, M., Ibáñez-Sánchez, S.: Understanding influencer marketing: the role of congruence between influencers, products and consumers. J. Bus. Res. **132**, 186–195 (2021)
26. Condello, I., Santarpino, G.: Considerations for influencer marketing in cardiac surgery and interventional cardiology. Ann. Thorac. Surg. **112**(2), 689 (2021)

Electronic Voting: Review and Challenges

Ghizlane Ikrissi(✉) and Tomader Mazri

National School of Applied Sciences Kenitra, University Ibn Tofail, Kenitra, Morocco
{ghizlane.ikrissi,tomader.mazri}@uit.ac.ma

Abstract. Electronic voting (e-voting) is the use of electronic systems and technologies in elections to cast and count votes. There are several types of electronic voting systems, including Direct Recording Electronic (DRE) systems, Optical Scan Systems, Internet Voting, and Remote Electronic Voting, among others. It is a means of improving and strengthening democratic processes in modern information societies.

E-voting has the potential to provide several advantages over traditional paper-based voting methods, including increased efficiency, accessibility, and accuracy. It does, however, raise some concerns and challenges that must be addressed in order to ensure the Transparency, privacy, integrity, and security of the voting process. This paper aims to present some requirements of the voting system and give a review of scholarly studies that proposed various schemes and systems to meet the conditions necessary for a highly secure electronic voting system. It offers also an overview of the experiences of various countries that have implemented electronic voting systems, highlighting their successes and challenges.

Keywords: Electronic voting · Secure voting · e-voting requirements

1 Introduction

Voting is an essential act of democracy because it allows people to express their preferences and make decisions on specific issues, usually through elections or decision-making processes. By voting, citizens can actively shape their government, influence policy decisions, express their opinions, and ensure that their elected representatives are accountable to them. As a result, voting is a crucial pillar of representative governance, allowing citizens to exercise their democratic rights and liberties.

There are numerous methods of voting all over the world. The availability and selection of a specific voting system can differ significantly across countries, regions, and organizations. Many factors influence this decision, including legal requirements, cultural preferences, technological infrastructure, and security concerns.

Classical voting, also known as paper-based voting, is a popular method of voting. It refers to the process by which eligible voters physically go to designated polling stations to vote during specific election periods. Each individual receives a separate voting ballot, then he makes an independent decision that is not revealed to others. Nonetheless, this method is heavily dependent on the human factor. The counting commission, for

M. Ben Ahmed et al. (Eds.): SCA 2023, LNNS 906, pp. 110–119, 2024.
https://doi.org/10.1007/978-3-031-53824-7_11

example, or those who printed voting ballots may have an impact on the voting results and can help determine how a specific participant voted [1]. Figure 1 depicts an example of a classical voting and e-voting flowchart.

Due to the rapid development of the Internet and information technologies, voting, like many other traditional offline services, is transitioning to online one. Online voting, also known as electronic voting or e-voting, is a digital method that allows eligible voters to cast ballots using electronic machines or internet-connected electronic devices such as smartphones, computers, or tablets. This operation typically entails accessing a secure online platform or voting system where voters authenticate their identities and make their selections electronically. Figure 2 depicts an example of the process in the e-voting system.

E-voting is currently in high demand due to its ability to address a slew of issues associated with traditional voting methods. It provides several significant advantages, which contribute to its popularity. These advantages include improved voter accessibility, cost savings, faster vote counting, on physical infrastructure, and the potential for more streamlined election administration. Furthermore, e-voting allows voters with disabilities and those living in remote areas to participate in the electoral process, increasing inclusivity [2].

However, electronic voting raises several issues and concerns. The voting process's security and integrity are critical, as any vulnerability in the electronic systems could be exploited. Ensuring voter anonymity and privacy, protecting against cyber threats, and keeping the voting process verifiable and auditable are all important considerations.

This paper is organized as follows: Sect. 2 introduces various electronic voting requirements. Section 3 examines previous studies on e-voting systems. Section 4 provides an overview of e-voting systems in different countries. Section 5 discusses the key focal point of our paper. Finally, Sect. 6 concludes our work.

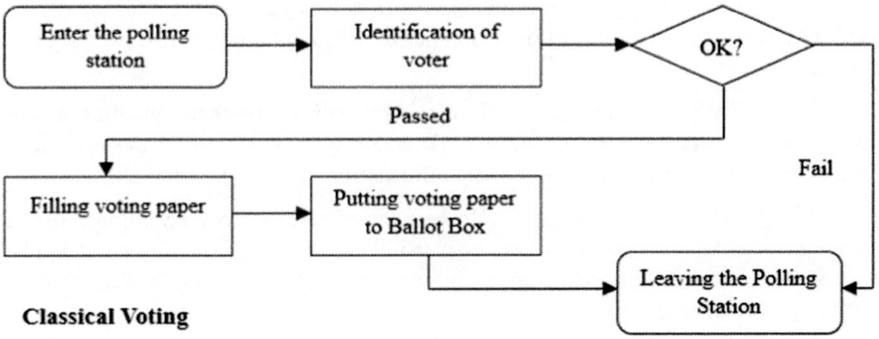

Fig. 1. The classical voting processes

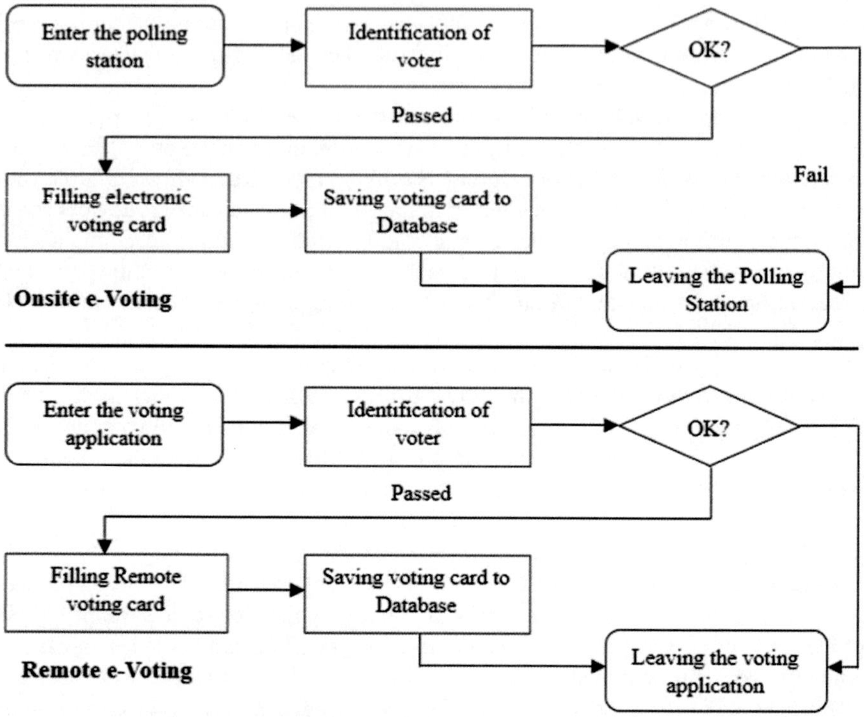

Fig. 2. The electronic voting processes.

2 Electronic Voting Requirements

Several conditions and requirements must be fulfilled when discussing traditional paper-based voting, voting via digital voting machines or online voting systems. Here are some common considerations:

Eligibility: Only legitimate individuals who meet certain predetermined criteria are eligible and can take part in the voting. To accomplish this, during the registration process, voters must provide all necessary information to be considered eligible.

Accessibility: The voting system should be accessible to all people eligible to vote, including those with language barriers or disabilities. Measures should be put in place to accommodate various needs, such as providing interfaces for visually impaired or motor-impaired people.

Privacy: E-voting systems should protect the privacy and anonymity of voters. It should be impossible to link individual votes back to specific voters. This can be achieved through anonymization techniques that ensure votes remain confidential.

Security: It is important to have strong security measures in place to prevent unauthorized access, tampering, or manipulation of votes. Implementing secure authentication mechanisms, and audit trails are examples of security measures that should be incorporated to ensure the security and integrity of the voting process.

Reliability: The voting system should be reliable, guaranteeing precise vote counting, minimal errors, and trustworthy infrastructure. This entails incorporating measures such as redundancy protocols to address technical failures and implementing backup systems to protect against data loss.

Transparency: It is essential for instilling trust and confidence in e-voting systems. To achieve this, the design and operation of the system should be transparent. It is crucial for voters to have a clear understanding and general knowledge of the voting process.

Verifiability: E-voting systems should allow for independent verification of the voting process. One way to achieve this is through the implementation of cryptographic protocols that allow voters to confirm the accurate recording and counting of their votes. Furthermore, conducting post-election audits is essential to ensure the integrity of the system and to detect any irregularities or anomalies that may have occurred.

3 Literature Review

Electronic voting systems can be implemented on the basics of three general schemes or approaches. The blind signature is one of these approaches that was introduced for e-voting in [3, 4], this protocol aims to protect the confidentiality and privacy of the voter by validating their identity without revealing the specific choices they made on their ballot.

Another approach used in e-voting systems is homomorphic encryption based on maintaining the privacy of voters' choices throughout the voting process. Votes can be encrypted, and computations, such as vote tallying, can be performed on the encrypted data directly. This eliminates the need for decrypting individual votes, reducing the risk of privacy breaches or tampering. This approach was first introduced in [5, 6], and it has since been improved by many authors [7, 8].

A popular scheme for ensuring secrecy and verifiability in electronic voting is the Mixnet-based approach, which is introduced and improved towards e-voting in [9, 10]. This scheme plays an essential role in hiding the origin of each ballot, effectively breaking the connection between a voter's identity and their specific vote. The objective is to preserve the privacy and confidentiality of the voting process by ensuring that there is no traceable relation between the voter and their ballot. Messages within mixnets can be mutated using a variety of techniques, including encryption, decryption, and re-encryption [11].

The adoption of e-voting has introduced new challenges that must be addressed in order to ensure the e-voting requirement such as security, privacy, trust, and acceptance of such systems. As a result, research efforts in the field of e-voting have been focused on developing highly secure systems that can effectively prevent fraudulent activities, such as attempts to rig elections by corrupt politicians and election officials.

In [12] the researchers proposed a biometric-secure cloud-based e-voting system for election processes. The study's motivation was to address issues like vote duplication and the high cost of producing ballot papers. The objectives were to develop and construct a secure electronic voting system based on biometric fingerprint methods, specifically Histogram Equalization and the Fourier Transform for fingerprint and iris identification.

While the research accomplished voter authentication through biometric identification, it appears that it fell short of other important e-voting system requirements such as secrecy, confidentiality, transparency, integrity, audibility, and convenience. Furthermore, the use of fingerprint and iris authentication may be economically costly due to the increased memory capacity required for data storage.

The researchers in [13] sought to address issues such as imposter voting and the slow collation of election results, which frequently result in delayed result announcements. The study's goals were to design, implement, and test a secure voting system based on IoT technology. In this research, a ridge and valley features extraction technique was used for authentication purposes. The election data was transferred to a central database for storage and processing. The proposed ridge extraction technique was utilized to achieve authentication as a means of meeting the e-voting system's requirements. However, the research did not adequately address certain important challenges for ensuring the overall reliability and trustworthiness of an e-voting system, such as confidentiality, secrecy, integrity, convenience, transparency, and audibility of e-voting functional and security requirements. Furthermore, results obtained over an insecure network are vulnerable to attack.

The researchers' aim in the paper [14] was to address the issues associated with paper-based voting systems, such as the volume of papers and delays in processing election results. The goals were to create, test, and implement a secure online voting system that would encourage remote voter participation. Multiple authentication methods were used, including one-time passwords (OTP), interactive voice response (IVR), and passwords. The registration and voting processes were designed to be convenient for users. Yet, this system may face some security challenges because OTPs and passwords were identified as potential vulnerabilities that could be stolen or detected, compromising the security of the system. Additionally, voice recognition could be difficult due to the negative effects of diseases on human voices, potentially impacting the accuracy of authentication.

In [15] the researchers designed an electronic voting system that allowed voters to vote via mobile apps. The goals of this study were to design and test an electronic voting system capable of dealing with issues such as multiple votes cast by a single voter and irregularities such as result manipulation at polling booths. The methodology required the use of Aadhaar cards (India's unique identification system) and mobile phones. The research presented techniques for matching a voter's fingerprint to their Aadhaar ID in order to achieve authentication and vote confirmation. However, it should be noted that voter confirmation of votes could potentially lead to vote buying and selling, which could have a negative impact on election outcomes.

A study by researchers in [16] was conducted to design and implement an electronic voting machine equipped with facial recognition and fingerprint sensors. The researchers' motivation stemmed from the issue of voter impersonation, which often results in unlawful votes and provides corrupt candidates with an opportunity to win. The research's goal was to implement and test a voting system that used fingerprint and facial recognition technologies. Face recognition was accomplished using the Support Vector Machine and Local Binary Pattern Histogram methods, while fingerprint recognition was accomplished using the High Sensitive Pixel Amplifier (HSPA). The voting system was implemented using the Visual Basic programming language. This successfully met the authentication challenge but did not address the other requirements such as transparency, confidentiality, secrecy, integrity, and convenience. It should be noted that lighting affects face recognition accuracy, and the error rate (FAR or FRR) of the less reliable biometric can have a significant impact on the overall effectiveness of the system.

4 Existing Voting Systems in Different Countries

Certainly, the use of e-voting systems has been a source of discussion and experimentation in a number of countries. While some countries have embraced e-voting as the future of elections, others are skeptical of its implementation. Here's a look at how e-voting systems have been implemented in a few countries:

4.1 Electronic Voting in Estonia

Estonia is regarded as a pioneer in e-voting. It was first used for local council elections in 2005, and for the first time in the world, e-voting was made available for parliamentary elections in 2007.

Estonia's e-voting system relies on a secure authentication process that combines digital signatures and national ID cards. Before casting a ballot, voters log into the system using their ID card or mobile ID during a designated pre-voting period. To protect voter privacy, the system removes the voter's identity from the ballot before it reaches the National Electoral Commission for counting. This ensures that the votes remain anonymous and prevents any identification of individual voters. [17] As a result, e-voting has gained substantial popularity in Estonia, with a significant portion of the population regularly utilizing this method.

4.2 Electronic Voting in Brazil

Since 1996, e-voting has been an integral part of the Brazilian electoral process. The Brazilian government converted to fully electronic voting in 2000, and over 400,000 kiosk-style machines were used in elections that year [18]. Brazil has been using Direct Recording Electronic (DRE) voting machines, known as urnas eletrônicas. These machines are designed to provide a secure and efficient method for casting votes.

The urnas eletrônicas are standalone units equipped with a keyboard, display, and a card reader for voter authentication. To cast a vote, a voter needs to enter the designated number for a particular candidate on the machine's keyboard. The candidate's picture then appears on the screen, allowing the voter to visually confirm, reject or change their selection.

The electronic voting process in Brazil has experienced consistent development and improvement across multiple aspects, such as software, hardware, interface, security, and efficiency. Notably, the Electoral Court has introduced the use of voters' biometric data as an additional means of identification, enhancing the process of voter authentication.

4.3 Electronic Voting in India

India is the world's largest democracy with more than 668 million electorates spread across 543 parliamentary constituencies. In the past, the manual counting of votes in India used to be a time-consuming process, often spanning three to four days. To address this issue and minimize errors in the voting process as well as expedite the vote counting process, the country has introduced electronic voting machines (EVMs) developed by Electronics Corp of India and Bharat Electronics.

The EVM is divided into two sections: one for poll workers and one for voters. In the balloting unit, voters can select their preferred candidate by pressing the button adjacent to the candidate's symbol and name. The control unit, managed by the polling staff, records the vote electronically. When a voter presses the button on the balloting unit, a light adjacent to the button illuminates, and a short beep sound confirms that the vote has been successfully cast. After each voter has cast their vote, the polling officer presses a switch to clear the EVM and prepare it for the next voter. This ensures that each vote is recorded independently and securely, maintaining the integrity of the electoral process [18].

4.4 Electronic Voting in Namibia

In order to address certain deficiencies in the previous election system, Namibia took a significant step in modernizing its electoral process by implementing e-voting using Electronic Voting Machines (EVMs) during the Presidential elections.

The Electoral Commission of Namibia (ECN) purchased 3,400 EVMs from India, which were specifically developed and designed for the electoral process in Namibia taking into account the country's unique requirements. Each EVM can record a maximum of 3,840 votes, providing an efficient means of collecting and tallying the votes [19].

4.5 Electronic Voting in the USA

In the United States of America (USA), various voting technologies and methods have been implemented for the voting process. These include Direct Recording Electronic (DRE) machines, which enable voters to make their selections electronically. Additionally, there are Optical Scan systems that utilize paper ballots and scanning machines for vote tabulation, Hybrid Voting Machines are also used, combining features of both DRE machines and Optical Scan systems [20].

The 2016 US presidential election did face allegations of cybersecurity interference and tampering with digital ballots. The US government accused the Russian government of being responsible for this interference, suggesting that they had the motive to favour the Republican nominee, Donald Trump. The protection of the US voters' registration database was deemed crucial to prevent fraudulent or abnormal activity.

To address the reliability issues encountered during the 2016 election, there were suggestions to integrate biometric technology into the US voting system [21]. Biometric technology, such as fingerprint or iris scans, can provide an additional layer of security and verification in the voting process.

5 Discussion

Electronic voting systems provide numerous benefits compared to traditional paper-based methods, including enhanced efficiency, reduced chances of human and mechanical errors, faster result tabulation, and the potential for cost savings. Nonetheless, the adoption of e-voting systems presents substantial challenges such as verifying voter identity and safeguarding voter privacy, ensuring auditability, preserving data integrity, promoting transparency, and addressing significant security concerns.

As presented in this paper, several studies have addressed various e-voting challenges. While some of these studies have succeeded in meeting certain e-voting requirements, such as authentication, it becomes apparent that they have not fully met other crucial e-voting system prerequisites like convenience, ballot secrecy, auditability, data confidentiality, and transparency.

Building e-voting systems that are secure and efficient requires the incorporation of diverse technologies and the selection of the most optimal e-voting method. Remote voting, a type of electronic voting, has the potential to shape the future of democracy by increasing voter engagement and inclusivity. However, its successful implementation demands a careful strategy due to system complexities.

To ensure its viability and uphold fundamental democratic principles, our forthcoming efforts will focus on examining modern technologies like blockchain and AI and studying their potential to enhance the e-voting system.

This combination of technologies aims to establish a remote voting system that not only encourages broader participation but also preserves essential attributes such as accessibility, security and effectively addresses potential voter fraud, authenticity, increasing transparency, and enhancing the system's overall integrity and accuracy.

6 Conclusion

This paper presents the electronic voting systems and discusses some solutions proposed by researchers to address related challenges of this system. Furthermore, we have provided examples of some electronic voting systems implemented in various countries worldwide. It should be noted that the availability and selection of a specific voting system can differ significantly across countries, regions, and organizations. Many factors influence this decision, including cultural preferences, technological infrastructure, security concerns and legal frameworks based on their requirements and available resources.

E-voting systems are inherently complex, which is why significant research and development efforts are currently focused on designing highly secure and reliable e-voting systems that strike a balance between leveraging technology to enhance the voting process and upholding trust, security, and inclusivity in democratic elections.

References

1. Dyachkova, I., Rakitskiy, A.: Anonymous remote voting system. In: 2019 International Multi-conference on Engineering, Computer and Information Sciences (SIBIRCON), October 2019, pp. 0850–0852 (2019). https://doi.org/10.1109/SIBIRCON48586.2019.8958064

2. Bokslag, W., de Vries, M.: Evaluating e-voting: theory and practice. arXiv, 08 February 2016. https://doi.org/10.48550/arXiv.1602.02509
3. Fujioka, A., Okamoto, T., Ohta, K.: A practical secret voting scheme for large scale elections. In: Seberry, J., Zheng, Y. (eds.) AUSCRYPT 1992. LNCS, vol. 718, pp. 244–251. Springer, Heidelberg (1993). https://doi.org/10.1007/3-540-57220-1_66
4. Okamoto, T.: Receipt-free electronic voting schemes for large scale elections. In: Christianson, B., Crispo, B., Lomas, M., Roe, M. (eds.) Security Protocols 1997. LNCS, vol. 1361, pp. 25–35. Springer, Heidelberg (1998). https://doi.org/10.1007/BFb0028157
5. Cohen, J.D., Fischer, M.J.: A robust and verifiable cryptographically secure election scheme. Presented at the 26th Annual Symposium on Foundations of Computer Science (SFCS 1985), pp. 372–382. IEEE Computer Society, October 1985 (1985). https://doi.org/10.1109/SFCS.1985.2
6. Distributing the power of a government to enhance the privacy of voters. In: Proceedings of the Fifth Annual ACM Symposium on Principles of Distributed Computing. https://dl.acm.org/doi/10.1145/10590.10595. Accessed 18 June 2023
7. Benaloh, J., Tuinstra, D.: Receipt-free secret-ballot elections (extended abstract). In: Proceedings of the Twenty-Sixth Annual ACM Symposium on Theory of Computing, in STOC '94, pp. 544–553. Association for Computing Machinery, New York, NY, USA, May 1994. https://doi.org/10.1145/195058.195407
8. Feng, C., Xin, Y., Yang, Y., Zhu, H.: Multi-integer somewhat homomorphic encryption scheme with China remainder theorem (2015). Accessed 18 June 2023. https://www.semanticscholar.org/paper/Multi-integer-Somewhat-Homomorphic-Encryption-with-Feng-Xin/fa831df884ce85be9cc78033dce2148119258db6
9. Sako, K., Kilian, J.: Receipt-free mix-type voting scheme. In: Guillou, L.C., Quisquater, J.-J. (eds.) EUROCRYPT '95. LNCS, vol. 921, pp. 393–403. Springer, Heidelberg (1995). https://doi.org/10.1007/3-540-49264-X_32
10. Jakobsson, M., Juels, A., Rivest, R.: Making mix nets robust for electronic voting by randomized partial checking (2002). https://eprint.iacr.org/2002/025. Accessed 18 June 2023
11. Islam, N., Alam, K.M.R., Rahman, A.: The effectiveness of mixnets – an empirical study. Comput. Fraud Secur. **2013**(12), 9–14 (2013). https://doi.org/10.1016/S1361-3723(13)70111-8
12. Publication, T.: A biometric-secure cloud based e-voting system for election process. Int. J. Electric. Electron. Eng. Res. (IJEEER) (2014). https://www.academia.edu/7312101/A_BIOMETRIC_SECURE_CLOUD_BASED_E_VOTING_SYSTEM_FOR_ELECTION_PROCES. Accessed 16 June 2023
13. Desai, M.M., Patoliya, J.J., Mewada, H.K.: Advanced secure voting system with IoT. Int. J. Eng. Comput. Sci. (2016). https://www.academia.edu/24110413/Advanced_Secure_Voting_System_with_IoT. Accessed 17 June 2023
14. Nadaph, A., Bondre, R., Katiyar, A., Goswami, D., Naidu, T.: An implementation of secure online voting system (2015). https://www.semanticscholar.org/paper/An-Implementation-of-Secure-Online-Voting-System-Nadaph-Bondre/1d4e4a386ff10c4b0c271505814653fc97d12ad3#cited-papers. Accessed 17 June 2023
15. Tamizhvanan, C., Chandramohan, S., Navfar, A.M., Kumar, P., Vinoth, R.: Electronic voting system using Aadhaar card (2018). https://www.semanticscholar.org/paper/Electronic-Voting-System-Using-Aadhaar-Card-Tamizhvanan-Chandramohan/5aa1d9ba933e1257be660a36eae07c9cb9c37829. Accessed 17 June 2023
16. Ganesh, B., Gokulprashanth, P., Udhayakumar, G.: Electronic voting machine with facial recognition and fingerprint sensors. Int. J. Adv. Res. Dev. **3**(3), 165–170 (2018)
17. Tsahkna, A.-G.: E-voting: lessons from Estonia. Eur. View **12**(1), 59–66 (2013). https://doi.org/10.1007/s12290-013-0261-7

18. Kumar, M., Walia, E.: Analysis of electronic voting system in various countries. Int. J. Comput. Sci. Eng. **3** (2011)
19. Mpekoa, N., van Greunen, D.: E-voting experiences: a case of Namibia and Estonia. In: 2017 IST-Africa Week Conference (IST-Africa), May 2017, pp. 1–8 (2017). https://doi.org/10.23919/ISTAFRICA.2017.8102303
20. Johnson, N., Jones, B.M., Clendenon, K.: E-Voting in America: current realities and future directions. In: Meiselwitz, G. (ed.) SCSM 2017. LNCS, vol. 10282, pp. 337–349. Springer, Cham (2017). https://doi.org/10.1007/978-3-319-58559-8_27
21. Adewale Olumide, S., BoyinbodeOlutayo, K., Adekunle, S.E.: A review of electronic voting systems: strategy for a novel. IJIEEB **12**(1), 19–29 (2020). https://doi.org/10.5815/ijieeb.2020.01.03

Smart Mobility Systems

Recommended LEED-Compliant Cars, SUVs, Vans, Pickup Trucks, Station Wagons, and Two Seaters for Smart Cities Based on the Environmental Damage Index (EDX) and Green Score

Osama A. Marzouk(✉)

University of Buraimi, Al Buraimi 512, Sultanate of Oman
osama.m@uob.edu.om

Abstract. An environment with reduced pollution from road vehicles and decarbonized transportation is one of the dimensions of smart cities. In this regard, new sales of vehicles intended for urban use should be oriented toward cleaner (greener) vehicles with less harmful environmental impacts. In the current study, two environmental rating variables provided by the American Council for an Energy-Efficient Economy (ACEEE) for model year 2023 vehicles (U.S. market) in 6 broad classes are employed to identify the best 10 models in each class. These classes are: two seaters (sports cars), cars, SUVs (sport utility vehicles), vans, station wagons (estate cars), and pickups (pickup trucks). The method used in these ratings is based on a combination of emissions life cycle assessment (LCA) and environmental economics. The first ACEEE rating variable is the environmental damage index (EDX), representing an estimated environmental damage cost (in U.S. cents per driving mile). The second ACEEE rating variable is the Green Score, which is a non-dimensional number (0–100 scale) derived from EDX. According to version 4 of the green building certification program LEED (Leadership in Energy and Environmental Design) of the U.S. Green Building Council (USGBC), green vehicles are defined as those having a Green Score of 45 or higher. In the current study, 85 selected top models were found to have a Green Score range from 41 to 67. Only 55 models of them (64.7% portion) are LEED compliant (classified as green vehicles), and thus are more recommended for use within smart cities than other models.

Keywords: ACEEE · EDX · Green Score · Vehicles · Smart City · LEED

1 Nomenclature (Alphabetical Order, Numbers Come First)

2wd	two-wheel drive
4×4	four-wheel drive
4dr	4 side doors (2 left doors and 2 right doors)
4wd	four-wheel drive

M. Ben Ahmed et al. (Eds.): SCA 2023, LNNS 906, pp. 123–135, 2024.
https://doi.org/10.1007/978-3-031-53824-7_12

6.2L 8 (example engine specification): total engine displacement (capacity) 6.2 liters, in 8 cylinders
20in 20-inch wheel diameter (tire rim)
21in 21-inch wheel diameter (tire rim)
22in 22-inch wheel diameter (tire rim)
ACEEE American Council for an Energy-Efficient Economy
auto automatic transmission
auto cvt continuously variable automatic transmission
auto stk manually adjustable automatic transmission
AWD all-wheel drive
Awd all-wheel drive
¢/mi United States cent per mile (United States dollar per 160.9 km)
[CNG] compressed natural gas (as a vehicle fuel)
[D] diesel (as a vehicle fuel)
EDX environmental damage index
FFV flexible fuel vehicle
FWD front-wheel drive
GHG greenhouse gas
HEV unplugged hybrid electric vehicle
kWh kilowatt-hour (unit of usable electric battery storage energy)
LEED Leadership in Energy and Environmental Design
Li-Ion lithium-ion battery pack
LWB long wheel base
manual manual transmission
[P] premium gasoline/petrol (as a vehicle fuel)
RWD rear-wheel drive
SRT street and racing technology
SUV sport utility vehicle

2 Introduction

2.1 Overview

Smart city solutions can facilitate more convenient and safer portability of personnel and goods through non-conventional transportation modes and heavy use of technology for planning and monitoring [1]. Smart city transportation can also help reduce harmful emissions through the use of greener (more sustainable, less polluting) vehicles and innovative mobility ideas [2].

Transportation emissions may be divided into three categories [3]:

1. species that cause health problem, such as particulate matter (PM)
2. species that cause global warming through their GHG (greenhouse gas) effect, such as carbon dioxide (CO_2)
3. species that cause both health hazards and global warming, such as carbon monoxide (CO)

Either health damages or environmental damages lead to economic losses [4], such as loss of labor hours, health care expenses, governmental spending on remedial actions, interrupted sectors due to climate change (like agriculture) or natural disasters.

By expressing the emissions impact in terms of economic penalties (using environmental economics), not only due to direct emissions but also due to indirect ones (using life cycle assessment, LCA, of emissions), it becomes possible to devise an indicator to assess the environmental harms by a vehicle during its estimated lifetime.

2.2 Study Objective

This study uses 2 indicators for identifying top 10 performing vehicle models (in the United States market, with model year 2023 only) in 6 broad classes, which are:

1. two seaters (sports cars)
2. cars
3. SUVs (sport utility vehicles)
4. vans
5. station wagons (estate cars)
6. pickups (pickup trucks)

The subset of these best models that meets LEED (Leadership in Energy and Environmental Design) criterion for "green vehicles" is determined for 3 of the 6 broad classes: two seaters, vans, and pickups. For the other 3 classes, all top 10 models are LEED compliant as green vehicles. Only top models (per broad vehicle class) that are also LEED green vehicles are recommended for smart city use.

2.3 Criteria for Selecting Vehicle Models

The ratings of vehicle models represent annual assessment published by ACEEE (American Council for an Energy-Efficient Economy) online at its web-based tool greenercars.org [5]. The original data have 14 narrow vehicles classes. However, similar narrow classes were grouped here for simplification, leading to 6 broad vehicle classes. For example, the two original narrow vehicle classes (minivans) and (large vans) were combined to form the broad vehicle class (vans).

Then, in each broad vehicle class, the models with largest 10 distinct EDX ratings were selected. Additional EDX ratings (rank 11th for example) may be added if there are models with an EDX rating that is very close to the 10th best (smallest) EDX rating. If two EDX ratings are assigned to the same vehicle model, only the better (lower) EDX rating is mentioned here (and typically the difference is small).

3 ACEEE's EDX and ACEEE's Green Score

The Green Score of the American Council for an Energy-Efficient Economy (ACEEE) ranges from 0 (theoretically worst case) to 100 (theoretically best case). This Green Score is a unitless value derived from another dimensional one called the environmental damage index (EDX), which represents an estimate of monetized impacts of air pollution and global warming in ¢/mi. Unlike the Green Score rating, the EDX rating is unbounded (does not have an upper limit). Both ratings are related as:

$$\text{Green Score (no unit)} = 100\,\frac{e^{-\frac{EDX\,(¢/mi)}{c\,(¢/mi)}}}{\left(1+\frac{EDX\,(¢/mi)}{c\,(¢/mi)}\right)^{3}} \tag{1}$$

The value of the parameter (c) in Eq. (1) is subject to update and may change over years. It has the same unit of EDX (i.e., ¢/mi). Its value was 6.83 ¢/mi in 2016. The value of 2023 is approximately 6.67 ¢/mi, because it satisfies published data of average 2023 EDX of 1.52 ¢/mi, corresponding to an average Green Score of 43 [6].

There are 5 air pollutants considered while computing EDX, which are:

1. carbon monoxide (CO)
2. hydrocarbons (HC) or volatile organic compounds (VOC)
3. nitrogen oxides (NOx)
4. sulfur oxides (SOx)
5. particulate matter with a diameter of 10 μm or smaller (PM_{10})

Each of these air pollutants is assigned health-damage cost factors (in ¢/mi), which are used to compute an EDX rating (and then a Green Score rating) for a vehicle.

The EDX rating also accounts for 3 greenhouse gas (GHG) emissions, despite not directly have health damage. These are:

1. carbon dioxide (CO_2)
2. methane (CH_4)
3. nitrous oxide (N_2O)

Three other air pollutant gases (CO, HC or VOC, and NOx) are added the above 3 GHGs, leading to a total of 6 GHGs contributing to the EDX estimation. The impacts from these 6 GHGs are quantified through assigned penalty cost factor (in ¢/mi). Combining the economic impacts of GHG effect (global warming) and health damage (air pollution) gives the EDX rating, with smaller values mean greener vehicles.

Electrified vehicles or vehicles using fuels other than gasoline (petrol) have special process for proper estimation of the EDX, such as accounting for additional embodied emissions due to an added large battery pack in battery-electric vehicles (BEV).

The EDX rating involves uncertainties in expressing the impact of emissions on health and climate, and in estimating the useful life of a vehicle. However, the EDX method does not account for direct tailpipe emissions only, but also for emissions from manufacturing the vehicle (referred to as "embodied emissions"), as well as from the production and distribution of its fuel or electricity (referred to as "upstream emissions") and also hydrocarbon vapors leaking from fuel tanks or lines or other non-fuel evaporating substances (both together are referred to as "evaporative emissions", and when added to the tailpipe emissions they are called "in-use emissions"). The EDX and Green Score ratings are for vehicles with a gross vehicle weight of up to 10,000 lb (4,536 kg) [7]. The published Green Scores are rounded as integers.

Vehicles with a Green Score rating of 45 or above are considered "green vehicles" according to version 4 (v4) of LEED [8].

4 Results

4.1 Extreme Cases Per Broad Vehicle Class

Figure 1 can be useful in visualizing the range of EDX for each broad vehicle class. A similar function is achieved by Fig. 2 for Green Scores. The overall best Green Score for the 2023 models is 67 (either cars or SUVs), while the worst is 17 (two seaters).

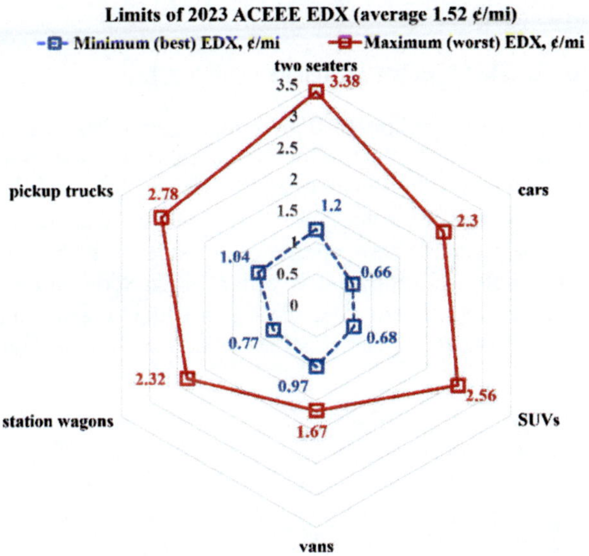

Fig. 1. Illustration of the minimum (best, indicated by the inner dashed-line blue loop) and maximum (worst, indicated by the outer solid-line red loop) value of the environmental damage index (EDX) rating of ACEEE for each of the 6 broad vehicle classes, for model year 2023.

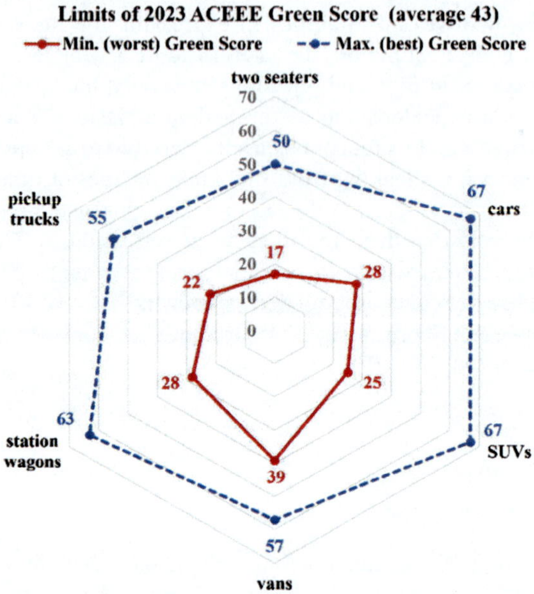

Fig. 2. Illustration of the minimum (worst, indicated by the inner solid-line red loop) and maximum (best, indicated by the outer dashed-line blue loop) value of the Green Score rating of ACEEE for each of the 6 broad vehicle classes, for model year 2023.

4.2 Lists of Top Vehicle Models Per Broad Vehicle Class

Table 1 gives a summary of the top-10 EDX ranks for the 2023 two seater vehicles. Similarly; Table 2 is for cars, Table 3 is for SUVs, Table 4 is for vans, Table 5 is for station wagons, and Table 6 is for pickup trucks.

It should be mentioned that for the vans, few models were actually available in the original data. Thus, the given list here represents all models (rather than top 10 ranks).

Shaded rows in 3 tables indicate models with a Green Score of 44 or below. Thus, these highlighted vehicle models are not qualified for being “green vehicles” according to LEED definition, despite being relatively best models within their broad class.

Table 1. Two seaters with model year 2023 having the 10 best (smallest) EDX ratings

index	Rank (by EDX)	EDX (¢/mi)	Green Score	Make	Model	Specification
1	1st	1.2	50	BMW	Z4 sDrive30i	2.0L 4, auto stk [P]
2	2nd	1.24	49	Toyota	Supra 2.0	2.0L 4, auto stk [P]
3	3rd	1.3	47	Toyota	Supra 3.0 (automatic)	3.0L 6, auto stk [P]
4	3rd	1.3	47	Audi	TT Roadster	2.0L 4, auto Awd
5	4th	1.32	47	BMW	Z4 M40i	3.0L 6, auto stk [P]
6	5th	1.46	43	Nissan	Z	3.0L 6, auto stk [P]
7	6th	1.48	43	Porsche	718 Cayman	2.0L 4, auto [P]
8	6th	1.48	43	Porsche	718 Boxster	2.0L 4, auto [P]
9	7th	1.49	43	Porsche	718 Boxster T (automatic)	2.0L 4, auto [P]
10	7th	1.49	43	Porsche	718 Cayman T (automatic)	2.0L 4, auto [P]
11	8th	1.51	42	Toyota	Supra 3.0 (manual)	3.0L 6, manual [P]
12	9th	1.52	42	McLaren	Artura	Electric (Li-Ion) / 3.0L 6, auto *
13	10th	1.53	42	Porsche	718 Cayman T (manual)	2.0L 4, manual [P]
14	10th	1.53	42	Porsche	718 Boxster T (manual)	2.0L 4, manual [P]

* This is a plug-in hybrid electric vehicle (has 2 specifications, for electricity and fuel).

Table 2. Cars with model year 2023 having the 10 best (smallest) EDX ratings

index	Rank (by EDX)	EDX (¢/mi)	Green Score	Make	Model	Specification
1	1st	0.66	67	Mini	Cooper SE Hardtop 2 Door	Electric (Li-Ion)
2	2nd	0.68	67	Nissan	Leaf	Electric (Li-Ion)
3	3rd	0.73	65	Hyundai	Elantra Hybrid Blue	1.6L 4, auto
4	4th	0.77	63	Toyota	Camry Hybrid LE	2.5L 4, auto CVT

(*continued*)

Table 2. (*continued*)

index	Rank (by EDX)	EDX (¢/mi)	Green Score	Make	Model	Specification
5	5th	0.79	63	Nissan	Leaf SV	Electric (Li-Ion)
6	6th	0.8	62	Hyundai	Elantra Hybrid	1.6L 4, auto
7	7th	0.81	62	Toyota	Corolla Hybrid AWD	1.8L 4, auto CVT Awd
8	8th	0.83	61	Hyundai	Sonata Hybrid	2.0L 4, auto
9	9th	0.86	60	Polestar	Polestar 2 Single Motor	Electric (Li-Ion)
10	10th	0.87	60	Honda	Accord Sport / Touring	2.0L 4, auto CVT
11	11th *	0.88	60	Hyundai	Ioniq 5 SE Standard Range	Electric (Li-Ion)

* This is an added rank (beyond the 10th) because its vehicle model has an equal "rounded" Green Score of (60/100) as that of the 10th-ranked vehicle model

Table 3. SUVs with model year 2023 having the 10 best (smallest) EDX ratings

index	Rank (by EDX)	EDX (¢/mi)	Green Score	Make	Model	Specification
1	1st	0.68	67	Mazda	MX-30	Electric (Li-Ion)
2	2nd	0.69	66	Toyota	bZ4X Limited	Electric (Li-Ion)
3	2nd	0.69	66	Toyota	bZ4X	Electric (Li-Ion)
4	2nd	0.69	66	Toyota	bZ4X AWD	Electric (Li-Ion)
5	2nd	0.69	66	Subaru	Solterra AWD	Electric (Li-Ion)

(*continued*)

Table 3. (*continued*)

index	Rank (by EDX)	EDX (¢/mi)	Green Score	Make	Model	Specification
6	3rd	0.7	66	Subaru	Solterra Limited/Touring AWD	Electric (Li-Ion)
7	3rd	0.7	66	Toyota	bZ4X Limited AWD	Electric (Li-Ion)
8	4th	0.8	62	Lexus	NX 450h + AWD	Electric (Li-Ion)/2.5L 4, auto CVT Awd *
9	5th	0.81	62	Hyundai	Kona Electric	Electric (Li-Ion)
10	6th	0.86	60	Hyundai	Tucson Plug-in Hybrid	Electric (Li-Ion) / 1.6L 4, auto Awd *
11	7th	0.87	60	Volvo	XC40 Recharge twin	Electric (Li-Ion)
12	7th	0.87	60	Volvo	C40 Recharge twin	Electric (Li-Ion)
13	8th	0.88	59	Mercedes-Benz	EQB 250+	Electric (Li-Ion)
14	9th	0.9	59	Mercedes-Benz	EQB 300 4MATIC	Electric (Li-Ion)
15	10th	0.91	59	Mercedes-Benz	EQB 350 4MATIC	Electric (Li-Ion)
16	10th	0.91	59	Hyundai	Santa Fe Plug-in Hybrid	Electric (Li-Ion) / 1.6L 4, auto Awd *

* This is a plug-in hybrid electric vehicle (has 2 specifications, for electricity and fuel)

Table 4. Vans with model year 2023 having the 10 best (smallest) EDX ratings

index	Rank (by EDX)	EDX (¢/mi)	Green Score	Make	Model	Specification
1	1st	0.97	57	Chrysler	Pacifica Hybrid	Electric (Li-Ion) / 3.6L 6, auto CVT *
2	2nd	1.12	52	Toyota	Sienna	2.5L 4, auto CVT
3	3rd	1.14	51	Toyota	Sienna AWD	2.5L 4, auto CVT Awd
4	4th	1.25	49	Mercedes-Benz	GLA 250	2.0L 4, auto [P]
5	5th	1.34	46	Ford	Transit Connect Wagon LWB FWD	2.0L 4, auto stk
6	6th	1.38	45	Ford	Transit Connect Van FWD	2.0L 4, auto stk
7	7th	1.48	43	Chrysler	Pacifica	3.6L 6, auto
8	7th	1.48	43	Chrysler	Voyager	3.6L 6, auto
9	8th	1.5	42	Honda	Odyssey FWD	3.5L 6, auto stk
10	9th	1.53	42	Kia	Carnival	3.5L 6, auto stk
11	10th	1.67	39	Chrysler	Pacifica AWD	3.6L 6, auto Awd

* This is a plug-in hybrid electric vehicle (has 2 specifications, for electricity and fuel).

Table 5. Station wagons with model year 2023 having the 10 best (smallest) EDX ratings

index	Rank (by EDX)	EDX (¢/mi)	Green Score	Make	Model	Specification
1	1st	0.77	63	Kia	Niro FE	1.6L 4, auto
2	2nd	0.83	61	Kia	Niro	1.6L 4, auto
3	3rd	0.84	61	Nissan	Ariya Engage FWD 63kWh	Electric (Li-Ion)
4	4th	0.86	60	Kia	EV6 Standard Range RWD	Electric (Li-Ion)
5	5th	0.92	58	Volvo	V60 T8 AWD Recharge ext. Range	Electric (Li-Ion) / 2.0L 4, auto stk Awd *
6	6th	0.93	58	Nissan	Ariya Venture + FWD 87kWh	Electric (Li-Ion)

(*continued*)

Table 5. *(continued)*

index	Rank (by EDX)	EDX (¢/mi)	Green Score	Make	Model	Specification
7	7th	0.94	58	Kia	Niro Electric	Electric (Li-Ion)
8	8th	0.97	57	Chevrolet	Bolt EV	Electric (Li-Ion)
9	8th	0.97	57	Nissan	Ariya EVO + /EMP + /PRM FWD 87 kWh	Electric (Li-Ion)
10	9th	0.98	56	Chevrolet	Bolt EUV	Electric (Li-Ion)
11	10th	1	56	Kia	EV6 Long Range RWD	Electric (Li-Ion)

* This is a plug-in hybrid electric vehicle (has 2 specifications, for electricity and fuel)

Table 6. Pickup trucks with model year 2023 having the 10 best (smallest) EDX ratings

index	Rank (by EDX)	EDX (¢/mi)	Green Score	Make	Model	Specification
1	1st	1.04	55	Ford	Maverick HEV FWD	2.5L 4, auto CVT
2	2nd	1.23	49	Ford	F150 Pickup Lightning 4WD	Electric (Li-Ion)
3	3rd	1.32	47	Ford	Maverick FWD	2.0L 4, auto
4	4th	1.37	45	Ford	Maverick AWD	2.0L 4, auto Awd
5	5th	1.38	45	Ford	F150 Lightning Extended Range 4WD	Electric (Li-Ion)
6	6th	1.41	45	Ford	F150 Lightning Platinum 4WD	Electric (Li-Ion)
7	7th	1.47	43	Ford	F150 Pickup 2WD HEV	3.5L 6, auto stk
8	7th	1.47	43	Ford	Ranger 2WD	2.3L 4, auto stk
9	8th	1.52	42	GMC	Sierra 2WD	3.0L 6, auto [D]

(continued)

Table 6. (*continued*)

in-dex	Rank (by EDX)	EDX (¢/mi)	Green Score	Make	Model	Specification
10	8th	1.52	42	Lordstown EV	Endurance	Electric (Li-Ion)
11	8th	1.52	42	Chevrolet	Silverado 2WD	3.0L 6, auto [D]
12	9th	1.53	42	Rivian	R1T 21in	Electric (Li-Ion)
13	9th	1.53	42	Ford	F150 Pickup 2WD	2.7L 6, auto stk
14	10th	1.55	41	Ford	F150 Pickup 2WD FFV	3.3L 6, auto stk
15	11th *	1.56	41	Rivian	R1T 22in	Electric (Li-Ion)
16	11th *	1.56	41	Ford	F150 Pickup 2WD	2.7L 6, auto stk
17	11th *	1.56	41	Ford	Maverick Tremor AWD	2.0L 4, auto Awd
18	12th *	1.57	41	RAM	1500 4x2	3.6L 6, auto
19	12th *	1.57	41	Ford	Ranger 4WD	2.3L 4, auto stk 4wd
20	13th *	1.58	41	Ford	F150 Pickup 2WD FFV	3.3L 6, auto stk
21	13th *	1.58	41	Rivian	R1T 20in	Electric (Li-Ion)
22	13th *	1.58	41	Toyota	Tacoma 2WD	2.7L 4, auto stk

* This is an added rank (beyond the 10th) because its vehicle model has an equal "rounded" Green Score of (41/100) as that of the 10th-ranked vehicle model.

5 Conclusions

Based on analysis of 2023 EDX and Green Score ratings for 85 top vehicle models in 6 broad classes, 55 models are "LEED green". In terms of maximum Green Score within a class, cars and SUVs (67/100) are better than station wagons (63/100), then come vans (57/100), pickups (55/100), and finally two seaters (50/100). The overall best EDX is 0.66 ¢/mi (cars), and the overall worst EDX is 3.38 ¢/mi (two seaters).

References

1. Hai, T., Wang, J.Y., Yaseen, Z.M., Mohammed, M.N., Zain, J.M.: Shrewd vehicle framework model with a streamlined informed approach for green transportation in smart cities. Environ. Impact Assess. Rev. **87**, 106542 (2021)
2. Zawieska, J., Pieriegud, J.: Smart city as a tool for sustainable mobility and transport decarbonisation. Transp. Policy **63**, 39–50 (2018)
3. Kontovas, C.A., Psaraftis, H.N.: Transportation emissions: some basics. In: Psaraftis, H. (ed.) Green Transportation Logistics. International Series in Operations Research and Management Science, vol. 226, pp. 41–79. Springer, Cham (2016). https://doi.org/10.1007/978-3-319-17175-3_2
4. Resosudarmo, B.P., Napitupulu, L.: Health and economic impact of air pollution in Jakarta. Econ. Rec. **80**, S65–S75 (2004)
5. Greener Cars (part of American Council for an Energy-Efficient Economy – ACEEE), Best by Class. https://greenercars.org/greenest-meanest/best-class. Accessed 06 Aug 2023
6. Greener Cars (part of American Council for an Energy-Efficient Economy – ACEEE), How We Determine Ratings. https://greenercars.org/greenercars-ratings/how-we-determine-ratings. Accessed 06 Aug 2023
7. Vaidyanathan, S., Slowik, P., Junga, E.: Rating the Environmental Impacts of Motor Vehicles: ACEEE's Greenercars.org Methodology (Report T1601), 2016 edn. American Council for an Energy-Efficient Economy (ACEEE), Washington, DC, USA (2016)
8. U.S. Green Building Council: LEED v4 - Location and Transportation - Green vehicles. https://www.usgbc.org/credits/new-construction-core-and-shell-retail-new-construction-data-centers-new-construction-2. Accessed 06 Aug 2023

Tracking and Tracing Containers Model Enabled Blockchain Basing on IOT Layers

Safia Nasih(✉), Sara Arezki, and Taoufiq Gadi

Faculty of Sciences and Techniques, Mathematics Computer Science and Engineering Laboratory, Hassan First University of Settat, 26000 Settat, Morocco
s.nasih@uhp.ac.ma

Abstract. Smart transportation nowadays is based on tracking vessels, assets, etc.. Tracking and tracing operation is an important services that can enhance the maritime trade around the world, minimize loss and increase gain in national economies, this is a field that is finding its place among others integrating new technologies such as Internet of things, Big Data, IOT and Blockchain. Blockchain is a pervasive technology that is building trust and transparency in sectors where it is difficult to ignore intermediaries' intervention. In this article, we will present a model in which we will study the integration of Blockchain technology with IOT in the maritime field, especially container tracking and tracing during their stay in the maritime territory. This model is important to build a system of tracking assets with transparent data using both technologies Blockchain and Internet of things.

Keywords: Tracking and Tracing · Blockchain · IOT · Containers

1 Introduction

In the last decades, maritime industry has known a disruptive development on the implementation of new technologies such as Artificial Intelligence, Blockchain, Big Data, Internet of Things, etc. Many ports, as a result, were transformed to smart ones especially with the rise of awareness about the importance of technologies in keeping their competitiveness in the trade era.

Considering a container terminal, there are many operations where services are digitalized then transformed to intelligent ones. However, it is still a fertile area to conduct researches especially tracking and tracing service. This was confirmed in a recent work, where we tried to extract services that were developed and transformed using these disruptive technologies during the last five years. On one hand, Traceability is defined in the United Nations Global compact and Business for Social Responsibility (2014) as "the ability to identify and trace the history, distribution, location and application of products, parts, and materials, to ensure the reliability of sustainability claims, in the areas of human rights, health and information security, the environment and anti-corruption".

On the other hand, Blockchain is considered as an innovative solution to keep trace of assets while ensuring security and data immutability to the supply chain area where

M. Ben Ahmed et al. (Eds.): SCA 2023, LNNS 906, pp. 136–147, 2024.
https://doi.org/10.1007/978-3-031-53824-7_13

visibility and transparency of product flows are the principal challenges. Blockchain-based traceability allows for safe information sharing, which aids in product quality control, real-time data collecting, transparency, and visibility throughout the supply chain [1, 2].

The aim of this paper, is to propose a model for tracking and tracing containers using Blockchain technology implemented in an IOT platform. This model traces the way of containers arriving by vessel to the port, going through handling, stacking and delivering to carriers then suppliers and clients, to help port information systems to maintain containers localization even out of the port territory.

In the first section, we will describe the background of this study, where we mention the most important works dealing with containerization, containers tracking and blockchain technology. In the second section, we describe the method used to create our model that we explain in the third section. The conclusion will contain a summary of our work with some perspectives for applying this model.

2 Background

Maritime transportation went through an important disruptive change with the notion of containerization. Up to1950s, Ships were loaded manually; Baskets of fruits might be stacked next to Industrial cables. In the 1950s, Malcom Mc Lean and other entrepreneurs introduced standardized containers that can be loaded to ships and passed on trucks and rail network. For holding containers, both ships and ports were redesigned. In the early 1970s, Ships could carry four times the cargo capacity of traditional ships, their speed becomes faster to make six round trips a year between Europe and Asia, compared to three and a half for the older ships.

Every container is identified by (code BIC). The International Bureau of Containers and Intermodal Transport (BIC) has published and maintained the BIC code register since 1970. The BIC code is a container code that allows identification of containers, their owners and enables border crossing without delay [3].

2.1 Containerization and Containers Tracking

Containerization is an intermodal freight transport system that employs shipping containers. These containers have standard dimensions, can be loaded and unloaded, stacked, efficiently transported over long distances, and converted from and to multiple modes of transport, such as container ships or rail transport. Containers are numbered under an international system of numbering so they can be automatically tracked. Containerization reduces also shipping time and minimizes damage and theft losses. In addition, it eliminates manual sorting of most shipments and the need for warehousing.

Tracing Supply chain process can reduce container transportation, administration and communication costs. All involved stakeholders (actors in ports or in customs) will have a mutual access to the supply chain data. [4] proposes a system that covers supply chain container linked with road, train and vessel. This system is built on an intermodal terminal with rail and road access that is linked to the sea via an inland waterway. The following are the system's primary components:

- Sensors that are used to track and trace containers on trains. Instead of using separate GPS links for each container, they communicate with a central unit within the locomotive, which drastically cuts cost and energy usage. In addition, other sensors are employed to detect any unforeseen activity inside the container, such as its opening or an abnormal temperature rise.
- The planning of drayage fleets utilized for the collection and delivery of containers can benefit from real-time information. An automatic signal box and a specially constructed planning unit make it easier for trains to enter the port terminal.
- A smart navigation system is used to direct ships, schedule events and crossings, control light buoys, and identify timing deviations.
- Finally, The container placed onto the vessel is tracked using a concentrator similar to those used on trains [4]. In [5], authors propose a state-of-art review progress on visual tracking methods. They classify these methods into different categories, as well as identifying future trends. The problem of visual tracking is difficult due to visual noise, complicated object motion, and information loss brought on by projecting a 3D scene onto a 2D image, etc. According to [5], visual tracking is based on some points: The object needs to be descripted so as to be tracked, this description can be a template, a shape, a texture or color model. This is a critical task because of the quality of the description directly related to the quality of the tracking process. Then, objects are inserted into certain context, this context is linked to the tracked object with very strong links, and integrating context information into a tracking framework will help on the research of visual tracking.

By [6], operational activities in Nagoya, Japan, specifically at the Nabeta Pier container terminal, are simulated to analyze the processing time and the bottlenecks of the operations flows. Electronic real-time tracking data is accumulated from the information systems. It is found that the obtained information by performing simulation is effective for analyzing the performance of the portal operation. Although, [7] deals with traceability and control of containers movement identification and monitoring by automatic data capture technologies by RFID technology. Using a web application, authors identified and measured time spent by containers in the selected center. The result of this study was the determination of conditions for the implementation of the RFID technology in relation to monitoring and identification of containers. In sum, t's a step into the identification and the visibility of the entire logistics chain. In [8], authors propose an effective container transport security system that includes subsystem such as container tracker real time container intrusion monitoring. The hypothesis of this paper propose that combination of advanced software and hardware solutions can reduce costs effectively in Intermodal and Maritime Transport Management by Electronic Container Tracking System. The results of this research can be beneficial for Practitioners, they can use the research results to increase competitiveness for carriers, forwarders, logistics and IT companies.

2.2 Blockchain and Its Use in Tracking

Blockchain technology is used in different purposes, whether in the maritime industry or outside supply chain processes: for communications between different stakeholders, for storing transaction history, and for tracking and tracing processes. For example, Chang et al. in [9] proposed a blockchain-based framework using smart contracts, for tracking information and encouraging a network for collaboration within supply chain members, with the use of smart contracts. This study gives conceptual guidance for practical business system design and implementation, as well as improved operational efficiency. Supply chain managers can watch the progress of logistics and cash flows using the suggested blockchain-based architecture with smart contracts, and build relevant strategies to prevent inefficiency.

According to [9], transparent tracking can help real time control of the supply chain. It was found that an alternative payment method using digital currency can reduce payment lead time. In [10], authors developed a system for COVID-19 data tracking. It proposed using Ethereum smart contracts and oracles, as a blockchain-based system for tracking data related to the number of new cases, deaths, and recovered cases. In this paper, [10] present algorithms that describe the interactions between stakeholders in the network. Otherwise, authors gave more details about the cost acquired with some security analysis. In [11], authors propose a systemic literature review dealing with an external dataset related to extracted Blockchain applications in supply chain management. They defined five case studies where technologies (such as Blockchain) are not used frequently. One of them is tracking and tracing. According to [12], a study about a cotton supply chain using blockchain, with specific smart contracts and transaction rules, is provided. In addition, the authors tested the constructed blockchain to verify its application and use. The suggested solution can help supply chain participants develop technology based on trust, while the distributed ledger can be used to store and authenticate supply chain transactions.

This study uses a case study approach along with a literature survey to investigate supply networks as part of a project connected to the energy sector in Finland. Using key performance indicators, real-time data acquired from the developed portal is used to measure the overall performance of project logistics planning and execution.

3 Containers Tracking in Portal Supply Chain

In this paper, we propose a model for tracking containers from the arrival of the ship to its leaving. The aim of this work is to build a system that can track containers in the portal territory as well as the way taken by them to suppliers then to clients, out of the portal territory, using IOT, Blockchain and RFID sensors. Figure 1 explains different stages from containers handling to delivering, parking, transporting (carriers) in the port territory CASA PORT, including the container journey in the exit part of the port, until its return to the port via carriers, suppliers and customers. Before the arrival of the vessel, a file named BAPLIE is sent to the harbormaster's office and the operator who is taking charge of containers handling. The operator sends then the concerned crew and logistics to bring intended containers to be treated. In the same time, the operator sends a notification to the client whose containers have arrived. Containers before being delivered are divided into three types: containers that are planned to be delivered to carriers directly (like containers…); containers that hold pharmaceutical or fresh food that need to be in a controlled temperature and mostly are refrigerated shipping containers or called “Reefer’s containers”. These containers will be sent to refrigeration zone that is equipped with power outlet. The last type is containers with no delays. They are placed in the storage zone, waiting for carriers.

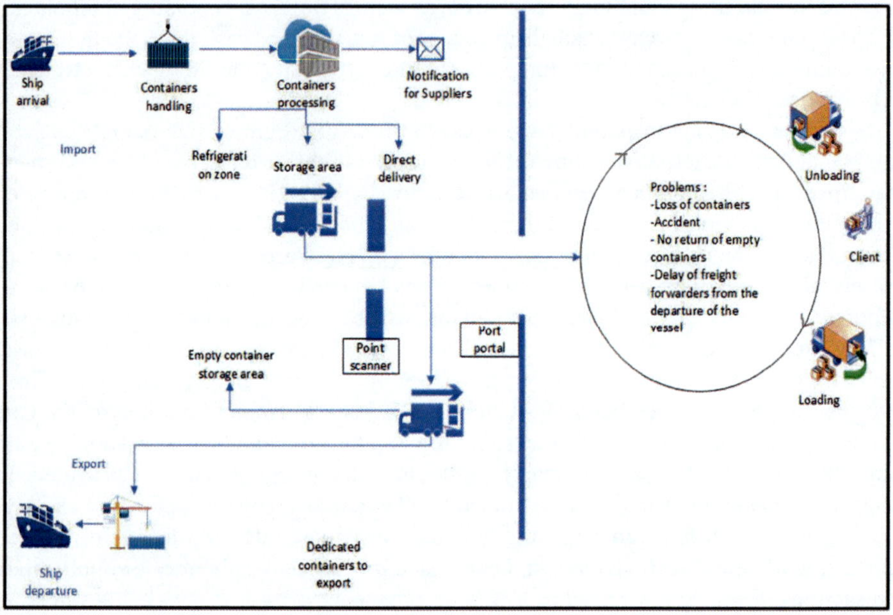

Fig. 1. Container journey from arrival to vessel departure.

Out of the port, there is three actors: Carriers, Suppliers and Clients. Carriers are companies responsible for transporting containers and taking them up to suppliers then to client; Suppliers who get notification message or call from the operator, inform the carrier, then unload or load the container truck; client in some cases receives goods in the container by the carrier. The carriers acquire some documents, at the port access he must show them to authorities, collect the containers, go through the scanner and customs to control the goods and containers. After leaving the port, the carrier transports the containers to the supplier who redirects them with the products within, to the customer, who unload them and return them to the supplier. In some cases, containers are returned to the port after being unloaded, in other cases, suppliers keep them for other transactions. Concerning container security, the supplier and the carrier fully take charge of container’s security. Carriers pay insurance to deliver these containers, and suppliers pay customs in case they will keep them. Containers are divided into two types: containers intended to export whether in the same vessel or another one, or containers are returned empty to the port and classified in the empty containers zone. In most of ports, containers in the port territory are controlled due to checking points between different stages, and Information systems used to control data in ports, and especially CASA PORT, records data regarding the location of containers, the BIC container’s code, the arrival of the ship transporting containers, the customer to whom the container is intended. The most important problems related to containers are: the delay of freights forwarders from the departure of the vessel, the container damage caused by accidents, this leads to think of tracing the way of containers as a necessity. Although, tracing items in general and containers in particular, require the integration of other technologies in addition to Blockchain technology: for capturing data, sensors or other technology with similar functions is needed, for connecting these sensors and collecting data and controlling the network and the communication between different layers, an IOT platform is required. Blockchain technology is used here to decentralize the processing of data, below more details will be given.

4 Technologies Used in the Tracking and Tracing Operation

In this paper, we will use disruptive technologies that are not frequently used in the tracking-tracing problem: Blockchain technology, Internet of things IOT, sensors.

4.1 Internet of Things IOT

The Internet of Things IoT is a network of interconnected devices that may share data and information via the Internet. IoT makes these devices smarter and processes more efficient by collecting and analyzing data from them. In 1999, Kevin Ashton coined the term "Internet of Things" to promote Radio Frequency Identification (RFID) technology. The Internet of Things concept did not acquire widespread acceptance until 2010. Gartner included the term "Internet of Things" on its "hype-cycle for emerging technologies" in 2011 [13].

IoT technologies are intended to connect a vast number of devices dispersed across large geographical areas. These technologies also strive for low power consumption so that battery-powered gadgets can operate for an extended period of time before requiring battery replacement. As a result, wearable IoT devices can be used to track animals' lives and behaviors over lengthy periods of time and throughout huge geographical areas [13].

It has been proposed that the Internet of Things (IoT) consists of a vast distributed network of devices with sensors and actuators. IoT devices are predicted to generate and exchange enormous volumes of data, offering previously unresolved security and scalability challenges. IOT devices can be used for measuring soil humidity in agriculture, or smoke sensors for detecting fire used in smart home installation, touch and motion sensors for security [15] etc. Although, there is no agreement on IOT architecture, the number of layers varies depending on business needs. For example, Cisco IBM and Intel introduced a reference model in 2014 in the IOT World Forum with seven layers presented in Fig. 2. Four layers among these layers, arepresented in Fig. 2: 1) physical devices layer, 2) connectivity layer, 3) processing layer 4) application layer. This structure is considered as the backbone of the IOT technology, Other layers are implemented in the architecture depending on the use case or business needs.

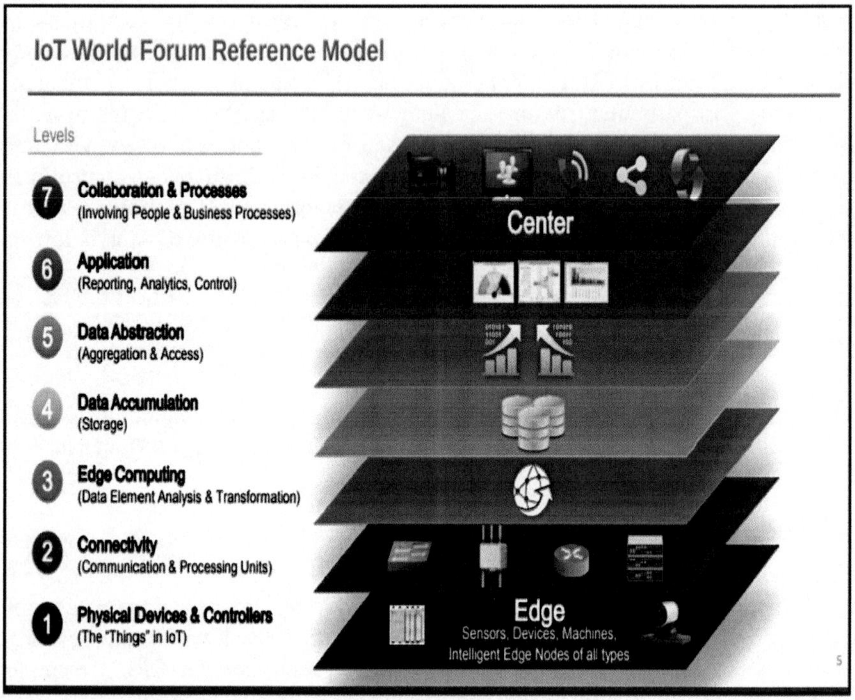

Fig. 2. The standardized architectural model proposed by IoT industry leaders. Source: Internet of Things World Forum

4.2 Blockchain Technology

Blockchain technology is a disruptive technology based on a distributed ledger used to build trust in transactions of assets, money, information etc. It is the basis technology behind many cryptocurrencies, such as Bitcoin, Ethereum etc… This technology appeared for the first time in a scientific paper authored by certain Satoshi Nakamoto and published in 2008, this paper discusses the problem of double spending problem in cash machines. Blockchain is an immutable ledger; once a transaction is recorded, it is difficult to change or delete it. For sending transaction, every node or element participating in the process has two keys: a public key and a private key, every transaction is signed by both of these keys, and verified. While sending a transaction, it should be verified, then validated by miners so as to be added to a block that will be added himself to the chain of blocks called Blockchain. This process is assured with cryptographic algorithms, and based on the notion of consensus.

The process through which a collection of nodes agrees on the validity and order of transactions included in blocks is known as consensus. The purpose of consensus is to reach an agreement on the order and validate the correctness of the block's set of transactions. Consensus is a mechanism through which a network of nodes offers a guaranteed ordering of transactions and validates the block of transactions. In reality, there is a list of consensus algorithms, each one can be useful for different use case, in the next lines some important algorithms are mentioned:

Proof of Work: This type of consensus is based on the computational resources used to propose a value to be accepted by the network. This is a term that is used in bitcoin and other cryptocurrencies. This method has been shown to be effective against sybil attacks.

Proof of Stake: This algorithm is based on the assumption that each node or user has a sufficient stake in the system. Another consideration is coin age, which is determined by the duration of time and the number of coins not spent in this algorithm, The likelihood of proposing and signing the next block increases with coin age [16].

Trust and transparency are key components in many distributed applications, and blockchain technology has proven to be a potential answer. It is not surprising, then, that both industry and academia are heavily debating how to effectively connect IoT systems with Blockchain technology. [14].

5 Results

To model a diagram for the problematic detailed above, we made interviews with experts from an operator in Casa PORT and with a supplier dealing with the same port. Below a diagram that represents the business process of containers carried in and out of the terminal. In this process, there are four stakeholders who interact with, receive, store or carry containers: the operator, the carrier, the supplier and the client. The shipowner is included in this process, but we limited our study since data brought by the shipowner is possible to get from the operator.

This model details the journey of containers from the arrival of the vessel carrying containers to its leaving or storing them in the container storage zone. The process is divided into two parts: in the hinterland and out of the port territory. In Fig. 3, the most important actors in the delivering containers process are mentioned as follows:

- **The terminal operator:** When the ship approaches the port, it sends a message to the authorities presented by the ANP, which takes charge of informing the terminal operator or operators concerned by the containers carried by the ship. Containers are handled as soon as they arrive at the port, and the operator sends the agents for the unloading, handling and checking of the containers.
- **The carrier:** Some of incidents discussed in this work, concerning containers, were found while being handled by carriers, whether in the terminal or in their way to the costumer. One of the benefits of the system proposed in this work, is to track containers in this part to get help as soon as possible (in case of accident, stealing or damage)
- **The supplier:** The supplier and the terminal operator are the most concerned on having the trace of containers in and out of the terminal to establish the delivery more efficient. As a result: the supplier should have the possibility to get the information in real-time.
- **The client:** The role of the client is receiving containers to get goods, then returning empty containers to the supplier, so he is not concerned about tracked containers.

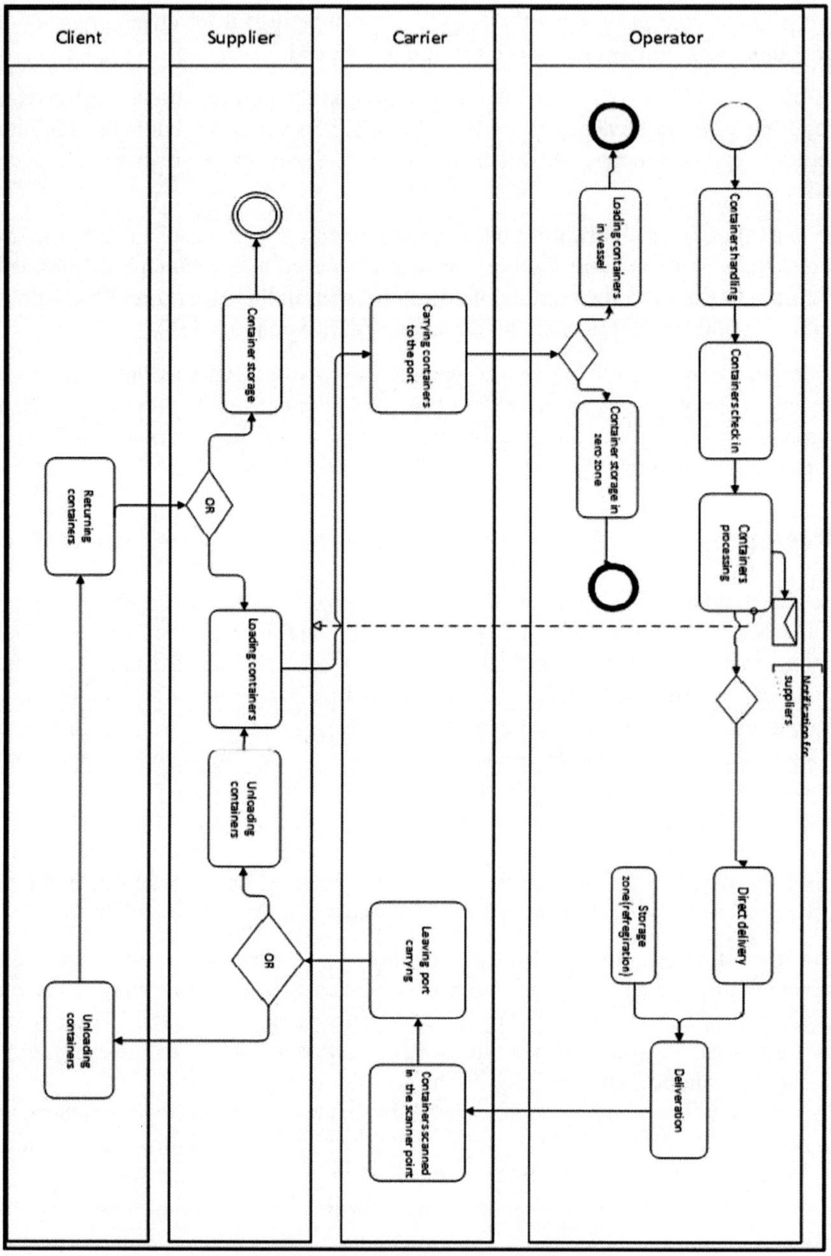

Fig. 3. BPMN model designs the process of containers in and out of the terminal.

As mentioned above, the operator information system controls movement and data concerning the container location in the port territory. Otherwise, as soon as the carrier quit the port territory, container's location is unknown to operators nor clients and sometimes even the supplier can't get his goods location. Figure 4 presents a schema about the framework layers proposed in this work:

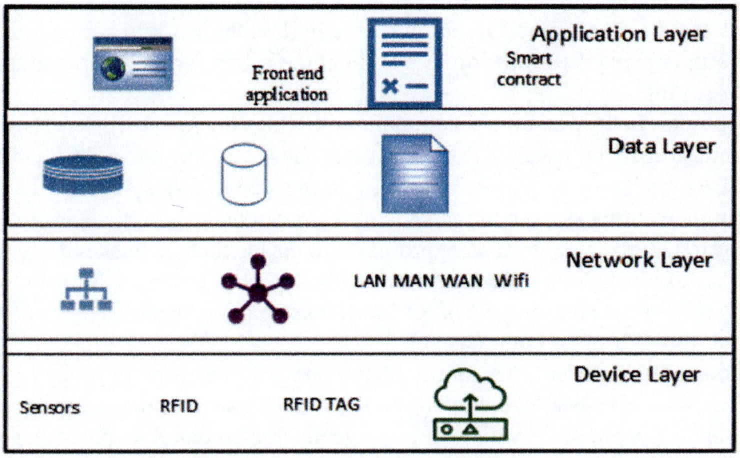

Fig. 4. Proposed system layers standing on Blockchain based IOT

In this figure, the architecture of the model proposed is organized under four layers. This structure is adopted from the IOT architecture:

- *Device layer*: Consists of Devices that are linked to the system. Smart devices are items or assets that have sensors, processors, actuators, and the ability to transfer data over the internet. They can collect data from their surroundings and exchange it with system operators, users, other smart devices, and apps. To collect data from their surroundings, intelligent devices might integrate many sorts of sensors. In this model, we employ location sensors mounted on containers to determine whether they are in or out of the port. Sensors transmit the location information to the network layer.
- *Network layer:* The Network layer is established on the IOT network. The network layer is responsible for encapsulating and forming the packets, then transferring them: from source to destination to accumulate and manage all data streams; in our case: data collected from RFID devices is stored in packets in the network IOT with the use of an IOT platform.
- *Data layer*: Data layer is a layer where data is processed analyzed then sent to elements concerned. IOT platform, in most of applications, belongs to data layer. In our proposition, blockchain technology will be integrated in this layer too, its role is to transform data formatted by the IOT platform to blocks of data validated then added to the chain of blocks.
- *Application layer:* The application layer is responsible for ensuring effective communication with external applications. In this proposition, a front-end application is

implemented with a smart contract to the application layer, this smart contract is sent to the user of the system depending on the need, the user of the system is getting his copy of blockchain data that is permanently updated.

6 Conclusion and Discussion

This article is a conception and a modeling for the problem of tracking and tracing containers in the port terminal. In our case, we limited the work to containers handled by carriers and tracing their location in and out of the hinterland, from unloading to their loading or storing.

Some studies in the literature recommends utilizing Blockchain for port access policies to impose trust between multiple partners, however owners must handle policy updates. This method is incapable of scaling to the size of IoT systems. According to [14], By implementing smart contracts, blockchain can make policy management more scalable by transferring policy update processes to the blockchain backend. By doing so, we make use of blockchain technology and IOT to open up further options for research. According to [17], the combination of IOT and Blockchain has benefits like: autonomous transactions through smart contracts [17, 18].

To summarize, both industry and academia are debating how to properly combine IoT devices with Blockchain technology [14]. This paper presents a model in which this combination is proposed with its advantageous implementation, particularly for the supply chain and different stakeholders; it tests the true potential of both technologies, particularly with the number of challenges encountered, which opens up more chances and opportunities for research; and it opens up more chances and opportunities for research.

References

1. Azzi, R., Chamoun, R.K., Sokhn, M.: The power of a blockchain-based supply chain. Comput. Ind. Eng. **135**, 582–592 (2019). https://doi.org/10.1016/j.cie.2019.06.042
2. Frizzo-Barker, J., Chow-White, P.A., Adams, P.R., Mentanko, J., Ha, D., Green, S.: Blockchain as a disruptive technology for business: a systematic review. Int. J. Inf. Manage. **51**, 102029 (2020). https://doi.org/10.1016/j.ijinfomgt.2019.10.014
3. EENI École d'Affaires Business, Conteneurs et transport international, EENI École d'Affaires/Business School. http://fr.reingex.com/Containers.shtml. Accessed 18 Dec 2021
4. Muñuzuri, J., Onieva, L., Escudero-Santana, A., Cortés, P.: Impacts of a tracking and tracing system for containers in a port-based supply chain. Brazilian J. Oper. Product. Manage. **13**, 352–359 (2016). https://doi.org/10.14488/BJOPM.2016.v13.n3.a12
5. Yang, H., Shao, L., Zheng, F., Wang, L., Song, Z.: Recent advances and trends in visual tracking: a review. Neurocomputing **74**(18), 3823–3831 (2011). https://doi.org/10.1016/j.neucom.2011.07.024
6. Jain, S.: American Statistical Association, Association for Computing Machinery, and IEEE Systems, Man, and Cybernetics Society, Eds., Modeling the materials handling in a container terminal using electronic real-time tracking data. IEEE, Piscataway, NJ (2011)
7. Madleňák, R., Madleňáková, L., Kolarovszká, Z.: System of management and traceability of logistic items through new technologies. Procedia. Soc. Behav. Sci. **230**, 128–135 (2016). https://doi.org/10.1016/j.sbspro.2016.09.016

8. Miler, R.K.: Electronic Container Tracking System as a Cost-Effective Tool in Intermodal and Maritime Transport Management no. 1, p. 13 (2015)
9. Chang, S.E., Chen, Y.-C., Lu, M.-F.: Supply chain re-engineering using blockchain technology: a case of smart contract based tracking process. Technol. Forecast. Soc. Chang. **144**, 1–11 (2019). https://doi.org/10.1016/j.techfore.2019.03.015
10. Marbouh, D., et al.: Blockchain for COVID-19: review, opportunities, and a trusted tracking system. Arab. J. Sci. Eng. **45**(12), 9895–9911 (2020). https://doi.org/10.1007/s13369-020-04950-4
11. Blossey, G., Eisenhardt, J., Hahn, G.: Blockchain technology in supply chain management: an application perspective. Presented at the Hawaii international conference on system sciences (2019). https://doi.org/10.24251/HICSS.2019.824
12. Agrawal, T.K., Kumar, V., Pal, R., Wang, L., Chen, Y.: Blockchain-based framework for supply chain traceability: a case example of textile and clothing industry. Comput. Ind. Eng. **154**, 107130 (2021). https://doi.org/10.1016/j.cie.2021.107130
13. Dian, F.J., Vahidnia, R.: British Columbia Institute of Technology, BC Open Textbook Project, and BCcampus, IoT use cases and technologies. (2020). [https://open.bccampus.ca/browse-our-collection/find-open-textbooks/?uuid=b71da430-8abb-45b3-ae10-b35e68bbb09b&contributor=&keyword=&subject=. Accessed 30 Oct 2021
14. Truong, H.T.T., Almeida, M., Karame, G., Soriente, C.: Towards secure and decentralized sharing of IoT data. In: 2019 IEEE international conference on blockchain (Blockchain), pp. 176–183. IEEE, Atlanta, GA, USA (2019). https://doi.org/10.1109/Blockchain.2019.00031
15. Nightingale, A.: A guide to systematic literature reviews. Surg. Infect. (Larchmt.) **27**(9), 381–384 (2009). https://doi.org/10.1016/j.mpsur.2009.07.005
16. Imran Bashir, "Mastering Blockchain - Second Edition [Book]. https://www.oreilly.com/library/view/mastering-blockchain-/9781788839044/. Accessed 13 Aug 2020
17. Kshetri, N.: Can blockchain strengthen the Internet of Things? IT Prof. **19**(4), 68–72 (2017). https://doi.org/10.1109/MITP.2017.3051335
18. Christidis, K., Devetsikiotis, M.: Blockchains and smart contracts for the Internet of Things. IEEE Access **4**, 2292–2303 (2016). https://doi.org/10.1109/ACCESS.2016.2566339

A Grid-Based and a Context-Oriented Trajectory Modeling for Mobility Prediction in Smart Cities

Hocine Boukhedouma[1,2](✉), Abdelkrim Meziane[1], Slimane Hammoudi[3], and Amel Benna[1]

[1] Department of Multimedia and Information Systems, CERIST, Algiers, Algeria
ghocine@gmail.com
[2] Department of Computer Science, USTHB, Algiers, Algeria
[3] ESEO-TECH - ERIS TEAM, Angers, France

Abstract. In the last decade, mobility prediction has played a crucial role in urban planning, traffic forecasting, advertising, and service recommendation. This paper addresses the *prediction of mobility* and emphasizes an essential step that is *trajectory modeling* (better the modelling is, better is the prediction). First, we propose a *context-based* and *prediction-oriented* trajectory model. Our model is based on a grid-oriented trajectory description technique that allows overcoming low precision and ambiguity issues. Second, our model is compared to some related trajectory models. Third, an application of the model in intelligent transportation domain is illustrated. Finally, to evaluate our model, we experiment it on a data mining-based prediction algorithm and show the results in terms of prediction accuracy.

Keywords: urban human mobility · mobility prediction · trajectory description and modeling · context

1 Introduction

Nowadays, with the use of new information and communication technologies in smart cities, big quantity of data about human mobility and its environment is provided. These data concern mobile's location and data associated to a smart city environment (date, weather, means of transportation…). Human mobility data are exploited to accomplish several tasks such as mobility prediction, which became a challenging problem, and therefore has an important role in smart cities. Indeed, to enhance the daily life of mobile citizens, mobility prediction is used in several domains such as mobile communication, intelligent transportation, and location-based service recommendation. In this perspective, a context-based mobility prediction approach is highly desirable.

According to the use of mobility traces, we distinguish two types of mobility prediction models: *historical-based models* [1] and *knowledge-based models* [2].

Moreover, to achieve better mobility prediction results, two main challenges are considered: *mobility modeling* that refers to the identification, representation and integration

M. Ben Ahmed et al. (Eds.): SCA 2023, LNNS 906, pp. 148–157, 2024.
https://doi.org/10.1007/978-3-031-53824-7_14

of all data required for prediction; and *prediction process* that defines steps, techniques and algorithms used for prediction.

Trajectory modeling of a mobile object requires data related to its different positions over time (GPS coordinates, date, direction…). Other contextual data of the object's movements are also essential (Our reference, 2020), since they can influence mobility.

The set-up of a trajectory description technique, which aims to define the main items constituting the trajectory, is the base for trajectory modeling. Many trajectory descrip-tion techniques have been proposed in the literature. For example, trajectory is described in terms of places [3] (e.g., cities, streets…), in terms of landmarks [4], etc. These techniques suffer from low accuracy or erroneous description of the trajectory.

In this paper, we focus on the trajectory modeling challenge and propose a *conceptual trajectory model* for mobility prediction which considers four main *context* parameters: *time, weather, means of displacements* and *road traffic*. Our model allows: i) the representation and the integration of data based on the *technique of the grid* for a trajectory *description*; and ii) the storage of data in a database for further analyses and visualization.

The rest of this paper is organized as follows: Sect. 2 focuses on related works of trajectory modelling. Section 3 describes our mobility prediction approach and emphasizes our trajectory model. Section 4 shows the use of our model in a specific domain and exposes an evaluation of our model using a data mining-based prediction algorithm. Section 5 concludes the paper and gives an overview of future works.

2 Related Works

In the context of mobility, trajectory modeling has been the subject of several works in different fields. In [5], authors propose a general framework for modeling trajectory patterns during the conceptual design of a database. In [4], authors propose a method to construct and analyze trajectories of mobile objects and apply it on ship and herbivore movements.

Some works [6–9] focus on the semantic aspect of trajectory. In [6], authors present a semantic trajectory conceptual data model named CONSTAnT, which defines the most important aspects of semantic trajectories. In [7] a semantic model for trajectories is proposed as well as relationships of trajectories with geographic and environment information. In this model, trajectory relationships are associated with vehicles and roads. Recently, authors propose in [8] a new conceptual model of trajectories, which accounts for semantic and indoor space information and supports the design and implementation of context-aware mobility data mining and statistical analytics methods. In [9], authors designed a multi-aspect and multi-level trajectory model that takes into account: (i) the description of sequences of hierarchical episodes, (ii) the definition of semantic aspects integrating spatial, temporal and thematic dimensions, and (iii) the association of such semantic aspects to positions or to trajectory episodes.

In [3] and [10] authors treat mobility patterns-based models. In [3], authors propose a software framework for modeling, querying and analysing an individual or group's behavior. In [10], authors propose to describe the trajectories, at the conceptual level, with their geometric, temporal and semantic aspects and their structural components: *starting point, ending point, stops and intermediate displacements.* The modeling approach is based on modeling patterns, which allow a generic solution to model the standard characteristics of the trajectories. This model named "*Stops and Moves*" constitutes the basis of several works, such as [5, 6, 11].

Other works [11–13] are oriented towards urban human mobility. In [12], authors survey a data driven modeling of urban human mobility. They distinguish three main categories of urban human mobility models: *Population Mobility Models (PMMs), Unified Mobility Models (UMMs) and Individual Mobility Models (IMMs).* PMMs include *gravity model, intervening opportunity model* and *radiation model.* IMMs include *Brownian motion, Levy flight, Continuous time random walk* and *Social-based model.* UMMs cover some issues of population and individual mobility models. Among these models, we cite the universal model proposed by [13]. In [11], authors focus on modeling spatiotemporal trajectories from digital traces of mobile phone in order to study the human mobility. For that, a conceptual data model allowing modeling spatiotemporal dimensions of GSM data is proposed.

Recently, authors in [14] propose to model trajectory data with a deep generative model such that: (a) the model is lightweight, (b) the model is of high utility to support the downstream trajectory applications, and (c) the model preserves individual privacy.

Compared to the proposed approaches in the literature, we aim to propose a context based conceptual trajectory model for mobility prediction, which should satisfy the following requirements:

- Avoid problems of existing trajectory description techniques, such as low precision and ambiguity of trajectory description due to the overlapping problem.
- General enough to be used by different user centred mobile applications in smart cities, and covers the main contextual entities proposed in the state-of-the-art of mobile applications in smart cities.

In the following, we describe our trajectory modeling approach that we plan to use in a mobility prediction solution to improve the results of Location-Based Services Recommendation (LBSR).

3 Trajectory Modeling For Mobility Prediction

Our mobility prediction approach contains three main phases (see Fig. 1). The first phase concerns *trajectory modelling and description.* The second phase includes *data acquisition, transformation, normalization and storage.* The last phase is related to the *prediction process (learning* and *prediction algorithms).*

3.1 Trajectory Description Technique: Technique of the Grid

Trajectory modeling is based on the technique used for the description of trajectory, allowing to answer the following question: *in what terms is the trajectory described*?

Many ways for trajectory description were used: trajectory can be described in terms of geographical places as in [3], in terms of landmarks as in [4], and in terms of cells of a cellular network as in [15]. However, some limits are observed with these techniques. The formers are not very precise and the latter suffers from the oscillation phenomenon due to the overlapping cells offering an erroneous description (see Fig. 2).

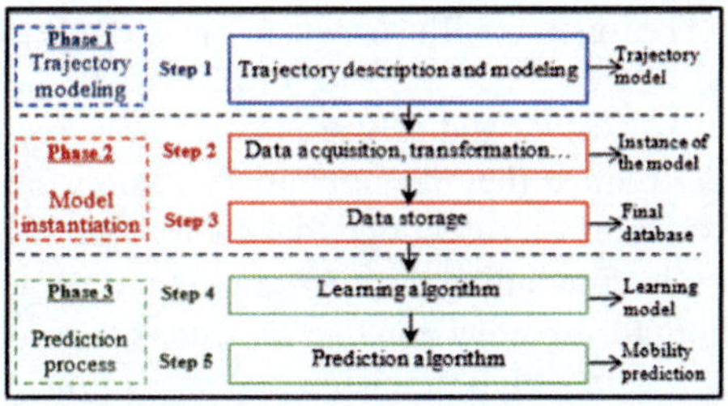

Fig. 1. Main steps of our mobility prediction approach

In cellular networks, theoretically, a network is represented in a hexagonal form where each cell corresponds to a *Base Transceiver Station* (*BTS*), known as an *antenna.* The reality, however, deviates from the theory because of the antennas radiate in a circular form. This causes the famous *overlap problem*, which causes the *oscillation phenomenon.* This phenomenon may imply a user movement even if the user does not move at all. A widely adopted solution is cell clustering, but at the expense of precision in mobility prediction, the oscillation deteriorates the prediction performance [15]. To overcome the cited limits, we propose a trajectory description technique, called *technique of the grid.*

This technique consists of breaking down the geographical study space into regions of identical shapes (a square) of a fixed size. The size of the regions will be defined based on the standards used for the definition of the sizes of cells in cellular networks, which depends on the nature of the geographic area. This size varies from a few tens of meters in urban areas to a few hundred meters or even kilometers in rural and deserted areas. Figure 3 illustrates the region division adopted.

In this technique, we distinguish two types of regions: *sojourn regions* and *transit regions. Sojourn region* is a region where the mobile spends a time greater than a given *threshold*. Departure and arrival regions are always considered as *sojourn regions.*

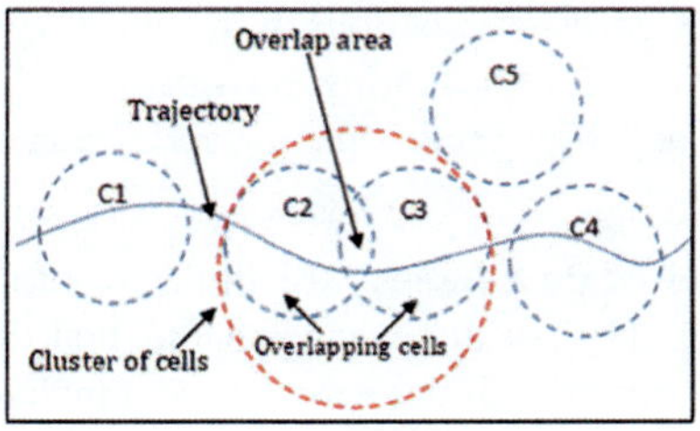

Fig. 2. Overlapping cells problem

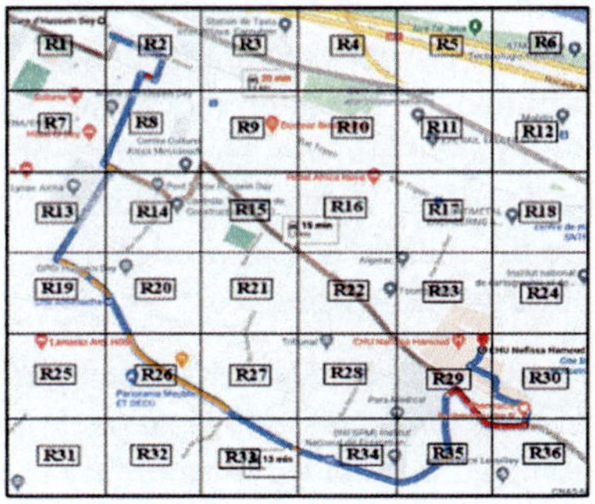

Fig. 3. Overview of a part of a city splitted using the technique of the grid

Illustrative Example. The trajectory illustrated on Fig. 3 is represented by the following sequence*: R2 R8 R7 R8 R14 R13 R19 R20 R26 R27 R33 R34 R35 R29 R35 R36 R30 R29*, where *Ri* denotes a *region, R2* and *R29* are two *sojourn regions* and the *remaining regions* are *transit regions* since they do not satisfy the condition of *sojourn region.* From this illustrative example, we can notice that each point of trajectory is associated to one region only. This allows avoiding the overlapping problem observed in the Id-Cell based descriptions without losing precision that depends on the requirements of the end application. If the latter requires high precision, the size of the grid elements (regions) needs to be reduced (augmentation of the spatial resolution). Thus, precision depends strongly on the "size" parameter that can be defined empirically.

In addition, our technique allows preserving the user's privacy since the position of the user is given in terms of regions instead of an exact position defined in terms of a GPS coordinates (latitude and longitude).

3.2 Trajectory Modeling

There are two alternatives of trajectory modeling. The first one is to design the trajectory model based on the needs and the available data; it concerns the data-driven mobility modeling [12]. The second one is to design the model based solely on needs. In this second case, the data will be collected according to the model after its creation. In our work, we adopt the second alternative and propose a trajectory model, which accounts the contextual data and which is oriented for mobility prediction.

Context-Based and Prediction-Oriented Trajectory Model for Location-Based Services Recommendation. In order to enhance results of location-based services recommendation, we aim to apply trajectory modeling to mobility prediction of a mobile object moving under variable context. Thus, we define three key parameters in our model: *context, prediction* and *location-based services (LBSs) of the city.* The *context* includes time, urban traffic, weather and means of displacement; it is a main parameter that influences mobility. The *LBSs* tacking place in the city are important because of the final objective attached to services recommendation. The prediction in the model represents prediction algorithm allowing to predict future position (s) based on history.

Our model allows also the identification of significant positions in trajectories, considered as PoIs, through the consideration of *stations*, representing, possibly, the places where the mobile object practices an activity; and *significant transit points*, related to the locations of LBSs.

Trajectory Model Description: Principle and Main Classes. Our trajectory model extends the *Stops* and *Moves* model, proposed in [10], with advanced contextual information, and is intended for mobility prediction. Figure 4 shows a meta-model including the main classes and links between them.

Smart City Mobility View (SCMV): Consists of an aggregation of trajectories of the mobile object and allows to express the different possible views on mobility in a city. Through this class, three main views of mobility in a city can be specified and visualized:

- The past mobility view: gives a view on the mobility of a person or a group of people during a period in the past. This view also allows to visualize the general mobility in the city in a given past period.
- The real-time mobility view: this view gives real-time mobility in the city. Mobility on a person, a group of people or a district of a city or the whole city can be visualized.
- The predicted mobility view: This view is used to calculate and display a mobility prediction in the city. It may concern either a person (trajectory prediction) or a neighbourhood in the city (group mobility prediction).

Mobile Object: Consists of the entity performing one or more displacements. The mobile object can be of two types: human and other.

Trajectory: It is composed of a set of displacements.

Displacement: Consists of an aggregation of transit regions. Each displacement is delimited by two consecutive stations (source and destination) and happens in a specific context that can be variable for the same displacement.

Context: Includes specific data that can influence mobile user displacements. We consider the following data: Time (as standard contextual data), weather, means of displacement and road traffic (as advanced contextual data). As we can see in Fig. 4, each movement of a mobile object is linked to a context through the association class *Context*.

Station: Represents a sojourn region where the mobile object stays a time superior to a given threshold.

Transit Region: Represents the regions through which the mobile object passes without staying in. The transit region contains two types of transit points: standard points and significant points. These latter are very important in our work because it concerns relevant locations related to specific location-based services taking place in the city.

Location-Based Service. Concerns all services of the city, for which the location is a key parameter. These services are relevant for a recommendation to specific profiles.

Prediction: Associates the mobile object to the SCMV. It is about making predictions according to context filtering, a current date and place, and opting for a defined prediction model.

3.3 Comparative Study

In this section, we compare our model with those of the cited works based on five criteria that are: *application domain, respect of the Stops and Moves principle, context and semantic consideration* and *items used for the description technique* (see Table 1).

Comparatively to the cited models, innovative ideas, such as the consideration of the technique of the grid and the explicit integration of the prediction, have been added to our model. The advantages derived from other models (consideration of the semantic aspect, adoption of the *Stops and Moves* principle, etc.) have been retained. As a result, we obtained a personalized model with respect to prediction purpose, but generic with respect to trajectory modeling in urban places and with respect to service recommendation, which is not limited to a particular type of service. Our model is easier to understand, but also more complete in terms of the diversity of data required to meet our needs (predict to recommend).

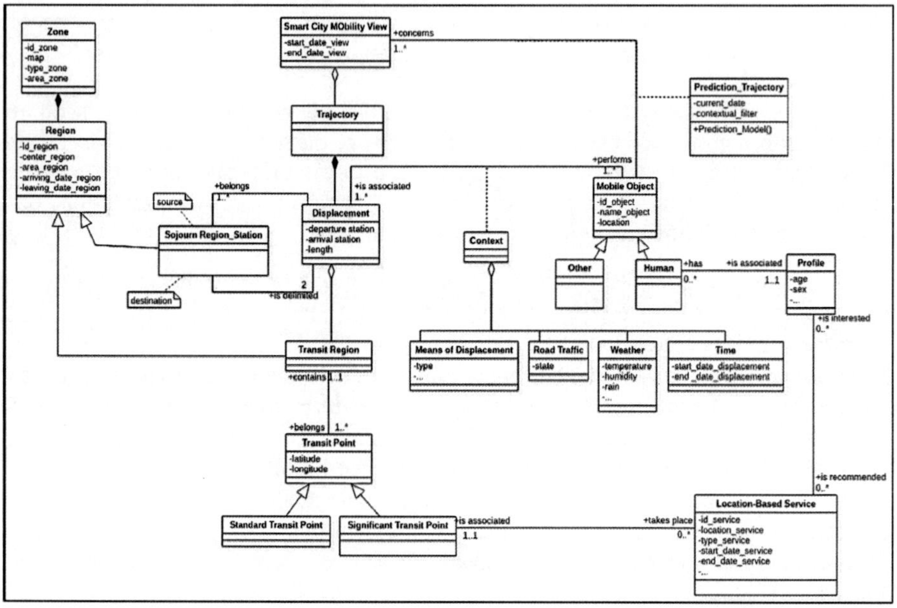

Fig. 4. Context-based and prediction-oriented trajectory model

4 Application of the Model in Road Traffic Field

In this section, we present an example related to the application of mobility prediction to intelligent transportation and road traffic systems. For that, we are working on the design of a *mobile application* for *road anomaly detection and notification*. This application serves to collect anomalies happening on roads and to notify citizens about those situated in their *future* trajectories. This latter will be obtained using our mobility prediction approach. We focus in this scenario on trajectory modeling and data acquisition steps, allowing construction of users' trajectories.

The construction and storage of a user's trajectory is performed based on the grid technique and according to the proposed trajectory model. In fact, first, the study area is delimited and broken down into regions. Subsequently, the necessary data will be collected, processed and projected onto the study area in order to determine the key elements of the trajectory (station, transit region, etc.). These data will then be transformed according to the model and finally recorded in the associated database.

At the mobile application level, a *data collection module* will collect all data required for the model. These data can be collected when users install the application on their mobile devices and agree to the terms of use. Positions, date and time will be extracted implicitly. Advanced contextual data can be extracted implicitly, explicitly or inferred.

When enough quantities of data is available, they will be recorded, in form of trajectories, in the database conformably to the trajectory model. These trajectories will be used in the prediction process (learning and prediction algorithms).

Table 1. Trajectory models comparison

Work	Application of the model / case study	Based on stops & moves model	Context consideration	Semantic consideration	Items used for description technique	Specificities
[4]	Movement of ships in maritime space	None	Not Mentioned (NM)	Yes	Existing landmarks (port)	-
	Herbivores moving	None	NM	Yes	Defined landmarks (PoIs positions)	-
[5]	Identification of touristic places	Yes	Yes	Yes	Landmarks (touristic places)	Pattern
[6]	Tourism Bird migration	Yes	Yes	Yes	Landmarks	-
[7]	Vehicle network	None	Yes	Yes	Road	-
[3]	Monitoring of a fleet of vehicles	None	None	Yes: Activities	Territorial partitions (zones)	Mobility patterns and interrogation patterns
[10]	Migration of storks	Yes	Yes	Yes	Countries and zones	Mobility patterns
[11]	Study of tourist movements	Yes	Time Means of displacement	Yes : Activities	Voronoï Cell	-
[13]	Statistical analysis	None	None	None	Cities (attractive places)	Individual and population mobility
[15]	Mobility prediction	NM	NM	NM	Cell-Id	Overlap problem Low precision
Our model	Mobility prediction for LBSR	Yes	Yes	Yes	Technique of the grid (zone & region)	Mobility prediction oriented

At this stage, the mobility prediction will serve to improve the dissemination of alerts to affected users whenever a traffic anomaly occurs in the city. The users concerned are those whose predicted trajectory, from their current position, contains significant waypoints corresponding to the location of anomalies stored in the LBS classe of the model.

4.1 Model Evaluation

The performance of our model is evaluated on a prediction algorithm using real user movements. For that, we developed a data mining-based prediction algorithm using *Generalized sequential pattern (GSP)* algorithm. To show the impact of our model on the prediction algorithm, we executed the algorithm according to two alternatives and we compared the prediction accuracy values obtained for each case. In the first alternative, we used our model to describe the trajectories with the consideration of standard contextual data (date and time) only. In the second alternative, no trajectory model was taken into account. We have described the trajectories as being a succession of positions with the consideration of the PoIs.

The evaluation was made on 33 persons. For each alternative, we calculated the following values: the average precision of all the users, the maximum precision value corresponding to the user U1, and the minimum precision value corresponding to the user U2. The precision is measured as the ratio of correctly predicted transitions to the number of computed predictions. Results showed that all values of precision of the first alternative outperforms the values of the second alternative. The best value of precision (82%) is obtained for the user U1 (having a regular mobility) with the first alternative.

5 Conclusion and Future Works

We proposed a model based on the grid technique to obtain more precise and unambiguous trajectories. Our model adopts the Stops and Moves approach, allowing to integrate the semantic aspect by identifying sojourn regions. The contextual aspect, reflecting the travel conditions (e.g., weather,) is also present in the model.

Compared to other models, data present in the model are sufficient for the construction of fairly rich trajectories, in terms of information on the travel conditions, as well as for the prediction process. The model is scalable as it tolerates the addition of other contextual data and takes into consideration any type of LBS belonging to any domain. Moreover, the model is intended for service recommendation, which became very popular among citizens, especially those moving in smart cities. An example of application of our model was presented. The results of evaluation showed the positive contribution of our trajectory model on the prediction algorithm.

Our future work aims to integrate all the contextual data present in our trajectory model (climate, means of transport, etc.) in the prediction phase of our prediction process to improve accuracy.

References

1. Boc, M., Amorim, D.M.D., Fladenmuller, A.: Near-zero triangular location through time-slotted mobility prediction. Wireless Netw. **17**(2), 465–478 (2011)
2. Samaan, N., Karmouch, A.: A mobility prediction architecture based on contextual knowledge and spatial conceptual maps. IEEE Trans. Mob. Comput. **4**(6), 537–551 (2005)
3. Chardonnel, S., Du Mouza, C., Fauvet, M.C., Josselin, D., Rigaux, P.: Patrons de mobilité : proposition de définition, de méthode de représentation et d'interrogation. In : Colloque Cassini'04–7ème conférence du GDR Sigma" Géomatique et Analyse Spatiale, pp. 19–23 (2004)
4. Buard, E., Devogele, T., Ducruet, C.: Trajectoires d'objets mobiles dans un espace support fixe. Revue Internationale de géomatique **25**(3), 331–354 (2015)
5. Bogorny, V., Heuser, C.A., Alvares, L.O.: A conceptual data model for trajectory data mining. In: Fabrikant, S.I., Reichenbacher, T., van Kreveld, M., Schlieder, C. (eds.) Geographic information science, pp. 1–15. Springer Berlin Heidelberg, Berlin, Heidelberg (2010). https://doi.org/10.1007/978-3-642-15300-6_1
6. Bogorny, V., Renso, C., de Lucca Siqueira, F., Alvares, L.O.: Constant–a conceptual data model for semantic trajectories of moving objects. Trans. GIS **18**(1), 66–88 (2014)
7. Brakatsoulas, S., Pfoser, D., Tryfona, N.: Modeling, storing and mining moving object databases, In Proceedings. International Database Engineering and Applications Symposium, IDEAS'04, pp. 68–77. IEEE (2004)
8. Kontarinis, A., Zeitouni, K., Marinica, C., Vodislav, D., Kotzinos, D.: Towards a semantic indoor trajectory model: application to museum visits. GeoInformatica **25**(2), 311–352 (2021)
9. Cayèré, C., Sallaberry, C., Faucher, C., Bessagnet, M.N., Roose, P.: Proposition d'un modèle de trajectoires multi-aspects et multi-niveaux appliqué au tourisme. In: IC, pp. 56–64 (2021)
10. Spaccapietra, S., Parent, C., Damiani, M.L., et al.: A conceptual view on trajectories. Data Knowl. Eng. **65**(1), 126–146 (2008)
11. Raimond, A.M.O., Ouronné, T., Fen-Chong, J., Smoreda, Z.: Le Paris des visiteurs étrangers, qu'en disent les téléphones mobiles-Inférence des pratiques spatiales et fréquentations des sites touristiques en Île-de-France. Revue internationale de géomatique **22**(3), 413–437 (2012)
12. Wang, J., Kong, X., Xia, F., Sun, L.: Urban human mobility: data-driven modeling and prediction. ACM SIGKDD Explor. Newsl. **21**(1), 1–19 (2019)
13. Yan, X.Y., Wang, W.X., Gao, Z.Y., Lai, Y.C.: Universal model of individual and population mobility on diverse spatial scales. Nat. Commun. **8**(1), 1–9 (2017)
14. Wang, Y., Li, G., Li, K., Yuan, H.: A deep generative model for trajectory modeling and utilization. Proc. VLDB Endowment **16**(4), 973–985 (2022)
15. Göndör, S., Uzun, A., Rohrmann, T., Tan, J., Henniges, R.: Predicting user mobility in mobile radio networks to proactively anticipate traffic hotspots. In: 2013 international conference on MOBILe wireless MiddleWARE, operating systems, and applications, pp. 29–38, IEEE (2013)

Real-Time Mapping of Mobility Restrictions in Palestine Using Crowdsourced Data

Hala Aburas(✉) and Isam Shahrour

Civil and Geo-Environmental Engineering Laboratory (LGCgE), Lille University, Rue Paul Duez, 59000 Lille, France
{hala.aburas.etu,isam.shahrour}@univ-lille.fr

Abstract. This study introduces the service of real-time mapping of mobility restrictions RT-MMR using crowdsourced data to improve people's mobility in Palestine. The service is part of the Smart and Resilient Mobility Services (SRMS) platform, a mobile web app that provides real-time road information and mobility services in areas with frequent road restrictions. The literature review highlights the adverse effects of mobility restrictions on the Palestinian population's daily life, employment opportunities, and the environment. It also notes the lack of research exploring strategies to mitigate the negative impacts of mobility restrictions, largely due to the difficulty in accessing real-time and historical data on the behavior and functionality of these restrictions. The research argues that common commercial apps such as Waze and Google Map are not effective solutions due to their limited coverage and inaccuracies in accounting for mobility restrictions. To address these challenges, the paper proposes the RT-MMR service, which uses ArcGIS Online and ArcGIS Survey123 to provide real-time and accurate information on mobility restrictions. The service aims to help interurban travelers make informed decisions and optimize their travel to reduce delays, congestion, time losses, and energy consumption. The paper also introduces the Restriction Notification System (RNS), which will notify users in real-time of any updates in specific restrictions, further improving the service's efficiency.

Keywords: Crowdsourcing · Real-time Mapping · Mobility Restriction

1 Introduction

Road traffic disruption due to man-made events can significantly impact the flow of traffic, causing delays, congestion, and in some cases, road closures [1], leaving severe socioeconomic and environmental impacts. These events can range from planned construction projects to unplanned incidents, such as checkpoints, protests, accidents, or acts of terrorism [2]. Kurth et al. [3] emphasized the impact of random road disruptions on GDP in various U.S. cities. Their findings reveal a direct link between travel time delays and GDP decline. In San Francisco, with disruptions on only 3% of road segments, travel time increased by 34%, causing a significant 6.64% GDP decrease.

Additionally, Chen et al. [4] quantified vehicle emissions during traffic congestion using real-world traffic data in China. The findings showed that typical congested traffic,

M. Ben Ahmed et al. (Eds.): SCA 2023, LNNS 906, pp. 158–167, 2024.
https://doi.org/10.1007/978-3-031-53824-7_15

with speeds below 5 km/h, can produce emissions 5 to 9 times higher than uncongested roads with speeds over 50 km/h. Without congestion, CO, HC, and NOx emissions dropped by 12 to 28%.

The West Bank (WB) area of Palestine has been subject to long-term mobility restrictions, causing significant difficulties for people traveling within and between cities. These restrictions began about thirty years ago with the establishment of permanent and temporary checkpoints [5, 6] and the construction of a separation wall [7]. Recently, a new type of mobility restriction has emerged that poses a threat to the safety of travelers, which is settlers-related violent incidents. The violent actions range from blocking roads, throwing stones at cars, physical assault against travelers, and using live ammunition [8]. According to OCHA report [8], the year 2022 witnessed an unprecedented increase in settlers' violence, with an average of 6.6 injuries per day. Around 21% of total settlers-related incidents are related to violence on vehicles, drivers and passengers, and road blockages [9]. So, settlers' violence is a dynamic risk threatening Palestinian mobility and could cause physical and human loss [10].

The mobility restriction system in WB caused severe disturbances in the daily life of the population, with such adverse effects as anxiety, increased physical risk, time losses, and decreased employment opportunities. They also significantly increased transport route distance and time, resulting in higher energy consumption and greenhouse gas emissions. Cali and Miaari [11] analyzed the impact of mobility restrictions on employment. They showed that checkpoints reduced employment opportunities, the number of working days, and working wages. For example, installing a checkpoint ten minutes from a Palestinian locality reduced employment opportunities and working days by 0.14 and 0.22 percentage points, respectively.

Aburas and Shahrour [12] assessed the environmental consequences of mobility restrictions in one governorate of the WB. The researchers found that these restrictions have a significant impact on energy consumption and CO_2 emissions. Gasoline vehicles, for instance, experienced a surge in energy consumption ranging from 67% to 731%, with an average of 275%, while CO_2 emissions increased similarly. For diesel vehicles, the increase in CO_2 emissions was even more pronounced, ranging from 101% to 881%, with an average of 358%.

While there is a vast amount of literature on the socioeconomic and environmental impacts of mobility restrictions in the WB, there is a notable lack of research exploring the strategies to mitigate their negative effects. This is largely due to the lack of access to real-time and historical data on the behavior and functionality of mobility restrictions. The difficulty in accessing this information has long been a challenge for travelers, government officials, and transportation authorities in the WB.

Accessing real-time information on mobility restrictions has become an urgent need in WB. However, relying on common commercial apps such as Waze is not an effective solution due to its limited coverage [13]. Waze may not provide accurate and up-to-date information for all areas in Palestine, particularly in Palestinian built-up areas. Additionally, the accuracy of the information may be questionable. The road network in the WB is complex, with many roads and checkpoints that are not well-marked or mapped. As a result, users may not be able to accurately report the location and severity of the restrictions. Moreover, data ownership is limited, as Waze is owned by an international company, meaning that the data generated by users may not be owned or controlled by local communities or transportation authorities in Palestine.

Travelers in WB have developed a self-coping mechanism to stay informed about mobility restrictions by creating customized groups and channels on social media platforms. However, this approach has limitations, including limited accessibility to geolocated data as most of the shared data is textual or image-based, limited coverage, as the data is limited to users who actively participate in the groups and chats, lack of structure, which makes it difficult to gather and analyze data consistently and meaningfully, and limited data analysis since social media groups and chats do not provide transportation authorities with a centralized platform to collect and analyze road traffic data. This makes it challenging to monitor traffic patterns and make informed decisions about traffic management.

Palestinian startups, like Doroob [14], have developed location-based applications for sharing road traffic data. Doroob offers navigation services based on traffic reports but doesn't specifically address incidents related to mobility restrictions like checkpoints or settler violence. To address the lack of historical and real-time data on mobility restrictions in the West Bank, this study provides a novel contribution by ensuring free access to real-time information on mobility restrictions and traffic conditions. It uses spatial crowdsourcing to map checkpoints, road gates, traffic congestion, and settler violence in real-time. Additionally, it introduces a real-time Restriction Notification System (RNS) to inform users of any changes in mobility restrictions.

Providing users with access to accurate and up-to-date information will enable them to make informed decisions about their travel, ultimately reducing the socioeconomic and environmental costs associated with mobility. Furthermore, this will create for the first time a valuable spatial database of mobility restrictions that helps in understanding the behavior of mobility restrictions, determining the weighted waiting time, and creating prediction models. This will maintain the performance of the mobility system and enhance its resilience [15].

2 Methods

This section outlines the methodology used to create real-time mobility restrictions service RT-MMR and build the restriction notification system RNS service. It includes three main steps: (i) identifying the data sources and collection methods; (ii) processing and analyzing the data, and (iii) publishing the service for public use.

2.1 Data Sources and Collection

Real-time mapping of mobility restriction services mainly depends on crowdsourced data transmitted from interurban travelers via the SRMS platform. The service also relies on the External Spatial Database (ESD) to verify crowdsourced data and provide a base map of interurban road networks and population communities. To accurately provide information about an event, the system requires an event description, location, and time. The crowd can report using the ArcGIS Survey123 embedded crowdsourcing tool in the SRMS platform. Survey123 was released by Esri in 2016 [16] and has been used sparingly as a GIS data collection tool ever since [16, 17]. It is a simple and intuitive form-centric data-gathering tool with the power of publishing results in real-time through the Web Feature Service (WFS) in the web GIS.

ArcGIS Survey123 stands out in data collection, analysis, and visualization when compared to other crowdsourcing apps like Ushahidi, Maptionnaire, Open Data Kit, and GIS Cloud [18]. The study [18] assessed various categories, including data input, management, analysis, visualization, and costs. Survey123 excels by offering web and mobile apps compatible with Android and IOS devices, making it unique. It also leads in data management, providing a built-in database, support for data entry removal, mass deletions, sorting, filtering, and single-entry editing, along with multiple supported format options. Survey123 also outperforms data analysis and offers robust visualization capabilities compared to other crowdsourced apps.

Besides the crowdsourced data needed to deliver event real-time mapping, the system relies on static data to verify the crowdsourced data. Hence, the platform depends on the External Spatial Database ESD. ESD is a repository for the processing and storing of open-source and authoritative data. The open-source data is presented in the monitoring information provided by NGOs to describe the mobility restrictions. The monitoring reports describe fixed mobility restrictions, locations, photos, and operations.

The authoritative data is presented mainly in the spatial data that forms the base map of SRMS, including interurban road networks and population communities. The authoritative data is obtained from a higher governmental transportation body, such as the Ministry of Transport (MoT) and the Ministry of Public Work and Housing (MPWH). The open-source and authoritative data was obtained in the Esri vector data storage format (Shapefile). ESD independently processed the spatial data using GIS capabilities, including georeferencing, digitizing, classifying, converting, geoprocessing, and removing duplication and missing data. Table 1 shows the datasets used in the real-time mapping of mobility restriction.

2.2 Data Processing

This phase concerns processing and preparing the crowdsourced data in ArcGIS Online, which is a cloud-based software-as-a-service (SaaS) platform for creating, sharing, and managing geographic information using cloud-based servers and infrastructure [19, 20]. The data transmitted from Survey123 will be stored in the ArcGIS Online cloud server, which is a certified trusted server [21, 22] that uses the cloud infrastructure of Microsoft Azure and Amazon Web Services (AWS); therefore, users' data will be stored within them.

The transmitted data will be stored as a hosted feature layer service, which is ideal for storing road information data because it enables adding, editing, and deleting data in real-time. The hosted feature service will process the reported data in real-time through configuring the instant filtering and validation rules [23]. For example, the reports with missing or manipulated or those transmitted from malicious users will be detected and ignored by the service. Validated reports from authenticated users will be visualized on the SRMS base map [24] with specific symbology based on the type of mobility restriction, and each report will include event type, time stamp, and audio data for further description.

Table 1. Data sources for mapping the mobility restrictions.

Dataset	Source	Data type	Description	Dataset purpose
Crowdsourced data	SRMS Application	Crowdsourcing	Citizen's entries with time, location, and description	Mapping
Mobility restriction reports	NGOs	Open-source data	Textual and spatial data contains a description of fixed restrictions, photos, location, and operation mechanisms	Mapping, verification
Mobility infrastructure	High governmental transportation bodies	Authoritative data	Spatial data of the transportation infrastructure and the built environment	Mapping, verification

Besides storing the data on a trusted cloud server to ensure the privacy of the user's data, the SRMS has developed a privacy-based policy [25], that explains the rules for using the users' data and sharing information and security. Furthermore, tightening the accessibility of sensitive user data, such as users' location, for reporting purposes only, so the live location of the users is not necessary for the platform's operation. Hence, prevents an attacker from tracking the user's trajectory. The reported data will be presented as anonymous users with random IDs without revealing the users' identities (Fig. 1).

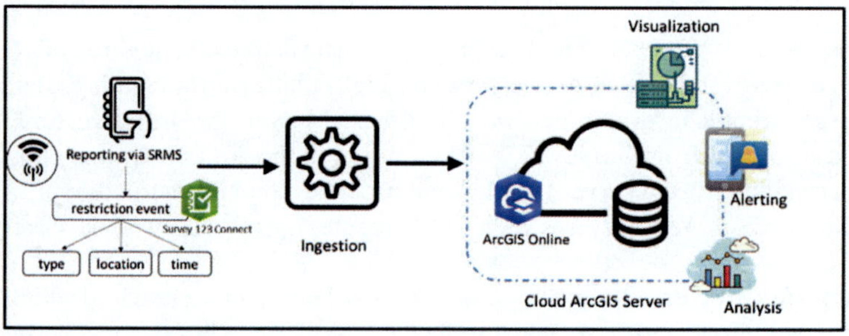

Fig. 1. Data Storage and Processing in ArcGIS Online.

The service of mapping restrictions is optimized by developing the Restriction Notification System (RNS). RNS aims to inform SRMS users of updates regarding one or multiple restriction types they have subscribed to. The RNS service sends email notifications to users when new reports related to mobility restrictions are added, deleted, or modified. The RNS service was developed using a Python script and a JSON configuration file [26]. The script uses the Requests Python module to send HTTP requests [27] to a feature service endpoint and retrieves the maximum date and time of edits in the feature service. It compares this with the last edit detected by the script and sends email notifications to recipients listed in the configuration file if there are any new edits.

The configuration file is a JSON object that contains information related to an email, a service, and a filename. The email section includes; the list of subscribed suers' emails that will be notified; (ii) the email of senders, which could be a person or organization; (iii) the server, which contains settings for configuring the email server connection, including the host of email server which can be specified by hostname or IP address, port of the server; (iv) The mail text and subject. The service section includes the feature service URL, the service username and password, the layer number, and the viewer URL and level. Filename contains the name of a JSON file that contains the layer information.

In order to maintain the continuous operation of the RNS service, it is necessary to host both the Python script and the configuration file on a machine that is constantly connected to the internet. The Windows operating system's Task Scheduler is then used to schedule the script to run automatically every 5 min. The development methodology for the RNS service is illustrated in Fig. 2.

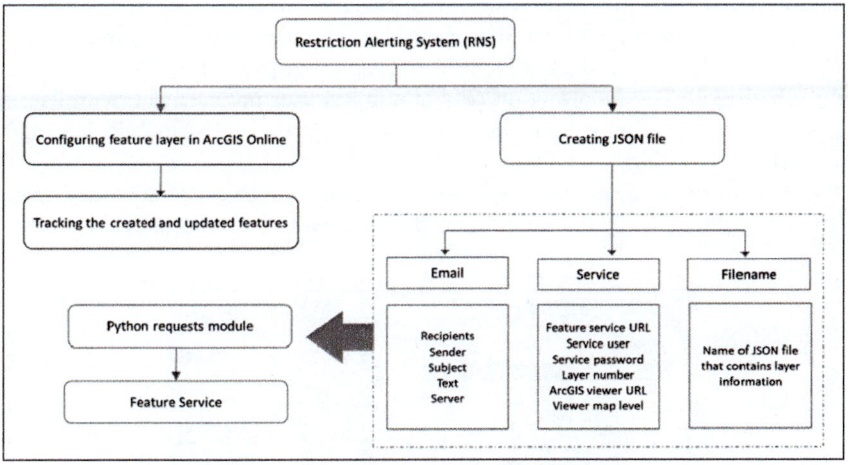

Fig. 2. Method of developing RNS Service.

2.3 Service Publishing

This section concerns disseminating and visualizing the processed data stored in the hosted feature layer for public use. This phase will enable users to access information about mobility restrictions and adjust their travel plans accordingly. The hosted feature service configuration is required for sharing the reported data with the public. It includes configuring the accessibility of the users to the data by assigning the feature service to support public data collection, which enables users to add or modify their data. The real-time mapping service will be published as a reporting widget on the SRMS web mobile application. The published data includes information about the location, time, descriptive audio, and type of mobility restriction or traffic congestion. Users can access this information through the SRMS mobile application, which displays an interactive map with icons representing the reported incidents. Users can access detailed information about the incident by clicking on these icons.

3 Application

RT-MMR service is provided via the SRMS platform web mobile application, a type of app that can be accessed through a mobile web browser without needing to be downloaded from an app store or installed on a device. This approach ensures compatibility across various devices and operating systems, making it cost-effective and flexible for developers seeking a broad user base [28].

RT-MMR service was implemented using Web Experience Builder (WEB), which is a user-friendly web development platform that allows users to create and share web applications, maps, and dashboards without extensive coding knowledge [29]. The reported data is visually presented with custom colors based on event types (Fig. 3). The map displays current events like checkpoints, road closures, and traffic jams reported in the WB. Users can zoom in, out, and filter events by type using the interactive legend on the right side. This action directs them to the reporting page (Fig. 3) to review data before submission. The reporting process minimizes user involvement, with date and time auto-filled as read-only fields, location auto-detected via GPS mobile data, and an optional voice recording.

The RNS service, developed in Python 3.11, resides on a continuously running desktop computer. Users can subscribe to the service by selecting an icon on the UI

Fig. 3. Reporting page and mapped checkpoints using RT-MMR.

and choosing one or more restriction types (Fig. 4). Subscribed users receive email notifications for updates, showing the type and location of updated restrictions (Fig. 4).

The RT-MMR service's validation involved cross-referencing information from WB road traffic social media channels with data reported through the SRMS platform. The outcomes confirm the RT-MMR service's effectiveness in providing real-time and precise data. Nevertheless, a more comprehensive discussion of service validation will be included in future research.

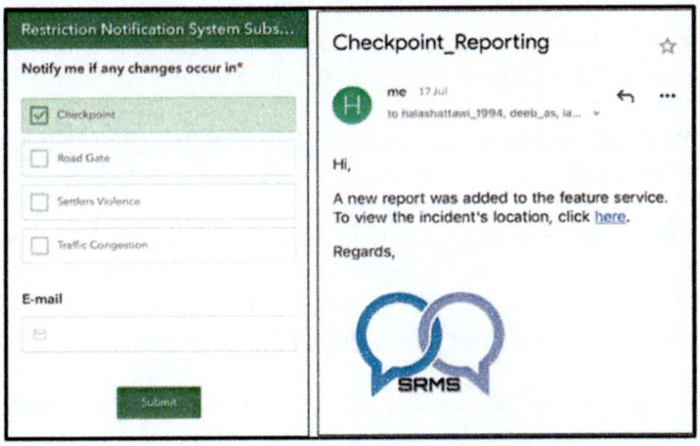

Fig. 4. RNS subscription system, and e-mail body sent to RNS subscribers.

4 Conclusion

Using spatial crowdsourcing to generate real-time maps of mobility restrictions has emerged as a promising solution to the challenges posed by these restrictions. This approach offers several potential benefits, including improved access to information for interurban commuters to make informed decisions about routes and timing, enhanced safety and security by alerting travelers to potential hazards or security risks, and optimization of resources for transportation authorities and emergency responders to respond more quickly to incidents and emergencies. This study proposes a novel approach employing spatial crowdsourcing to map real-time road restrictions and traffic conditions. This will create for the first time a real-time and historical database documenting the devour of the restrictions, which will help develop coping strategies to mitigate the adverse impacts of mobility restrictions on the daily lives of Palestinians.

References

1. Huang, X., Wang, X., Pei, J., Xu, M., Huang, X., Luo, Y.: Risk assessment of the areas along the highway due to hazardous material transportation accidents. Nat. Hazards **93**, 1181–1202 (2018). https://doi.org/10.1007/s11069-018-3346-4

2. Arrighi, C., Pregnolato, M., Castelli, F.: Indirect flood impacts and cascade risk across interdependent linear infrastructures. Nat. Hazards Earth Syst. Sci. **21**, 1955–1969 (2021). https://doi.org/10.5194/nhess-21-1955-2021
3. Kurth, M., et al.: Lack of resilience in transportation networks: economic implications. Transp. Res. Part D Transp. Environ. **86**, 1–14 (2020). https://doi.org/10.1016/j.trd.2020.102419
4. Chen, X., et al.: Quantifying on-road vehicle emissions during traffic congestion using updated emission factors of light-duty gasoline vehicles and real-world traffic monitoring big data. Sci. Total Environ. **847**, 157581 (2022). https://doi.org/10.1016/j.scitotenv.2022.157581
5. Griffiths, M., Repo, J.: Women and checkpoints in Palestine. Secur. Dialogue **52**, 249–265 (2021). https://doi.org/10.1177/0967010620918529
6. Rijke, A., Minca, C.: Inside checkpoint 300: checkpoint regimes as spatial political technologies in the occupied palestinian territories. Antipode **51**, 968–988 (2019). https://doi.org/10.1111/anti.12526
7. Habbas, W., Berda, Y.: Colonial management as a social field: The Palestinian remaking of Israel's system of spatial control. Curr. Sociol. 2–28 (2021). https://doi.org/10.1177/00113921211024695
8. OCHA: Data on casualties. https://www.ochaopt.org/data/casualties. Accessed 13 Feb 2023
9. B'Tselem: Settler Violence in the WB. https://www.btselem.org/settler_violence_updates_list?f%5B2%5D=nf_district%3A181&f%5B3%5D=nf_type%3A173&f%5B4%5D=date%3A%28min%3A1640995200%2Cmax%3A1672444800%29&page=1. Accessed 22 Jan 2023
10. UNHRC: UN experts alarmed by rise in settler violence in occupied Palestinian territory. https://www.ohchr.org/en/press-releases/2021/11/un-experts-alarmed-rise-settler-violence-occupied-palestinian-territory. Accessed 14 March 2023
11. Calì, M., Miaari, S.H.: The labor market impact of mobility restrictions: evidence from the West Bank. Labour Econ. **51**, 136–151 (2018). https://doi.org/10.1016/j.labeco.2017.12.005
12. Aburas, H., Shahrour, I.: Impact of the mobility restrictions in the palestinian territory on the population and the environment. Sustain. **13**, (2021). https://doi.org/10.3390/su132313457
13. Miriam Berger: When Waze Won't Help, Palestinians Make Their Own Maps. https://www.wired.com/story/palestine-jerusalem-mapping/. Accessed 24 June 2023
14. Doroob. http://www.doroob.net. Accessed 20 Sept 2022
15. Wang, T., Xie, X., Cao, X., Pedersen, T.B., Wang, Y., Xiao, M.: On efficient and scalable time-continuous spatial crowdsourcing. In: 2021 IEEE 37th International Conference on Data Engineering (ICDE), pp. 1212–1223 (2021). https://doi.org/10.1109/ICDE51399.2021.00109
16. Esri: ArcGIS Survey123. https://doc.arcgis.com/en/survey123/reference/whatissurvey123.htm#:~:text=Survey123. Connect provides the survey, the survey and its behavior. Accessed 10 March 2023
17. Jordan, E.J., et al.: Current Issues in Tourism Does tourism really cause stress ? A natural experiment utilizing ArcGIS Survey123. Curr. Issues Tour. 1–15 (2019). https://doi.org/10.1080/13683500.2019.1702001
18. Jay, B., Alberta, S., Royal, M.: The tools of citizen science: an evaluation of map-based crowdsourcing platforms Chickadee Technology. Spat. Knowl. Inf. Canada. **7** (2019)
19. Kholoshyn, I.V., Bondarenko, O.V., Hanchuk, O.V., Shmeltser, E.O.: Cloud ArcGIS Online as an innovative tool for developing geoinformation competence with future geography teachers. CEUR Workshop Proc. **2433**, 403–412 (2019). https://doi.org/10.55056/cte.401
20. Esri: ArcGIS Trust Center. https://trust.arcgis.com/en/security/cloud-options.htm#:~:text=ArcGIS. Online is a cloud, being mobile in the field. Accessed 15 March 2023
21. Yuan, D., Li, Q., Li, G., Wang, Q., Ren, K.: PriRadar: a privacy-preserving framework for spatial crowdsourcing. IEEE Trans. Inf. Forensics Secur. **15**, 299–314 (2020). https://doi.org/10.1109/TIFS.2019.2913232

22. Wang, Z., et al.: Privacy-preserving crowd-sourced statistical data publishing with an untrusted server. IEEE Trans. Mob. Comput. **18**, 1356–1367 (2019). https://doi.org/10.1109/TMC.2018.2861765
23. Esri: Hosted feature layers. https://developers.arcgis.com/documentation/mapping-apis-and-services/data-hosting/hosted-feature-layers/. Accessed 7 April 2023
24. Esri: ArcGIS Online. https://www.esri.com/en-us/arcgis/products/arcgis-online/overview. Accessed 20 Dec 2022
25. SRMS: SRMS Privacy Policy. https://app.notice.studio/editor?workspace=25406a57-a1d7-471d-9d93-758a388f5fdf&project=6c2f34d6-9ec4-4013-b176-950c76172e89&page=6c2f34d6-9ec4-4013-b176-950c76172e89. Accessed 18 June 2023
26. Esri: A simple e-mail notification system for Survey123 for ArcGIS. https://community.esri.com/t5/arcgis-survey123-blog/a-simple-e-mail-notification-system-for-survey123/ba-p/895031. Accessed 28 April 2023
27. Pypi.org: Requests: HTTP for Humans. https://docs.python-requests.org/en/latest/index.html. Accessed 23 May 2023
28. Tandel, S.S., Jamadar, A.: Impact of progressive web apps on web app development. Int. J. Innov. Res. Sci. Eng. Technol. **7**, 9439–9444 (2018). https://doi.org/10.15680/IJIRSET.2018.0709021
29. Esri: ArcGIS Experience Builder. https://www.esri.com/en-us/arcgis/products/arcgis-experience-builder/overview. Accessed 22 Feb 2023

Comparative Analysis of ITS-G5 and C-V2X for Autonomous Vehicles with an Improved Algorithm of C-V2X

Kawtar Jellid(✉) and Tomader Mazri

National School of Applied Sciences, Kenitra, Morocco
kawtarjellid@gmail.com

Abstract. Autonomous vehicles rely on sensors, radars, and lidar entertainment devices to operate. They provide a safer solution for traffic safety, both for the driver and nearby cars. However, they also pose a challenge, as ensuring safety necessitates taking all necessary precautions to guarantee flawless functioning and communication of the autonomous vehicle. Consequently, numerous research investigations have been conducted on the communication protocols employed by self-driving cars. These studies have examined the vulnerabilities of each protocol to propose enhancements and refine existing ones.

In our prior research papers, we have conducted comparative analyses among various communication protocols using software such as OMNET++. The critical factor that determines the resilience of a protocol is its response time, which ensures effective communication between the protocols. Some of the protocols examined include C-V2X, DSRC, ITS-G5, and 5G. This article aims to conduct a comparative study between the C-V2X and ITS-G5 protocols based on criteria such as latency, synchronization, resource selection, and response time. This comparison is crucial for improving the performance of both protocols and ensuring efficient, reliable, and smooth communication between vehicles and the environment.

Keywords: ITS-G5 · C-V2X · ITS · V2V · V2I · Autonomous car

1 Introduction

Automated technology has long been an appealing concept. It holds the potential to revolutionize our commuting experiences and long-distance travel, extracting individuals from hazardous work environments while streamlining our industries. It plays a vital role in the construction of safer cities and travel. In this context, we refer to self-driving vehicles as automobiles that enable autonomous driving without human intervention. These vehicles are expected to drive the next transformative shift in the transportation field. Although the benefits and challenges associated with their implementation are still under critical evaluation and discussion, major technology companies and car manufacturers have been actively engaged in a race to develop the first operational vehicle for many years now [1]. Autonomous vehicles are likely to substantially increase travel

M. Ben Ahmed et al. (Eds.): SCA 2023, LNNS 906, pp. 168–177, 2024.
https://doi.org/10.1007/978-3-031-53824-7_16

for non-drivers. By enhancing passenger comfort and productivity, autonomous operation can make long-distance journeys, including commutes, more tolerable, resulting in increased vehicle travel and urban expansion [2].

The intelligent transportation infrastructure facilitates the provision of essential digital automation services that can be applied and utilized in real-time to fulfill the operational requirements of autonomous vehicles. These services and automations are capable of adapting autonomously to infrastructure enhancements and the specific conditions of each vehicle and element, taking into account their environmental context [3]. Vehicular ad hoc networks (VANETs) are indispensable for autonomous cars as they contribute to safe driving by optimizing traffic flow and reducing car accidents. This advantage is achieved through the sharing of pertinent information. However, any alteration in this real-time information can lead to system failure, compromising the safety of road users. This highlights the critical importance of the response time criterion [4]. Thus, VANETs establish a network of mobile devices and vehicles, enabling vehicle-to-infrastructure (V2I) and vehicle-to-vehicle (V2V) communication. The primary objective of this technology is to enhance road safety by facilitating effective communication, for instance, in scenarios such as traffic congestion and accidents, where vehicles can exchange vital and essential information with one another and with the network [5].

For efficient communication, autonomous cars rely on communication protocols such as DSRC (Dedicated Short-Range Communications), C-V2X (Cellular Vehicle-to-Everything), 5G Generation, and ITS-5G (5G Intelligent Transportation System). In this article, we conduct an in-depth study of the two protocols, C-V2X and 5G-ITS, with a particular focus on the response time criterion, which represents the most crucial aspect for prompt and efficient information reception, facilitating communication within autonomous cars and between two vehicles.

2 CV2X

In this initial segment, we will introduce the notion of the C-V2X Cellular Vehicle-to-Everything protocol, a robust technology for delivering mobility and safety services. Additionally, we will examine the various categories of C-V2X applications and the services facilitated by the C-V2X protocol architecture.

2.1 Concept of C-V2X

The standardization of Cellular Vehicle-to-Everything (C-V2X) technology by 3GPP was completed in 2017. This cellular technology, which is built upon LTE, enables vehicles to establish connections with one another, road infrastructures, cloud-based services, and other road users. C-V2X has been designed to be compatible with future 5G mobile technologies and is based on the 3rd Generation Partnership Project (3GPP) version 14. This communication technology has recently gained significant attention due to its ability to meet V2X communication requirements. For the successful operation of C-V2X [6], 3GPP identifies four distinct types of C-V2X applications: V2V (Vehicle-to-Vehicle), V2I (Vehicle-to-Infrastructure), V2P (Vehicle-to-Pedestrian), and

V2N (Vehicle-to-Grid). These applications are depicted in Fig. 1 [7]. The C-V2X protocol ensures robust and reliable communication for vehicle connectivity. With the increasing support from an extensive stakeholder ecosystem that includes the automotive industry, system-on-a-chip (SoC) manufacturers, and Tier 1 vendors, C-V2X emerges as a formidable competitor to DSRC (Dedicated Short Range Communication Protocol), which has been developed based on the IEEE 802.11p standard [8].

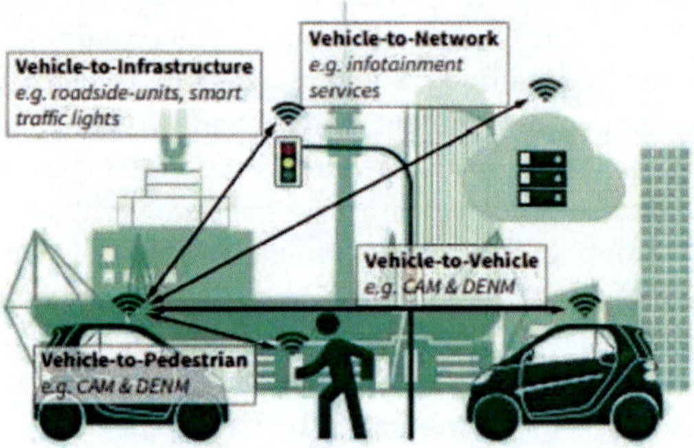

Fig. 1. Caption types of C-V2X applications.

2.2 C-V2X Services

- It extends the capabilities of existing cellular network components and functionalities to ensure minimal delay.
- This protocol facilitates reliable and robust communication between different communication systems, encompassing vehicle-to-pedestrian (V2P), vehicle-to-vehicle (V2V), vehicle-to-network (V2N), and vehicle-to-infrastructure (V2I) communications.
- It provides a user-friendly understanding of the comprehensive system, supporting diverse V2X applications and a wide range of V2X scenarios.
- C-V2X plays a crucial role as an enabling technology, transitioning from individual vehicle intelligence to connected intelligence [8].

2.3 C-V2X Architecture

The fundamental structure of the C-V2X protocol consists of two main components: the physical layer and the 3GPP Rel-14 V2X Scope. Additionally, the protocol leverages other standards, as depicted in the figure. A notable addition introduced by this protocol is the PC5 interface, which works alongside the existing Uu interface in LTE. This interface enables direct communication between devices, regardless of the presence of an eNodeB (eNB) interface. V2X communications rely on the side link period, which ranges from 40 ms to 320 ms, corresponding to 40-320 subframes. During this sidelink subframe period, any vehicle can transmit data twice, with one blind re-transmission on two selected subframes within the time domain (Fig. 2).

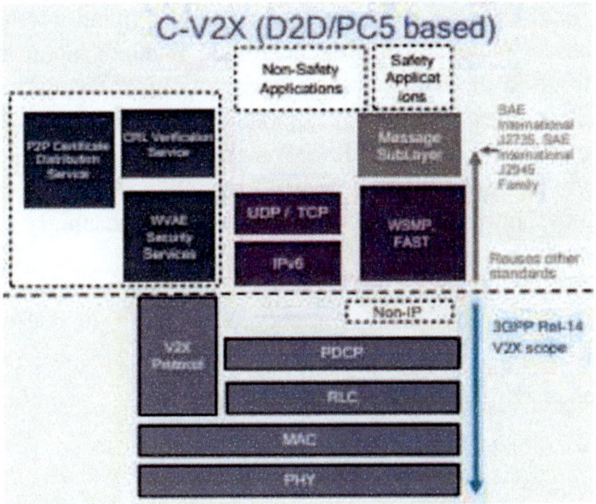

Fig. 2. Basic architecture of the cv2x protocol.

Dedicated resource pools are allocated for sidelink transmission in each sidelink subframe. The vehicle dynamically selects a subset of these physical resource blocks (PRB) resource pools for transmission [9].

To summarize this section, in order to analyze the weaknesses of the C-V2X protocol, it is essential to understand its operation, historical background, evolution, modes of communication, and particularly its architecture, including the physical architecture.

3 C-ITS

A wireless variant known as ITS-G5, primarily used in Europe, has been developed. In this subsequent section, we will delve into the fundamental aspects of the ITS-G5 protocol, including its principles, services, and architecture.

3.1 Concept of C-ITS

Recently, cooperative intelligent transportation systems (ITS-C) have emerged as a solution to enhance vehicular communication systems. These systems aim to provide a range of services that optimize traffic management and ensure road safety. They achieve this by facilitating data exchanges between vehicles, as well as between vehicles and other road users, including pedestrians and centralized infrastructure. Several wireless technologies are currently available to support communication in vehicular networks. Among the most effective wireless technologies for ITS services are ITS-G5 and LTE-V2X.

ITS-G5 serves as the European standard for vehicular communications, introduced by the European Telecommunications Standards Institute (ETSI). It is based on the IEEE standard [10]. ITS-G5 is an access technology specifically designed by ETSI for vehicle-to-vehicle and vehicle-to-infrastructure communications. Messages from ITS services, whether related to safety or non-safety applications, are encapsulated in a Cooperative

Awareness Message (CAM) and a Decentralized Environmental Notifications (DENM) message. These messages are further encapsulated in Geographic Networking messages and transmitted via the Basic Transport Protocol (BTP) at the access layer. The transmission is facilitated through decentralized congestion control (DCC) [11].

C-V2X plays a crucial role in the development of cooperative intelligent transport systems (C-ITS), which aim to address challenges such as pollution and congestion. By utilizing C-ITS, cities can become smarter and support increasingly secure, automated, and efficient transportation systems.

3.2 ITS-5G Services

Amidst the services provided by STI-C:

- Alerting about road hazards
- Enhancing driving assistance

The aim of these services is to provide real-time information to vehicles regarding incidents, similar to variable message signs installed on roads. However, they offer the flexibility of delivering personalized and dynamic information to selected geographic areas. One notable example is the Traffic Ahead service, which sends messages about detected traffic incidents to specifically targeted regions using DEN messages.

The availability of these services enables the Traffic Management Center (TMC) to send C-ITS notifications to connected and autonomous vehicles. Irrespective of the automation level of connected vehicles, these services provide information that surpasses the visual horizon limitations faced by vehicle sensors and human drivers. This, in turn, ensures safety and enhances the efficiency of road transportation across all levels [12, 13].

3.3 ITS-G5 Architecture

The frequency band defined by ITS-G5A is used by ITS road safety applications, with the Control Channel (CCH). Other services use other channels, such as ServiceChannel (SCH). Today, the ITS-G5 technology makes it possible to broadcast traffic messages called CAM (Cooperative Awareness Message) and event messages called (DENM: Decentralized Environmental Notification Message) [14] (Fig. 3).

Different situations are studied to guarantee the associated services and road safety according to the use cases [15], The IEEE 802.11p MAC layer is based on the OCB Offset code book mode of operation where association, authentication and data confidentiality services are not used. Ocb is a mode that adapts to the rapid dissemination of short messages and it allows great mobility [16].

Finally, ITS-G5 defines the MAC and PHY layers of the open systems interconnection (OSI) architecture which is based on orthogonal frequency division multiplexing (OFDM) carrier sense multiple access with collision avoidance (CSMA/CA).

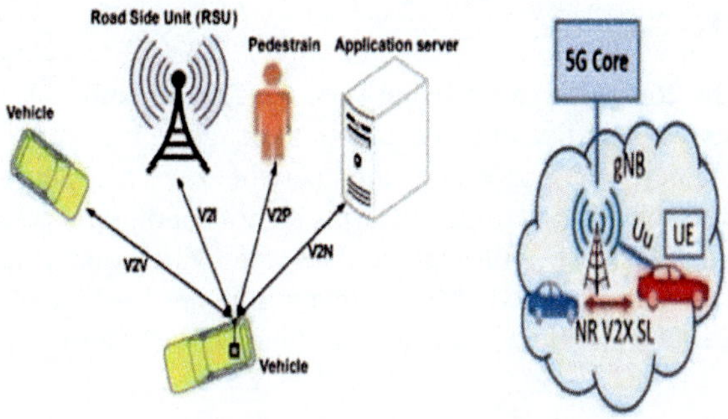

Fig. 3. This figure representing the ITS-5G architecture.

4 Comparison ITS-G5 VS CV2X

In this section we have carried out a comparative study between the two protocols C-V2X and ITS-G5 based on main criteria such as latency and Data rate also a study on the possibility of coexistence between these two protocols on the same band.

4.1 Coexistence Between c-v2x and ITS-G5

C-V2X has the capability to transmit a message over a longer period compared to ITS-G5. It offers greater flexibility in terms of the code rate, resulting in improved efficiency and robustness with a lower effective transmission rate [17]. Furthermore, the organization of radio resources into orthogonal slots introduces a higher bounded access delay, which contributes to the overall performance efficiency of C-V2X and is dependent on the allocation process [18] (Fig. 4).

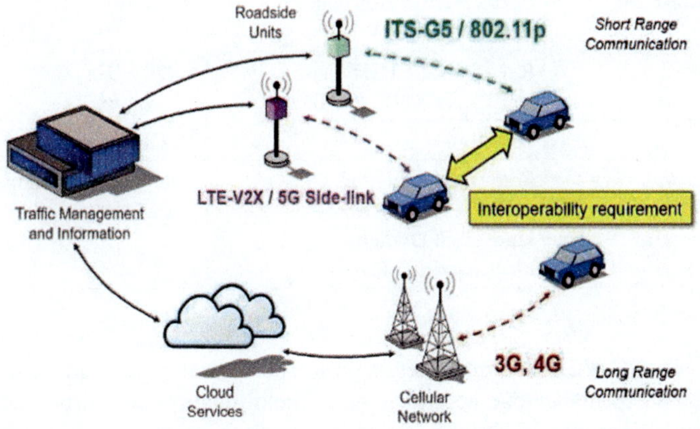

Fig. 4. C-V2X and IEEE 802p standards coexistence in the 5.9 GHz band.

4.2 C-V2X vs. ITS-G5

In this section of the comparative study between the ITS-G5 and C-V2X protocols, we will focus on several key indicators to evaluate their respective performances. Firstly, we will examine latency, which measures the delay between message transmission and reception. Reduced latency is crucial for ensuring smooth communication in autonomous vehicle applications. Next, we will analyze data rate, which quantifies the amount of information that can be transmitted within a given time unit. A higher data rate enables faster exchange of information between vehicles and infrastructures (Table 1).

Table 1. Comparison study between C-V2X and IT5-G5 based on parameters.

Parameters	C-V2X	ITS-G5
Waveform	Single-Carrier Frequency Division Multiple Access SC-FDM	Orthogonal Frequency Division Multiplex OFDM
Modulation support	UP to 64 QAM	Up to 64 QAM
Mimo-Support	Rx: diversity for 2 antennas mandatoryTx: diversity for 2 antennas supported	No support standardized
Ressource-selection	Semi persistent transmission with relative energy-based selection	CSMA-CA(Carrier Sense Multiple Access with Collision Avoidance)
Data channel coding	Turbo	Convolution
Latency	20 ms	<5 ms
synchronization	Synchronous	Asynchronous
Harq Retransmission mechanism	Harq (Hybrid Automatic Repeat Request)	No Harq
Chqnnel Size	Rel 14:10,20 MHz Rel 15:Nx20,10,20 MHz	10/20 MHz
Data Rate	150Mbps	10 Gbps
Ressource multiplexing across vehicles	Frequency-division multiplexing (FDM)and Time Division Multiplexing (TDM)	Time Division Multiplexing (TDM)

In conclusion, C-V2X demonstrates superior performance at high vehicle speeds. This improved performance can be attributed to the utilization of a turbo coding mechanism, which provides optimal channel estimation by employing a higher number of DMRS symbols. It is worth noting that ITS-5G offers low latency, which is a crucial criterion for seamless communication. Additionally, there is increased flexibility in the code rate, further enhancing the system's adaptability.

5 Improvement of C-V2X Algorithm

The main disadvantage of the C-V2X protocol is its dependence on cellular network infrastructure. Unlike ITS-G5, which operates on a dedicated frequency band, C-V2X relies on existing cellular networks for communication. This reliance on cellular infrastructure introduces potential challenges in terms of network coverage, reliability, and latency, especially in areas with limited cellular connectivity or high network congestion. Additionally, the use of cellular networks for C-V2X communication may incur additional costs and dependencies on service providers.

In order to enhance communication reliability and efficiency, an algorithm is proposed that intelligently selects the optimal communication channel based on factors such as network congestion, signal quality, and available bandwidth. This algorithm employs dynamic channel switching, allowing seamless transitions between different channels or frequencies. By continuously monitoring network conditions and evaluating relevant metrics, such as congestion levels, signal strength, and available bandwidth, the algorithm dynamically adapts the communication channel to maximize performance. This approach ensures that the system can intelligently and automatically select the most suitable channel for reliable and efficient communication. The algorithm is implemented using C++ programming language, providing a flexible and efficient solution for channel selection in communication systems.

```
struct Channel {
    int id;
    double signalQuality;
    int bandwidth;
    int congestionLevel;};
int main() {
    std::vector<Channel> channels;
    // Populate the channels with sample data
    // Initialize the optimal channel to the first channel
    Channel optimalChannel = channels[0];
    // Iterate through the channels and find the optimal one
    for (const auto& channel : channels) {  // Calculate a score for the channel based
on the factors
        double score = channel.signalQuality * channel.bandwidth / channel.conges-
tionLevel;
        // Check if the current channel has a higher score than the optimal channel
        if (score > optimalChannel.signalQuality * optimalChannel.bandwidth / opti-
malChannel.congestionLevel) {
            optimalChannel = channel}  }
    // Print the optimal channel
    std::cout << "Optimal Channel: " << optimalChannel.id << std::endl;
    return 0;}
```

6 Conclusion

C-V2X offers significant advantages over other vehicle connectivity technologies, supporting a wide range of use cases and ensuring navigation, security, and integrated transport systems. It utilizes direct communication for exchanges without cellular coverage and provides redundancy as needed. While ITS-G5 achieves lower latency in core network traversal, research is exploring the coexistence of these protocols. The proposed C++ algorithm intelligently selects the optimal communication channel based on factors like network congestion, signal quality, and available bandwidth. It dynamically switches channels for reliable and efficient communication in diverse network conditions, optimizing system performance and enhancing overall reliability and efficiency.

References

1. Narayanana, S., Chaniotakisb, E., Antonioua, C.: Shared autonomous vehicle services: a comprehensive review, vol. 111, pp. 255–293 (2020)
2. Litman, T.: Autonomous Vehicle Implementation Predictions, Implications for Transport Planning 9 May (2022)
3. Lee, L.: The Future of Transportation A system of systems perspective on the future of thecar, NextCurve, 15 August (2021)
4. Hasrounyab, H., Ellatif, A., Bassilc, S.C., Laouitia, A.: VANet security challenges and solutions: a survey. Vehicul. Commun. **7**, 7–20 (2017)
5. Nadeem Ahangar, M., Ahmed, Q.Z., Khan, F.A., Hafeez, M.: A survey of autonomous vehicles: enabling communication technologies and challenges. Sensors **21**(3), 706 (2021). https://doi.org/10.3390/s21030706
6. Shah, G., Saifuddin, Md, Fallah, Y.P., Gupta, S.D.: RVE-CV2X: a scalable emulation framework for real-time evaluation of CV2X-based connected vehicle applications. In: 2020 IEEE vehicular networking conference (VNC) (2020)
7. Eckermann, F., Kahlert, M., Wietfeld, C.: Performance analysis of C-V2X Mode 4 communication introducing an open-source CV2X simulator. In: 2019 IEEE 90th vehicular technology conference (VTC2019-Fall) (2019)
8. Saifuddin, Md., Zaman, M., Toghi, B., Fallah, Y.P., Rao, J.: Performance analysis of Cellular-V2X with adaptive & selective power control. In: 2020 IEEE 3rd connected and automated vehicles symposium (CAVS) (2020)
9. Chen, S., Hu, J., Shi, Y., Zhao, L., Li, W.: A vision of C-V2X: technologies, field testing and challenges with Chinese development. IEEE Internet Things J. **7**(5) (2020)
10. Sattiraju, R., Weinand, A., Schotten, H.D.: Channel estimation in C-V2X using deep learning. In: IEEE international conference on advanced networks and telecommunications systems (ANTS), (2019)
11. Maaloul, S., Aniss, H., Kassab, M., Berbineau, M.: Classification of C-ITS services in vehicular environments. IEEE Access **9**, 117868–117879 (2021). https://doi.org/10.1109/ACCESS.2021.3105815
12. Karoui, M., Freitas, A., Chalhoub, G.: LIMOS-CNRS - University Clermont Auvergne, Performance comparison between LTE-V2X and ITS-G5 under realistic urban scenarios. In: 2020 IEEE 91st vehicular technology conference (VTC2020-Spring) (2020)
13. Naranjo, J.E., Jiménez, F., Gómez, A., Valle, A., González, J., Cruz, A.: Integration of C-ITS in Autonomous Driving, 2020 IEEE Intelligent Vehicles Symposium (IV) October 20–23, Las Vegas, USA (2020)

14. Ahmad, M.: Cellular V2X vs. ITS-G5: who will dominate autonomous car connectivity?, embedded computing (2018)
15. ETSI, EN 302 637 v1.2.1: Intelligent Transport Systems (ITS)– Vehicular Communications; Basic Set of Applications; Part2: Specification of Decentralizes Environmental Notification Basic Service, ETSI Std. (2014). http://www.etsi.org/deli-ver/etsi_en/302600_302699/30263703/01.02.01_30/en_302637 03v010201v.pdf
16. Randriamasy, M., Cabani, A., Chafouk, H., Fremont, G.: Formally validated of novel tolling service with the ITS-G5. IEEE Access **7**, 41133–41144 (2019). https://doi.org/10.1109/ACCESS.2019.2906046
17. Mannoni, V., Berg, V., Sesia, S., Perraud, E.: A comparison of the V2X communication systems: ITS-G5 and C-V2X. In: IEEE 89th vehicular technology conference (VTC2019-Spring) (2019)
18. Sattiraju, R., Wang, D., Weinand, A., Schotten, H.D.: Link Level Performance Comparison of C-V2X and ITS-G5 for Vehicular Channel Models. arXiv:2002.11450 [eess.SP] (2020)

Evaluation of Resilience Based on Resources and Adaptation Level in Critical Transport Infrastructures

N. K. Stamataki[1](✉), D. V. Achillopoulou[2], and N. Makhoul[3]

[1] Laboratory of Reinforced Concrete and Seismic Design of Structures, Democritus University of Thrace, Xanthi, Greece
nstamata@civil.duth.gr
[2] James Watt School of Engineering, University of Glasgow, Glasgow, UK
dimitra.achillopoulou@glasgow.ac.uk
[3] École Spéciale des Travaux Publics, ESTP, Paris, France
nmakhoul@estp.fr

Abstract. Adaptation to climate change is crucial for Critical Transport Infrastructures (CTIs) due to their proximity to completing their life cycle and the lack of design considerations for weather extremes. Considering the insufficient financial resources during the ageing periods of such infrastructure, investment is necessary to support their resilient and sustainable maintenance. The potential need for investment is examined in this paper by utilizing a novel data-driven methodology and a Life Cycle Cost (LCC) analysis to assess the CTIs' resources resilience. Further to this, the effectiveness of the mitigation measures on the infrastructures' adaptative capacity is estimated with the Adaptation Index (AdI), which considers resilience and sustainability levels. Resilience is accounted for three different components related to structural integrity, functionality, and resources. Sustainability is indirectly assessed by weighting the distinct ageing periods of the asset's life cycle and the calculated resilience levels. To document the proposed methodology, a case study Reinforced Concrete (RC) highway bridge in the Netherlands is considered. The results of this research aim to inform owners and stakeholders about the operation level and the effectiveness of adopted policies for the bridge's management as well as assist in decision-making for a rapid and reliable data-driven adaptation analysis of the asset.

Keywords: Critical Transport Infrastructures · Adaptation Index · Resources Resilience

1 Introduction

According to the Centre of Research on the Epidemiology of Disasters, the natural hazards in Europe increased by approximately 37% in 2021 compared to 2020, resulting in a significant worldwide impact, affecting 193.4 million people and leading to a $52.2 billion USD loss to the European economy, 20.7% of the global economic loss [1].

M. Ben Ahmed et al. (Eds.): SCA 2023, LNNS 906, pp. 178–187, 2024.
https://doi.org/10.1007/978-3-031-53824-7_17

Meanwhile, the effects of climate change are expected to worsen and directly impact the deterioration of Critical Transport Infrastructures (CTIs), such as highway bridges [1, 2]. Alongside the ongoing effects of climate change, population growth and commute habits will affect the condition of bridges worldwide leading to further damages [2, 3].

Considering these challenges, it is crucial to develop climate adaptation and resilience strategies that quantify the resilience level and adaptive capacity of CTIs, particularly those approaching the end of their life cycle. By quantifying adaptive capacity, stakeholders can gain awareness of an infrastructure's ability to withstand long-term consequences, such as damages and disruptions [2, 3]. Efficient monitoring of the structural and functional condition of the assets is needed for minimizing the high level of uncertainties [2] and providing real-time information. Key tools in this regard include Structural Health and Functionality Monitoring (SHFM), wireless sensors of Computer Vision-based (CV-based) monitoring techniques and emerging technologies of Artificial Intelligence (AI). These tools facilitate the development of reliable data-driven resilience methods that improve the bridge's operability, safety, and management for future generations [4–6].

To ensure future quality investment and adaptation in CTIs, it is crucial to establish proper metrics that enhance the knowledge of adaptive transformability. In case proper metrics are used resilience and sustainability can be combined to assist in the allocation of resources throughout the assets' life cycle. In this research, an already developed methodology of Achillopoulou et. al [8] and Stamataki & Achillopoulou [9] that utilizes Structural Health Monitoring (SHM) data to estimate the resilience and sustainability levels is used. A Life Cycle Cost (LCC) analysis is also employed to examine the resilience of resources, and the computed indices of resilience and sustainability levels serve as the basis for estimating the Adaptation Index (AdI). The AdI is utilized to evaluate the effectiveness of the adopted mitigation measures and/or policies. The research outcomes derived from this study can be utilized to enhance the CTIs' disaster preparedness levels, ultimately minimizing damages to the CTIs and the surrounding communities.

2 CTIs Data-Driven Investment

Nowadays, there is a growing global demand for CTIs investment to fulfil the Sustainable Development Goals of the United Nations [10]. According to the Global Infrastructure Outlook, it is projected that $94 trillion will be needed by 2040 to meet the demands of resilience and sustainable infrastructures [11]. However, the economic risk and level of uncertainties resulting from climate change, population growth and vulnerable countries, pose significant challenges that can hinder or even prevent infrastructure investment [7, 10]. This shortfall of economic alternations is expected to worsen the condition of poorly maintained CTIs. Therefore, the proper allocation of resources throughout the infrastructures' life cycles is vital for ensuring the acceptable operation and maintenance levels and enhancing resilience [3, 10].

Data-driven investment in CTIs plays a pivotal role in fostering dynamic economic resilience by incorporating sustainable criteria and facilitating dynamic adaptation pathways. By leveraging data-driven approaches, decision-makers gain valuable insights into the performance and condition of CTIs, enabling them to allocate resources effectively and enhancing economic resilience. Through the continuous analysis of data, CTIs can be proactively adapted to evolving needs and mitigating risks. Alongside, data-driven approaches can provide the foundation for dynamic adaptation pathways, allowing stakeholders to anticipate and respond to changing environmental conditions and emerging challenges.

2.1 Dynamic Economic Resilience

Economic resilience expresses the ability of an asset to recover after the occurrence of disruptive events by evaluating the effectiveness of the adopted mitigation measures and policies. Microeconomic and macroeconomic are the two types of economic resilience that are linked to the minimization of the negative impact and financial losses at local and global level respectively [12–14].

Dynamic economic resilience is characterized by the rapidity recover after an event occurrence [13]. Investment can accelerate financial dynamic resilience, but it necessitates increased levels of reliability [10, 13]. Even though there is a wide range of limitations to enhancing CTIs' quality investment, dynamic data-driven policies based on infrastructures' life cycle can enable it [8, 11]. Monitoring of the resources and conducting Life Cycle Cost (LCC) analysis can be implemented to estimate the cost of different mitigation actions [15, 16]. Based on the LCC analysis the various possible solutions can be compared to identify the more economical option [8, 11]. Under this prism, dynamic economic resilience is of high importance.

2.2 Adaptation and Sustainable Criteria

Decision-making strategies that promote flexibility, innovation, and proactive planning can enhance the economic robustness and adaptive capacity of CTIs [17, 18] and can be assisted by the development of dynamic adaptative pathways. The adaptive pathways can enhance the preparedness levels by identifying the uncertainties of hazardous events at present and future [18, 19]. The utilization of emerging technologies further strengthens asset management and maintenance practices, facilitating the development and evaluation of adaptive pathways [4, 19, 20].

The incorporation of sustainable criteria and limitations, stemming from safety, functionality, environmental and resource requirements, into the decision-making process of CTIs' maintenance is crucial to ensure their rapid recovery and adaptation. Sustainable criteria encompass the socioeconomic and environmental impacts to ensure that individual short-cost benefits do not exist [20, 21]. These criteria may be related to energy efficiency and reduction of emissions during maintenance works as well as with the decrease of unemployment levels, human well-being, and environmental protection.

3 Data-Driven Methodology

The proposed methodology builds upon the methodology of Achillopoulou et al. [8] for quantifying the resilience and sustainability of CTIs, as well as the adaptive pathways framework introduced by Makhoul et al. [19]. Figure 1 provides a schematic representation of the adopted methodology and framework, illustrating the occurrence of natural or man-made hazards throughout the life cycle of the asset. The risk associated with these hazards varies based on the asset's ageing period (e.g., exposure, idle time, or recovery period) and the frequency of hazard occurrence, as indicated by the colored ribbon. High-risk levels are depicted in orange to red colors, while moderate-risk levels are shown in yellow. For example, an anticipated hazard event may be associated with a moderate-risk level, though the risk level is expected to increase in cases where another hazard takes place before (idle time) or after mitigation actions (recovery period).

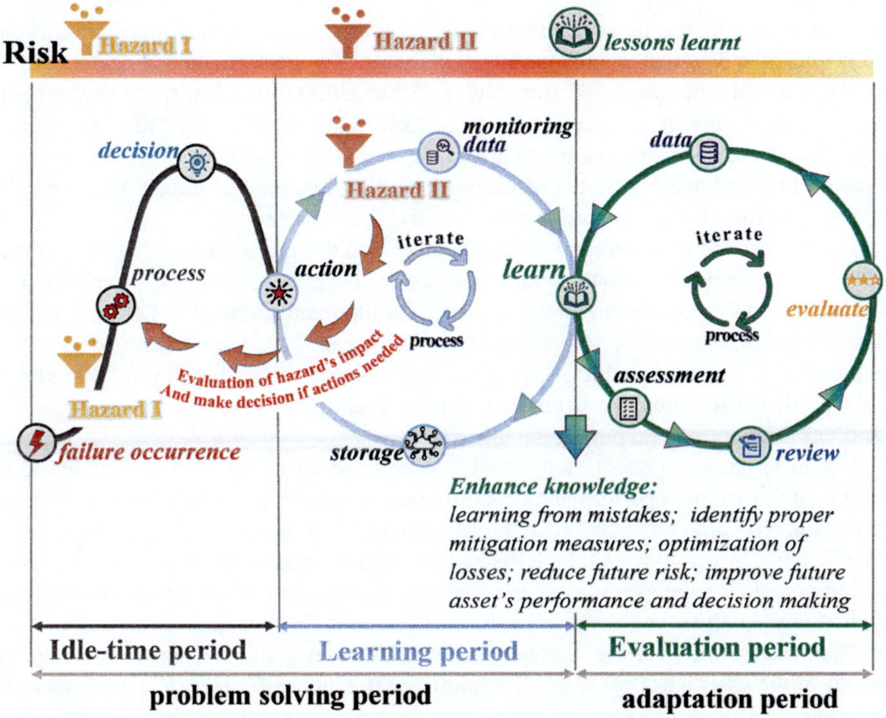

Fig. 1. Schematic representation of the adopted framework.

To evaluate properly the risk level, engineering decisions are supported by up-to-date information from contemporary and smart SHM systems (CV-based systems, AI, and machine learning). During the problem-solving period, decision-making is guided by the iterative process of collecting, evaluating, and storing monitoring data obtained through continuous and/or periodic monitoring of the asset. Lessons learned from already implemented actions and policies are also utilized to minimize both risk and mitigation cost while developing possible adaptive pathways.

Once the pathways are developed they are evaluated by estimating the resilience and sustainability indices. The development and evaluation period of the pathways is considered to have a crucial role in decision-making leaning period. The pathway characterized with the highest resilience level, the lowest resources cost, and the most beneficial socioeconomic and environmental impact is the chosen one. The progress of the selected pathway is monitored to ensure the completion of the adaptation period. This monitoring phase, referred to as the evaluation period, involves data collection to assess the pathway's capacity, considering the possibility of sudden events. The assessment and review of the pathways progress are continuous and contribute to enhancing the knowledge for the asset's management in the future.

3.1 Resilience and Sustainability Quantification

The quantification of resilience and sustainability levels is a data-driven multicriteria approach where the different ageing periods of the asset (Key Performance Indicators/KPIs) and the selected parameters (Performance Indicators/PIs) of the resilience components (Group of Performance Indicators/GPIs) are factored [8, 9]. The key difference of this research and the one proposed by Achillopoulou et al. [8] is that incorporates an additional GPI related to the required resources, resulting in a total of three GPIs: i) asset's life cycle performance, ii) functionality, and iii) resources needed.

The PIs related to the resources' component and the preliminary estimation of the component's participation to the final resilience level are based on an LCC analysis. The total number of PIs considered for the resources' component is seven (7). Different PIs are activated through the ageing periods and each one is evaluated with the consideration of an importance (L_i) and reliability (γ_i) factor as has already introduced by Achillopoulou et al. [8, 9]. For instance, the parameter of investment is expected to activate when the resources eliminating and during the idle time.

The different components are strongly correlated as a lack of functionality may result from insufficient performance condition or resources. Therefore, the components are considered complementary with the use of participation factors (β_c). The values of these factors can vary through the KPIs and are further used to identify the criticality of the components on the overall resilience level. The selection of the factors results from a Sensitivity Analysis as Stamataki and Achillopoulou [9] have already described.

The Sustainability level is indirectly estimated by weighting (W_f) the distinct ageing periods of the asset's life cycle and the estimated Resilience Indices (RI). In this paper, the Sustainability Index (SI) estimation for a certain time derives as the relative sum of the SI between two successive periods according to the equation presented in Fig. 2.

3.2 Adaptation Level Assessment

The adaptive capacity of CTIs expresses the ability of them to effectively adjust and cope with the impacts of a constantly changing environment [19]. In this research, the adaptive capacity of CTIs is measured under certain conditions and mitigations measures with the use of the Adaptation Index (AdI). The AdI is employed for evaluating the adaptation level of the assets throughout their life cycles as well as the effectiveness of the various mitigation measures and policies. The adaptation of CTIs is considered that should encompass both resilience and sustainability demand, and therefore is taken into account as their intersection. In accordance with this, the AdI is a function of the calculated resilience and sustainability indices, as Fig. 2 represents.

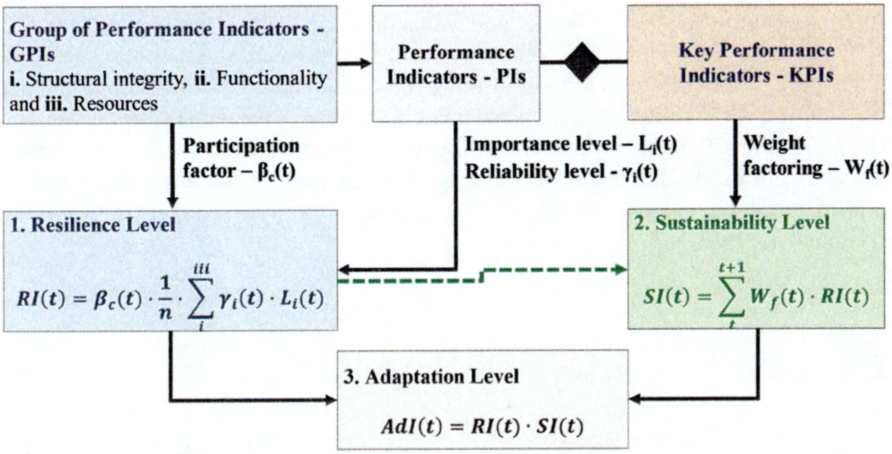

Fig. 2. Diagrammatic Illustration of the employed methodology.

4 Case-Study

The case-study of this research is the oldest prestressed RC highway bridge in the Flevopolder region of the Netherlands, known as "Hollandse Brug" [8, 9]. The bridge is a critical asset to the region's transport network, as it is the primary route connecting the town of Almere, the capital city Amsterdam, and the Schiphol Airport. Since its construction in 1969, the Hollandse Brug has dealt with numerous functionality issues prompting the implementation of various mitigation measures over time. In 2007, the bridge underwent refurbishment, and an SHFM system was installed [8, 9] enabling the continuous measurements of strains, deflections, and temperatures of the bridge's deck. The manipulation of this open-source monitoring data can also assist to the quantification of the resilience and sustainability levels of the bridge as well as to the prediction of those levels for future hazard scenarios by using emerging technologies of AI algorithms and ML methods [8, 9].

5 Results

The resilience and sustainability indices of the case-study bridge have already been estimated by Achillopoulou et al. [8] and Stamataki and Achillopoulou [9] for the components of structural integrity and functionality neglecting the resources component. The recourses required are elaborate in this research through an LCC analysis to assist in the assessment of the resources resilience and update the overall resilience quantification of the asset. Further to this, the AdIs are calculated to evaluate the effectiveness of the various mitigation measures considered through the bridge's life cycle.

5.1 LCC Analysis

The LCC analysis of the case-study bridge for the estimation of resources required for the asset's maintenance is performed in accordance with the proposed approach of the

Dutch Directorate for Public Works and Water Management in the Netherlands [22, 23]. The cost of the planned maintenance actions is estimated based on the age of the bridge through the different periods. Among the asset's ageing periods the mitigation actions differ, while routine maintenance takes place every year with its cost estimating to be equal with 5000€ per year. The mitigation measures considered in this research include except for routine maintenance, inspections as well as maintenance actions for the primary and secondary elements of the superstructure and substructure of the bridge, as outlined in Fig. 3.

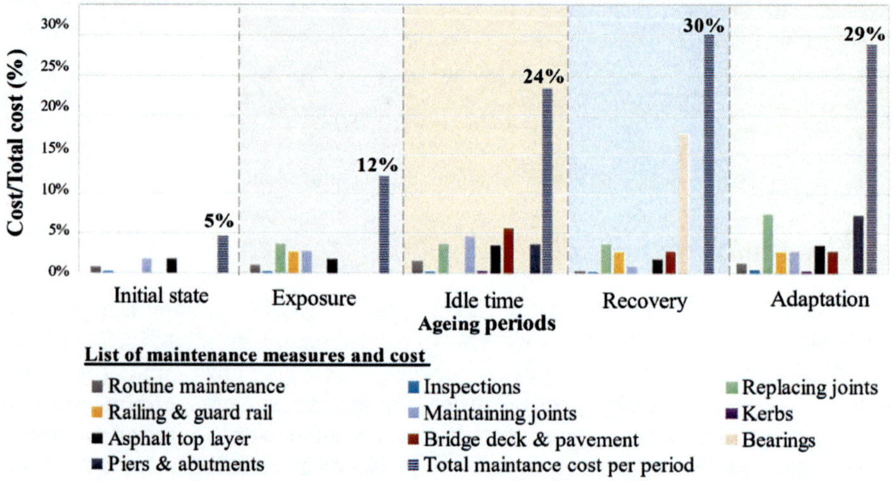

Fig. 3. Proportionate cost of maintenance actions and cumulative cost over the ageing periods of the bridge's life cycle.

Figure 3 presents a bar chart diagram illustrating the proportionate cost of each maintenance action to the total maintenance cost throughout the bridge's life cycle. The diagram also includes the cumulative cost of the various ageing periods that occur over time. The hatched bars with horizontal lines represent the percentage of the total maintenance cost of each period compared to the overall cost of the ageing periods. These bars reflect also, the participation factor considered for the resilience component of the resources, indicating that the criticality of the component increases after the exposure period. The highest participation level of resources is observed during the recovery period (30%) for the case study bridge while, the small difference in the participation values from the idle period and on, can be attributed to the interconnection between the periods of idle time and mitigation efforts (1993 to 2011).

5.2 Adaptation Level

The implementation of the proposed methodology for the quantification of the resilience, sustainability and adaptation levels of the case-study bridge derives from the estimation of the RI, SI and AdI indices described in Sect. 3. The overall calculated RI and SI through time are outlined in Fig. 4.

Figure 4 presents a 3D surface diagram providing a representation of the RI and SI changes over time. The horizontal axes of the diagram represent the progression of those indices through time, while the vertical axis represents the adaptation level of the asset. The adaptation level is expressed as a percentage of the relative AdI to the maximum potential AdI derived from the novel methodology considering the maximum RI and SI values. The diagram allows for a comprehensive understanding of the relationship among the indices through the ageing periods of the asset's life cycle.

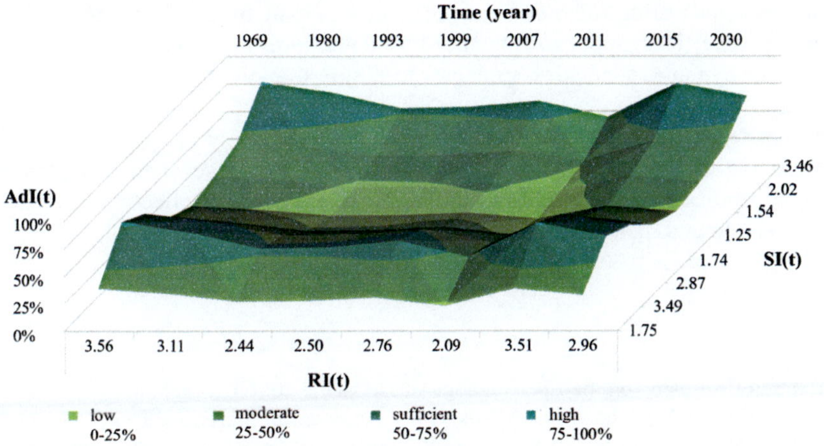

Fig. 4. 3D representation of the correlation among the RI, SI and AdI indices over time.

It is remarkable that the highest SI value (3.49) is observed at the completion of the initial state period, while the RI experiences a slight decrease of 13% (from 3.56 to 3.11) indicating the start of the exposure period. At this stage, the adaptation level stands at a moderate level of 39%. The subsequent mitigation periods from 1993 to 2011 exhibit varying levels of low to moderate adaptation, ranging from 20% to 44%. The lowest AdI value of 20% corresponds to 2007 when traffic restrictions were imposed on the bridge. During the adaptation periods the AdI reaches levels of 44% and 64%, denoting sufficient to high levels of adaptation. In fact, in 2030 the AdI corresponds to high adaptation levels due to the bridge reconstruction in 2015, despite the 16% decrease in the RI (from 3.51 to 2.96). This reduction in the RI value is correlated with the significant increase of the traffic volume in 2030, while the rise of the AdI is affected by the SI.

6 Conclusions

The findings of this research emphasize the significance of integrating resilience and sustainability considerations for evaluating the effectiveness of the maintenance measures and management policies for CTIs like bridges. The AdI serves as a useful metric for the evaluation and enhances the knowledge of the asset's adaptive capacity under certain conditions. The key findings of this study can be summarized as follows:

- The highest participation factors of the resource's resilience component are observed from idle time to adaptation periods, with the recovery period exhibiting the highest level of participation.
- The continuous monitoring of the asset from 2007 enhances the reliability of the proposed metrics, especially during mitigation periods where the various changes in the bridge's performance are recorded and evaluated with the AdI.
- The AdI proves to be a valuable metric for evaluating the bridge's ability to adapt to changing conditions. It provides a comprehensive understanding of the asset's adaptive capacity, enabling informed decision-making regarding mitigation measures and policies.
- The case-study bridge demonstrates the dynamic nature of resilience and sustainability over time. This emphasizes the need for continuous monitoring and the development for dynamic adaptation strategies to ensure the long-term performance of the bridge.
- Further research is necessary to explore the correlation between the AdI and the estimation of the adaptive space to enhance the knowledge of the bridge's adaptive capabilities and contribute to the development of more effective strategies for maintaining and managing CTIs.

References

1. Centre of Research on the Epidemiology of Disasters. 2021 Disasters in Numbers. CRED Publication (2022). https://shorturl.at/kpQW9. Accessed 27 March 2023
2. Bellon, M.M., Massetti, E.: Economic Principles for Integrating Adaptation to Climate Change into Fiscal Policy. International Monetary Fund (2022)
3. UN Office for Disaster Risk Reduction. Principles for resilient infrastructure (2022). https://shorturl.at/cemR3. Accessed 5 June 2023
4. Makhoul, N., Achillopoulou, D.V., Stamataki, N.K., Kromanis, R.: The resilience of computer vision-based monitoring systems and their measurement. Data Driven Methods for Civil Structural Health Monitoring & Resilience: Latest Developments & Applications. Taylor & Francis (2023, in press)
5. Pimenov, D.Y., Bustillo, A., Wojciechowski, S., Sharma, V.S., Gupta, M.K., Kuntoğlu, M.: Artificial intelligence systems for tool condition monitoring in machining: analysis and critical review. J. Intell. Manuf. **34**(5), 2079–2121 (2023)
6. Ozer, E., Malekloo, A., Ramadan, W., Tran, T.T., Di, X.: Systemic reliability of bridge networks with mobile sensing-based model updating for post event transportation decisions. Comput.-Aided Civil Infrastruct. Eng. **38**(8), 975–999 (2023)
7. Organisation for Economic Co-operation and Development. OECD Compendium of Policy Good Practices for Quality Infrastr. Investment (2020). https://shorturl.at/BGLNT. Accessed 5 June 2023
8. Achillopoulou, D.V., Stamataki, N.K., Psathas, A., Iliadis, L., Karabinis, A.: Resilience Quantification based on Monitoring & Prediction data Using Artificial Intelligence (AI). IABSE Congress. Nanjing 2022-Bridges Structures. Connect. Integration & Harmon. Nanjing, 21–23 September 2022 (2022)
9. Stamataki, N., Achillopoulou, D.: Sensitivity analysis on resilience components throughout the lifecycle of an asset. In: IALCCE 2023 8th international symposium on life-cycle civil engineering (2023)

10. Outlook, G.I.: Infrastructure investment needs 50 countries, 7 sectors to 2040. Global Infrastructure Hub (2017)
11. The World Bank: Well Maintained: Economic Benefits from More Reliable & Resilient Infrastructure (2021). https://shorturl.at/gltPV. Accessed 5 June 2023
12. European Commission: Guidance on Public Investment into Sustainable Infrastructure Projects (2022). https://shorturl.at/pwN48. Accessed 5 June 2023
13. Dormady, N.C., Rose, A., Roa-Henriquez, A., Morin, C.B.: The cost-effectiveness of economic resilience. Int. J. Product. Econ. **244**, 108371 (2022)
14. Rose, A., Dormady, N.: Advances in analyzing & measuring dynamic economic resilience (2018)
15. Almeida, J.O., Teixeira, P.F., Delgado, R.M.: Life cycle cost optimisation in highway concrete bridges management. Struct. Infrastr. Eng. **11**(10), 1263–1276 (2015)
16. Leiva-Maldonado, S.L., Bowman, M.D.: Superstructure bridge selection based on Life-Cycle cost analysis for short and medium span bridges. Eng. Struct. **285**, 116031 (2023)
17. González-Ruiz, J.D., Botero-Botero, S., Duque-Grisales, E.: Financial eco-innovation as a mechanism for fostering the development of sustainable infrastructure systems. Sustainability **10**(12), 4463 (2018)
18. Badr, A., Li, Z., El-Dakhakhni, W.: Dynamic resilience quantification of hydropower infrastructure in multihazard environments. J. Infr. Syst. **29**(2), 04023012 (2023)
19. Makhoul, N., Achillopoulou, D.V., Stamataki, N.K., Kromanis, R.: Adaptive pathways using emerging technologies Applications for Critical Infrastructures. Heliyon Physical & Applied Science (under review) (2023)
20. Cheng, M., Frangopol, D.M.: Efficient scenario analysis for optimal adaptation of bridge networks under deep uncertainties through knowledge transfer. Struct. Saf. **100**, 102278 (2023)
21. Yang, D.Y., Frangopol, D.M.: Life-cycle management of deteriorating civil infrastructure considering resilience to lifetime hazards: a general approach based on renewal-reward processes. Reliab. Eng. Syst. Saf. **183**, 197–212 (2019)
22. Klatter, H.E., Van Noortwijk, J.M.:. Life-cycle cost approach to bridge management in the Netherlands. In: Proceedings of the 9th Intern. Bridge Management Conference, 28–30 April 2003, pp. 179–188. Orlando, Florida, USA (2003)
23. Klatter, H.E., Vrouwenvelder, A.C.W.M., van Noortwijk, J.M.: Life-cycle-cost-based bridge management in the Netherlands. In: Proceedings of the second international conference on bridge maintenance, safety & management, pp. 18–22 (2004)

Optimizing Station Selection and Routing Efficiency Using the Pickup and Delivery Problem Method with A-Star and Genetic Algorithm

Soukayna Abibou[1(✉)], Dounia El Bourakadi[1], Ali Yahyaouy[1], Hamid Gualous[2], and Hussein Obeid[2]

[1] LISAC Laboratory, Department of Computer Sciences, Faculty of Sciences Dhar El Mahraz, Sidi Mohammed Ben Abdellah University, 30003 Fez, Morocco
{soukayna.abibou,dounia.elbourakadi,ali.yahyaouy}@usmba.ac.ma

[2] LUSAC Laboratory, University of Caen Normandy, Caen, France
{hamid.gualous,hussein.obeid}@unicaen.fr

Abstract. As the popularity of hydrogen vehicles grows, so does the demand for an efficient and reliable refueling station infrastructure. Hydrogen vehicle (HV) owners face the challenge of finding the best refueling site to meet their specific requirements in terms of comfort, accessibility, and price. Finding the best refueling station for HVs is becoming increasingly important as hydrogen stations expand. This requires analysis of variables such as hydrogen availability, cost, and distance. In this piper, we have two global objectives. In the first one, we drew inspiration from the Pickup and Delivery Problem (PDP) to find the most optimal Hydrogen Refueling Station (HRS) for our HV using a Genetic Algorithm (GA). Secondly, we want to trace the shortest path between the found station and the current location of the vehicle, for which we use the A-Star Algorithm. The approaches proposed in this document have been tested on real data and lead to the conclusion that customers would be exponentially better served by saving time and energy through optimal selection of hydrogen station services rather than the traditional method. The study results indicate the validity of these methods.

Keywords: Hydrogen Refueling Station · Hydrogen Vehicle · Pickup and Delivery Problem · Genetic Algorithm · Shortest Path · A-Star Algorithm

1 Introduction

In recent times, the automotive sector has undergone a notable transition towards sustainable transportation options. Among these advances, the rise of HVs and the associated infrastructure supporting them. HVs present a considerably more environmentally friendly and cleaner substitute for conventional gasoline-driven vehicles, thereby curtailing carbon emissions and diminishing reliance on fossil fuels. To accommodate the increasing community of HV users, a network of refueling stations has been established, providing convenient and easily accessible refueling choices.

M. Ben Ahmed et al. (Eds.): SCA 2023, LNNS 906, pp. 188–198, 2024.
https://doi.org/10.1007/978-3-031-53824-7_18

The demand for an efficient and dependable refueling station infrastructure is growing in tandem with the surge in HV popularity. HV owners encounter the task of locating an optimal refueling spot that aligns with their specific preferences regarding comfort, accessibility, and affordability. As the deployment of hydrogen stations increases, the quest for the finest refueling station for HVs gains even greater significance. It becomes imperative to assess factors such as hydrogen availability, cost, and distance. By doing so, HV owners can save time and energy by selecting the most convenient refueling station and minimizing the distance covered. By resolving this challenge, we can accelerate the transition to a sustainable transportation future and ensure the seamless integration of HVs into our daily routines.

Worthy research effort has been focused on HVs and HRSs, including hydrogen stations and delivery costs [1], offering valuable insights into potential risks and necessary safety measures for hydrogen storage and utilization [2]. Another study focuses on variables such as station capacity, customer demand, and waiting times to optimize hydrogen station operations and enhance consumer satisfaction [3]. Additionally, in order to understand the pivotal factors contributing to the success of HRSs, a thorough evaluation of the operational and performance aspects of public hydrogen refueling stations is conducted [4]. In a comprehensive review of hydrogen internal combustion engine (ICE) vehicles, the authors delve into the current state of hydrogen ICE technology, discussing its advantages, challenges, and potential applications [5].

The Pickup and Delivery Problem (PDP) stands as a formidable logistics challenge, persistently employed for devising the most efficient routes for a vehicle fleet. These routes encompass collecting goods from predefined sites and subsequently delivering them to specified destinations [6–8]. Curiously, to our awareness, there exists a paucity of research dedicated to refining station selection and route efficiency by harnessing the principles of the PDP approach. This void propels us to intricately adapt the PDP methodology to the quest for the optimal HRS.

We face two distinct tasks. Firstly, our objective is to identify the utmost optimal station for refueling our HV. To accomplish this, we employ the Genetic Algorithm (GA), a proficient technique for navigating intricate search spaces to unearth high-caliber solutions for optimization predicaments. Secondly, our focus shifts to delineating the briefest route connecting the selected station with the vehicle. This endeavor guarantees a thorough optimization of time, distance, and expenditure. Fulfilling this second task is the realm of the A-Star algorithm, a potent tool renowned for tackling diverse search and optimization challenges. Its prowess in efficiently discovering optimal pathways gives a significant advantage to our challenge.

Continuing forward, this paper's organization unfolds as follows: Sect. 2 furnishes a succinct literature review. Section 3 furnishes a structured exposition of our approach along with its pertinent parameters. Elaborating on our Genetic Algorithm, Sect. 4 is dedicated to its depiction. The accrued outcomes are laid out in Sect. 5. Lastly, Sect. 6 encapsulates the paper's essence and charts the course for future investigations.

2 Literature Review

Within this section, we unveil two distinct facets. We start with delving into the spectrum of existing research pertinent to HVs and HRSs. Subsequently, we present a state of the art of the different methods used to solve the different scenarios of the PDP in its conventional format.

Within the realm of hydrogen supply chain network life cycle optimization, a study has directed its efforts towards formulating an enhanced framework aimed at curbing environmental repercussions and enhancing overall efficiency, through a meticulous analysis of multiple facets encompassing production, storage, transportation, and distribution [9]. Furthermore, a distinct investigation contributes to augmenting the operational efficacy and sustainability of hydrogen refueling stations, through a holistic consideration of variables spanning station placements, delivery paths, and vehicle capacities [10]. In a separate endeavor, scrutiny is cast upon the financial dimensions underpinning hydrogen refueling stations within Shanghai [1]. Amidst these explorations, the paramount aspect of safety in handling hydrogen across storage, transportation, and utilization stages remains of paramount importance. A resolute commitment is made to minimizing risks and ensuring the secure management and utilization of hydrogen [2]. Further enriching this tapestry, an endeavor strives to heighten operational efficiency and enhance the overall customer experience [3]. Noteworthy is the presentation of a quantitative analysis of a thriving public hydrogen station, shedding light on its performance attributes [4]. These works collectively address diverse dimensions, including economic, safety, operational, and customer-centric perspectives. Their contributions extend the boundaries of knowledge pertaining to the advancement and cultivation of hydrogen refueling infrastructure.

Several articles have focused on applying different metaheuristic algorithms to address optimization challenges in pickup and delivery scenarios. These efforts encompass diverse aspects, including tackling multi-objective PDP using hybrid genetic algorithms [6]. Additionally, a novel metaheuristic method for the selective PDP was introduced [7], or by a particle swarm optimization (PSO) [8]. Collectively, these articles contribute to the advancement of optimization methods and algorithms tailored to solve complex pickup and delivery problems in their conventional contexts.

3 The Mathematical Modeling

Solving complex problems often demands a systematic and organized methodology. One commonly strategy used is mathematical modeling, enabling the representation of problems through a set of essential components. Within this paper, we delve into the foundational aspects of mathematical modeling, namely sets, parameters, decision variables, cost function and constraints.

3.1 Sets

$c \in C$	OD (Origin-Destination) pairs (Vehicle-Station)
$s \in S$	Set of HRSs

3.2 Parameters

P_{s_i}	Hydrogen price in the HRS s_i
V	HV customer
CP	Current Position of vehicle
D	Destination
Nodes	$S \cup V \cup CP \cup D$
d (a, b)	Euclidean distance between a and b

3.3 Decision Variables

HS_i	1, If the Si station has hydrogen; 0 otherwise
$X^{V}{}_{i\_j}$	1, If vehicle v has chosen the route from i to j; 0 otherwise
$X^{V}{}_{i}$	1, If vehicle v moves to i; 0 otherwise

3.4 Cost Function

A cost function defines metrics for evaluating system performance and guiding the optimization process. The objective revolves around minimizing or maximizing this function based on decision variables and constraints, in order to obtain.

$$MaximizeOBJ = 1/(Total\_Distance + Total\_Cost + 1) \tag{1}$$

Subject to:

$$\sum_{i \in \mathbb{S}} X_i^V \leq 1 \tag{2}$$

$$\sum_{i=1}^{Nodes} X_{CP}^V \leq 1 \tag{3}$$

$$\sum_{i=1}^{Nodes} X_D^V \leq 1 \tag{4}$$

$$HS_i \leq 1 \tag{5}$$

$$X_{i\_j}^V \leq HS_i \tag{6}$$

Here, (1) includes the total distance, the total cost and a constant 1 to avoid division by zero. Constraint (2) guarantee that each node of HRS can be visited at most once. Constraint (3) ensures that the vehicle route start at the current position. Constraint (4) ensures that the vehicle route finish at the destination. Constraint (5) guaranteeing hydrogen availability. Constraint (6) ensures that all HRSs with hydrogen are on route in the same direction as the vehicle, so that the vehicle doesn't roll back. In addition, if we have two HRSs within the same distance or 10 km difference, we choose the HRS with the lowest price.

4 Our Approach

Our approach is based on two key objectives that are intricately linked, jointly enhancing the efficacy of our solution. The initial objective is to seek out the most optimal HRS that aligns with all predefined constraints using the PDP method. Subsequently, the secondary objective pertains to identifying the shortest route connecting the chosen station to the vehicle.

4.1 PDP Method

PDP, is a well-popular combinatorial optimization problem in the literature [6]. Since ancient times, this method has been known as having a set of collection locations where items have to be collected and a set of delivery locations where these items have to be delivered. The crux of the challenge lies in devising a plan that is optimal or near-optimal that can minimize the total cost or time requisite for transportation.

We adopt this approach within the context of our specific problem, aiming to find the most optimal HRS.

In our scenario, the PDP can be employed by leveraging the information presented in Table 1:

Table 1. Description of how to model PDP for our problem.

Literature	Our Modeling
Starting from depot	Starting from the current position of the vehicle
Back to depot	Reaching the destination
Collection Point	Current Position
Delivery Points	HRSs
Limited Capacity	Hydrogen Availability

To identify the most optimal station, our strategy involves the utilization of the PDP method, for which we've laid out the key principles of this method. Drawing insights from the literature, the PDP process starts at the depot and ends there, and it tries to determine the optimal routes for transporting objects from pick-up points to delivery destinations. In our case, we consider a single collection point with multiple delivery points, and our goal is to determine the optimal route connecting these points. Given that the PDP method yields a sequence of delivery points to visit, our intention is to designate the initial point of this sequence as the recommended station.

4.2 Genetic Algorithm

In recent years, challenging real-world problems spanning diverse sectors such as economics, politics, management, and engineering have found resolutions through the application of metaheuristic algorithms [11]. One well-known metaheuristic algorithm is the GA, which takes its cues from the biological evolution process [12]. At the core of the GA are fundamental components: chromosome representation, fitness selection, and biologically-inspired operators that encompass selection, mutation, and crossover.

Initial Population Operator

GAs tend to exhibit optimal performance with a certain initial population. The quality of the solution is also impacted by population size [13]. The researchers contend that the approach requires more computing time when a huge population is taken into account. The limited population could lead to inadequate solutions [14]. In our case, this operation starts by obtaining a list of all available stations that have hydrogen, then calculates the direction between the vehicle's current position and the designed destination. This directional vector serves as a filter, excluding stations that deviate from the vehicle's trajectory to prevent backtracking to ensure that the vehicle doesn't go back, and stores them in a list. Should this list turn out to be empty, a new list is generated, encompassing all the stations that have hydrogen. This operation returns a list of filtered stations sorted by distance, which represents the initial solution for the GA.

Crossover Operation
Crossover operators are utilized to produce offspring by merging the genetic information of two or more parents. For us, this operation takes two parent solutions as input, then selects a random cut-off point in the parent solutions, and then swaps the segments of the parent solutions after the cut-off point to create two descendants.

Mutate Operation
Mutation is a mechanism that preserves genetic variation from one population to the next. Our mutate operation takes a solution (a list of stations) as input, then selects two random stations and switches their positions to explore new possibilities.

Fitness Operation
The fitness function is the determinant of the most suitable individual in each iteration of the algorithm. A costly fitness function can be altered if there aren't many iterations. In order to maintain diversity, we use the shared formula (OBJ) to rank individuals according to their fitness values. The basic idea of the sharing method is to penalize the fitness of solutions located over long distances or at high prices, thus reducing their probability of survival for the next generation.

4.3 A-Star Algorithm

The shortest path problem is a widely recognized and classical topic [15, 16], although it can be challenging, particularly in large-scale networks [17, 18]. It can be found in a variety of real-world contexts, including social networks [19], isometric feature mapping [20], pattern mining [21, 22], RDF clustering [23], biological network analysis [24], and transportation networks [25, 26]. The A-Star algorithm is one of the ways to solve this problem. It is a powerful search algorithm using the following evaluation function:

$$f(c) = dist(s, c) + h(c) \tag{7}$$

where f(c) denotes the evaluation function, dist (s, c) denotes the value of the shortest path between the source s and the current vertex c, and h(c) is the heuristic value that calculates the value of the shortest path from the current vertex c to the destination. The choice of h(c) must be estimated in advance, which strongly affects the efficiency of the A-Star algorithm. Whilst there are a variety of calculation methods, such as Manhattan distance, diagonal distance, and Euclidean distance, they are all based on known vertex coordinates.

5 Results

In this section, we present some experimental results obtained using the GA and A-Star algorithms. Our experimental environment is Windows 10, and the computer has an Intel Core i5 1.90GHz CPU and 8GB of RAM. The implementation of these algorithms was carried out using the Python programming language. The parameters employed in this study were established through rigorous testing and experimentation.

Starting with the GA outcomes, the parameter values employed are as follows: maximal population = 50, maximal number of iterations = 100, mutation rate = 0.01. We conducted tests on modest instances with 10 h nodes and a single vehicle. The results are shown in Tables 2 and 3. Table 2 gives an overview of the fitness value, while Table 3 shows the execution time for each distinct test. Notably, these tests are conducted in isolation from one another.

Table 2. Fitness value for each Test

Test	Fitness value
1	0.006
2	0.0007
3	0.001
4	0.0013
5	0.00132
6	0.00112
7	0.000828
8	0.000993
9	0.000956
10	0.001209

After using the PDP method with the GA, we obtained promising results. By plotting the fitness value for each test, we tried different scenarios (a station closer but further back, no station in front, stations at the same distance or with a 10 km difference, etc.) and each test found the best HRS satisfying our predefined constraints. This demonstrates the effectiveness of the GA. Furthermore, by examining the execution time graph for each test, we found that the GA was able to find high-quality solutions within a reasonable time. These results suggest that the use of GAs in PDP to find the optimal HRS may be a feasible approach to optimizing paths and time.

Table 3. Timing of execution

Test	Time
1	29.594
2	43.841
3	13.876
4	13.402
5	13.094
6	13.220
7	16.297
8	16.028
9	16.031
10	13.497

In the second part, we present the result of the A-Star algorithm in Table 4, which we have set the real distance and the distance optimized by the algorithm.

Table 4. The result by A-Star Algorithm

Real Distance	Optimized Distance
20.885630677503038	17.165634258127838
29.756766564076635	26.351525355044892
30.595491343140466	27.592691021405709
51.22952750075645	47.112798501030921
89.49000366794757	84.4225614270492

Using the A-star algorithm with Euclidean distance as the heuristic function, we obtained very satisfactory results. This algorithm can efficiently determine the shortest path between origin and destination. The Euclidean distance is a good estimate of the actual distance and helps to ensure the accuracy of the result. Therefore, using the A-star algorithm with a heuristic function is an effective strategy for solving my path-finding problem.

6 Conclusion

The primary aim of this study was to determine the most efficient route between the current position of the HV and the optimal HRS. To achieve this goal, we leveraged the PDP method, and to trace the shortest path we used the A-Star algorithm, which is the most efficient path-finding algorithm to the present day. It's important to acknowledge that further endeavors are required to enhance the proposed approach and to attain more complete results.

References

1. Weinert, J., Shaojun, L., Ogden, J., Jianxin, M.: Hydrogen refueling station costs in Shanghai. Int. J. Hydrogen Energy **32**(16), 4089–4100 (2007). https://doi.org/10.1016/j.ijhydene.2007.05.010
2. Abohamzeh, E., Salehi, F., Sheikholeslami, M., Abbassi, R., Khan, F.: Review of hydrogen safety during storage, transmission, and applications processes. J. Loss Prev. Process Ind. **72**, 104569 (2021). https://doi.org/10.1016/j.jlp.2021.104569
3. Brown, T., Kisting, H.: Analysis of customer queuing at hydrogen stations. Int. J. Hydrogen Energy **47**(39), 17107–17120 (2022). https://doi.org/10.1016/j.ijhydene.2022.03.211
4. Brown, T., Shane Stephens-Romero, G., Samuelsen, S.: Quantitative analysis of a successful public hydrogen station. Int. J. Hydrogen Energy **37**(17), 12731–12740 (2012). https://doi.org/10.1016/j.ijhydene.2012.06.008
5. Wróbel, K., Wróbel, J., Tokarz, W., Lach, J., Podsadni, K., Czerwiński, A.: Hydrogen internal combustion engine vehicles: a review. Energies **15**(23), 8937 (2022). https://doi.org/10.3390/en15238937
6. Al Chami, Z., Manier, H., Manier, M.-A., Fitouri, C.: A hybrid genetic algorithm to solve a multi-objective pickup and delivery problem. IFAC-PapersOnLine **50**(1), 14656–14661 (2017). https://doi.org/10.1016/j.ifacol.2017.08.1906
7. Al Chami, Z., El Flity, H., Manier, H., Manier, M.-A.: A new metaheuristic to solve a selective pickup and delivery problem. In: 2018 4th International Conference on Logistics Operations Management (GOL), pp. 1–5. IEEE, Le Havre (2018). https://doi.org/10.1109/GOL.2018.8378089
8. Peng, Z., Al Chami, Z., Manier, H., Manier, M.-A.: A particle swarm optimization for selective pickup and delivery problem. In: 2018 IEEE 30th International Conference on Tools with Artificial Intelligence (ICTAI), pp. 947–952. IEEE, Volos (2018). https://doi.org/10.1109/ICTAI.2018.00146
9. Li, L., Feng, L., Manier, H., Manier, M.-A.: Life cycle optimization for hydrogen supply chain network design. Int. J. Hydrogen Energy **52**, 491–520 (2024). https://doi.org/10.1016/j.ijhydene.2022.03.219
10. Li, L., Chami, Z.A., Manier, H., Manier, M.-A., Xue, J.: Incorporating fuel delivery in network design for hydrogen fueling stations: Formulation and two metaheuristic approaches. Transp. Res. E Logistics Transp. Rev. **152**, 102384 (2021). https://doi.org/10.1016/j.tre.2021.102384
11. Kumar, V., Chhabra, J.K., Kumar, D.: Parameter adaptive harmony search algorithm for unimodal and multimodal optimization problems. J. Comput. Sci. **5**(2), 144–155 (2014). https://doi.org/10.1016/j.jocs.2013.12.001
12. Michalewicz, Z., Schoenauer, M.: Evolutionary algorithms for constrained parameter optimization problems. Evol. Comput. **4**(1), 1–32 (1996). https://doi.org/10.1162/evco.1996.4.1.1
13. Piszcz, A., Soule, T.: Genetic programming: optimal population sizes for varying complexity problems. In: Proceedings of the 8th Annual Conference on Genetic and Evolutionary Computation, pp. 953–954. ACM, Seattle Washington USA (2006). https://doi.org/10.1145/1143997.1144166
14. Pelikan, M., Goldberg, D.E., Cantu-Paz, E.: Bayesian OptiamnidzaTtiiomneAtolgo-Croitnhvmer,gPenocpeulation Sizing.

15. Sommer, C.: Shortest-path queries in static networks. ACM Comput. Surv. **46**(4), 1–31 (2014). https://doi.org/10.1145/2530531
16. Zhang, D., Liu, Y., Liu, A., Mao, X., Li, Q.: Efficient path query processing through cloud-based mapping services. IEEE Access **5**, 12963–12973 (2017). https://doi.org/10.1109/ACCESS.2017.2725308
17. Akiba, T., Iwata, Y., Yoshida, Y.: Fast exact shortest-path distance queries on large networks by pruned landmark labeling. In: Proceedings of the 2013 ACM SIGMOD International Conference on Management of Data, in SIGMOD 2013, pp. 349–360. Association for Computing Machinery, New York, NY, USA (2013). https://doi.org/10.1145/2463676.2465315
18. Zhu, A.D., Ma, H., Xiao, X., Luo, S., Tang, Y., Zhou, S.: Shortest path and distance queries on road networks: towards bridging theory and practice. In: Proceedings of the 2013 ACM SIGMOD International Conference on Management of Data, in SIGMOD 2013, pp. 857–868. Association for Computing Machinery, New York, NY, USA (2013). https://doi.org/10.1145/2463676.2465277
19. Rezvanian, A., Meybodi, M.R.: Sampling social networks using shortest paths. Phys. A Stat. Mech. Appl. **424**, 254–268 (2015). https://doi.org/10.1016/j.physa.2015.01.030
20. Chen, C.-H.: A semi-supervised feature selection method using a non-parametric technique with pairwise instance constraints. J. Inf. Sci. **39**(3), 359–371 (2013). https://doi.org/10.1177/0165551512456502
21. Han, J., Zhenhui Li, L., Tang, A.: Mining moving object, trajectory and traffic data. In: Kitagawa, H., Ishikawa, Y., Li, Q., Watanabe, C. (eds.) Database Systems for Advanced Applications, pp. 485–486. Springer, Berlin, Heidelberg (2010). https://doi.org/10.1007/978-3-642-12098-5_56
22. Youxi, W., Wang, L., Ren, J., Ding, W., Xindong, W.: Mining sequential patterns with periodic wildcard gaps. Appl. Intell. **41**(1), 99–116 (2014). https://doi.org/10.1007/s10489-013-0499-4
23. Khosravi-Farsani, H., Nematbakhsh, M., Lausen, G.: SRank: shortest paths as distance between nodes of a graph with application to RDF clustering. J. Inf. Sci. **39**(2), 198–210 (2013). https://doi.org/10.1177/0165551512463994
24. Kuperstein, I., Grieco, L., Cohen, D.P.A., Thieffry, D., Zinovyev, A., Barillot, E.: The shortest path is not the one you know: application of biological network resources in precision oncology research. Mutagenesis **30**(2), 191–204 (2015). https://doi.org/10.1093/mutage/geu078
25. Sun, Y., Xinpei, Y., Bie, R., Song, H.: Discovering time-dependent shortest path on traffic graph for drivers towards green driving. J. Network Comput. Appl. **83**, 204–212 (2017). https://doi.org/10.1016/j.jnca.2015.10.018
26. Quercia, D., Schifanella, R., Aiello, L.M.: The shortest path to happiness: recommending beautiful, quiet, and happy routes in the city. arXiv (2014). Consulté le: 19 juin 2023. Disponible sur: http://arxiv.org/abs/1407.1031

Spatio-Temporal Clustering for Optimal Real-Time Parking Availability Estimation

Hanae Errousso[1,2], Youssef Filali[1,4], Nihad Aghbalou[1], El Arbi Abdellaoui Alaoui[3(✉)], and Siham Benhadou[2]

[1] EIGSI, La Rochelle-Casablanca, 17041-20410 Casablanca, Morocco
[2] National and High School of Electricity and Mechanic, HASSAN II University, 8118 Casablanca, Morocco
[3] IEVIA Team, IMAGE Laboratory, Department of Sciences, ENS, Moulay Ismail University of Meknes, Meknes, Morocco
e.abdellaouialaoui@umi.ac.ma
[4] Faculty of Sciences Dhar-Mahraz, 30000 Fez, Morocco

Abstract. Urban transport systems represent a major infrastructure asset in contemporary cities, enabling many millions of people to commute and travel every day. Transport systems are increasingly complex because of rapid urbanization and rising vehicle ownership. Effectively predicting parking availability across a city means more efficient parking management, better urban planning, smoother traffic flow, lower fuel wastage and, ultimately, less environmental pollution. To this end, several research studies have proposed predictive approaches to parking space availability based on supervised machine learning, mainly regression algorithms. But if real-time information on parking space availability is lacking, these approaches become useless, since their driving force is historical data. What's more, many city zones simply don't require exact information on parking space occupancy; all that's needed is a global view of whether there are any spaces available or not. That's why, in this paper, we outline an approach to predicting parking availability that combines both clustering and classification. Our aim is to model parking availability through a number of typical days. A typical day is defined by a profile characterizing parking space occupancy. We start by determining precisely how many typical days per spatial cluster are required to map parking availability, and then cluster data to form groups typified by each typical day, using six clustering algorithms of ascending difficulty. Obtained results show that this number varies between 3 and 4, depending on which algorithm is applied. Thus, model evaluation reveals that distance between cluster elements is short and separation between clusters is high, in other words, clusters are far apart and not very dispersed.

Keywords: Urban transportation · parking availability · prediction model · clustering · classification

1 Introduction

Machine learning algorithms fall into four principal categories: Supervised, Unsupervised, Semi-supervised and Reinforcement Learning. Supervised learning

M. Ben Ahmed et al. (Eds.): SCA 2023, LNNS 906, pp. 199–214, 2024.
https://doi.org/10.1007/978-3-031-53824-7_19

involves training an objective function on annotated examples. That is, data already labelled with the correct label (class, continuous value...). The objective function, a model expression describing data, predicts the dependent or output variable value, from a collection of independent variables or characteristics. Unlike supervised learning, unsupervised learning consists in feeding the function with a set of instances having no corresponding labels. The system seeks to reveal the concealed data structure or associations between variables. The third category combines the previous two. It makes use of labelled as well as unlabelled data. It operates in principle like unsupervised learning, but takes advantage of tagging part of the data [21]. In the last learning mode, algorithms attempt to predict a problem's output according to a set of tuning parameters. A new output is then calculated taking the previously determined output as an input parameter, and this operation is repeated until the optimum output is reached [18].

This paper deals with parking space availability by proposing a new approach founded on supervised and unsupervised machine learning algorithms. Our present work relies on clustering algorithms to separate a dataset into clusters, such that instances pertaining to a cluster are as similar as possible, and instances belonging to different clusters differ as much as possible. Each group point is characterized by a feature vector representing a typical day of occupancy for a car park group. Parking lots with similar spatial characteristics are grouped together in a single cluster, thanks to spatial clustering. Before partitioning data, it's necessary to determine the number of typical days (resp. the number of spatial groups) needed to cover the variance of every parking history (resp. to map a city's spatial heterogeneity). To do this, these numbers are estimated by means of clustering stability technique, elbow method and silhouette analysis. Subsequently, six algorithms from two fundamentally different types of clustering (graph-based and individual-based methods) are trained. These algorithms are selected considering two parameters: computational intensity and processing speed. In our case, real-time analysis is required, which is why the retained algorithm must have a high computational speed to keep up with changes in the input data and produce the targeted results on time. The idea, therefore, is to identify the best model in terms of computation time and accuracy, by comparing multiple clustering metric values.

Our primary contribution is to provide global information on parking availability, rather than a more detailed view of it. This is intended to adapt the information level to users' needs and avoid delivering information that goes beyond the expectations and requirements of drivers visiting a given area. Most parking management systems inform users either of available spaces number, or of car park occupancy rate. However, users only want to know if there are available spaces in the desired parking lot. Consequently, these systems are overdimensioned, resulting in unnecessarily high costs and investment. Other parking solutions, meanwhile, are undersized in cases when drivers serving an area require to know all available parking space coordinates or identifiers. The same is true for parking administrators who manage special-status spaces, such as delivery bays, where carriers with very high requirements in terms of time and walking distance park their vehicles. This shortcoming can also easily be addressed with our proposal, by simply reducing the modelling scale from a parking lot to a single space. It is also possible to increase or decrease decision point frequencies according to urban area criticality. All of which highlights our proposed method's modularity and scalability.

The rest of this paper is structured in four sections. Section 2 reviews research into parking availability prediction, focusing particularly on cluster-based approaches. Section 3 outlines our systematic framework and explains the principle of each machine learning algorithm employed. Section 4 summarizes results obtained by applying the proposed approach, and evaluates the efficiency of developed models. Section 5 concludes this document and discusses possible future work.

2 Related Works

According to a study conducted by the International Parking Institute [28], 40% of traffic jams in urban areas are due to cruising cars, i.e. cars looking for a parking space, representing around 30% of cars in circulation [24]. Finding a parking spot is thus recognized as a challenge that many authors have sought to overcome by predicting this infrastructure's availability. Previous works fall into two key categories: contextual data- and real-time data-based prediction. Each category is divided further into two sub-categories depending on parking facility type: on- and off-street parking. Parking organization strongly influences features to include in models, like weather conditions, events, urban renewal projects...

Regardless of datatype or parking form, most papers have applied machine or deep learning techniques to large datasets. Vlahogianni et al. [26] predict short-term parking lot occupancy rates by means of a neural network-based model. In [3], the number of available parking spaces in urban garages equipped with barriers is estimated employing four different techniques (Bayesian regularized neural network, support vector regression, recurrent neural network and autoregressive integrated moving average). Other algorithms are tested for modelling the same parameter in [11], namely XGBoost, Random forest and K-nearest neighbours. The authors of [31] evaluated three well-known methods for their performance (regression tree, support vector regression and neural network), with three feature sets composed of carefully selected parameters to illustrate their importance in parking space occupancy modelling. This issue is also addressed in [8] by dedicated approaches, chiefly Shap Values. Other than the proposed explanation model, four algorithms of different principle are implemented with four feature sets to predict parking occupancy rates. Camero et al. [5] sought to improve recurrent neural network performance by introducing two metaheuristics approaches (with genetic algorithm or with evolutionary strategy) to optimize the deep neuro-evolutionary hyperparametric architecture. In line with this perspective, a semi-supervised hierarchical recurrent neural network is proposed in [30]. Particular to this work is its modelling of non-Euclidean spatial autocorrelation between parking lots by a hierarchical graph convolution module. Additionally, it captures local and global spatial dependencies between parking lots by means of a contextual graph convolution block and a multi-resolution soft clustering graph convolution block. Other authors have thought differently about enhancing conventional algorithms' efficiency as they tested several stacking architectures, differing in number of levels, basic regressor types and meta-learner deployed [10].

Our key focus lies in leveraging clustering algorithms to improve parking management. To date, many approaches targeting parking lot characterization with spatial and temporal information have been put forward by the scientific community. Bock et al. [4] modelled each parking meter's occupancy derived from ticket payment time stamps and parking ticket validity period. They calculated parking meter similarity as the mean squared deviation of average daily parking space occupancy trends. They performed hierarchical clustering according to this similarity measure. The study carried out in [15] set out to generalize models developed for areas equipped with sensors to areas without them. It is based on calculating similarity values between city neighbourhoods using geographic information system data. It concludes by clustering parking lot spatial characteristics employing K-means algorithm. In line with sharing parking lot occupancy prediction models across multiple parking lots, a new two-stage clustering technique based on parking lots' spatio-temporal characteristics is described in [22]. It identifies all relevant spatial and temporal features of each parking lot, and then aggregates these profiles to accurately predict occupancy for a set of parking lots. It reduces the number of prediction models, requiring individual tuning for every car park. Similarly, the authors of [7] have exploited recurring trends in parking dynamics to develop a new processing pipeline. In turn, they seek to reduce both the number of models required and the number of training examples needed. They deployed hierarchical clustering technique to first cluster road segments with high similarity in parking availability patterns. They predicted future parking availability by training a shared prediction model on each of these clusters. They pruned the training set, retaining just the most representative instances through the Kennard-Stone algorithm. Another alternative to expensive sensor and camera investment costs is presented in [27]. This software solution estimates real-time parking availability for a given street block by gathering data from cell phones. It further calculates historical parking availability profiles for an arbitrary street block with algorithms incorporating weighted statistical schemes and Kalman filters.

Capitalizing on this state of art, many related works have tackled this challenge of parking occupancy prediction with advanced machine or deep learning techniques to obtain good predictions, with strong generalization properties. While these methods have proven their effectiveness and relevance, they require significant computing resources, meaning that deploying these solutions on a large-scale is often unfeasible. What's more, they call for large data sets for each parking lot, which makes their implementation even more arduous, given the high cost of deploying data collection infrastructures. Added to the development costs of predictive models, their maintenance costs for any given car park are not negligible. This is why several authors have attempted to reduce an entire city's number of parking availability prediction models. Admittedly, this approach is very ambitious, especially in terms of feasibility and adoption, but it can greatly impact prediction accuracy. Extending a parking occupancy prediction model over several car parks can lead to a drop in shared model performance if urban areas do not show a high degree of similarity on several levels. This is indeed true in most cities, where parking trends differ greatly from one zone to another due to particular characteristics (attractiveness, sociology, anthropology, typology...) and special events (vacations, unusual weather conditions, festivals or sporting events...). No matter how similar developers think two city zones are, the algorithms don't always concur. Another limitation of this alternative is that generalized models may be over-adjusted. To overcome all these shortcomings, we first propose to divide cities into homogeneous zones along spatial features. Next, we designate a number of typical parking occupancy models over a day. By grouping a city's zones into similar groups, fewer models will be needed to represent parking availability on an urban scale, without impacting on predictive accuracy and parking trend modelling, since there will be no loss of parking event information or assumptions about driver habits.

3 Methodology

Our real-time parking availability prediction approach is driven by two clustering principles. Spatial clustering involves grouping parking lots into homogeneous clusters in accordance with their spatial profile. This time-independent profile is defined by the car park's geographic coordinates, parking capacity, surrounding infrastructure, number and proximity. Temporal clustering brings together parking occupancy trends per spatial cluster, in relation to spatial cluster occupation rate, day of week and time of day. These temporal features evolve strongly over time and depict parking occupancy dynamics.

For time-based clustering, six algorithms are selected: K-means, Fuzzy c-Means, Bisecting k-means, Hierarchical agglomerative clustering, BIRCH clustering and mean shift clustering. These algorithms are mainly picked because their inner mechanisms are different, their underlying objectives are dissimilar and their settings are distinct.

K-means is a typical clustering algorithm that classifies given data objects into k different clusters via an iterative process converging on a local minimum. This unsupervised learning algorithm initially chooses k centroids at random, then assigns each data vector to cluster with the closest centroid. It determines distances between each data point and cluster centroids, usually by Euclidean distance. When it finishes mapping data points to clusters, it recalculates centroids of formed clusters. It repeats this process until the criterion function reaches a minimum [23]. The criterion function is expressed as the sum of all database objects' squared errors.

$$\sum_{i=1}^{n}\sum_{j=1}^{k} p_{ij} \left| x_i - c_j \right|^2 \tag{1}$$

where n is the database size, k is the number of clusters, p_{ij} is a binary variable denoting membership of data point x_i to cluster of centroid c_j.

With K-means, data is divided into distinct clusters, where each data element belongs to only one cluster. Fuzzy c-means (FCM) assigns each data point to each cluster out of k clusters created, but each element is linked to a set of membership levels. This membership factor reflects the strength of a data element's association with a particular cluster. It varies between 0 and 1: 1 means that the data vector belongs entirely to the cluster, and 0 implies that it is entirely excluded from the cluster. The degree u_{ij} to which point i belongs to cluster j is calculated by the Lagrange multiplier method [17]. This algorithm's iterative process begins by initializing the fuzzy partition matrix and the number of clusters. It then calculates the initial fuzzy cluster centroids. It continuously updates cluster centroids and membership weights: if a point is very close to a cluster centroid, its membership degree will be greater, and conversely. It stops if the maximum number of iterations is reached, or if the objective function variation with respect to the previous value is less than the specified threshold value [16].

Bisecting k-means (BKM) algorithm is an optimized and extended version of k-means. This hierarchical clustering technique does not partition data into k clusters at each iteration, but divides a cluster into two sub-clusters at each bisecting step using k-means until obtaining k clusters. It first splits the dataset into two clusters, then selects one of these clusters to divide, and so on. It chooses which cluster to split in two, either by considering size, sum squared error or both [19]. For this work, the cluster picked for splitting is larger in size and greater in sum squared error.

To diversify tested clustering types, hierarchical agglomerative clustering (HAC) is implemented. This approach initially treats each point as a single cluster, then repeatedly merges the two closest clusters until only one overall cluster remains in the active set. This active set is that set containing all clusters to be merged. When two clusters are merged, each one is removed from the active set and their union is added to it [1]. The clusters chosen for merging are selected such that distances between resulting clusters are monotonically increasing. This means that two merged clusters must always be at least as far apart as the previously merged sub-clusters that formed them.

Another important clustering algorithm is BIRCH (Balanced Iterative Reducing and Clustering using Hierarchies) [29]. This method constructs a height-balanced hierarchical cluster tree taking into account the branching factor (the maximum number of a node's children), the number of clusters k and the threshold for a cluster's radius [20]. Each node symbolizes a cluster within the cluster hierarchy, intermediate nodes represent superclusters and leaf nodes correspond to final clusters. Every node contains the most relevant characteristics of its respective cluster. To build the cluster feature tree, a point begins at the root and traverses the tree recursively, always joining in the sub-cluster whose centroid is closest, until it reaches a given leaf node. At this stage, it is added to this leaf cluster, unless the cluster radius exceeds the specified threshold. If not, a new cluster is formed with this point as sole member. If, following this creation, the parent's number of child nodes is greater than the branching factor value, the parent is split. Once the tree has been elaborated, the leaf cluster centroids are globally grouped by k-means algorithm to reduce their number and improve their quality.

Mean shift is a general non-parametric procedure for grouping data. Unlike the classic K-means method, it makes no assumptions about data distribution or cluster counts. Its basic principle entails expressing all points in the d-dimensional feature space as an empirical probability density function, knowing that dense regions of this space correspond to local maxima or modes of the subjacent distribution [6]. It conducts, for each data vector within the feature space, a gradient ascent iterative process on the estimated local density till convergence. It deems data points associated at least approximately with the same distribution mode to be members of the same cluster. Effectively, each data point is iteratively shifted by the kernel bandwidth until it converges on a local mode of the estimated kernel density. A data point's number of shifts to a peak depends on its bandwidth and its distance from the mode.

Spatial clustering of parking lots considers both categorical and numerical features as input. Therefore, a mixed-input clustering algorithm is required. Our choice fell on applying k-prototypes, a popular generalization of k-means to mixed data. It's a combination of K-means and K-modes clustering algorithms [14]. It makes use of Euclidean and Manhattan distances for numerical data and matching distance for categorical data. This latter distance metric is the 0–1 indicator function that checks whether two categories match [25]. Its objective function minimizes the following weighted sum:

$$\sum_{i=1}^{n}\sum_{j=1}^{k} p_{ij}\,\text{dist}\left(x_i, c_j\right) \tag{2}$$

where dist (x_i, c_j) is the dissimilarity measure given by:

$$\sum_{r=1}^{m_r}\sqrt{\left(x_{ir}-c_{jr}\right)^2} + L_j\sum_{t=1}^{m_t} H\left(x_{it'}c_{jt}\right) \tag{3}$$

where m_r is the number of numerical attributes, m_t is the number of categorical attributes, x_{ir} is the value for the numerical attribute r at point x_i, x_{it} is the value of the categorical attribute t for the same point, c_{jr} is the mean of the numerical attribute r for j, c_{jt} is the most current value of the categorical attribute t within the same cluster, L_j is a user-specified weight for categorical attributes in cluster j and $H\left(x_{it'}c_{jt}\right)$ is the matching distance between two categorical attributes, which equals 0 if $x_{it} = c_{jt}$ otherwise equals 1.

Objective function optimization is achieved by means of an iterative process that alternates between finding optimal cluster centroids and calculating distance between data points and new cluster centroids [13]. Convergence is achieved when cluster centroids stabilize after two consecutive iterations.

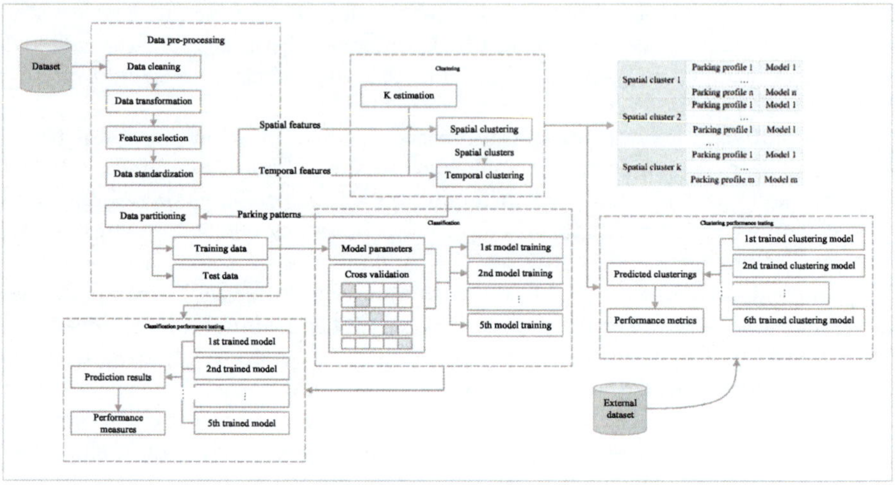

Fig. 1. Our proposal's methodological framework

The flowchart shown in Fig. 1 outlines our solution's sequence of operations and highlights the methodology adopted to conceive our framework. It is composed of five main phases, each comprising several stages briefly described as follows:

- Database pre-processing to eliminate noisy values impairing collected data, restructure columns, adapt variable types and standardize selected characteristics;
- Grouping data first by their spatial characteristics to determine spatial profiles representing a city, and then by each spatial cluster's temporal features to define its parking profiles;
- Performance measurement to evaluate clustering models by calculating multiple metrics (adjusted mutual info score, completeness score, homogeneity score...) and exploiting the results of [9], where clusters are inspected qualitatively;
- Classification model training that performs ten learning and validation iterations for each algorithm to deploy, so that the training set is randomly split into ten equally sized folds using cross-validation;
- Testing accuracy of predictions made by classifiers on the test set with various performance indices (Balanced accuracy, Precision, F1-Score, Jaccard...).

4 Implementation and Results

Our proposed approach is deployed with Python 3 thanks to a web-based interactive development environment for notebooks, known as JupyterLab. It is assessed by capitalizing on historical parking data from Aarhus, Denmark. This parking data stream covers eight parking lots spanning a five-month timeframe, from May 22, 2014 to November 4, 2014. These car parks are kitted out with sensors to indicate how many spaces are vacant [2]. Parking events are marked by a code indicating car park name and location, total number of spaces (i.e. maximum parking lot capacity), number of vehicles (i.e. number of spaces occupied in a given parking lot at a given time), update time (since data is updated every 30 min) and stream time.

Accessible from this link[1], the database contains a total of 55 264 rows. Before deploying our selected machine learning models, we first undertake a data pre-processing phase to clean and restructure data, perform any necessary calculations and fill in missing values. Details of all pre-processing operations conducted and a graphical representation of the pre-processed database can be found in [8]. This database is supplemented by commodity information collected using OpenStreetMap and Google Maps.

Our approach begins with spatial clustering, precisely by determining the number of spatial profiles representing Aarhus city. For this purpose, the clustering stability method is employed. It involves dividing data into training and validation subsets, then comparing stability of the resulting clusters. It begins by applying the clustering algorithm to the training subset for different values of k (in this paper $k = \{2, 3, 4, 5\}$). It then runs it a second time on the validation subset, utilizing the same cluster centers obtained from the training subset. It subsequently calculates a stability index, defined in our case as the percentage of points falling in the same cluster in both runs. The optimal number of clusters producing the highest stability (93%) for our dataset and algorithm is 3. In other words, this value of k yields three similar and robust spatial profiles over several runs.

K-prototype algorithm parameterized with found number of clusters groups all data according to their spatial characteristics. Its performance is measured by computing multiple metrics and manually completing the database with the true labels. It obtains an adjusted mutual information value of 0.964237, meaning that there is a good match between clusters and true labels. Calculated entropy reveals that both partitions (predicted clusters and real classes) share a very large amount of information. This is confirmed by the adjusted rand index score (0.958143), which measures similarity between predicted clusters and true labels. It indicates that the proportion of sample pairs in the same clusters (or in different clusters) relative to true labels and predicted clusters is very high. Likewise, the proportion of same-class members grouped together in the same cluster is 0.923854. This completeness score implies that the clusters are coherent and capture the true label structure well. Unlike completeness, which evaluates clustering's ability to recover all a class's members, homogeneity rates the purity of clusters, ensuring that each cluster contains only points pertaining to a single class. Our spatial clusters are well separated by class and correctly reflect the real label pattern, given the homogeneity metric's high value (0.979468). These two

[1] http://iot.ee.surrey.ac.uk:8080/datasets.html#parking.

metrics are combined in V-Measure to provide a more comprehensive overview of clustering quality, emphasizing both homogeneous and complete clusters. It uses conditional entropy of true labels knowing predicted clusters and conditional entropy of predicted clusters knowing true labels to compute a compound measure. With a V-Measure of 0.964241, our clustering algorithm succeeded in grouping points belonging to the same class together in distinct clusters, and in forming homogeneous clusters containing only points of a single class. Fowlkes-Mallows score combines precision and recall using their product's square root. This measure equals, for this work, 0.964152. Clusters thus have both high precision (i.e. the number of false positives is low) and high recall (i.e. the number of false negatives is also low).

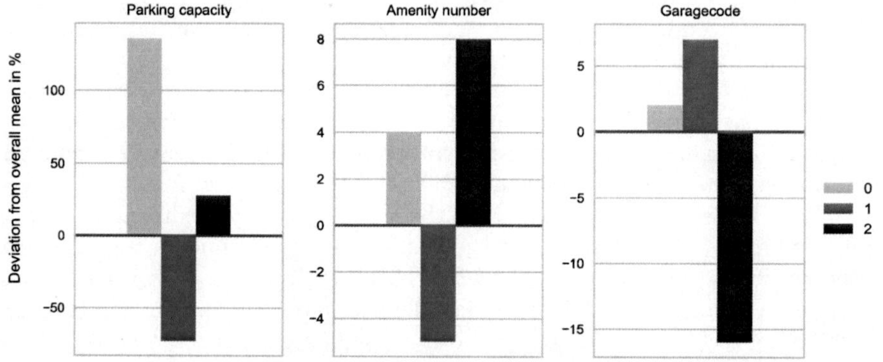

Fig. 2. Comparison of the mean per cluster to the overall mean in %

The bar graphs in Fig. 2 allow to compare each feature's mean deviation per cluster to its overall mean. It appears that there's a big difference between cluster 0's average value for parking capacity and the global average over the entire data set. Cluster 0 may represent a particular segment of data points that have specific characteristics regarding this feature. In cluster 2, however, its low standard deviation indicates that its values spread little around the mean value. Interestingly, this observation implies that the data points in Cluster 2 are relatively homogeneous in terms of parking capacity. The mean value of the two features Parking capacity and Amenity number are lower in cluster 1 than the overall mean. This suggests that data points in cluster 1 tend to have fewer available parking spaces and fewer amenities than the average for all data. Garage code (or parking geographical coordinates) in cluster 2 represents a distinctive feature that sets this cluster apart from the others.

For temporal clustering, presented results concern parking profiles for cluster 2. Before identifying this urban area's parking patterns, it is necessary to determine precisely how many clusters are sufficient to cover parking data variation. To do this, two different methods are employed, according to each applied clustering algorithm's particularity. Elbow method helps to determine the elbow point, at which adding more data samples does not greatly alter cluster membership. It implies that increasing cluster counts can reduce total intra-cluster variance. It is based on minimizing sum squared deviations within clusters.

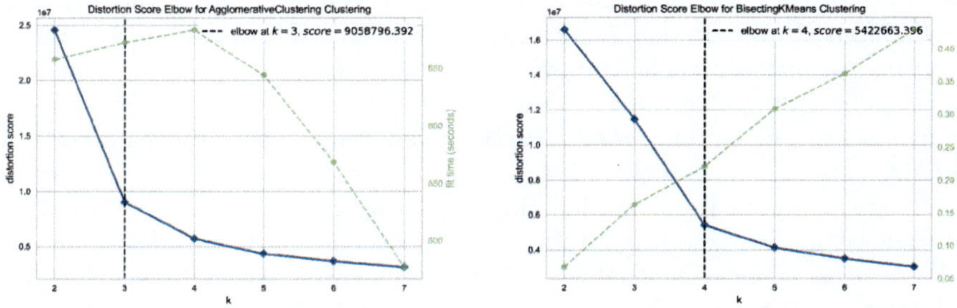

Fig. 3. Sum squared error plots to find the optimum number of clusters

The value retained for k when applying agglomerative clustering is equal to 3, as this marks onset of curve flattening: for lower values, clustering quality is significantly poorer, while for higher values, quality does not improve appreciably (Fig. 3). The elbow when using k-means and BIRCH is at k always equal to 3, i.e. distances between data points and their respective centroids no longer vary rapidly after this inflection. Distortion metric values, which calculate sum of squared distances between each point and its assigned cluster centroid, show that clusters determined by k-means are more homogeneous and those returned by agglomerative clustering are the least refined. This score decreases further with bisecting k-means, as the curve's elbow lies at $k = 4$. Assuming that the higher the number of clusters, the closer the points belonging to a cluster are to its centroid, and consequently the greater the similarity between cluster members.

Elbow method is in fact mainly utilized for distance-based clustering algorithms, as it relies on how much the sum of squared distances (inertia) between data points and cluster centroids varies. Since fuzzy c-means is based on a fuzzy clustering approach where each data point is assigned to each cluster with a degree of membership, inertia is not a suitable measure for determining the optimal number of clusters. Therefore, another method is invoked, namely silhouette analysis. This method measures a data point's similarity to its own cluster versus other clusters. The silhouette coefficient is calculated from average intra-group distance and average distance between nearest groups for each sample. It varies between -1 and 1: a value close to 1 indicates that the point is well clustered within its cluster and is far from other clusters, while a score approaching -1 means that the point could be better placed in another cluster and is not well matched to its current cluster, and a value close to 0 denotes that the point is close to the boundary between two clusters. Average silhouette for all data points is calculated by averaging silhouette values for each point. It is an overall measure of clustering quality. An average silhouette closer to 1 reflects better clustering quality. In our case, the average silhouette value is 0.6 for $k = 3$, and this value decreases as the number of clusters increases. It means that clusters are separated and that each point is properly clustered.

Mean shift is a non-parametric clustering algorithm. Unlike other clustering methods, it does not require prior specification of the number of clusters. It is able to automatically identify the number of clusters in a dataset without needing explicit input. However, it is important to note that this technique is sensitive to certain parameters, such as moving window radius (bandwidth), which can affect significantly cluster number and size.

Once a suitable number of clusters has been determined for each algorithm, the dataset is clustered to identify parking occupancy patterns. Next, several metrics are analysed to determine this clustering quality. It is clear from our performance indicator values (Table 1) that k-means is the worst-performing model, with an adjusted mutual information score equal to 0.495827, an adjusted rand score equal to 0.517256 and a V-measure score equal to 0.495835. FCM provides the best results, achieving an adjusted mutual info score greater than 0.949547, an adjusted rand score higher than 0.975348 and a V-measure score above 0.949548. Mean shift and BKM rank second by having a mean adjusted mutual info score of 0.806525, a mean adjusted rand score of 0.781252 and a mean V-measure score of 0.8065285.

Table 1. Metric values measuring models' performance

	K-means	FCM	HAC	BIRCH	Mean shift	BKM
Adjusted mutual info score	0.495827	0.949547	0.502900	0.683445	0.845833	0.767217
Adjusted rand score	0.517256	0.975330	0.475070	0.728112	0.900090	0.662414
Completeness score	0.550479	0.952073	0.490017	0.669008	0.855986	0.62382
Homogeneity score	0.451060	0.947036	0.516493	0.698528	0.835922	0.996231
Fowlkes mallows score	0.803768	0.988598	0.747319	0.870165	0.954462	0.823956
V-measure score	0.495835	0.949548	0.502907	0.683449	0.845835	0.767222

Considering adjusted mutual information metric values, FCM outperforms all other models, followed by Mean shift, BKM and BIRCH, with drops in performance of 10.92%, 19.20% and 28.02% respectively. Based on adjusted rand score values, FCM is the most efficient method, 7.71% better than mean shift, 25.34% better than BIRCH and 32.08% better than BKM. Measured against Fowlkes mallows scores, FCM performs consistently best, beating mean shift by around 3.45%, BIRCH by approximately 11.97% and BKM by roughly 16.65%. Looking at V- measure scores, FCM obtains an excellent rating, ahead of mean shift, BKM and BIRCH in order, where score variations range from 10.92% to 28.02%. Models' performance when examining the remaining two evaluation parameters doesn't really differ, with FCM staying on top, pursued by mean shift, BKM or BIRCH, whose scores fluctuate in a similar order as the other metrics.

Undoubtedly, FCM far surpasses all other models in terms of excellent results and short runtime. So obviously, it's these outputs that will feed our classification algorithms. But this work is detailed in [12]. Below, parking profiles generated by this method for spatial cluster 2 are presented.

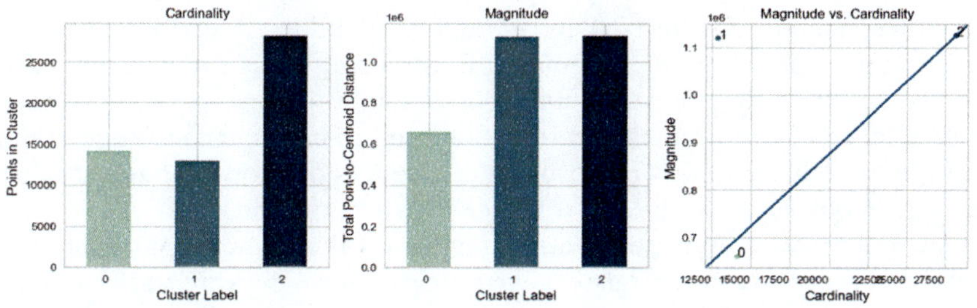

Fig. 4. Temporal cluster cardinality and magnitude

Clusters are diagnosed with regard to magnitude and cardinality (Fig. 4). The cluster cardinality graph, or cluster size graph, plots the number of data points per cluster. Data point distribution for clusters 0 and 1 appears to be correct. Their membership counts are very close, as cluster 0 contains around 14100 vectors and cluster 1 has about 12500 members. Cluster 2, however, holds many more data points than the other clusters, totalling 28700 points - twice as many as clusters 0 and 1. This size suggests that cluster 2 probably contains important structures or dominant classes in the data. The cluster magnitude graph illustrates the total distance between the centroid of each cluster and its points. Cluster 0 has the smallest point dispersion, meaning that the points are relatively close to their centroids and the cluster is therefore compact. Clusters 1 and 2, despite their different sizes, are characterized by very close, almost equal, total distances. High total cluster distances reveal that points within a cluster are distributed over a wide range of distances from its centroid. This suggests that the cluster is spread out or extended, and points may be relatively far apart. So, cluster 1 may contain sub-groups or diverse structures and consequently heterogeneous points. The last plot in Fig. 4 compares the cardinality of clusters with their magnitude. Clusters featuring a higher cardinality also tend to exhibit a higher magnitude. Clusters 0 and 2 lie on or very close to the 45-degree line, and are therefore normal clusters. Cluster 1, on another hand, is very far from this line, and hence most probably an abnormal cluster. A fact that proves our previous remark based on the second graph.

5 Conclusion

By combining spatial and temporal analyses, our dual-level clustering approach is proving powerful for modelling urban parking availability holistically. It provides a global view of urban parking availability, and detects variations in time and space. Spatial clustering groups parking lots into homogeneous clusters, according to their geographical proximity and infrastructural similarity. Next, within each spatial cluster, temporal clustering segments parking days and hours into different availability patterns. Some clusters are distinguished by high availability throughout daylight hours, probably indicating under-utilized parking lots. Other clusters are marked by limited availability at certain peak times, suggesting parking lots in high demand. Recognizing these profiles, our solution identified specific trends for each urban area, facilitating better resource allocation and parking management strategies tailored to each cluster. Understanding parking availability schemes can prompt concrete action for more efficient urban planning, smoother mobility and a better liveability for citizens.

This approach's first perspective involves taking more parameters into account for either spatial patterns (i.e. city regulations, parking tariffs and urban development projects) or temporal patterns (i.e. road traffic, weather conditions and special events). It is also worth applying transfer learning to improve the performance of models classifying parking profiles using cluster information. Besides, our dual spatial and temporal clustering methodology can be adapted to other complex urban issues, such as traffic management, energy resource optimization or territorial planning. This approach offers a flexible and scalable framework for analysing spatial and temporal data in a wide range of domains.

References

1. Adams, R.P.: Hierarchical agglomerative clustering. In: Proceedings of Ninth SIAM Data Mining Conference (SDM 2009), $\Sigma\lambda$, pp. 510–516 (2009)
2. Ali, M.I., Gao, F., Mileo, A.: CityBench: a configurable benchmark to evaluate RSP engines using smart city datasets. In: Arenas, M., et al. (eds.) ISWC 2015. LNCS, vol. 9367, pp. 374–389. Springer, Cham (2015). https://doi.org/10.1007/978-3-319-25010-6_25
3. Badii, C., Nesi, P., Paoli, I.: Predicting available parking slots on critical and regular services by exploiting a range of open data. IEEE Access **6**, 44059–44071 (2018). https://doi.org/10.1109/ACCESS.2018.2864157
4. Bock, F., Xia, K., Sester, M.: Mapping similarities in temporal parking occupancy behavior based on city-wide parking meter data. Proc. ICA **1**, 1–5 (2018). https://doi.org/10.5194/ICA-PROC-1-12-2018, https://doi.org/10.5194/ica-proc-1-12-2017
5. Camero, A., Toutouh, J., Stolfi, D.H., Alba, E.: Evolutionary deep learning for car park occupancy prediction in smart cities. In: Battiti, R., Brunato, M., Kotsireas, I., Pardalos, P.M. (eds.) LION 12 2018. LNCS, vol. 11353, pp. 386–401. Springer, Cham (2019). https://doi.org/10.1007/978-3-030-05348-2_32
6. Derpanis, K.G.: Mean shift clustering. Lect. Notes **32**, 1–4 (2005)
7. Di Martino, S., Origlia, A.: Exploiting recurring patterns to improve scalability of parking availability prediction systems. Electronics **9**, 838 (2020). https://doi.org/10.3390/ELECTRONICS9050838, https://www.mdpi.com/2079-9292/9/5/838/htm
8. Errousso, H., Abdellaoui Alaoui, E.A., Benhadou, S., Medromi, H.: Exploring how independent variables influence parking occupancy prediction: toward a model results explanation with SHAP values. Prog. Artif. Intell. **11**(4), 367–396 (2022). https://doi.org/10.1007/S13748-022-00291-5/METRICS, https://link.springer.com/article/10.1007/s13748-022-00291-5

9. Errousso, H., Abdellaoui Alaoui, E.A., Benhadou, S., Medromi, H.: Intelligent parking space management: A binary classification approach for detecting vacant spots. Multimedia Tools and Applications (2024)
10. Errousso, H., El Ouadi, J., Benhadou, S., Medromi, H., Malhene, N.: Improving delivery conditions by dynamically managing the urban parking system: Parking availability prediction. In: 2020 13th International Colloquium of Logistics and Supply Chain Management, LOGISTIQUA 2020 (2020). https://doi.org/10.1109/LOGISTIQUA49782.2020.9353890
11. Errousso, H., Malhene, N., Benhadou, S., Medromi, H.: Predicting car park availability for a better delivery bay management. Procedia Comput. Sci. **170**, 203–210 (2020). https://doi.org/10.1016/J.PROCS.2020.03.026
12. Filali, Y., Errousso, H., Aghbalou, N., Abdellaoui Alaoui, E.A., Benhadou, S.: Real-time parking availability classification on a large-area scale. In: The 8th Smart City Applications International Conference. Springer (2023)
13. HajKacem, M.A.B., N'cir, C.-E.B., Essoussi, N.: An accelerated Mapreduce-based K-prototypes for big data. In: Milazzo, P., Varró, D., Wimmer, M. (eds.) STAF 2016. LNCS, vol. 9946, pp. 13–25. Springer, Cham (2016). https://doi.org/10.1007/978-3-319-50230-4_2
14. Huang, Z.: Extensions to the k-means algorithm for clustering large data sets with categorical values. Data Min. Knowl. Discov. **2**(3), 283–304 (1998). https://doi.org/10.1023/A:1009769707641/METRICS, https://link.springer.com/article/10.1023/A:1009769707641
15. Ionita, A., Pomp, A., Cochez, M., Meisen, T., Decker, S.: Where to park? Predicting free parking spots in unmonitored city areas. ACM Int. Conf. Proc. Ser. **22**, 1–12 (2018). https://doi.org/10.1145/3227609.3227648, https://dl.acm.org/doi/10.1145/3227609.3227648
16. Jipkate, B.R., Gohokar, V.: A comparative analysis of fuzzy c-means clustering and k means clustering algorithms. Int. J. Comput. Eng. Res. **2**(3), 737–739 (2012)
17. Johnson, T., et al.: Clustering with Fuzzy C-means and common challenges. J. Phys. Conf. Ser. **1453**(1), 012137 (2020). https://doi.org/10.1088/1742-6596/1453/1/012137, https://iopscience.iop.org/article/10.1088/1742-6596/1453/1/012137
18. Kubat, M.: An Introduction to Machine Learning, pp. 1–348 (2017). https://doi.org/10.1007/978-3-319-63913-0/COVER
19. Liu, G., Huang, T., Chen, H.: Improved bisecting k-means clustering algorithm. Comput. Appl. Softw. **32**(2), 261–277 (2015)
20. Lorbeer, B., Kosareva, A., Deva, B., Softić, D., Ruppel, P., Küpper, A.: Variations on the Clustering Algorithm BIRCH. Big Data Res. **11**, 44–53 (2018). https://doi.org/10.1016/J.BDR.2017.09.002
21. Mohammed, M.: Machine learning : Algorithms and applications. CRC PRESS (2020). https://www.routledge.com/Machine-Learning-Algorithms-and-Applications/Mohammed-Khan-Bashier/p/book/9780367574673
22. Mufida, M.K., Ait El Cadi, A., Delot, T., Trépanier, M., Zekri, D.: Spatiotemporal clustering of parking lots at the city level for efficiently sharing occupancy forecasting models. Sensors **23**(11), 5248 (2023). https://doi.org/10.3390/S23115248, https://www.mdpi.com/1424-8220/23/11/5248/htm
23. Shi, N., Liu, X., Guan, Y.: Research on k-means clustering algorithm: an improved k-means clustering algorithm. In: 3rd International Symposium on Intelligent Information Technology and Security Informatics, IITSI 2010, pp. 63–67 (2010). https://doi.org/10.1109/IITSI.2010.74

24. Shoup, D.C.: Cruising for parking. Transp. Policy **13**(6), 479–486 (2006). https://doi.org/10.1016/J.TRANPOL.2006.05.005
25. Tobin, J., Zhang, M.: Clustering of big data with mixed features (2020). arXiv preprint arXiv:2011.06043
26. Vlahogianni, E.I., Kepaptsoglou, K., Tsetsos, V., Karlaftis, M.G.: A real-time parking prediction system for smart cities. J. Intell. Transp. Syst. **20**(2), 192–204 (2016). https://doi.org/10.1080/15472450.2015.1037955
27. Xu, B., Wolfson, O., Yang, J., Stenneth, L., Yu, P.S., Nelson, P.C.: Real-time street parking availability estimation. Proc. - IEEE Int. Conf. Mob. Data Manag. **1**, 16–25 (2013). https://doi.org/10.1109/MDM.2013.12
28. Zhang, J., Wang, F.Y., Wang, K., Lin, W.H., Xu, X., Chen, C.: Data-driven intelligent transportation systems: a survey. IEEE Trans. Intell. Transp. Syst. **12**(4), 1624–1639 (2011). https://doi.org/10.1109/TITS.2011.2158001
29. Zhang, T., Ramakrishnan, R., Livny, M.: BIRCH: a new data clustering algorithm and its applications. Data Min. Knowl. Discov. **1**(2), 141–182 (1997). https://doi.org/10.1023/A:1009783824328/METRICS, https://link.springer.com/article/10.1023/A:1009783824328
30. Zhang, W., Liu, H., Liu, Y., Zhou, J., Xu, T., Xiong, H.: Semi-supervised city-wide parking availability prediction via hierarchical recurrent graph neural network. IEEE Trans. Knowl. Data Eng. **34**(8), 3984–3996 (2022). https://doi.org/10.1109/TKDE.2020.3034140
31. Zheng, Y., Rajasegarar, S., Leckie, C.: Parking availability prediction for sensor-enabled car parks in smart cities. In: 2015 IEEE 10th International Conference on Intelligent Sensors, Sensor Networks and Information Processing, ISSNIP 2015 (2015). https://doi.org/10.1109/ISSNIP.2015.7106902

Real-Time Parking Availability Classification on a Large-Area Scale

Youssef Filali[1,4], Hanae Errousso[1,2], Nihad Aghbalou[1], El Arbi Abdellaoui Alaoui[3(✉)], and My Abdelouahed Sabri[4]

[1] EIGSI, La Rochelle-Casablanca, 17041-20410 Casablanca, Morocco
[2] National and High School of Electricity and Mechanic, HASSAN II University, 8118 Casablanca, Morocco
[3] IEVIA Team, IMAGE Laboratory, Department of Sciences, ENS, Moulay Ismail University of Meknes, Meknes, Morocco
e.abdellaouialaoui@umi.ac.ma
[4] Faculty of Sciences Dhar-Mahraz, 30000 Fez, Morocco

Abstract. In urban areas around the world, drivers face the daily challenge of finding a parking space. Unfortunately, these sought-after spaces, located close to their destination, are often either impossible to find, or excessively expensive, resulting in longer search times and increased congestion in city centers. The answer to this persistent problem is an intelligent parking solution. They provide drivers with real-time access to information on parking space availability, gathered through various sensing techniques such as crowdsourcing, parking meters and sensors. Some of these systems also offer opportunistic services, such as forecasts, to adapt to unforeseen dynamic situations. Drivers' biggest concern is find ing a spot, not knowing the exact number of available spaces or availability rate. Typically, these two parameters are estimated using regression or image processing techniques. While such solutions guarantee high predictive accuracy, their large-scale deployment is hampered by computational and data collection costs. This paper therefore proposes a new approach combining clustering and classification models to predict parking availability. Our aim is to test new methods that are relatively simple and less expensive in terms of both processing costs and amount of training data. Experimental results have proved promising, with accuracy predictions exceeding 0.84.

Keywords: Urban transportation · parking availability · prediction model · clustering · classification

1 Introduction

The rate of population growth is currently, on the rise. It is projected that by 2030 there will be around 8.5 billion people. The number is expected to increase to 9.7 billion by 2050 and reach a 11.2 billion by the year 2100 [19]. As a result, we are witnessing a concentration of people in cities with urban areas

M. Ben Ahmed et al. (Eds.): SCA 2023, LNNS 906, pp. 215–228, 2024.
https://doi.org/10.1007/978-3-031-53824-7_20

being home to half of humanity. However, this massive migration from areas to cities poses challenges for urban infrastructures and services in terms of environmental impact, scalability and safety [12]. In order to address these challenges municipalities are actively seeking solutions that ensure growth while improving energy efficiency and optimizing the use of natural resources. Additionally, they aim to manage infrastructure and reduce carbon emissions levels. To achieve these goals leveraging advancements in information and communication technologies (ICT) becomes imperative as cities move towards adopting a "city" framework. A "smart city" utilizes ICT to optimize systems and enables municipalities to cater to the increasing needs of their residents [16]. By implementing interconnected infrastructures and leveraging time data collection and analysis capabilities informed decisions can be made for efficient resource management leading to an improved quality of life for citizens across various aspects such as governance, mobility, environment sustainability lifestyle enhancement, among others. Where a smart city is a place where the various aspects of technology, physical structures, social interactions and commercial establishments are interconnected to utilize the knowledge and capabilities of the city. Numerous cities, in Asia, Europe and North America have taken steps to implement their smart city projects [22].

These and similar projects have been made possible by the advent of the Internet of Things (IoT) [3]. The IoT enables easy connection and interaction with a multitude of devices such as household appliances, surveillance cameras, control sensors, parking meters, actuators, displays, vehicles and so on. As a result, it generates a huge amount of varied data that can be exploited in the development of diverse applications such as smart grids, healthcare [13,17], traffic management and wearables. Where smart cities take advantage of this interconnected infrastructure and the data collected to improve their operation, efficiency and quality of life for their citizens. They aim to provide more efficient services, better manage available resources, reduce carbon emissions, improve urban mobility, and create a sustainable and pleasant environment for residents. It represents a major evolution in the way urban environments are designed and managed, using Internet of Things technologies to meet the current and future challenges of fast-growing cities.

In this article, our main contribution consists in outlining an approach for streamlining parking availability prediction using clustering and classification algorithms. This solution guarantees both accurate predictions and significant time savings, by eliminating any need to individually adjust prediction models for each parking lot. To achieve this, it begins by carefully pre-processing input data to guarantee its quality and reliability. Next, clustering algorithms are applied to determine each class's appropriate labels, which serve as independent variables for our various classification models. To improve predictive performance, several algorithms are combined using Majority Vote metaclassifier, where each classifier's weight is parameterized according to its accuracy. By modeling parking space availability as a binary variable (0 if there are available spaces in the area, 1 if there aren't), parking management systems could be revolutionized, enabling administrators to better anticipate parking availability and optimize resource allocation across multiple sites.

Adding a clustering phase is truly an invaluable addition to this work, as it enables tabular data intended for regression problems to be adapted to classification problems requiring qualitative data. This means that users are provided with a level of information that depends on their needs. If drivers only require zone-level information, they can check whether or not parking spaces are available. If, however, drivers have extreme requirements and need to know exactly which spaces are available, then classification will be carried out at parking space level.

The article is structured as follows: Sect. 2 provides a background on predicting the availability of parking spaces. In Sect. 3, the proposed approach is described. Section 4 covers the database used, along with the metrics and results. Lastly, Sect. 5 presents the conclusion and future perspectives of the study.

2 Related Work

Nowadays, predicting the availability of parking spaces has gained in recognition and has been widely studied, although it remains a major challenge. This complex process is closely linked to the interaction of various factors, including users' parking habits, current parking regulations, the type of parking facilities available, parking pricing policy, and drivers' individual behavior and preferences. However, despite the progress made, parking availability forecasting remains an active area of research, facing ongoing challenges. Variable driver behavior, changes in parking policies and unpredictable environmental factors all contribute to the complexity of this prediction, keeping it in constant evolution.

Most studies of parking occupancy prediction, regardless of data type or parking configuration, have adopted machine learning or deep learning techniques exploiting large datasets. For example, Vlahogianni et al. [26] developed a neural network-based model to predict short-term parking occupancy rates. In another study [2], four different techniques were used to estimate the number of available parking spaces in urban garages equipped with barriers, using a Bayesian regularized neural network, support vector regression, a recurrent neural network and an autoregressive integrated moving average. Other algorithms were also evaluated to model this parameter in [12], such as XGBoost, Random forest and K-nearest neighbors. Finally, in [28], the authors compared the performance of three well-known methods, namely regression tree, support vector regression and neural network, using three feature sets composed of parameters carefully selected to illustrate their importance in modeling parking space occupancy.

The issue of predicting parking lot occupancy rates is also addressed in [9], using specific approaches, mainly focused on shape values. In addition to the proposed explanation model, four different algorithms are implemented, each using four distinct feature sets to make these predictions. In another study, Camero et al. [7] focused on improving the performance of recurrent neural networks by introducing two metaheuristic approaches, namely the genetic algorithm and the evolutionary strategy, to optimize the hyperparametric architecture of deep neural networks. As part of this research, they proposed a semi-supervised hierarchical recurrent neural network in [27]. This approach stands out by modeling non-Euclidean spatial autocorrelation between parking lots using a hierarchical graphical convolution module. In addition, it captures local and global spatial dependencies between parking lots using a contextual graph convolution block and a multi-resolution hierarchical graph convolution module.

Other authors have taken a different approach to improving the efficiency of conventional algorithms by testing several stacking architectures, varying the number of levels, types of base regressors and meta-learners deployed [10]. Several papers have explored the problem of parking space availability using vision techniques. In paper [4], the authors present a vision-based intelligent parking framework. This framework retrieves video from surveillance cameras, segments each parking zone into blocks using calibration, then classifies each block to identify cars and inform drivers about the status of the parking lot (reserved or free). Although this image processing cannot predict future events, it can collect real-time data on parking occupancy.

Similarly, in paper [24], Tătulea et al. proposed a multi-stage system for video detection of parking space occupancy. These steps include image pre-processing, adaptive background subtraction, measurement, history creation, and fusion of results for final classification and parking space status. This approach can also determine the number of available parking spaces and their location, but does not predict future parking space availability.

In paper [20], the authors applied region-based convolutional neural networks to recognize available parking spaces by analyzing video data streams from city cameras. They implemented a chatbot for Telegram as a communication channel between users and their system, with a content management system developed in PHP, following the Model-View-Controller architectural scheme. These studies use computer vision and deep learning to solve the problem of parking space availability by providing drivers with real-time information on parking occupancy. However, they focus mainly on the current detection and classification of parking spaces, rather than on predicting future parking space availability. In this article, the authors have chosen the MySQL database management system for their work. Although this study shares the same limitations as the two previous articles in not allowing the prediction of future parking space availability, it nevertheless offers a comprehensive solution to help drivers quickly find parking spaces in a selected parking lot. Another approach to spatio-temporal detection of on-street parking space availability is analyzed in article [5]. They consider six different cab fleet sizes to estimate detection coverage in a real-life large-city scenario. They also evaluate the impact of detection errors on the detected availability information and propose a Kalman filter-based recovery strategy. The results indicate that just 300 cabs are already sufficient to achieve coverage comparable to that of stationary sensors.

According to reviewed literature, there are still unexplored research areas into parking space availability. Most of today's data collection architectures deliver continuous data, which is more suited to regression problems. Besides, it's rare to find a dataset of labeled parking spaces ready to drive classification algorithms. That's why scientists address this issue by resorting to regression algorithms when dealing with tabular data. What's more, adding a binary variable representing a zone's parking space availability, either manually or using an iterative algorithm, can be very time-consuming. Hence our idea of a methodology involving both clustering and classification algorithms.

3 Methodology

Our approach's methodological framework, shown in Fig. 1, represents the operational workflow for classifying parking availability within urban areas. This process takes place in tree main phases, each consisting of several intermediate steps briefly described as follows:

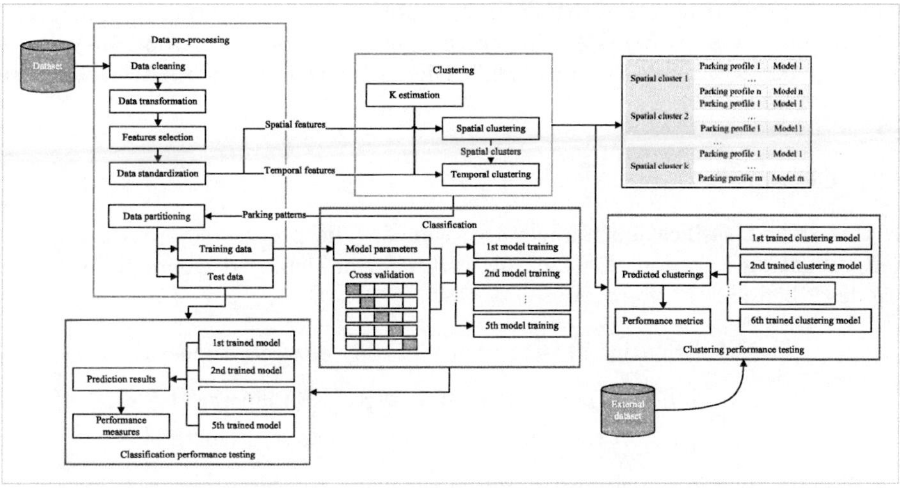

Fig. 1. Flowchart of our proposed parking classification approach

3.1 Pre-processing and Clustering

Each step in this process must be properly handled to avoid distorting the classification result. Poor preprocessing will have an impact on the classification results, resulting in a poor classification rate. For that the first step involves data preprocessing to eliminate noisy values that could falsify obtained results. That contains data cleaning, data transformation, features selection and data normalization. In the clustering phase, the data are first grouped according to their spatial characteristics to determine spatial profiles representing a city. Next, the data are clustered according to each spatial group's temporal characteristics, enabling parking profiles to be defined for each cluster:

- Spatial clustering: This involves grouping parking lots into homogeneous clusters based on their spatial profiles. The spatial profile is defined by various characteristics such as the car park's geographic coordinates, parking capacity, surrounding infrastructure, and the number and proximity of nearby facilities.
- Temporal clustering: This step brings together parking occupancy trends for each spatial cluster, considering factors such as the spatial cluster's occupation rate, the day of the week, and the time of day. Temporal factures are dynamic and evolve significantly over time, allowing us to depict the parking occupancy dynamics.

By combining spatial and temporal clustering, our approach can effectively predict parking availability in real-time, considering both the static characteristics of the parking lots and the dynamic patterns of parking occupancy over time. We tested several algorithms and selected the best performing model, that is Fuzzy C-Means (FCM) [25]. The clusters generated by this method are used as an independent variable for our classification algorithm. For more details, please refer to [11]

3.2 Classifiers

In this step, classification models are trained. The process comprises training and validation iterations for each algorithm to be deployed where the variable are described as:

$$X = \{ \text{ Spatial Cluster Identifier, day of week, time } \}$$

$$Y_i = \begin{cases} 0, & \text{if there are available spaces in cluster } i \\ 1, & \text{if there are no available spaces in cluster } i \end{cases}$$

To estimate the availability of car parks seven machine learning algorithms are selected:

Logistic Regression (LR) [21] is a linear classification technique primarily used for binary classification tasks. It establishes a connection between the dependent variable (binary outcome) and independent variables (features) using a logistic function. By estimating probabilities of an instance belonging to a specific class, LR makes predictions based on a predefined threshold.

Decision Tree (DT) [14] is a non-linear classification algorithm that recursively divides the data based on the most influential feature at each node. It creates a hierarchical structure resembling a tree, where internal nodes represent decisions based on features, and leaf nodes correspond to class labels. Decision trees are easily interpretable and handle both categorical and numerical data.

K-Nearest Neighbors (KNN) [18] is a learning algorithm suitable for classification and regression tasks. It classifies an instance based on the majority class of its k-nearest neighbors in the feature space. KNN is simple to implement and doesn't require training, but its performance can vary depending on the distance metric and the value of k.

Complement Naive Bayes (Complement NB) [23] extends the Naive Bayes algorithm to handle imbalanced datasets, especially when one class dominates the data. It assumes that each class complements the others, making it suitable for imbalanced data classification tasks. Bagging Classifier (BC) [6] is an ensemble learning method that combines multiple base classifiers, such as decision trees, by training them independently on different random subsets of the training data. It aggregates the predictions from individual classifiers to make the final prediction, resulting in improved generalization and reduced variance.

Histogram Gradient Boosting Classifier (HGBC) [15] efficiently implements the Gradient Boosting algorithm. It builds multiple weak learners (typically decision trees) sequentially, with each learner correcting the errors of its predecessors. The algorithm uses histograms to accelerate computations and is known for its high performance and ability to handle large datasets.

Majority Vote [8] another type of ensemble learning. This classifier combines multiple base classifiers through voting, such as majority voting for classification, to make the final prediction. Voting classifiers leverage the strengths of different models, leading to enhanced overall performance.

4 Result

The implementation of all the algorithms in this work was conducted using Python 3 on a computer equipped with an Intel Core^{TM}i7 processor, NVIDIA® GeForce® GTX 1660 Ti with Max-Q design (6 GB GDDR6 dedicated), and 8 GB memory. Our proposed approach will undergo evaluation and validation on the dataset 2014.

4.1 Dataset

The database used in this study contains a total of 55,264 data points and is accessible through reference [1]. Before constructing the proposed models, the data underwent several corrective operations to enhance its quality and coherence. These operations included the following steps:

- Data Cleaning: Rows with negative numbers of occupied parking spaces were deleted, and irrelevant columns (features) were removed. The data was then aggregated per parking lot to facilitate analysis.
- Car Park Naming: Each car park was assigned a distinct integer number between 1 and 8 for identification purposes.
- Time Information: The "Update time" variable was utilized to extract the day, month, year, and time information. Additionally, the day of the week was derived from the same variable and represented as an integer (1 for Monday, 2 for Tuesday, ... to 7 for Sunday).

- Weekend Indicator: A new binary column was added to indicate whether the day of the week was a weekday or a weekend.
- Occupancy Percentage: The percentage of parking spaces occupied was calculated by dividing the number of vehicles parked in a parking lot by its capacity and then multiplying the ratio by 100.
- Limiting Occupancy Rates: Parking lot occupancy rates exceeding 100% were capped at 100% to ensure consistency.
- Handling Missing Data: In case of missing data for a particular hour, the previous hour's value was replicated. If data for the entire hour was missing, the corresponding value from the same day of the preceding week was used.

After these data pre-processing and transformation steps, the database was reduced to 11,053 records related to 8 car parks, spanning 167 d. The pre-processing step is an important step that can affect the clustering quality and then the classification rate.

In our previous work [11], Data are first grouped according to their spatial characteristics using K-prototype algorithm with a cluster number of 3. Multiple performance measures, such as adjusted mutual information, are computed to evaluate how well clusters match true labels. They have demonstrated its effectiveness in forming coherent and pure clusters. Next, temporal clustering per spatial cluster with Fuzzy C-means generates very good results with an adjusted Rand index of 97.33, completeness of 95.20, Fowlkes mallows index of 98.85 and V-measure of 94.95%. These results are used as input for our classification algorithms. Correct pre-processing and clustering will inevitably lead to good parking lot classification.

4.2 Metrics

To evaluate our approach, we use different metrics as Accuracy, Balanced accuracy, F1-score, precision, Recall, Jaccard index and Roc-AUC, see Table 1.

With: TP, TN, FP, and FN are respectively: The number of true positives; the number of true negatives; the number of false positives; the number of false negatives.

4.3 Discussion

From Table 2, we can see the results of different metrics (Accuracy, Balanced accuracy, F1-score, precision, Recall, Jaccard index) obtained with six algorithms (LR, DT, KNN, CNB, BC, HGBC) and the Vote technique, we can conclude the following:

The Logistic Regression achieves average to bad performance in most metrics. Its best performances are in the Accuracy and Balanced accuracy, where it obtains values of 0.6612, 0.5 respectively. However, LR performs very poorly in the Recall and Jaccard where it obtains a value of zero or very close to zero.

Decision Tree shows average to good performance in most metrics. Its best performances are in the Accuracy, Balanced accuracy, F1 and Precision metrics,

Table 1. Performance metrics description

Metrics	Formula	Objectif of the evaluation
Accuracy (ACC)	$\frac{TP+TN}{TP+TN+FP+FN}$	Measure the ratio of accurate predictions to the overall number of evaluated instances
Specificity	$\frac{TN}{TN+FP}$	Calculate the percentage of negative instances that are correctly classified
Sensitivity or Recall	$\frac{TP}{TP+FP}$	Calculate the percentage of positive instances that are correctly classified.
Balanced accuracy (B_ACC)	$\frac{\text{Sensitivity} + \text{Specificity}}{2}$	Evaluate the effectiveness of a classification model by the half of the sum of sensitivity and specificity
Precision	$\frac{TP}{TP+FP}$	Calculate the ratio between the number of correctly predicted positive instances and the total number of predicted instances in a positive class
F1- Score (F1)	$\frac{(\text{Precision} \times \text{Recall}) \times 2}{(\text{Precision} + \text{Recall})}$	Calculate the fraction of the harmonic average between precision and recall
Random accuracy	$\frac{(FP+TN) \times (FN+TN) + (TP+FN) \times (TP+FP)}{\text{Total}^2}$	Measure the difference between the number of positive and negative instances divided by the total of instances
Jaccard	$\frac{\text{Accuracy} - \text{radom accuracy}}{1 - \text{random accuracy}}$	Estimate the similarity between two sets

where it obtains values of 0.7103, 0.6143, 0.4255 and 0.6480 respectively. However, like LR, DT performs poorly in the Recall and Jaccard metrics, with values close to zero. K-Nearest Neighbors (KNN) shows the same result that get with decision tree. In another hands, Complement Naive Bayes (CNB)shows mixed performance in the various metrics, without particularly standing out. Its best performances are in the recall metrics, where it obtains values 0.5138.

Bagging Classifier performs well overall in all metrics. Its best performances are in the Accuracy, Balanced accuracy, F1 and Jaccard metrics, where it achieves values of 0.7112, 0.6174, 0.4340 and 0.27717 respectively. Finally, Histogram Gradient Boosting Classifier is the algorithm with the most variable performance among the metrics. It achieves performance in the Recall metric, with a value of 0.3167. However, it performs relatively poorly in the Accuracy, Balanced accuracy and F1 metrics, compared with the other algorithms. Figures 2 and 3 shows the confusion matrix and ROC for each algorithm.

In summary, each algorithm has strengths and weaknesses in the various metrics evaluated. That's why we came up with the idea of exploiting the strengths of the six algorithms by using majority voting. The majority voting model could be used as a global solution if the objective is to obtain the best overall performance without focusing on a specific metric. For the "Accuracy" (ACC) metric, the majority voting model performs best, with a value of 0.8407, the "Balanced

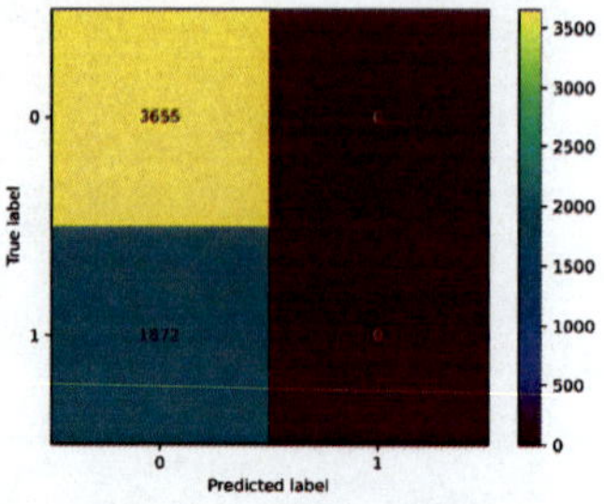

(a) Logistic Regression

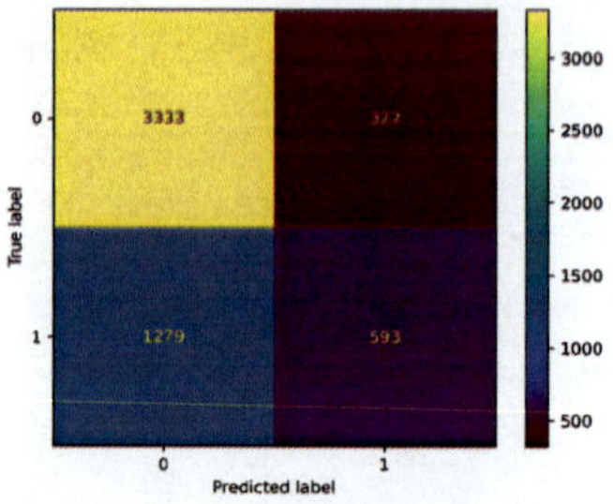

(b) Decision Tree

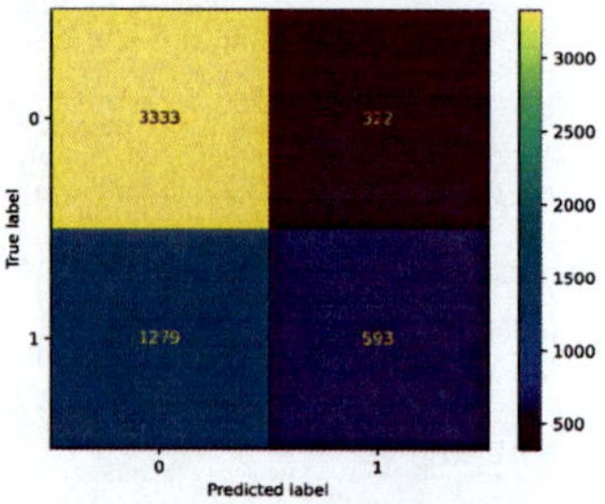

(c) K-Nearest Neighbours

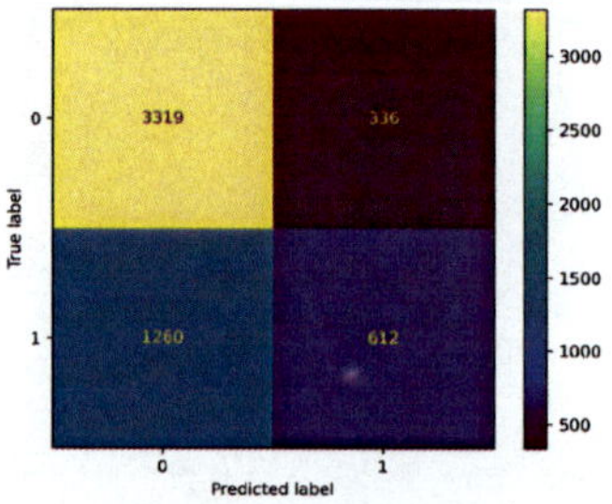

(d) Bagging Classifier

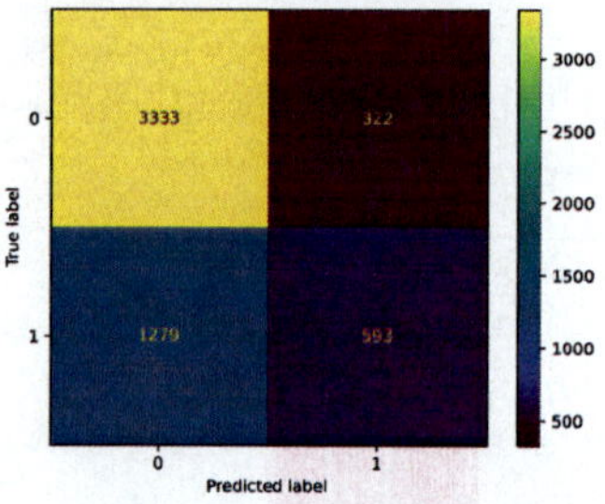

(e) Histogram Gradient Boosting

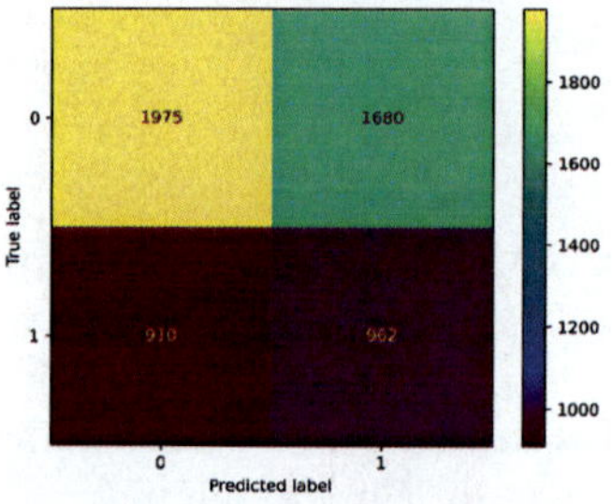

(f) Complement Naive Bayes

Fig. 2. Confusion matrix of the different classifiers

Accuracy" (B_ACC) metric, the majority voting model also performs best, with a value of 0.8264, the "F1" metric shows that the majority voting model performs best, with a value of 0.4491, the "Precision" metric shows that the majority voting model has the best performance with a value of 0.7820 also "Jaccard" metric shows that the majority voting model performs best, with a value of 0.624. As for the "Roc_auc" metric, the values for the majority voting model and the BC algorithm are identical (0.827). The majority voting model generally shows the best performance for most of the metrics evaluated, suggesting that it is more robust and capable of performing well under different conditions. The confusion matrix and ROC are showed in Fig. 4.

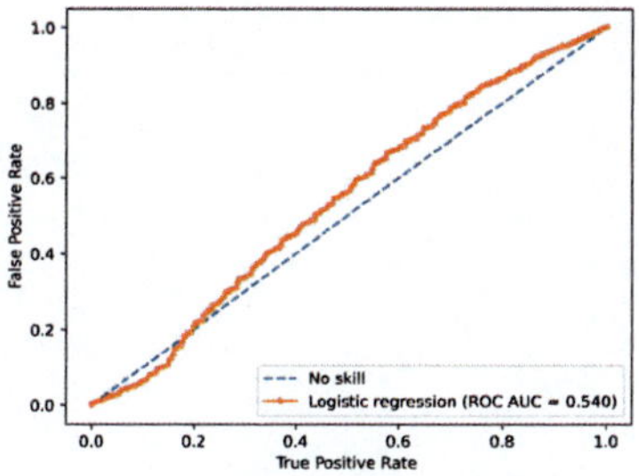

(a) Logistic Regression

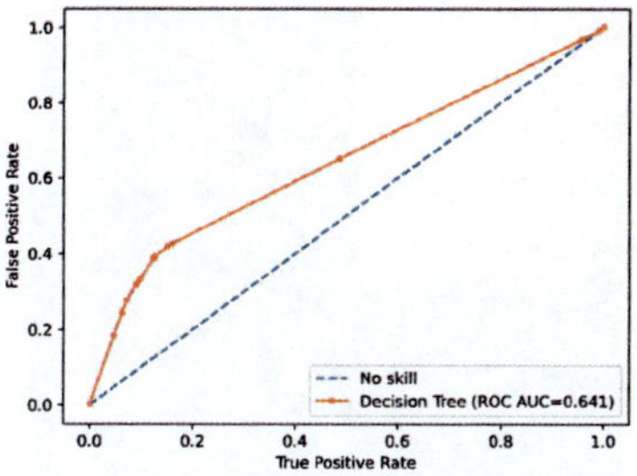

(b) Decision Tree

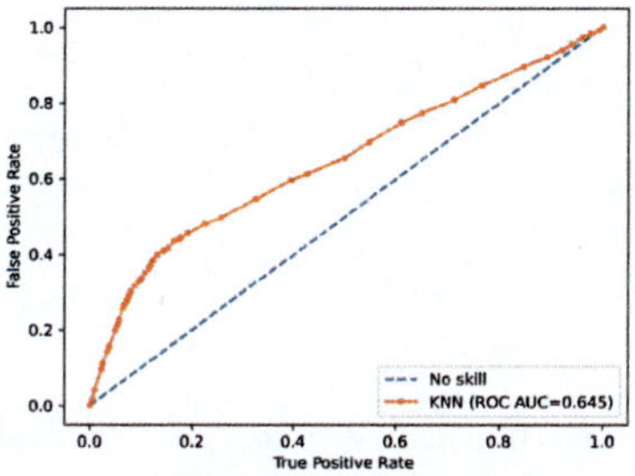

(c) K-Nearest Neighbours

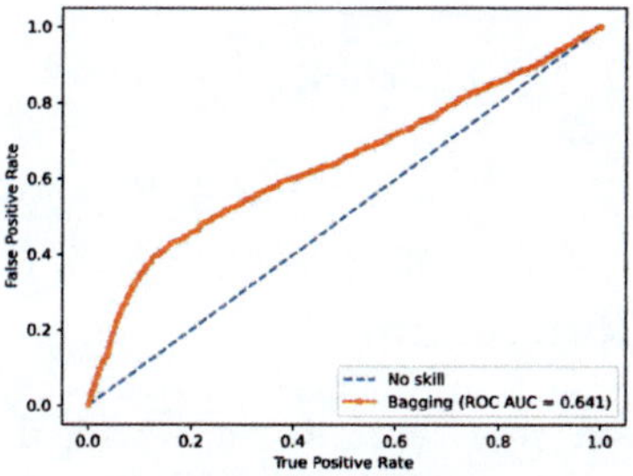

(d) Bagging Classifier

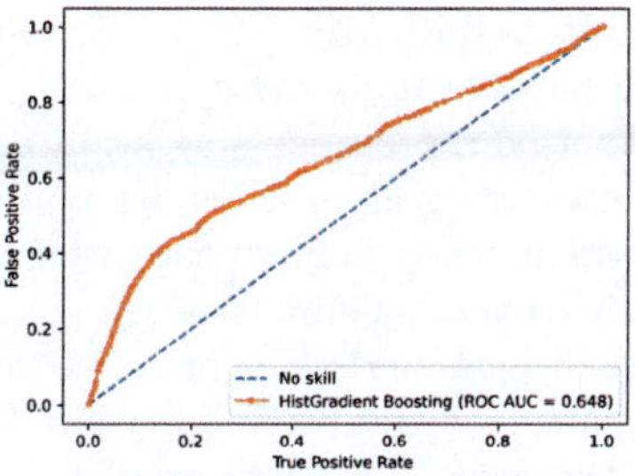

(e) Histogram Gradient Boosting

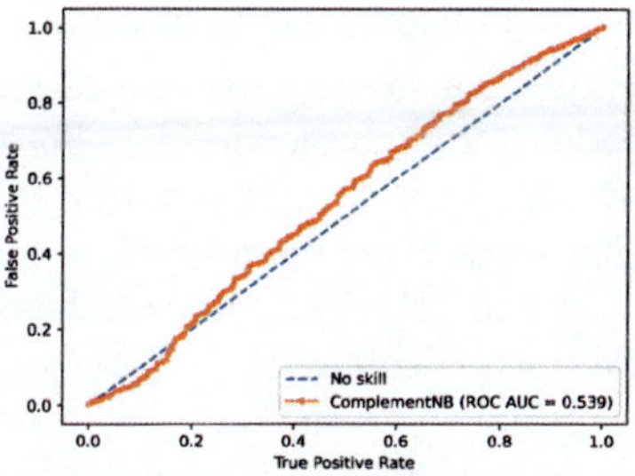

(f) Complement Naive Bayes

Fig. 3. Classification ROC curve of different classifiers

Table 2. Metrics result of different classifiers

Metric	LR	DT	KNN	CNB	BC	HGBC	Vote
ACC	0.6612	0.7103	0.7110	0.5313	0.7112	0.7103	0.8407
B_ACC	0.5	0.6143	0.6150	0.5271	0.6174	0.6143	0.8264
F1	0.0	0.4255	0.4255	0.4262	0.4340	0.4255	0.4491
Precision	0.0	0.6480	0.6480	0.3641	0.6455	0.6480	0.7820
Recall	0.0	0.3167	0.3167	0.5138	0.32692	0.3167	0.3150
Jaccard	0.0	0.2702	0.2702	0.2708	0.27717	0.2702	0.624

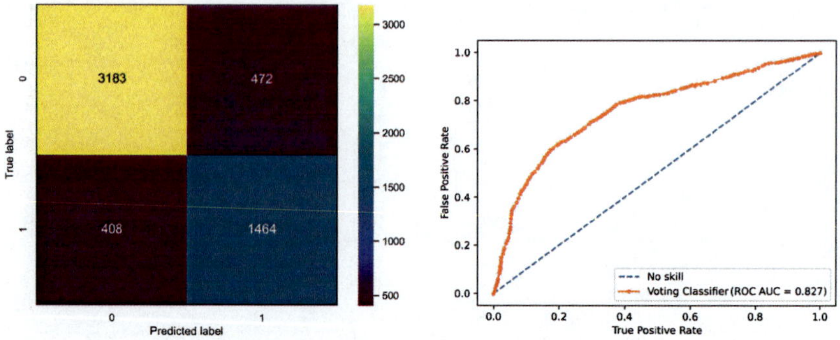

(a) Confusion matrix (b) ROC curve

Fig. 4. Result of voting classification

5 Conclusion

In urban areas around the world, drivers face the daily challenge of finding a parking space. In this article, our main contribution consists in outlining an approach for streamlining parking availability prediction using clustering and classification algorithms. This solution guarantees both accurate predictions and significant time savings, by eliminating any need to individually adjust prediction models for each parking lot. For that we used a preprocessing input data to guarantee its quality and reliability. Next, clustering algorithms are applied to determine each class's appropriate labels, which serve as independent variables for our various classification models. Finally, several algorithms are combined using Majority Vote meta-learner, where each classifier's weight is parameterized according to its accuracy. By modeling parking space availability as a binary variable (0 if there are available spaces in the area, 1 if there aren't), Experimental results have proved promising, with accuracy predictions exceeding 0.84. As a future work we will use deep learning classifier to improve the classification rate.

References

1. Babar, M., Arif, F., Jan, M.A., Tan, Z., Khan, F.: Urban data management system: towards big data analytics for internet of things based smart urban environment using customized hadoop. Futur. Gener. Comput. Syst. **96**, 398–409 (2019)
2. Badii, C., Nesi, P., Paoli, I.: Predicting available parking slots on critical and regular services by exploiting a range of open data. IEEE Access **6**, 44059–44071 (2018)
3. Bellini, P., Nesi, P., Pantaleo, G.: IoT-enabled smart cities: a review of concepts, frameworks and key technologies. Appl. Sci. **12**(3), 1607 (2022)
4. Bibi, N., Majid, M.N., Dawood, H., Guo, P.: Automatic parking space detection system. In: 2017 2nd International Conference on Multimedia and Image Processing (ICMIP), pp. 11–15. IEEE (2017)

5. Bock, F., Di Martino, S., Origlia, A.: Smart parking: using a crowd of taxis to sense on-street parking space availability. IEEE Trans. Intell. Transp. Syst. **21**(2), 496–508 (2019)
6. Breiman, L.: Bagging predictors. Mach. Learn. **24**, 123–140 (1996)
7. Camero, A., Toutouh, J., Stolfi, D.H., Alba, E.: Evolutionary deep learning for car park occupancy prediction in smart cities. In: Battiti, R., Brunato, M., Kotsireas, I., Pardalos, P.M. (eds.) LION 12 2018. LNCS, vol. 11353, pp. 386–401. Springer, Cham (2019). https://doi.org/10.1007/978-3-030-05348-2_32
8. Dogan, A., Birant, D.: A weighted majority voting ensemble approach for classification. In: 2019 4th International Conference on Computer Science and Engineering (UBMK), pp. 1–6. IEEE (2019)
9. Errousso, H., Abdellaoui Alaoui, E.A., Benhadou, S., Medromi, H.: Exploring how independent variables influence parking occupancy prediction: toward a model results explanation with shap values. Prog. Artifi. Intell. **11**(4), 367–396 (2022)
10. Errousso, H., El Ouadi, J., Benhadou, S., Medromi, H., Malhene, N.: Improving delivery conditions by dynamically managing the urban parking system: Parking availability prediction. In: 2020 IEEE 13th International Colloquium of Logistics and Supply Chain Management (LOGISTIQUA), pp. 1–6. IEEE (2020)
11. Errousso, H., Filali, Y., Aghbalou, N., Abdellaoui Alaoui, E.A., Benhadou, S.: Spatio-temporal clustering for optimal real-time parking availability estimation. In: The 8th Smart City Applications International Conference. Springer (2023)
12. Errousso, H., Malhene, N., Benhadou, S., Medromi, H.: Predicting car park availability for a better delivery bay management. Procedia Comput. Sci. **170**, 203–210 (2020)
13. Filali, Y., El Khoukhi, H., Sabri, M.A., Aarab, A.: Analysis and classification of skin cancer based on deep learning approach. In: 2022 International Conference on Intelligent Systems and Computer Vision (ISCV), pp. 1–6. IEEE (2022)
14. Fletcher, S., Islam, M.Z.: Decision tree classification with differential privacy: a survey. ACM Comput. Surv. (CSUR) **52**(4), 1–33 (2019)
15. Guryanov, A.: Histogram-based algorithm for building gradient boosting ensembles of piecewise linear decision trees. In: van der Aalst, W.M.P., et al. (eds.) AIST 2019. LNCS, vol. 11832, pp. 39–50. Springer, Cham (2019). https://doi.org/10.1007/978-3-030-37334-4_4
16. Hamamurad, Q.H., Jusoh, N.M., Ujang, U.: Adapting cities name with ICT and countries interested in the smart city. Int. J. Urban Civ. Eng. **17**(1), 24–30 (2023)
17. Ibrahim, K., Filali, Y., Sabri, M.A., Aarab, A.: Embedded skin cancer detection and classification on raspberry Pi. In: Masrour, T., El Hassani, I., Cherrafi, A. (eds.) A2IA 2020. LNNS, vol. 144, pp. 297–306. Springer, Cham (2021). https://doi.org/10.1007/978-3-030-53970-2_28
18. Kataria, A., Singh, M.: A review of data classification using k-nearest neighbour algorithm. Int. J. Emerg. Technol. Adva. Eng. **3**(6), 354–360 (2013)
19. Khatoun, R., Zeadally, S.: Smart cities: concepts, architectures, research opportunities. Commun. ACM **59**(8), 46–57 (2016)
20. Levkivskyi, V., Marchuk, D., Lobanchykova, N., Pilkevych, I.A., Salamatov, D.I.: Available parking places recognition system. In: CEUR Workshop Proceedings 4th Workshop for Young Scientists in Computer Science & Software Engineering. vol. 3077, pp. 123–134 (2022)
21. Maalouf, M.: Logistic regression in data analysis: an overview. Int. J. Data Anal. Tech. Strat. **3**(3), 281–299 (2011)

22. Richter, M.A., Hagenmaier, M., Bandte, O., Parida, V., Wincent, J.: Smart cities, urban mobility and autonomous vehicles: how different cities needs different sustainable investment strategies. Technol. Forecast. Soc. Chang. **184**, 121857 (2022)
23. Seref, B., Bostanci, E.: Performance comparison of naïve bayes and complement naïve bayes algorithms. In: 2019 6th International Conference on Electrical and Electronics Engineering (ICEEE), pp. 131–138. IEEE (2019)
24. Tătulea, P., Călin, F., Brad, R., Brâncovean, L., Greavu, M.: An image feature-based method for parking lot occupancy. Future Internet **11**(8), 169 (2019)
25. Tilson, L., Excell, P., Green, R.: A generalisation of the fuzzy c-means clustering algorithm. In: International Geoscience and Remote Sensing Symposium,'Remote Sensing: Moving Toward the 21st Century'. vol. 3, pp. 1783–1784. IEEE (1988)
26. Vlahogianni, E.I., Kepaptsoglou, K., Tsetsos, V., Karlaftis, M.G.: A real-time parking prediction system for smart cities. J. Intell. Transp. Syst. **20**(2), 192–204 (2016)
27. Zhang, W., Liu, H., Liu, Y., Zhou, J., Xu, T., Xiong, H.: Semi-supervised city-wide parking availability prediction via hierarchical recurrent graph neural network. IEEE Trans. Knowl. Data Eng. **34**(8), 3984–3996 (2020)
28. Zheng, Y., Rajasegarar, S., Leckie, C.: Parking availability prediction for sensor-enabled car parks in smart cities. In: 2015 IEEE Tenth International Conference on Intelligent Sensors, Sensor Networks and Information Processing (ISSNIP), pp. 1–6. IEEE (2015)

A Review of a Research in Autonomous Vehicles with Embedded Systems

Fulya Akdeniz, Mert Atay, Şule Vural, Burcu Kır Savaş(✉), and Yaşar Becerikli

Kocaeli University, Umuttepe Campus, Kocaeli 41001, Turkey
{fulya.akdeniz,180202090,190201062,burcu.kir,
ybecerikli}@kocaeli.edu.tr

Abstract. Designing an embedded system for practical applications needs objectives such as high accuracy, low power consumption and cost, and secure environment. NVIDIA's Jetson is the most popular platform for embedded system applications which promises to achieve a balance between all these objectives. This paper focused on autonomous vehicles and embedded system studies evaluating the Jetson card features used in smart cars. The characteristics of autonomous vehicles and presents the data was compared for researchers categorized by author names, vehicle names, platform, camera, purpose and environment. In paper shows that most of studies had different purposes and tested area in different environments. We also developed a cost-effective lane detection model for developing and testing the autonomous vehicle JetRacer AI Kit and detail the overall structure of the system in this work. We tested the lane detection performance for system feasibility. In addition, we presented our ongoing work to further develop the features of this vehicle based on the proposed framework.

Keywords: Nvidia Jetson Nano · Autonomous Vehicle · Lane Detection · Embedded Systems · Hough Transform · JetRacer AI Kit

1 Introduction

The world population has been growing steadily [1]. Recent studies indicate that there are many magnificent developments in smart and autonomous embedded systems and such systems are extensively used in autonomous vehicles such as drones, underwater vehicles and robots etc. With the Subterranean Challenge for autonomous vehicles, organized by the Defense Advanced Research Projects Agency (DARPA) in 2017 and 2021 respectively, robotic systems and their virtual implementation have been tested for mapping, navigating, and autonomously searching subterranean environments by both academia and industry. These implementations bring about other developments and global competition by pioneering companies such as Google, Nissan, BMW, and Tesla by using both real and virtual tests [2, 3]. Generally, autonomous vehicles use different sensors such as, external/internal camera, GPS/radar, lidar, proximity and position sensors, CPU, GPU, etc. Moreover, these vehicles combine various image/signal processing algorithms, machine learning algorithms, encode/decode and decision mechanisms. The

M. Ben Ahmed et al. (Eds.): SCA 2023, LNNS 906, pp. 229–239, 2024.
https://doi.org/10.1007/978-3-031-53824-7_21

self-determination level of these vehicles is specified by the integration of all different algorithms and devices at different grades of capability [4, 5]. Autonomous vehicles use real-time image processing, audio signal processing and control techniques to achieve safe driving [6–8]. Although autonomous vehicles have sensing, planning, and control capabilities, they are among the embedded systems that have gained popularity in the last decade. Recent developments in deep learning methods have contributed significantly to the development of these systems. These methods also facilitate the processing of data from camera images and pave the way for the advancement of other approaches such as learning hardware commands and autonomous driving ability [9]. Lane and line detections are highly useful for autonomous portable platforms such as cars, robots, and drones although it is a difficult task to achieve high accuracy and meet the essential requirements demanded by autonomous vehicles in real-time. The basic principles of conventional image processing algorithms [10] were based on principles such as color and edge processing to detect lane and line. Deep learning algorithms become a popular tool to detect and classify lanes. Also, the success in the accuracy of these systems is promising. JetRacer is an open-source platform that supports Donkey Car projects for industrial and academic areas. It is implemented with the Nvidia Jetson Nano hardware system. Generally, this system is suggested by researchers to improve embedded implementations. In this study, we tested the lane detection performance of this vehicle for system feasibility. The system's flowchart is given in Fig. 1. Contributions of this study,

- This paper was the first comprehensive review of research in autonomous vehicles with embedded systems such as JetRacer AI Kit.
- We developed an end-to-end lane detection system for autonomous vehicles.
- We performed image processing and lane detection processes using a HOG Transform model.

In the sections that follow, we explain related work in Sect. 2. After describing Hough transform for lane detection in Sect. 3, we detail experimental results in Sect. 4. We discuss our conclusions in Sect. 5.

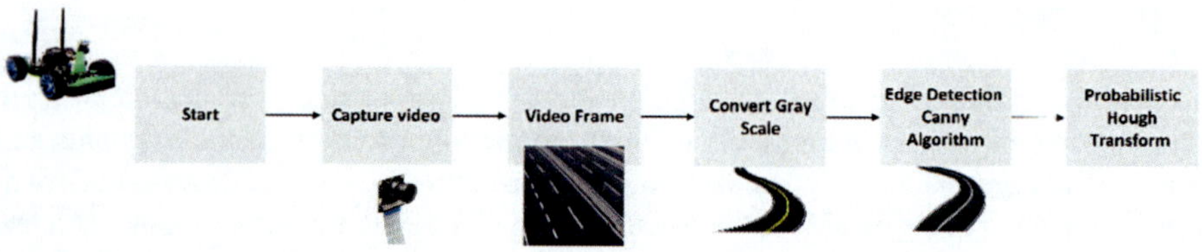

Fig. 1. System's flowchart

The first module, images were obtained from the camera. The second module, images were divided into frames. Framed images were converted into gray format in the third module. The fourth module, Canny algorithm was used to edge of images. In the last module, Probabilistic Hough Transform method was developed to track the lines.

2 Material Method

Roberts et al. [11] studied how to use cyber security test environments (cyber-physical testbed) to test autonomous vehicles. They used the MIT Duckiebot autonomous vehicle for testing in their study. Raspberry Pi embedded card and the 5-megapixel camera are integrated into this vehicle. Also, the whole system has a 10400 mAh battery. Kawakura et al. [12] designed a smart and flexible self-driving autonomous vehicle for smart agriculture. They developed a deep learning-based system that finds and avoids objects. They also utilized an autonomous vehicle named the JetBot AI kit, which has embedded system hardware developed by NVIDIA. This kit contains NVIDIA Jetson Nano Developer Kit and has an operating system using Linux4Tegra. This kit has been integrated with a Sony IMX-219 camera module which is 8 megapixel 160 FOV with a resolution of 3280 × 2464 pixels. Using CNN and LSTM deep learning networks, Bingöl et al., [13] examined the learning of autonomous vehicles from human driving based on data obtained from drivers. They utilized a Radio-controlled and remote-controlled Traxxas Slash 4 × 4 Platinum vehicle produced by Traxxas company. In the test stage, they have integrated Vedder Electronic Speed Controller (VEHK) into this autonomous vehicle. Also, this vehicle contains NVIDIA Jetson TX2 and has the operating system using Linux4Tegra. This system is integrated with ZED stereo camera which is 110 FOV and 2B-LIDAR for image processing. Yuhas et al. [14] improved a deep neural network-based out-of-distribution (OOD) detector. They utilized an autonomous vehicle named the Duckiebot for the embedded system part. This vehicle contained Raspberry Pi 3B + with a 64-bit CPU operating at 1.4GHz. Also, this vehicle has been integrated with a 160 FOV camera. Castro et al. [10] worked on the detection of lane in a circuit. They developed a color-based lane detection system. This autonomous vehicle contained Raspberry Pi 3 hardware which included a camera for image processing progress. Loeminto et al. [15] developed a traffic light detector using a Duckiebot in this study. This vehicle contained Raspberry Pi 3 hardware system. Also, this vehicle has been integrated a fisheye lens camera to get better visual surveillance. Kumar et al. [16] examined path and curve line using white pixels and center from frames. They utilized an autonomous vehicle named the Auto Car Mk II. In this vehicle, Arduino Uno and Arduino Nano hardware systems are used to establish electronic circuits and to communicate between modules and sensors. Also, this vehicle contained IP Webcam, L298N motor driver, NRF24L01 2.4GHz wireless modules and HC-SR04 Ultrasonic Sensor. Seth et al. [17] demonstrated an autonomous vehicle's self-driving learning using CNNs. They used real 1/10th scaled autonomous race vehicles. This autonomous vehicle contained Raspberry Pi 4 hardware system. Also, this vehicle has been integrated with SainSmart Wide Angle 5 MP Fish-Eye Camera, 2,000mAh battery, 10,000 mAh Li-on power portable power bank, Ultrasonic HC-SR04 sensor, PCA 9685 servo driver, DC motor and step motors. Sandha et al. [18] examined Deep Reinforcement Learning performance, the parameters that influence Deep RL and sampling rates alternation. They utilized an autonomous vehicle named DeepRacer for this performance analysis. This vehicle had 1/18th scaled autonomous and this vehicle contained an Intel Atom processor and an embedded GPU. Also, a front facing camera was added to this vehicle. Stocco et al. [4] examined autonomous vehicle movements for both real environment and simulation platform. They analyzed the similarity of the simulation environment to the real environment and the transferability. They also compared real environment results with simulation environment results and calculated similarity results. They utilized an autonomous vehicle named Donkey platform which is 1:16 scale for this similarity results. Katyal et al. [19] developed various algorithms. These algorithms are used to predict areas out of range of the vehicle sensor for reliable planning at high speeds. They utilized GANs for areas prediction. They utilized GANs for future prediction. They also used an autonomous vehicle named MIT race vehicle. This vehicle was built on 1/10-scale Traxxas Rally Car platform and it contains Nvidia Jetson TX2 hardware system. Also, this vehicle has been integrated with Intel Real sense D435 and T265 cameras. O'Kelly et al. [20] examined the research and education needs that will be required for future autonomous systems and automotive Cyber-Physical Systems. They explained the F1/10 test bed environment and how his environment could be used in various studies. In their study, they used an autonomous vehicle that contains Jetson TX2 embedded hardware system. Also, this vehicle included USB web cam, a ZED depth camera, Hokuyo 10LX scanning LIDAR modules. O'Kelly et al. [21] indicated F1TENTH autonomous racing platform properties in their study. They utilized 1/10th scaled DeepRacer autonomous vehicle for this requirement. This vehicle contained NVIDIA Jetson TX2 embedded hardware system and has been integrated with Hokuyo UST-10LX LIDAR, depth camera. Liu et al. [22] proposed a navigation system for self-driving autonomous vehicles using many cameras and a wheel odometer. Kalman Filter Based Sensor Fusion algorithm used for this multi-camera and wheel odometry. They used Quanser's Qcar. The autonomous vehicle contains NVIDIA Jetson TX2 embedded hardware system. Hu et al. [23] proposed a new trajectory tracking control system based on a kinematic model. They utilized an autonomous vehicle named Quanser self-driving car (QCar) in this study. The Qcar contains NVIDIA Jetson TX2 embedded hardware system. Sukhil et al. [24] designed a greedy algorithm for optimizing race metrics such as average lap speed, lap time, total deviation from the reference trajectory. They utilized F1/10 autonomous race vehicles platform for testing. This vehicle contains NVIDIA Jetson TX2 embedded system, Lidar and Depth camera for vehicle control. Also, the vehicle has a 5000 mAh li-po battery, electronic speed controller and various sensors. Balaji et al. [25] examined Reinforcement Learning algorithm for self-driving vehicles with monocular camera. The authors used a 1/18th scale DeepRacer autonomous vehicle. This autonomous vehicle contains an Intel Atom Processor. Also, this vehicle has been integrated with 4GB RAM, 4 MP Camera. Chiba et al. [26]. Examined the activity of Reinforcement Learning in autonomous vehicles. The authors employ 1/18th-scale AWS DeepRacer Car for the testing phase. This vehicle has an Intel Atom processor. Also, this vehicle included a 4MP Camera, 4GB RAM, Accelerometer, Gyro sensor. Karaman et al. [27], examined pedagogical implications of high-school STEM program in robotics organized by MIT Beaver Works Summer Institute Robotics Program. They used MIT Racecar which is a 1/10-scale Traxxas RC Rally Car in this program. The car contained Nvidia Jetson TX1 embedded hardware system. Also, this vehicle included a high-resolution ZED 3D stereo camera, Hokuyo UST 10-LX 2D Lidar Scanner and electronic speed control for image and pattern detection. Katyal et al. [19], studied robust occupancy map prediction at higher speeds. They examined Generative Neural Networks for predicting future occupancies. This vehicle is built on a 1/10-scale Traxxas Rally Car platform. The autonomous vehicle contained Nvidia Jetson TX2 embedded hardware system and Intel Realsense D435 and T265 cameras. Scheffe et al. [28], explained how a Racecar can be transformed into a networked autonomous vehicle. They used μCar which is 1:18 model-scale autonomous vehicle for the testing phase. This vehicle has been integrated with Raspberry Pi Zero W hardware and this vehicle does not have any a camera. Verschueren et al. [29] proposed a nonlinear model predictive control (NMPC) method to minimize the lap time. They used an autonomous vehicle named the Kyosho dNaNo. A comparison of research is given Table 1.

3 Probabilistic Hough Line Transform

The HOG is a method frequently used in object and pattern recognition. HOG features provide high performance under different and complex conditions and environments. During edge extraction phase, HOG method utilizes histograms for target shape. The main purpose of the HOG method is to describe an image as a set of local histograms. The numerical and graphical distribution of color values in an image is called a histogram. With the histogram values of the image, the brightness and tone information of the image can be resolved. One of the biggest problems of HOG Transform (HT) is that the method complexity is highly dependent on the size of the input data. Unlike the HT method, the Probability HT (PHT) method is independent of the size, shape and arrangement of the accumulator array. The PHT method also has a lower bandwidth and is much more robust/extra strong/resistant to noise [30]. PHT transformation greatly reduces the memory consumption and computational complexity [31]. The HT takes all points into account in lane detection, while the PHT transform does not consider all points. Instead of these points, it picks and uses a random sample of these points. The equation of lane detection is given in Eq. 1 and Eq. 2 [32, 33]. X_k feature values, a specific point in the z parameter space. p(y|Xm) is the probability density function.

$$H(z) = \ln\left[p(z|X_k)\right] \tag{1}$$

$$\mathrm{H}(z) = \Sigma_{t=1}^{k} \ln\left[p(x_t|z)\right] + \ln\left[p(z)\right] + C \tag{2}$$

4 Exerimental Analysis

This study consists of 4 stages: (1) The first stage was the integration part; the Jetson Nano embedded system kit mounted on the JetRacer vehicle was activated. System-required packages installed on this embedded kit using calculation and processing methods for image algorithms, and the basic setup is carried out. (2) The second stage was the software part; the video dataset was analyzed and all frames in the video preprocessed for lane detection. The image turned into a gray format because the color separation makes line detection difficult. Hence, it was studied in a single color in the study. (3) The third stage was the filter part; the Canny filter was applied to the image to determine the edges in the grayscale processed image. Cropping was implemented to remove the parts and noises

Table 1. A comparison of research.

Authors	Vehicle Name Platform	Camera	Field	Environment
Roberts et al	MIT Duckiebot/ Raspberry Pi Model3B	5 MP Raspberry Pi	automotive Cyber-physical test beds	real
Kawakura et al	JetBot AI kit NVIDIA Jetson Nano	Sony IMX 219	agri-workplace	real
Bingöl et al	Traxxas NVIDIA TX2	ZED stero, A2M6 2B-Lidar	to learn autonomous self-driving	real
Yuhas et al	Duckiebot Raspberry Pi 3B +	–	out-of-distribution detection	real
Castro et al	Duckiebot Raspberry Pi 3	–	lane following	real
Loeminto et al	Duckiebot Raspberry Pi 3	fisheye lens	traffic Light detection	real
Kumar et al	Auto Car Mk II Arduino Nano	IP Webcam	path selection and curve detection	real
Seth et al	– Raspberry Pi 4	5 MP fisheye lens	learn self-driving	real
Sandha et al	DeepRacer –	DeepRacer	they showed the efficacy of Time-inState RL on HalfCheetah	Simulation & real
Stocco et al	Donkey Car –	–	Simulated vs real environment	Simulation & real
Katyal et al	MIT Racecar NVIDIA JetsonTX2	Intel Realsense D435 and T265	robust occupancy map prediction for higher speeds	real
O'Kelly et al	F1/10 NVIDIA JetsonTX2	ZED depth cam, Hokuyo 10LX Lidar	automotive Cyber-Physical Systems	real
O'Kelly et al	F1/10 NVIDIA JetsonTX2	ZED depth cam, Hokuyo 10LX Lidar	define autonomous racing platform	Simulation & real
Liu et al	Quanser's Qcar NVIDIA JetsonTX2	CSI	a navigation system	–

(*continued*)

Table 1. *(continued)*

Authors	Vehicle Name Platform	Camera	Field	Environment
Hu et al	Qcar NVIDIA JetsonTX2	–	trajectory tracking control system	Simulation & real
Sukhil et al	F1/10 autonomous vehicle NVIDIA JetsonTX2	Lidar and depth cam	optimizing race metrics	Simulation & real
Balaji et al	DeepRacer –	4 MP	self-driving learn	Simulation &real
Karaman et al	DeepRacer NVIDIA JetsonTX1	ZED 3D stereo, Hokuyo UST 10-LX 2D Lidar	examine pedagogical implications	real
Verschueren et al	Kyosho dNaNo –	PointGrey Flea 3 Infrared	minimize the lap time	–
Chiba et al	AWS DeepRacer Car –	4MP	activity of reinforcement learning for autonomous vehicles	Simulation & real
Scheffe et al	μCar Raspberry Pi Zero	–	transformed into a networked autonomous vehicle	–

of the filtered image not related to the stripe. (4) Fourth stage was the lane detection part; the "Probabilistic Hough Line Transform" algorithm was analyzed and performed according to the polar coordinate system. The system overview was demonstrated in Fig. 2.

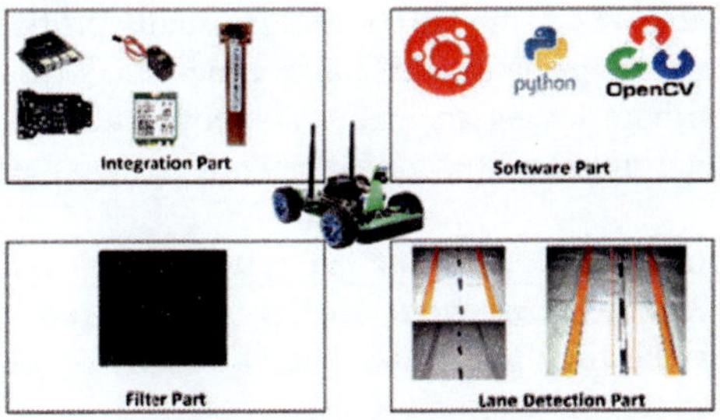

Fig. 2. System Overview

4.1 JetRacer AI KIT

JetRacer is a flexible and open-source platform that supports Donkey Car projects for industrial and academic scope. It is one of the autonomous vehicles implemented with the Nvidia Jetson Nano hardware system. This vehicle is recognized for its low-cost and energy-saving system as it utilizes Jetson Nano [34]. The flexibility of JetRacer is useful for users. Users use the default configuration and simply code. One of the greatest advantages is supporting deep learning algorithms. Thus, the vehicle learns autonomous driving and lane tracking automatically. Researchers can develop this system by making their own contributions to the JetRacer AI Kit, which has improved with updated features and alternative implementations.

4.2 Implementation Details and Experimental Setup

This study coded with Python v3.6.9 and v2.7.17 and OpenCV v4.1.1 versions. A GPU with NVIDIA's 128-core Maxwell architecture integrated into the JetRacer tool with Ubuntu v18.04 LTS operating system. The vehicle has Jetson Nano 2GB, Raspberry Pi CSI Camera, DC Motor-Metal Reducer Motor 2PCS to be used for forward and backward movement of the vehicle, Servo motor MG996R to move the vehicle to the right and left, Wireless-AC8265 used for communication. In this study, we focused on 3 experimental phases. These stages; Ensuring the movement of the vehicle with the Jetson Nano embedded system kit: By using Nvidia's Jetson Nano embedded system card, the forward/backward movement and right/left turn commands of the vehicle ensured to move correctly. The NVIDIA Jetson Nano Developer Kit was developed with artificial intelligence (AI) algorithms. Developers run AI software development environments and models such as image classification, object detection, segmentation and language processing on this embedded system kit. The Jetpack SDK is used across the entire NVIDIA Jetson product family and is fully supported by NVIDIA's world-leading model training and model deployment platform software. The 4-core 64-bit ARM CPU offers 4GB of memory and simultaneously processes several high-resolution sensors. Providing lane tracking with the real-time image from the camera and enabling the autonomous vehicle to proceed on the determined route: First, video or camera input was devided into frames. These frames were then converted to grayscale format. The Canny algorithm is based on a step-by-step algorithm to detect edges on frames. The output image consists of white and black colors, i.e. the edges in the image are colored white and the rest of the image is colored black. The Canny algorithm is usually proposed for finding edges. One of the advantages of the Probabilistic Hough Line transform is that it increases the computational speed without losing accuracy. Thereby, autonomous vehicles tracking lines speedly for the determined routes more easily. Route on the Racer pad and Racer pad is given in Fig. 3.

Lane detection from the camera of an autonomous vehicle traveling on a predetermined route: PHT features were used to analyze the images and tested on re-recorded videos and real-time hardware. The autonomous vehicle more easily and quickly follows the lines for the defined routes.

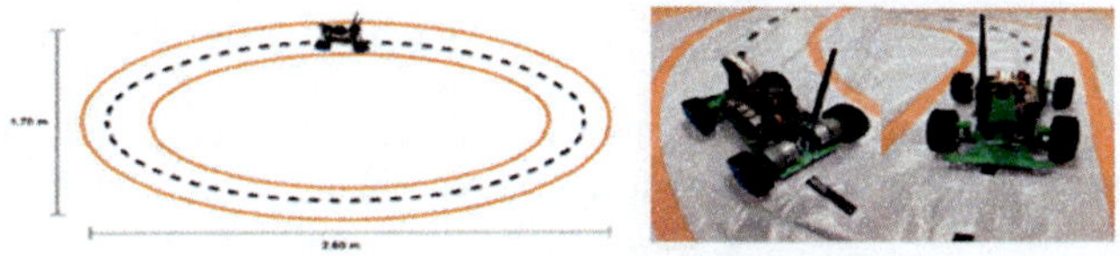

Fig. 3. Route on the Racer pad and Racer pad

5 Discussion and Conclusion

Recent research shows that autonomous vehicles are a popular area of research. Due to the growing interest in these vehicles, having a cost-effective framework to implement various applications, algorithms and try them out on a supervised platform further aid the development of this technology. In this context, we presented a procedure that uses a visual image as a precursor for motion to follow a lane using a cost-effective robot called JetRacer AI Kit. We observed a small performance loss in our system due to poor lighting conditions. As future work, we will develop a deep learning algorithm for robust lane detection that is not sensitive to lighting conditions. Furthermore, other embedded systems such as drones are intended to be driven simultaneously on the JetRacer AI Kit.

Acknowledgements. The study was carried out in the Image Processing and Computer Vision Laboratory of the Computer Engineering Department of Kocaeli University, the study is supported by the BAP research and development center with Project ID 2337 and also supported by the TÜBİTAK 2209-A Project with project no:1919B012112773.

References

1. Akdeniz, F., Kayikçioğlu, İ., Kaya, İ., Kayikçioğlu, T.: Using WignerVille distribution in ECG arrhythmia detection for telemedicine applications. In: 2016 39th International Conference on Telecommunications and Signal Processing (TSP), pp. 409–412. IEEE (2016)
2. DARPA Grand Challenge. https://en.wikipedia.org/wiki/DARPA_Grand_Challenge 2018–2021. Accessed 10 Jun 2022
3. Savaş, B.K., Becerikli, Y.: Real time driver fatigue detection system based on multi-task ConNN. IEEE Access **8**, 12491–12498 (2020)
4. Stocco, A., Pulfer, B., Tonella, P.: Mind the Gap! A Study on the Transferability of Virtual vs Physical-world Testing of Autonomous Driving Systems. arXiv preprint arXiv:2112.11255 (2021)
5. Shi, W., Alawieh, M.B., Li, X., Yu, H.: Algorithm and hardware implementation for visual perception system in autonomous vehicle: a survey. Integration **59**, 148–156 (2017)
6. Bonnefon, J.F., Shariff, A., Rahwan, I.: The social dilemma of autonomous vehicles. Science **352**(6293), 1573–1576 (2016)
7. Akdeniz, F., Becerikli, Y.: Performance comparison of support vector machine, k-nearest-neighbor, artificial neural networks, and recurrent neural networks in gender recognition from voice signals. In: 2019 3rd International Symposium on Multidisciplinary Studies and Innovative Technologies (ISMSIT), pp. 1–4, IEEE
8. Kır Savaş, B., Becerikli, Y.: Behavior-based driver fatigue detection system with deep belief network. Neural Comput. Appl. 2022 (2019)

9. Karabulut, G.: Mini autonomous car architecture for urban driving scenarios (Master's thesis, Middle East Technical University) (2019)
10. Castro, O., Céspedes, A., Ubaldo, R., Ramos, O.E.: Lane following with a duckiebot vehicle using visual feedback. In: 2019 IEEE Sciences and Humanities International Research Conference (SHIRCON), pp. 1–4, IEEE (2019)
11. Roberts, A., Maennel, O., Snetkov, N.: Cybersecurity test range for autonomous vehicle shuttles. In: 2021 IEEE European Symposium on Security and Privacy Workshops (EuroS&PW), pp. 239–248, IEEE (2021)
12. Kawakura, S., Shibasaki, R.: Deep learning-based self-driving car: JetBot with NVIDIA AI board to deliver items at agricultural workplace with object-finding and avoidance functions. Eur. J. Agric. Food Sci. **2**(3) (2020)
13. Bingöl, M.S., Kaymak, Ç., Ayşegül, U.Ç.A.R.: Derin Öğrenme Kullanarak Otonom Araçların İnsan Sürüşünden Öğrenmesi". Fırat Üniversitesi Mühendislik Bilimleri Dergisi **31**(1), 177–185 (2019)
14. Yuhas, M., Feng, Y., Ng, D.J.X., Rahiminasab, Z., Easwaran, A.: Embedded out-of-distribution detection on an autonomous robot platform. In: Proceedings of the Workshop on Design Automation for CPS and IoT, pp. 13–18 (2021)
15. Loeminto, E.S., Khoswanto, H., Lim, R.: Pengenalan Traffic Light Pada Robot Mobil Duckietown. Teknika **7**(2), 89–93 (2018)
16. Kumar, C.S., Anjali, T., Rao, S.N.: A cost-effective framework for developing and testing autonomous RC cars. In: 2021 Sixth International Conference on Wireless Communications, Signal Processing and Networking (WiSPNET), pp. 26–29, IEEE (2021)
17. Seth, A., James, A., Mukhopadhyay, S.C.: 1/10th scale autonomous vehicle based on convolutional neural network. Int. J. Smart Sens. Intell. Syst. **13**(1), 1–17 (2020)
18. Sandha, S.S., Garcia, L., Balaji, B., Anwar, F.M., Srivastava, M.: Sim2Real transfer for deep reinforcement learning with stochastic state transition delays. In: 4th Conference on Robot Learning, Cambridge MA, USA.CoRL (2020)
19. Katyal, K.D., Polevoy, A., Moore, J., Knuth, C., Popek, K.M.: Highspeed robot navigation using predicted occupancy maps. In: 2021 IEEE International Conference on Robotics and Automation (ICRA), pp. 5476–5482, IEEE (2021)
20. O'Kelly, M., et al.: F1/10: An open-source autonomous cyber-physical platform (2019). arXiv preprint arXiv:1901.08567
21. O'Kelly, M., Zheng, H., Karthik, D., Mangharam, R.: F1tenth: an open-source evaluation environment for continuous control and reinforcement learning. Proc. Mach. Learn. Res. **123**, 77–89 (2020)
22. Liu, Y., Schofield, H., Shan, J.: Navigation of a self-driving vehicle using one fiducial marker. In: 2021 IEEE International Conference on Multisensor Fusion and Integration for Intelligent Systems (MFI), pp. 1–6, IEEE (2021)
23. Hu, J., Zhang, Y., Rakheja, S.: Adaptive trajectory tracking for car-like vehicles with input constraints. IEEE Trans. Ind. Electron. **69**(3), 2801–2810 (2021)
24. Sukhil, V., Behl, M.: Adaptive Lookahead Pure-Pursuit for Autonomous Racing (2021). arXiv preprint arXiv:2111.08873
25. Balaji, B., et al.: DeepRacer: educational autonomous racing platform for experimentation with sim2real reinforcement learning (2019). arXiv preprint arXiv:1911.01562
26. Chiba, S., Sasaoka, H.: Effectiveness of transfer learning in autonomous driving using model car. In: 2021 13th International Conference on Machine Learning and Computing, pp. 595–601 (2021)
27. Karaman, S., et al.: Project-based, collaborative, algorithmic robotics for high school students: programming self-driving race cars at MIT. In: 2017 IEEE Integrated STEM Education Conference (ISEC), pp. 195–203, IEEE (2017)

28. Scheffe, P., et al.: Networked and autonomous model-scale vehicles for experiments in research and education. IFAC-PapersOnLine **53**(2), 17332–17337 (2020)
29. Verschueren, R., De Bruyne, S., Zanon, M., Frasch, J.V., Diehl, M.: Towards time-optimal race car driving using nonlinear MPC in real-time. In: 53rd IEEE Conference on Decision and Control, pp. 2505–2510, IEEE (2014)
30. Hassanein, A.S., Mohammad, S., Sameer, M., Ragab, M.E.: A survey on Hough transform, theory, techniques and applications (2015). arXiv preprint arXiv:1502.02160
31. Chen, M.: Recognition and localization of target images for robot vision navigation control. J. Robot. **2022**, 8565913 (2022)
32. Macdonald, I.: Probabilistic Hough Transform. Notes, 1–4 (2016)
33. Stephens, R.S.: Probabilistic approach to the Hough transform. Image Vis. Comput. **9**(1), 66–71 (1991)
34. Cass, S.: Nvidia makes it easy to embed AI: the Jetson nano packs a lot of machine-learning power into DIY projects-[Hands on]. IEEE Spectr. **57**(7), 14–16 (2020)

A Comparative Analysis of MANET Routing Protocols Using NS2 and NS3 Simulators

Boudhir Anouar Abdelhakim(✉) and Ben Ahmed Mohamed

SSET Research Team, C3S Laboratory, FSTT, UAE, University of Tetouan, Tétouan, Morocco
{aboudhir,mbenahmed}@uae.ac.ma

Abstract. The use of simulation software is crucial for determining the expected outcomes of a practical hardware arrangement, which is expensive and time-consuming to change on a regular basis. In the field of communication, there are many different types of network simulators. However, selecting the appropriate simulator for a given task can be challenging due to the wide variations in operating systems, hardware specifications, programming software needs, and scalability. The four mobile ad hoc routing protocols—DSDV, AODV, DSR, and OLSR—are analyzed using packet delivery ratio, throughput, average end-to-end delay, and energy in this study, which compares the results of NS3 and NS2.

Keywords: NS2 · NS3 · MANET · Network simulators · Routing Protocols · AODV · DSDV · OLSR · DSR

1 Introduction

A group of wireless mobile nodes linked by multi-hop wireless links that do not have any pre-existing infrastructure or centralized management is called a mobile ad hoc network, or MANET. Routing through a server that serves as an intermediary and router is necessary for communication between nodes. The creation of an ideal routing protocol for MANET is complicated, requiring effective algorithms for network connectivity, organization, link planning, and dynamic routing scenarios. In static networks, shortest path algorithms work well; but, in mobile area networks (MANNETs), several elements must be taken into account, including power, wireless link strength, path losses, fading, interference, and topological changes.

The purpose of this study is to evaluate the effectiveness of the network simulators, NS3 and NS2, using metrics such as packet delivery ratio, throughput, routing overhead, average end-to-end delay, and path optimality. For the comparison analysis, three different routing protocols were used: proactive, reactive, and hybrid.

The free, open-source discrete event network simulator NS3 has features like a freely available source code, a free network topology, and it's designed entirely in C++ with some elements that may be scripted in Python. Socket-based simulation and emulation are supported by NS3. An open-source simulator called NS2 is frequently used for ad hoc networking and assessing both new and old network protocols. It is written in C++ and OTCL and supports a number of IP network simulations.

M. Ben Ahmed et al. (Eds.): SCA 2023, LNNS 906, pp. 240–252, 2024.
https://doi.org/10.1007/978-3-031-53824-7_22

The article provides a summary of pertinent study findings for each. It then describes the simulation environment and the methods and strategies used in the research. The study ends with a presentation of the findings from simulations conducted in NS2 and NS3.

2 Related Works

In their study, Rakesh Kumar Jha and Pooja Kharga compared three routing protocols—Ad-hoc On Demand Distance Vector Routing (AODV), Optimized Link State Routing (OLSR), ET Dynamic Source Routing (DSR), and Distance Sequenced Distance Vector (DSDV)—using NS3 Simulator. The comparison was based on performance metrics, which included throughput, packet delivery ratio, packet dropped, jitter, and end-to-end delay measured after system simulation.

According to their research, AODV performs significantly better in terms of throughput than the other protocols for networks with different numbers of nodes. After taking into account the packet delivery ratio, packet lost, jitter, and end-to-end delay, they concluded that the OLSR protocol performs better than the other two protocols. By using the ns-2 simulator to simulate various routing protocols (DSDV, DSR, and AODV), authors in [2] provided performance evaluation studies based on QoS metrics (packet delivery ratio, average end-to-end delay, throughput, jitter, normalized routing overhead, and normalized MAC overhead). By altering the mobility speed, node count, and data rate, the performance comparison is carried out. Ultimately, they came to the conclusion that a variety of criteria, like the size of the network, load, mobility requirements, etc., influence whether a specific routing protocol is used in a mobile ad hoc network. Furthermore, simulation findings demonstrate that, rather than being the best, AODV performs maximally well for all sorts of net-works. This protocol is reliable for usage in mobile ad hoc networks, despite a few drawbacks.

Along with the protocols mentioned in the earlier study [1], Dilpreet Kaur and Naresh Kumar covered the features of ad hoc routing protocols in the research study [3], including Dynamic Source Routing (DSR) and Temporally Ordered Routing Algorithm (TORA). Based on performance parameters such packet delivery fraction, throughput, average latency, normalized load, and jitter under low mobility and low network conditions as well as high mobility and high traffic conditions, they described these routing protocols. In the end, they determine that DSDV has throughput in high traffic situations and AODV has maximum throughput in low traffic situations. While TORA performs well in dense networks in terms of packet delivery fraction, OLSR, DSR, and DSDV perform well since the denser network has a higher throughput than AODV. In addition, TORA's Normalized Routing burden is the highest of all the protocols in both networks. In both low and high traffic situations, DSDV has the lowest Normalized Routing burden. In both networks, OLSR and DSDV have the lowest average latency and jitter.

Using simulations carried out with the OPNET Modeler simulator, Ahmed and Alam [4] examine three routing protocols: Temporally Ordered Routing Algorithm (TORA), Ad-hoc On Demand Distance Vector (AODV), and Dynamic Source Routing (DSR). According to the simulation results, given particular simulation conditions, TORA performs better than AODV and DSR.

By comparing the simulation results of the NS3 and OMNET++ simulators, this study assesses the three MANET routing protocols (DSDV, AODV, and ZRP) based on

QoS metrics (packet delivery ratio, average end-to-end delay, throughput, routing overhead, and path optimality). Additionally, the best open-source simulator for research purposes was determined through analysis.

Routing protocols like AODV and OLSR were employed in the research study [5] and contrasted with the OMNET++ simulation environment. Ultimately, the research conducted by Reetika KaS-ingla [5] demonstrated that the AODV routing protocol produces superior performance. Better throughput is also provided by OLSR through throughput. As a result, while selecting routing protocols, network selection is crucial.

3 Methods and Technics

In this section, two network simulators are used to compare various MANET routing techniques. Numerous protocols, including DSR, OLSR, DSDV, and AODV. We provide simulation settings for NS3 and NS2. The virtual computer running Linux (Ubuntu 20.04) utilizes both the NS3 and NS2 simulation tools. We can determine which network simulator is best for MANET routing, such as reactive routing (AODV or DSR) or proactive routing (DSDV or OLSR), based on the simulation. We are able to compare which open-source network simulator is superior for MANET routing based on the simulation results.

Routing Protocols Overview

Every node in the proactive routing protocol keeps its data in tables even in the event of a change in the network architecture. Routing tables are modified in response to topological changes in the network. Proactive routing protocol, however, comes at a much higher cost than reactive routing protocol. One of the main benefits of proactive routing protocol is that it doesn't postpone discovery while establishing a new route.

Reactive or on-demand routing protocols find the route just when they're needed. That is to say, the route will only be determined by the reactive routing protocol, depending on demand. This lowers the cost of higher latency and the overhead of control messages when finding new routes. Reactive routing systems do not require messages to be sent from source to destination, and they consume less bandwidth while transmitting messages. Reactive and proactive routing approaches are combined in the hybrid routing protocol (Fig. 1).

Simulation Environment

Our simulation tests are carried out between proactive routing protocol (DSDV or OLSR) and reactive routing protocol (AODV or DSR) on NS3 and NS2 network simulators. The table below is a list of the parameters taken into account during NS3 simulation (Table 1).

Using two distinct simulators, NS2 and NS3, we compare four routing protocols in the simulation: two proactive and two reactive. We are utilizing 15, 25, 35, 45, and 50 nodes, among other numbers. First, we alter the density, or the number of nodes, and set the nodes' speed at 15 mps. Next, we set the number of nodes at 15, and we vary their speed between 15 and 30 mps. Since our simulation has a pause period of 0 s, our motion is continuous. Within the 600 by 600-m coverage region, the nodes are free to move at will. During packet transmission, we employ Manet Constant Bit Rate (CBR)

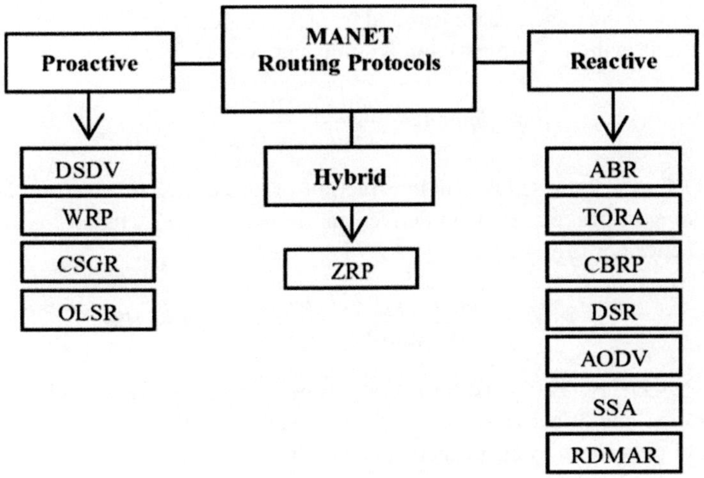

Fig. 1. Routing protocols of MANET

Table 1. Simulation Structure

Parameter	Values
Number of nodes	15,25,35,45,50
Interface type	Phy/WirelessPhy
Channel	Wireless Channel
Mac type	Mac/802_11
Interface	Phy/WirelessPhy
Movement Model	Random
Size of packet	64 bytes
Protocol	AODV, DSR, DSDV, OLSR
Traffic	CBR
Simulation area	600M*600M
Node speed	15,20,25,30 m/s

for traffic management. The evenly distributed IEEE 802.11 MAC protocol makes up the packet, which has a size limit of 64 bytes. The Random Waypoint Mobility Model, which may be developed and implemented by the NS2 and NS3 simulators, determines how the nodes in our simulation move.

Performance Metrics:
Our comparative simulation analyzes the performances of the MANET routing protocols using a variety of measures. The metrics listed below are considered in our simulation-based analysis [7]:

1. Throughput: is the total data transmitted from the source node to the destination node in a time unit, which is expressed in Kilobits per second (Kbps).

$$Throughput = \frac{(Total\ Received\ Bytes \times 8)}{(Simulation\ Time \times 1024)} \quad (1)$$

2. Packet Delivery Ratio (PDR): is the fraction of amount of received packets to the amount of sent packets. PDR is derived in % (percentage). Higher values of PDR provide better performance

$$PDR = \frac{Amount\ Of\ Received\ Packets}{Amount\ Of\ Packets\ Sent} \times 100 \quad (2)$$

3. End-to-End Delay (EED): is the average time interval between packets generated at the source and effective delivery of these packets at the destination. EED is the fraction of delay sum to the packets received (ms).

$$EED = \frac{Delay\ Sum}{PAckets\ Received} \quad (3)$$

4. Energy: is the fraction of the sum of energy consumed by the node to the number of nodes. Energy is derived in % (percentage).

$$Energy = \frac{\sum_{i=1}^{n} energy\ consumed\ by\ the\ node\ i}{Number\ Of\ Nodes} \times 100 \quad (4)$$

4 Results and Discussions

Performance on NS3

Simulation-based studies and performance comparisons of mobile ad hoc network routing algorithms were conducted using two different scenarios: one with changing node density and the other with changing node speed. The simulation scenarios and results are displayed in the tables and graphs below (Table 3 and Figs. 2, 3).

Table 2. Packet Delivery Ratio vs Number of Nodes

Packet Delivery Ratio				
Nbr Nodes	AODV	DSDV	OLSR	DSR
15	24	5	80	43
25	56	45	67	44
35	72	23	23	44
45	90	31	31	39
50	81	22	22	42

Table 3. Packet Delivery Ratio vs Speed

Packet Delivery Ratio				
Speed Node (m/s)	AODV	DSDV	OLSR	DSR
15	24	5	80	36
20	32	27	80	48
25	39	20	72	42
30	67	30	50	31

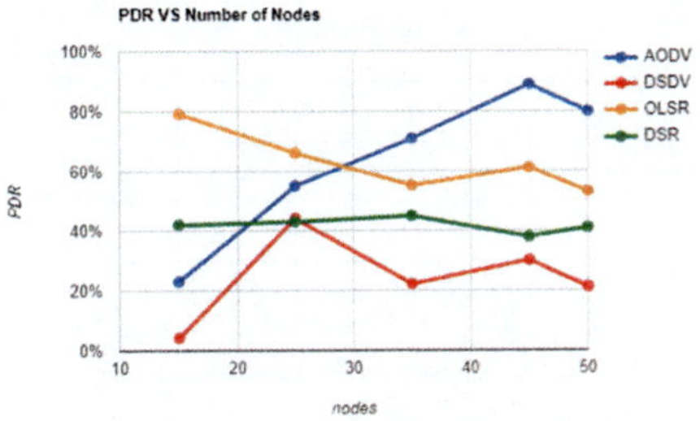

Fig. 2. PDR vs Number of Nodes

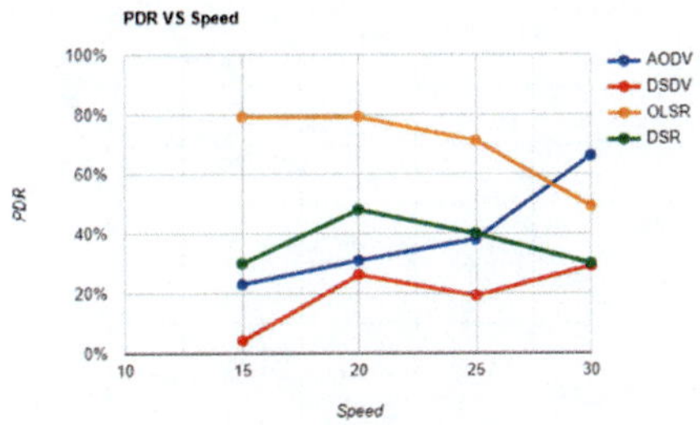

Fig. 3. PDR vs Speed

Table 4. Throughput vs Number of Nodes

Throughput				
Speed Node (m/s)	AODV	DSDV	OLSR	DSR
15	7,03	1,15	4	46
20	69	1,8	3	54
25	57	1,64	3,27	65
30	41	2,07	2,1	48

Table 2 from the NS3 simulation demonstrates that, in comparison to other protocols, the AODV routing protocol yields a greater percent of packet delivery ratio, followed by OLSR. In terms of speed, OLSR performs better when we reduce the number of nodes whereas AODV performs well when we increase the number of nodes (Figs. 4 and 5).

The throughput values of AODV, DSDV, DSR, and OLSR in relation to the number of nodes and its speed are shown in Tables 4 and 5. It demonstrates that in both scenarios, AODV had the highest throughput (Figs. 6 and 7).

The range of average end-to-end delays, or the amount of time it takes a packet to go from source to destination via the preferred path, is shown in Tables 6 and 7. The minimal average delay for DSDV and DSR is both speed- and node-related.

Table 5. Throughput vs Speed

Throughput				
Nbr Nodes	AODV	DSDV	OLSR	DSR
15	7,03	1,15	4	8
25	77,6	2,4	3,6	56
35	71	2,11	2,44	49
45	13,08	2,3	2,78	57
50	11,3	1,8	2,72	53

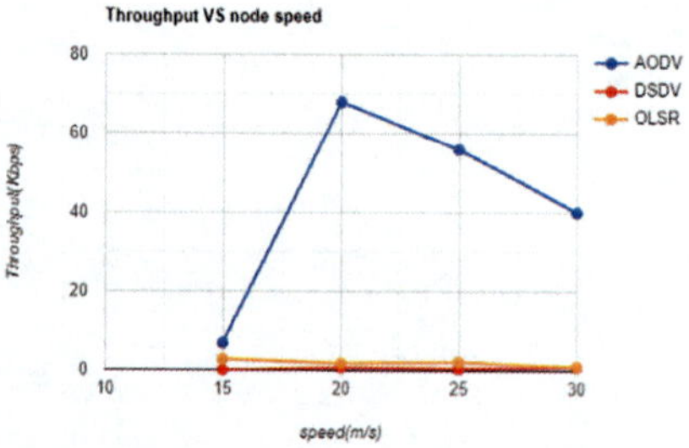

Fig. 4. Throughput vs Number of Nodes

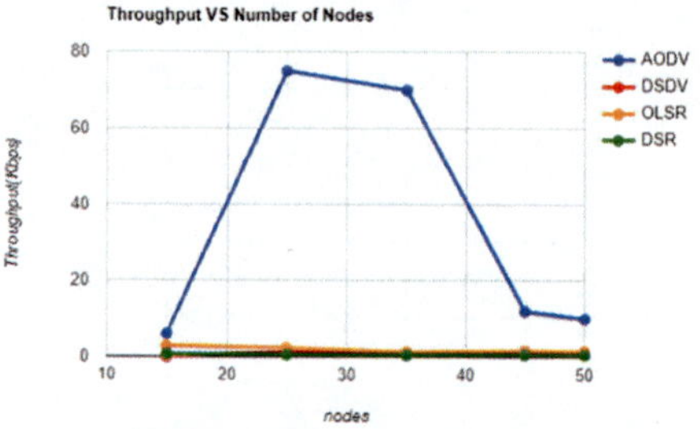

Fig. 5. Throughput vs Speed

Table 6. Average End-End Delay vs Number of Nodes

Nbr Nodes	AODV	DSDV	OLSR	DSR
15	229	0.05	0.08	0.03
25	46.09	14.5	0.32	0.1
35	69	18.6	3.58	0.12
45	168	5.89	62.6	0.15
50	223	1.2	4.01	0.06

Table 8 demonstrates that energy is highest for the OLSR protocol compared to the number of nodes, and it has little variety for the other protocols (Fig. 8).

Performance on NS2

We used the NS2 simulation environment to carry out the same scenario. The following tables and graphs show the simulation results and their analysis (Tables 9, 10 and Figs. 9, 10).

The NS2 simulator indicates that we achieved a higher packet delivery rate with the DSDV routing protocol than with other protocols, and that the DSR protocol performs better than other protocols as the number of nodes increases.

Table 7. Average End-End Delay vs Speed

Average End-End Delay				
Speed (m/s)	AODV	DSDV	OLSR	DSR
15	227	0.05	0.08	0.04
20	20.2	0.13	0.09	0.05
25	26	0.05	0.1	0.08
30	75	2.43	0.09	0.07

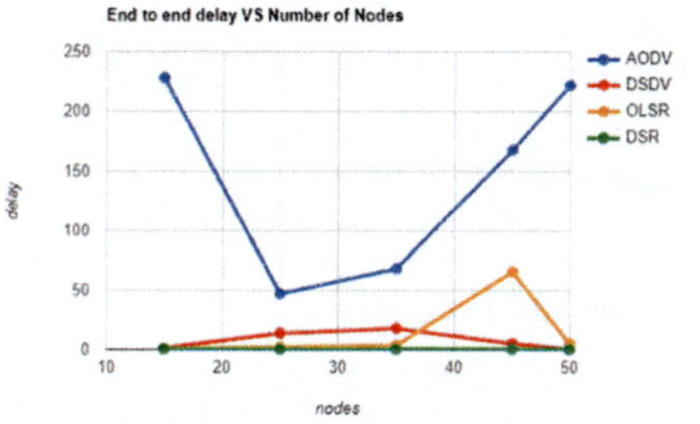

Fig. 6. Average End-End Delay vs Number of Nodes

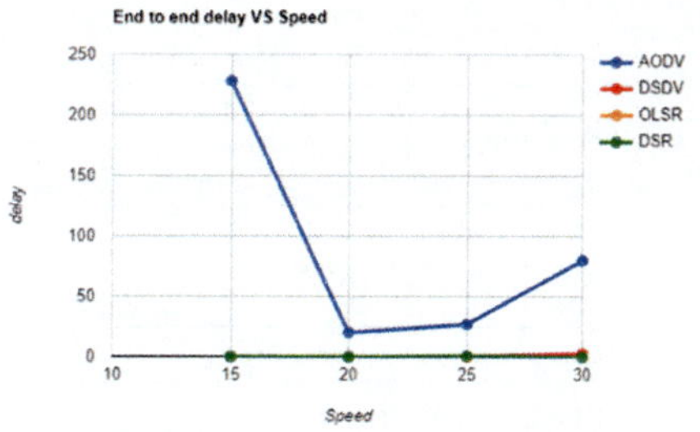

Fig. 7. End-End Delay vs Speed

Table 8. Energy vs Number of Nodes

Energy(Joul)			
Number of nodes	AODV	DSDV	OLSR
15	.8725	.8716	.8732
25	.8777	.8718	.8775
35	.8739	.8708	.8776
45	.8780	.8723	.8990
50	.8728	.8717	.8753

We observed that even when a node's speed is raised, the PDR essentially stays the same by fixing the number of nodes and adjusting their speed this time. When comparing the four protocols, we found that DSDV performs the best (Table 11 and Fig. 11).

The findings obtained indicate that the DSDV protocol is the best routing protocol in terms of throughput, even when compared to the other protocols. The AODV protocol will have the highest throughput when the number of nodes increases to more than 45.

By altering the node's speed, we concentrate on the changes in throughput (Table 12, Fig. 12). We may conclude that there is little correlation between throughput and node speed based on the results, which almost completely reveal that throughput does not change. And it remains the greatest of the four DSDV procedures (Table 13 and Fig. 13).

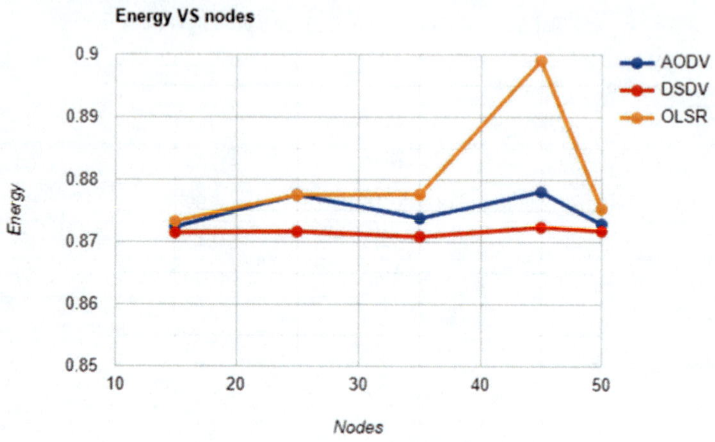

Fig. 8. Energy vs Number of Nodes

Table 9. PDR vs Number of Nodes

Packet Delivery Ratio				
Nbr Nodes	AODV	DSDV	OLSR	DSR
15	.67	.89	.57	.41
25	.38	.66	.53	.42
35	.24	.52	.31	.42
45	.28	.35	.32	.37
50	.18	.33	.36	.40

Table 10. PDR vs Speed

Packet Delivery Ratio				
Node Speed (m/s)	AODV	DSDV	OLSR	DSR
15	.65	.8	.46	.51
20	.65	.90	.56	.51
25	.64	.90	.61	.51
30	.67	.88	.57	.49

The findings show that while the two OLSR and DSR protocols have larger end-to-end delays, the other two protocols, AODV and DSDV, have the least end-to-end delays, with DSDV having essentially no delay at all.

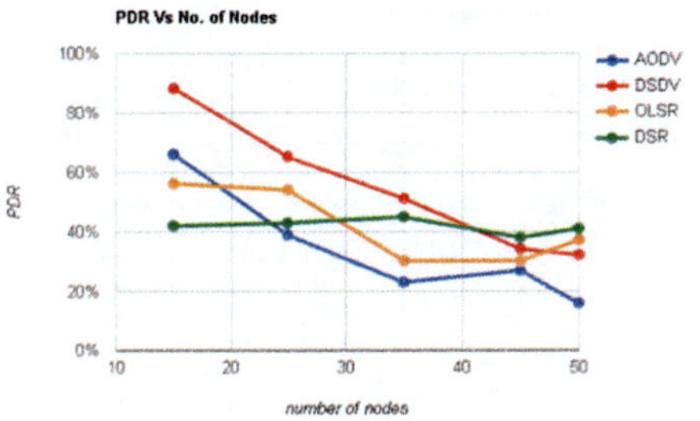

Fig. 9. PDR vs Number of Nodes

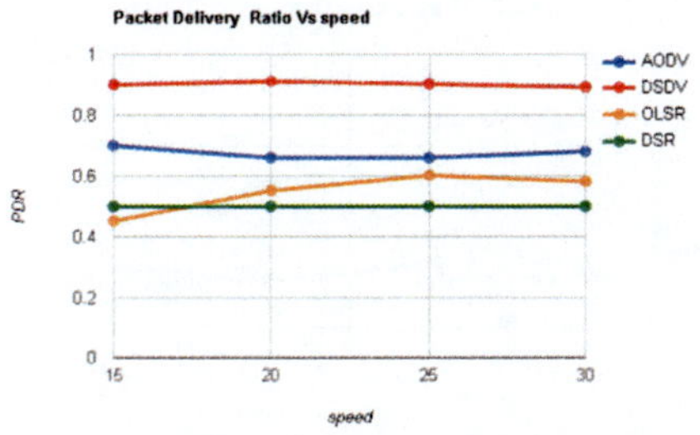

Fig. 10. PDR vs Speed

Table 11. Throughput vs Number of Nodes

Throughput				
Nbr Nodes	AODV	DSDV	OLSR	DSR
15	1.01	1.51	0.87	0.6
25	1.27	1.46	0.65	0.54
35	1.10	1.20	0.58	0.46
45	0.73	1.02	0.44	0.55
50	1.05	0.98	0.44	0.51

Table 12. Throughput vs Speed

Throughput				
Node Speed (m/s)	AODV	DSDV	OLSR	DSR
15	0.83	0.93	0.66	0.46
20	0.84	0.95	0.63	0.47
25	0.86	0.92	0.63	0.45
30	0.82	0.93	0.62	0.48

According to Table 14 and Fig. 14, the end-to-end latency of the DSDV protocol is nearly zero and is not affected by changes in node speed. In contrast, the end-to-end delay of the DSR protocols is greater and is increased by changes in node speed.

When comparing the energy consumption of these four protocols, DSR consumes nearly all of the energy and OLSR uses less, although overall energy consumption is relatively high when compared to the number of nodes (Table 15 Fig. 15).

Figure 16 in Table 16. The four protocols have very high overall energy consumption, which fluctuates depending on the speed. However, we have found that the OLSR protocol uses less energy at low speeds while DSR performs better at high speeds.

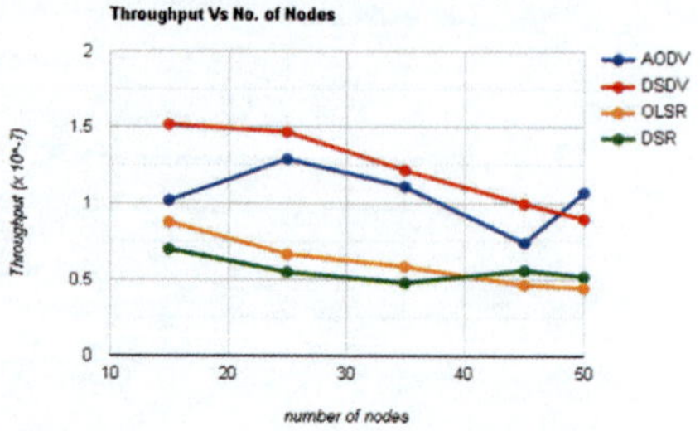

Fig. 11. Throughput vs Number of Nodes

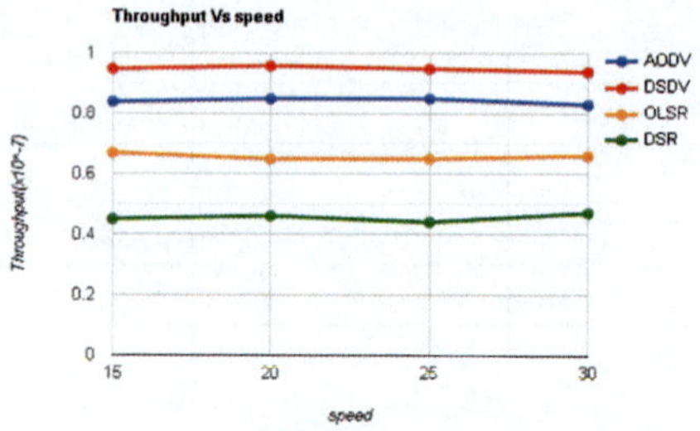

Fig. 12. Throughput vs Speed

Table 13. End-End Delay vs Number of Nodes

Average End-End Delay				
Nbr Nodes	AODV	DSDV	OLSR	DSR
15	0.09	0.0085	0.73	0.81
25	0.25	0.0085	0.60	0.66
35	0.14	0.0065	0.64	0.75
45	0.09	0.053	0.75	0.78
50	0.25	0.033	0.63	0.65

Table 14. End-End Delay vs Speed

Average End-End Delay				
Node Speed (m/s)	AODV	DSDV	OLSR	DSR
15	0.03	0.01	0.09	0.63
20	0.13	0.004	0.24	0.64
25	0.05	0.0009	0.06	0.75
30	0.15	0.004	0.17	0.766

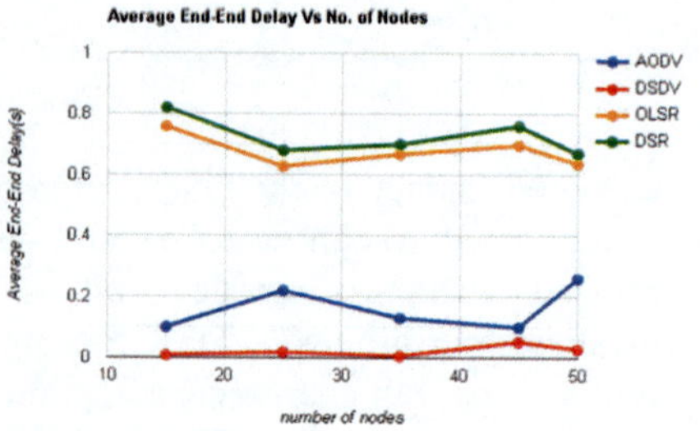

Fig. 13. End-End Delay vs Number of nodes

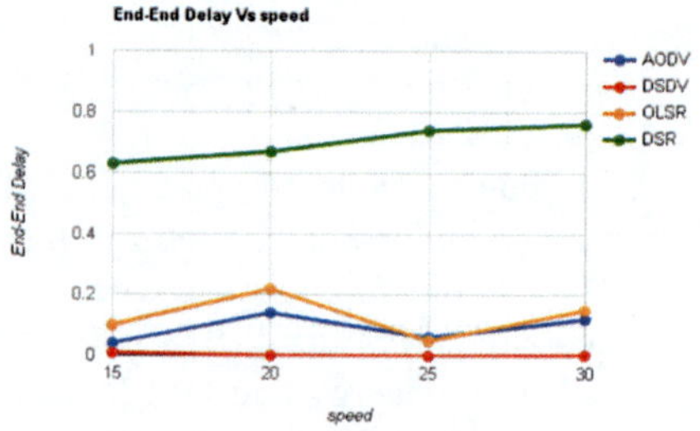

Fig. 14. End-End Delay vs Speed

Table 15. Energy vs Number of Nodes

	Energy (Joul)			
Nbr Nodes	AODV	DSDV	OLSR	DSR
15	0.9798	0.9969	0.9535	0.9942
25	0.9799	0.9727	0.9329	1
35	0.9809	0.9698	0.9413	0.9993
45	0.9802	0.9859	0.9124	0.9919
50	0.9925	0.9834	0.9638	1

Table 16. Energy vs Speed

Energy (Joul)				
Speed Node (m/s)	AODV	DSDV	OLSR	DSR
15	0.9886	0.9966	0.9813	0.9798
20	0.99545	0.9942	0.9921	0.9972
25	0.9992	0.9951	0.9938	0.9919
30	0.9898	0.9971	0.9901	0.9824

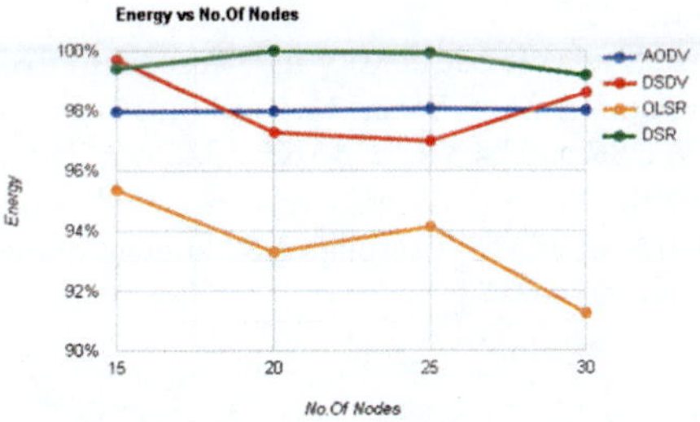

Fig. 15. Energy vs Number of Nodes

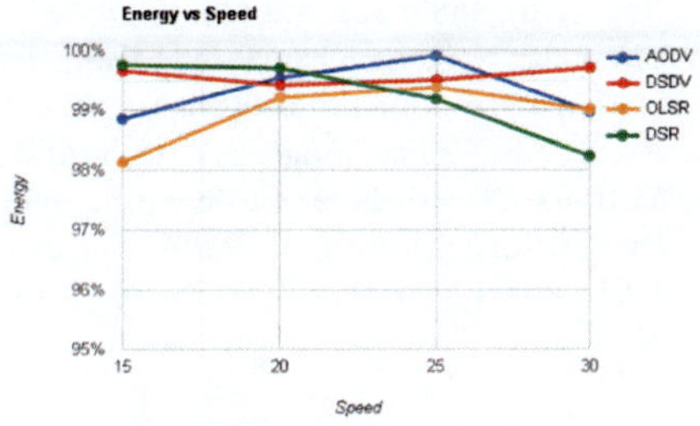

Fig. 16. Energy vs Speed

5 Conclusion and Future Work

In order to compare the protocols AODV, DSDV, OLSR, and DSR, we employed specific parameters and two simulators, NS2 and NS3. Our primary attention was on four parameters: end-to-end delay, packet delivery ratio, energy consumption, and throughput. First, we compared the results from the same simulator, but we altered one of the variables in our case. In this scenario, the DSDV protocol is the most efficient among the other protocols for the NS2 simulator, taking into account the node's speed or number of nodes based on all available data. The AODV protocol performs well in terms of packet delivery ratio and throughput even after modifying the two parameters, the node's speed and number, by looking at the output of the NS3 simulator; however, this is not the case in terms of end-to-end delay, where the two protocols, DSDV and DSR, are the most efficient. Furthermore, the DSDV protocol consumes less energy than competing protocols. In terms of power, performance, and adaptability, NS3 beats NS2. Therefore, in order to achieve the necessary results, it is recommended that you use the NS3 simulator. Future research endeavors could explore the commonalities between the MANET routing protocols and two other open-source simulators, namely OPNET and OMNET++.

References

1. Jha, R.K., Kharga, P.: A comparative performance analysis of routing protocols in MANET using NS3 simulator. Comput. Netw. Inf. Secur. **2015**(4), 62–68 (2015)
2. Gulatil, M.K., Kumar, K.: Performance comparison of mobile ad hoc network routing protocols. Int. J. Comput. Netw. Commun. **6**(2),127–142 (2014)
3. Kaur, D., Kumar, N.: Comparative analysis of AODV, OLSR, TORA, DSR and DSDV routing protocols in mobile Ad-Hoc networks. Comput. Netw. Inf. Secur. **5**(3), 39–46 (2013)
4. Ahmed, S., Alam, M.S.: Performance evaluation of important ad-hoc network protocols. EURASIP J. Wirel. Commun. Netw. **2**, 42 (2006)
5. Singla, R., Sharma, S., Singh, G., Kaur, R.: Performance evaluation of routing protocols in vanets by using Tcp variants on Omnet++ simulator. International Journal of Engineering Research and Applications (IJERA)
6. Bhatia, D., Sharma, D.P.: A comparative analysis of proactive, reactive and hybrid routing protocols over open source network simulator in mobile Ad Hoc network. Int. J. Appl. Eng. Res. **11**(6), 3885–3896 (2016). ISSN 0973–4562
7. Lakshman Naik, L., Khan, R.U., Mishra, R.B.: Analysis of node density and pause time effects in MANET routing protocols using NS-3. Comput. Netw. Inf. Secur. **12**, 9–17 (2016)
8. Ali, S., Ali, A.: Performance Analysis of AODV, DSR and OLSR in MANET. Masters Thesis, M.10:04, COM/School of Computing, BTH (2010)
9. Pandey, A., Srivastava, A.: Performance evaluation of MANET through NS2 simulation. Int. J. Electron. Electr. Eng. **7**(1), 25–30 (2014). ISSN 0974–2174

Sustainable Cities

Enhancing Sustainability: Leveraging Sensor Technology in Smart Bins for Real-Time Data Analysis

A. Idrissi(✉), R. Benabbou, J. Benhra, and M. El Haji

Optimization of Industrials and Logistics Systems Team OSIL, Laboratory of Advanced Research in Industrial and Logistic Engineering LARILE, Department of Industrial and Logistic Engineering GIL, National High School of Electricity and Mechanical Engineering (ENSEM), University Hassan II, Casablanca, Morocco

aya.idrissi@ensem.ac.ma

Abstract. The escalating challenges posed by urbanization and rapid population growth are exacerbating the global waste crisis. The World Health Organization underscores the risks of inadequate waste management, as it contributes to the spread of diseases.

To address these multifaceted issues, our research is driven by two primary objectives. Firstly, we are dedicated to designing cutting-edge smart garbage bins equipped with nine sensors, including waste level sensors to monitor bin capacity, Humidity sensors to measure moisture content, Temperature sensors for climate insights, Carbon dioxide (CO_2) sensors for air quality assessment, Methane (CH4) sensors to detect potential hazards, and ambient temperature sensors to gauge environmental conditions.

Secondly, our research establishes a technological backbone for real-time monitoring and informed decision-making through the seamless integration of ThingSpeak and WiFi ESP8266 technologies.

In addition, we will explore three usage scenarios for this intelligent waste bin. The initial scenario (Day 1) entails the collection of sensor data throughout the first day, with measurements taken every 15 min. The second scenario (Day 5) involves the collection of sensor data on the fifth day, also at a 15-min interval. Finally, the third scenario involves continuous monitoring of the smart bin over the course of one week.

In essence, our research represents a pioneering effort to harness the potential of smart bins and advanced data analysis techniques. It aspires to revolutionize waste management by introducing data-driven approaches that promise not only to improve the efficiency of waste collection and treatment but also to advance the cause of cleaner, more sustainable urban environments.

Keywords: Smart Bin · IOT · Arduino · MQ135 · MQ4

1 Introduction

Garbage presents a significant environmental and health risk [1, 2], with the potential to cause flooding, groundwater, soil, and atmospheric pollution, and the release of toxic gases such as Hydrogen Sulfide, Methane, and Ammonia that contribute to respiratory

M. Ben Ahmed et al. (Eds.): SCA 2023, LNNS 906, pp. 255–269, 2024.
https://doi.org/10.1007/978-3-031-53824-7_23

disorders and overall health issues [3, 4], highlighting the urgent need for a gas detection system to monitor and assess their concentrations in the air.

We have opted to prioritize methane (CH_4) and carbon dioxide (CO_2) emissions in our waste monitoring system due to their substantial environmental impact and the ready availability of sensors in the local market, making the implementation of our system more feasible. These gases constitute the primary anthropogenic (human-caused) sources of greenhouse gas emissions.

Methane is produced during the decomposition of organic waste, while CO_2 is primarily emitted through human activities [5].

Sensor devices have been utilized in diverse fields [6–11], such as waste management, where they assume a vital function in enhancing efficiency and mitigating environmental hazards.

This paper proposes a sensor-based garbage monitoring system that utilizes various sensors, including capacitive soil moisture sensors, DS18B20 sensors, hc-sr04 ultrasonic sensors, MQ135 and MQ4 sensors, and a DHT11 sensor, integrated with an Arduino Mega 2560 microcontroller, to explore the benefits of smart bins in waste management systems, optimize waste management practices, achieve cost savings and operational efficiencies, reduce pollution and harmful gas emissions, and contribute to a cleaner and more sustainable environment.

The remainder of this paper is organized as follows: Sect. 2 provides an overview of existing studies related to garbage monitoring and sensor-based systems. In Sect. 3, we present the hardware and software design and implementation of our proposed garbage monitoring system. We discuss the simulation results in Section 4. Finally, we conclude the paper in Section 5, summarizing our findings and outlining future research directions.

2 Literature Review

The critical problem of improper waste management, compounded by urbanization and population growth, necessitates efficient waste management systems. The conventional Solid Waste Management (SWM) approach is labor-intensive and resource-demanding, requiring innovative solutions. Integrating the Internet of Things (IoT) and Wireless Sensor Network (WSN) in waste management practices enables real-time monitoring and continuous surveillance of waste bins, offering a robust framework for efficiency improvement [12–14].

Smart bins have emerged as a compelling solution in waste management, integrating IoT and WSN technologies [15, 16]. These bins incorporate advanced components, including RFID readers, ultrasonic sensors, weight sensors, solar cells, and LEDs [17–21]. RFID tags enable identification, ultrasonic sensors measure distance, weight sensors determine waste quantity, solar cells provide power, and LEDs serve as indicators, collectively enhancing the functionality of the system.

The monitoring system developed by Mustapha et al. [22] utilizes GSM technology to create an electronic system that sends SMS notifications to the supervisor when the dustbin is filled, prompting the collection truck to be dispatched. Additionally, another SMS is sent to inform the supervisor once the waste has been collected. The system described in the paper employs an ultrasonic sensor to detect the level of trash in the dustbin, while a GSM module is utilized for messaging purposes. This module provides real- time information about the status of the dustbin, indicating whether it is filled or has been cleaned.

In the study by Morajkar et al. [23], a system has been implemented in the city or campus that involves the use of dustbins equipped with low-cost embedded devices. These devices help monitor the garbage levels in the bins, and each dustbin is assigned a unique ID for easy identification when they become full. The project module is divided into two sections: the Transmitter and the Receiver. In the Transmitter section, 8051 microcontrollers, RF Transmitter, and sensors are utilized and connected to the dustbins. The sensors within the dustbins detect the level of garbage, determining whether the bin is full or empty. When the sensor senses the content of the dustbin, it sends a signal or data to the 8051 microcontrollers.

The design of intelligent waste bins in this paper focuses on providing real-time visibility to stakeholders regarding fill level, temperature, humidity, and gas emissions, offering valuable advantages. Precise monitoring of fill levels enables route optimization and cost reduction, while tracking temperature and humidity levels ensures proper waste decomposition and mitigates health risks and environmental hazards. Continuous monitoring of gas emissions enables the proactive identification and resolution of air quality issues, contributing to a cleaner environment and helping mitigate the effects of climate change. Analyzing data trends enables stakeholders to discern patterns, predict fluctuations, and optimize waste management strategies strategically.

3 Methodology

The methodology of this study involves a detailed description of the smart bin system, the integration of level sensors, Humidity sensors, Temperature sensors, CO2 sensors, and CH4 sensors, and the calibration and validation procedures implemented to ensure accuracy and reliability. The calibration considers environmental conditions and waste composition, while the validation process compares sensor measurements with manual observations or reference standards to evaluate precision and consistency.

3.1 Selection and Integration of Physical Parameters

The comprehensive approach for the smart bin system involves identifying challenges in waste management and defining specific objectives to optimize waste management strategies and resource allocation. Real-time monitoring of fill levels facilitates efficient collection routes, reducing costs and greenhouse gas emissions (GHG). Monitoring CO2 and CH4 emissions during waste decomposition allows for the assessment and reduction of environmental impact [24]. Public health and safety are prioritized through appropriate waste monitoring practices, while resource optimization ensures effective allocation [25]. Key physical parameters for measurement include Fill Level, CO2 Emissions, CH4 Emissions, Humidity, and Temperature of waste and the ambient environment.

3.2 Selection and Integration of Appropriate Sensors

Ambient Temperature and Humidity: DHT11 Sensor

The DHT11 sensor is employed to monitor ambient Humidity and Temperature, crucial factors in waste management. This sensor offers temperature measurements within the range of 0 to 50 °C and detects relative humidity (RH) between 20% and 90% [26]. These ranges align well with the temperature and humidity conditions observed in Morocco over the past five years [27], making the DHT11 sensor suitable for monitoring the environmental parameters in this context.

The Fig. 1 illustrates the pinout graph of the DHT11 sensor.

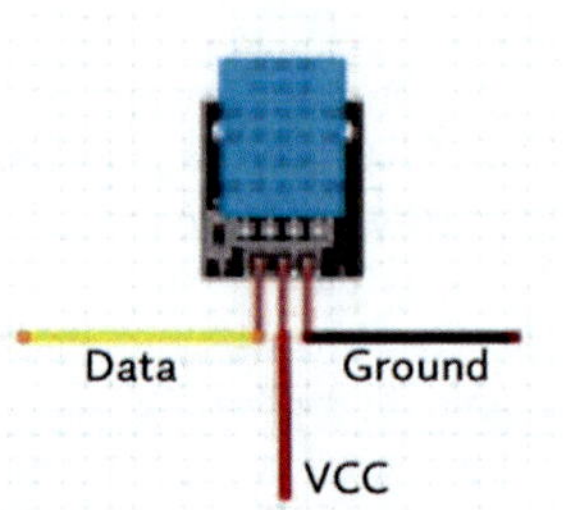

Fig. 1. DHT11 sensor pinout

Fill Level: HC-SR04 Ultrasonic Sensor

Ultrasonic sensors utilize sound pulses beyond human hearing range, emitting them in a conical shape at the speed of sound. When the pulse reflects off an object, it returns to the sensor as an echo, which is interpreted to calculate the time difference between transmission and reception. Using this time interval, the distance to the object can be computed using formula (1) [28].

$$\text{D} = (\text{Speed of sound} * \text{Time interval})/2 \quad (1)$$

In general, the speed of sound is often assumed to be a constant value, but it is important to recognize that it is influenced by temperature variations. Changes in temperature directly impact the propagation of sound waves through a medium, resulting in fluctuations in the speed of sound. To account for this effect, we can adjust the value used in the distance calculation formula by replacing it with the adjusted speed of sound based on the ambient temperature measured by the DHT11 sensor.

The adjusted speed of sound can be calculated using formula (2) [29].

$$\text{Adjusted Speed of Sound} = 331.4 + (0.6 * \text{Temperature}) \quad (2)$$

To overcome uneven filling, we use multiple strategically placed ultrasonic sensors to calculate an average fill level.

The Fig. 2 shows the pinout of an ultrasonic sensor with four pins.

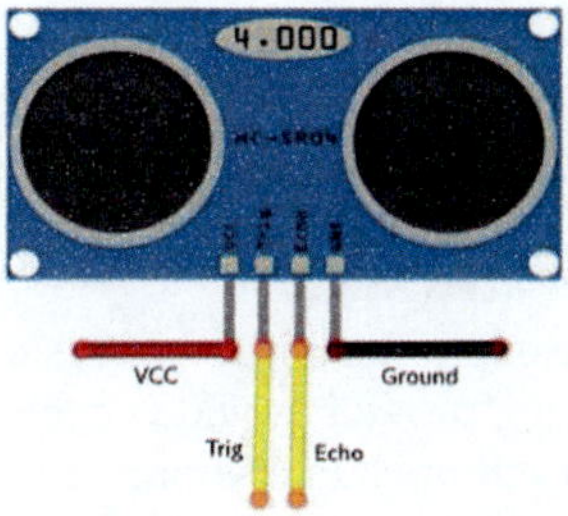

Fig. 2. HC-SR04 sensor

CO2 Emissions: MQ135 Sensor
The MQ135 gas sensor is widely used to detect and measure CO2 emissions in the air. It operates based on changes in electrical resistance when exposed to air containing CO2 [30]. By measuring the resistance variation, the sensor can determine the concentration of CO2, which can be displayed or utilized in various applications such as air quality monitoring, gas leak detection, and environmental monitoring.

CH4 Emissions: MQ4 Sensor
The MQ-4 gas sensor module detects methane (CH4) based on changes in gas conductivity, utilizing a heating element and a gas sensing element [31].

Waste Temperature: DS18B20 Sensor
The DS18B20 is a digital temperature sensor that is widely used in various applications for accurate temperature measurement. It is a versatile sensor known for its simplicity, high accuracy, and ease of integration into different electronic systems [32].

Two DS18B20 temperature sensors are strategically placed within the bin to monitor temperature accurately and calculate the average temperature.

Waste Humidity: Capacitive Soil Moisture Sensor
Capacitive soil moisture sensors measure moisture content in soil or waste based on capacitance changes between conductive plates [33]. They offer advantages such as increased sensitivity, corrosion resistance, calibration stability, and non-destructive measurement, making them ideal for precise and long-term monitoring in smart bins instead of resistive ones [34]. To monitor waste humidity, multiple capacitive sensors should be placed strategically within the bin. Averaging the readings from these sensors provides valuable insights into moisture conditions throughout the bin.

Microcontroller: Arduino Mega 2560
Given the current configuration, which involves the use of nine distinct sensors, including 2 HC-SR04 ultrasonic sensors, 1 DHT11 sensor, 2 DS18B20 temperature sensors for waste, 2 capacitive soil moisture sensors for waste humidity, 1 MQ135 sensor for CO2

emissions, and 1 MQ4 sensor for CH4 emissions, the Arduino Mega 2560 stands out as the optimal selection for seamlessly integrating these sensors. This choice is underpinned by its extensive array of digital and analog pins, substantial program memory, and robust RAM capabilities [35].

The Fig. 3 showcases the wiring and architecture of the smart bin system, illustrating the integration of various sensors with the Arduino Mega 2560 microcontroller.

Data transmission: WiFi ESP8266 Module

To transmit data from the sensors in the smart bin system, the WiFi ESP8266 module is used for wireless connectivity, providing wide coverage, fast data transfer, network compatibility, and data security through encryption [36].

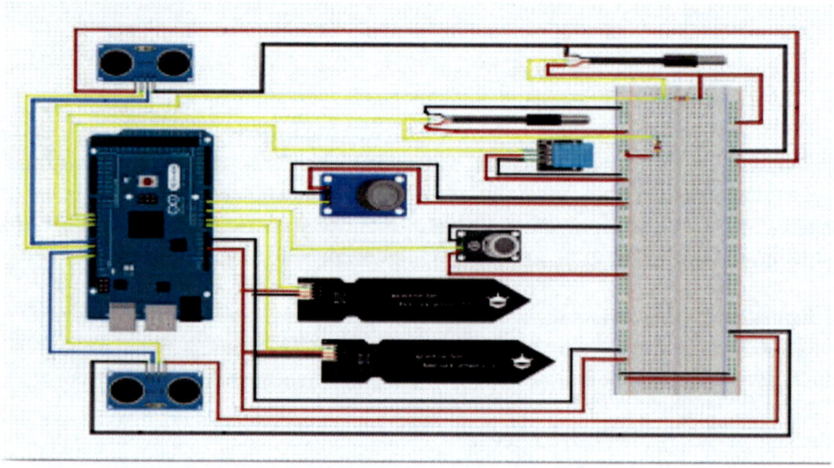

Fig. 3. Wiring configuration of the smart bin system with Arduino Mega 2560

Power Supply: Rechargeable 18650 Batteries

The smart bin system operates autonomously with rechargeable 18650 batteries, promoting sustainability and reducing waste, without the need for a constant external power supply [37].

The Fig. 4 shows the implemented smart bin system with integrated sensors, including those submerged in waste, mounted securely on a stainless-steel bracket for durability and corrosion resistance.

Fig. 4. Photo of the smart bin prototype and stainless-steel sensor mounting bracket.

4 Results and Discussion

The implemented system consists of nine sensors connected to the Arduino Mega, which acts as the central control unit. Sensor data is securely transmitted to ThingSpeak through the integrated WiFi ESP8266 module for efficient wireless data transmission. ThingSpeak provides a user-friendly web-based interface for real-time data visualization and analysis, allowing stakeholders to monitor and analyze the collected data easily. To gain deeper insights into the smart bin system, we analyzed three specific scenarios recorded on ThingSpeak, allowing for a comprehensive understanding of waste level and quantity variations over time.

First Scenario: Day 1

The first-day data analysis reveals significant trends in waste characteristics and environmental factors, as seen in the Fig. 5. Waste levels gradually increase from 14.03% to 36.28% by 10:00:00, with spikes at 12:30:00 (55.34%) and 13:00:00 (62.04%), followed by stability at around 94.17%. Ambient temperature remains at 19 °C but briefly peaks at 20 °C and 23 °C. In contrast, waste temperature, primarily organic, starts at 19.23 °C and rises to 29.8 °C, consistently exceeding ambient temperature, suggesting ongoing biological activity or decomposition in organic waste. Humidity levels drop from an initial 73% to 65% (ambient) and 66% (waste), with waste consistently slightly more humid due to organic materials. CO_2 remains stable at 23 ppm but rises to 37 ppm, indicating potential waste composition shifts, while CH4 stays at 15 ppm, slightly increasing to 20 ppm, implying shifts in anaerobic conditions or microbial activity.

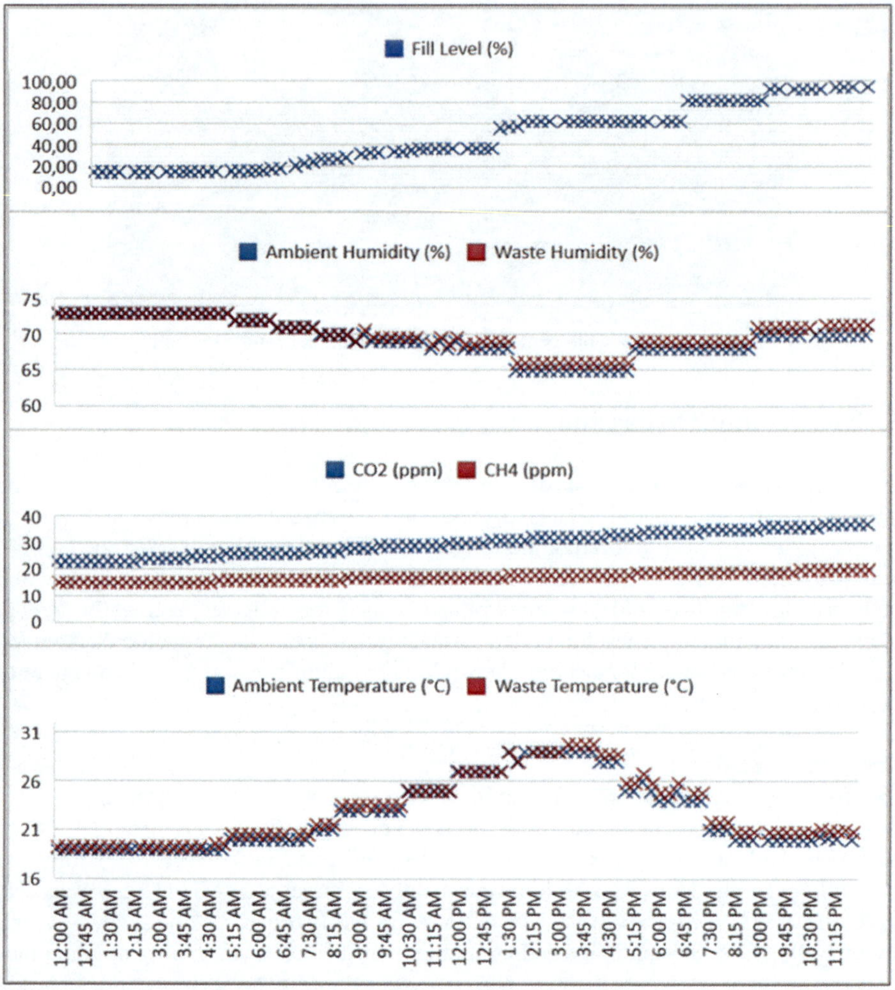

Fig. 5. Sensor Data Visualization for Day 1

Second Scenario: Day 5

On the fifth day of our smart waste bin observations, a significant transformation in the organic waste decomposition process became evident. The data as shown in the Fig. 6, revealed a substantial reduction in waste volume within the sealed container, dropping from 92.50 to 68.00% over the observation period. This remarkable decline can be attributed to the active microbial activity responsible for breaking down complex organic compounds, leading to a considerable decrease in waste mass. Moreover, this microbial activity generates heat and gases as byproducts, highlighting the dynamic nature of organic waste decomposition.

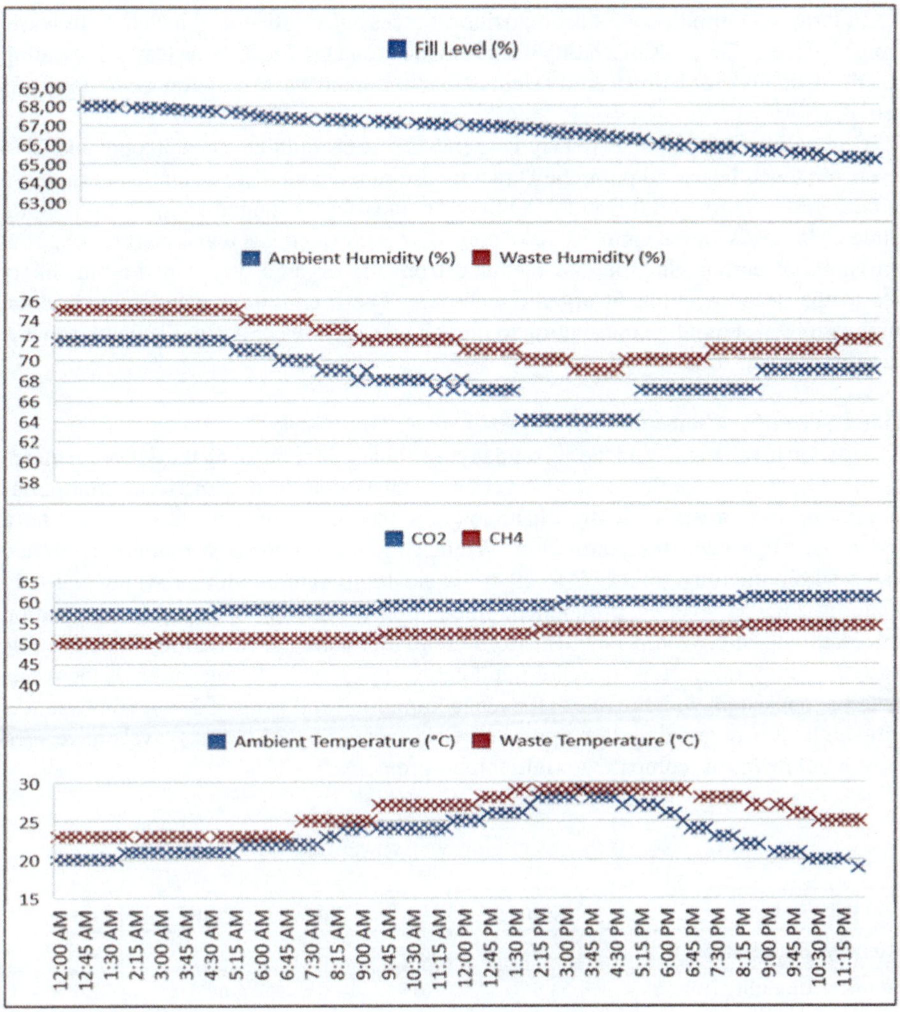

Fig. 6. Sensor Data Visualization for Day 5

Regarding temperature, the ambient temperature remained relatively constant at around 20–22 °C during the observation period, with the waste temperature closely tracking the ambient temperature, indicating minimal heat generated by the decomposition process initially. However, around the 7th hour, there was a notable increase in waste temperature, surpassing the ambient temperature and peaking at 29 °C around the 15th hour. This suggests a shift towards more active decomposition, likely due to increased microbial activity.

In terms of humidity, the decomposition process also influenced a change in waste moisture levels. The waste humidity went from 64% on Day 1 to 75% on Day 5, indicating that the decomposition process contributed to making the waste material wetter over this period.

Comparing these findings to Day 1 when CO2 levels didn't exceed 30 ppm and CH4 levels remained below 20 ppm, the Day 5 measurements indicate a significant increase in both gases. Throughout Day 5, CO2 levels consistently ranged from 57 to 61 ppm, while CH4 levels varied from 50 to 54 ppm. These elevated gas levels suggest ongoing emissions of carbon dioxide and methane from the organic waste in the bin, likely due to the decomposition of organic materials. These emissions have environmental implications, potentially contributing to greenhouse gas concentrations if they persist at these levels over time.

Third Scenario: Continuous Monitoring Over One Week
As depicted in the Fig. 7, and the accompanying Table 1 summarizing the data with mean, standard deviation, minimum, and maximum values, the third scenario of continuous monitoring over a week vividly illustrates how the smart bin's waste dynamics have undergone significant transformation. While ambient temperature remains relatively stable within the range of 19 °C to 29 °C, waste temperature, with an average slightly higher at 24.5 °C, exhibits pronounced fluctuations, reflecting ongoing decomposition processes within the bin, as evident from the higher standard deviation. Concurrently, waste humidity levels have increased, indicating a progressive moistening of the waste material, consistent with the rise in the mean humidity level from 66% to 75%. Remarkably, both CO_2 (averaging 45.8 ppm) and CH_4 (averaging 34.6 ppm) levels have seen substantial increases compared to initial measurements.

Table 1. Overview of Sensor Data

Variable	Mean	Std. Deviation	Min	Max
Ambient Temperature (°C)	22,9	3,18	19	29
Ambient Humidity (%)	68,9	2,63	64	73
Waste Temperature (°C)	24,5	3,27	19,2	30,7
Waste Humidity (%)	71,1	2,34	66	75
CO_2 (ppm)	45,8	10,54	23	61
CH_4 (ppm)	34,6	12,88	15	54

The elevated standard deviation of both CO2 and CH4 levels within the smart bin indicates a substantial degree of variability in these gases over the course of a week. This variability is consistent with the dynamic nature of waste decomposition processes, especially in the presence of predominantly organic waste. As organic materials decompose, microbial activities generate fluctuating levels of CO2 and CH4. The wide range of concentrations, from relatively low to relatively high, suggests ongoing changes in waste composition and microbial metabolic rates within the bin. Such significant fluctuations emphasize the need for close monitoring to maintain optimal environmental conditions and prevent the build-up of excessive gas emissions, which could have potential implications for both the bin's performance and environmental quality in its vicinity.

This comprehensive data analysis underscores the pivotal role of smart bins in providing real-time insights, facilitating informed decisions, and advancing sustainable waste management practices.

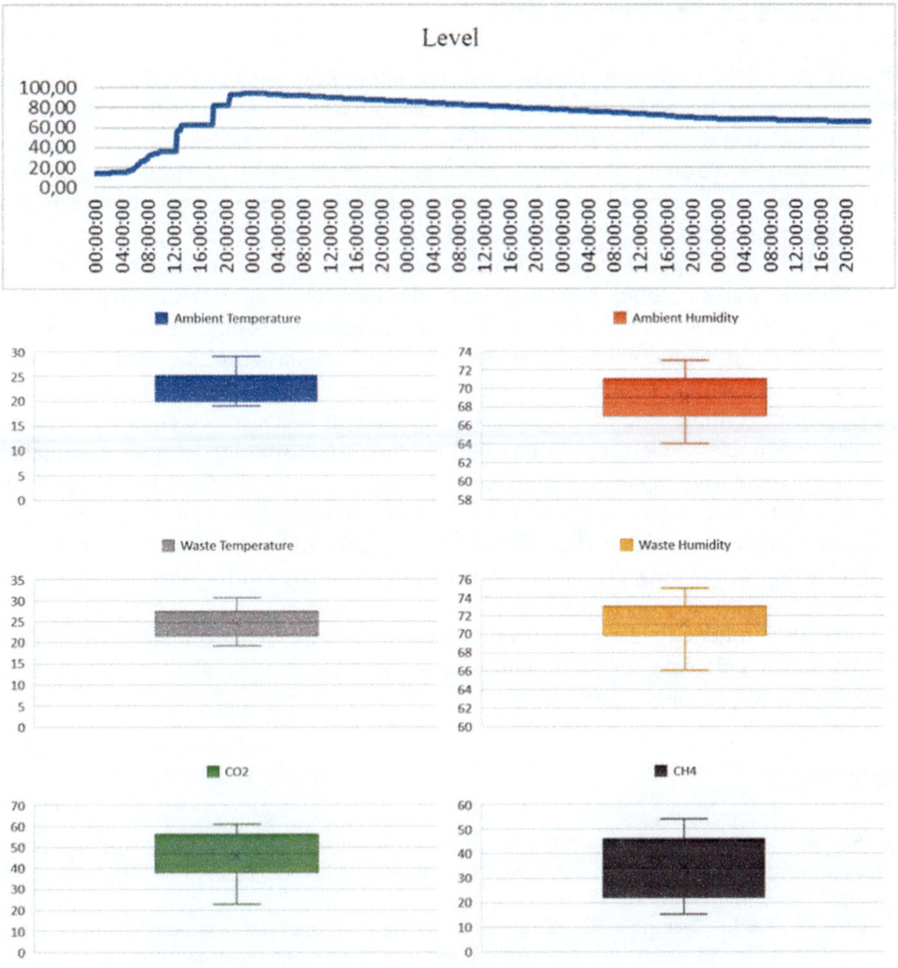

Fig. 7. Continuous Monitoring of Smart Bin Over One Week

The elevated standard deviation of both CO2 and CH4 levels within the smart bin indicates a substantial degree of variability in these gases over the course of a week. This variability is consistent with the dynamic nature of waste decomposition processes, especially in the presence of predominantly organic waste. As organic materials decompose, microbial activities generate fluctuating levels of CO2 and CH4. The wide range of concentrations, from relatively low to relatively high, suggests ongoing changes in waste composition and microbial metabolic rates within the bin. Such significant fluctuations emphasize the need for close monitoring to maintain optimal environmental conditions and prevent the build-up of excessive gas emissions, which could have potential implications for both the bin's performance and environmental quality in its vicinity.

5 Conclusion and Perspectives

The analysis of the three scenarios of smart bin monitoring captured in this study illustrate the invaluable importance of deploying smart bin technology in waste management. They demonstrate the ability of smart bins to provide real-time insights into waste characteristics and environmental factors, enabling more informed decision-making. The initial data in Day 1 revealed trends in waste levels, Temperature, Humidity, and greenhouse gas emissions (GHG), suggesting ongoing biological activity in organic waste. By Day 5, the decomposition process had become more evident, leading to substantial reductions in waste volume, increased waste Temperature, higher Humidity, and elevated levels of CO_2 and CH_4 emissions. Finally, the continuous monitoring over a week highlighted the dynamic nature of waste dynamics, emphasizing the need for proactive waste management strategies to minimize environmental impacts. These scenarios underscore the critical role of smart bins in optimizing waste management processes, reducing waste related GHG emissions, and ultimately contributing to more sustainable and environmentally responsible practices. It is important to conclude by recognizing that controlling and managing greenhouse gas emissions is not only pivotal for effective waste management but also a vital step in mitigating the impact of climate change.

Furthermore, utilizing the data obtained from the smart bin for fleet dimensioning and routing will serve as the focus of our future work. By leveraging advanced algorithms and predictive models, we aim to develop intelligent waste management systems that can dynamically allocate resources, optimize collection routes in real-time, and adapt to changing waste generation patterns.

References

1. Abubakar, I.R., et al.: Environmental sustainability impacts of solid waste management practices in the global south. Int. J. Environ. Res. Public Health **19**(19), 12717 (2022). https://doi.org/10.3390/ijerph191912717
2. Chen, S., Yu, L., Zhang, C., Wu, Y., Li, T.: Environmental impact assessment of multi-source solid waste based on a life cycle assessment, principal component analysis, and random forest algorithm. J. Environ. Manag. **339**(117942), 117942 (2023). https://doi.org/10.1016/j.jenvman.2023.117942
3. Unegg, M.C., Steininger, K.W., Ramsauer, C., Rivera-Aguilar, M.: Assessing the environmental impact of waste management: a comparative study of CO2 emissions with a focus on recycling and incineration. J. Clean. Prod. **415**(137745), 137745 (2023). https://doi.org/10.1016/j.jclepro.2023.137745

4. Li, H., Tian, Y., Li, Z., Wu, C., Liu, Y.: Life-cycle economic and environmental impacts of municipal solid waste reverse logistics in residential areas. Waste Management (New York, N.Y.), vol. 164, pp. 200–208 (2023). https://doi.org/10.1016/j.wasman.2023.04.016
5. Chan, Y.C., Sinha, R.K., Wang, W.: Emission of greenhouse gases from home aerobic composting, anaerobic digestion and vermicomposting of household wastes in Brisbane (Australia). Waste Manag. Res. **29**(5), 540–548 (2011). The Journal of the International Solid Wastes and Public Cleansing Association, ISWA.https://doi.org/10.1177/0734242X10375587
6. Sanger, J.B., Sitanayah, L., Ahmad, I.: A sensor-based garbage gas detection system. In: 2021 IEEE 11th Annual Computing and Communication Workshop and Conference (CCWC). IEEE (2021)
7. Abdullah, N., Alwesabi, O.A., Abdullah, R.: IoT-Based Smart Waste Management System in a Smart City. In: Advances in Intelligent Systems and Computing, pp. 364–371. Springer International Publishing, Cham (2019). https://doi.org/10.1007/978-3-319-99007-1_35
8. Babu, K.R.R., George, S.J., Samuel, P.: Optimal sensor selection from sensor pool in IoT environment. In: 2016 2nd International Conference on Applied and Theoretical Computing and Communication Technology (ICATccT). IEEE (2016)
9. Glass, T., Ali, S., Parr, B., Potgieter, J., Alam, F.: IoT enabled low cost air quality sensor. In: 2020 IEEE Sensors Applications Symposium (SAS). IEEE (2020)
10. Haque, K.F., Zabin, R., Yelamarthi, K., Yanambaka, P., Abdelgawad, A.: An IoT based efficient waste collection system with smart bins. In: 2020 IEEE 6th World Forum on Internet of Things (WF-IoT). IEEE (2020)
11. Chen, W.-E., Wang, Y.-H., Huang, P.-C., Huang, Y.-Y., Tsai, M.-Y.: A smart IoT system for waste management. In: 2018 1st International Cognitive Cities Conference (IC3). IEEE (2018)
12. Kumar, N.M., Parimala, S.: Jyothi Monitoring the smart garbage bin filling status: an IoT application towards waste management Int. J. Int J Civ Eng Technol **9**(6), 373–381 (2018)
13. Chaudhari, M.S., Patil, B.: Raut IoT based waste collection management system for smart cities: an. In: Overview Proceedings of the International Conference on Computing Methodologies and Communication, pp. 802–805. IEEE (2019)
14. Murugesan, S., Ramalingam, S., Kanimozhi, P.: Theoretical modelling and fabrication of smart waste management system for clean environment using WSN and IOT. Mater. Today: Proc. **45**, 1908–1913 (2021). https://doi.org/10.1016/j.matpr.2020.09.190
15. Al-Jabi, M., Diab, M.: IoT-enabled citizen attractive waste management system. In: 2017 2nd International Conference on the Applications of Information Technology in Developing Renewable Energy Processes & Systems (IT-DREPS). IEEE (2017)
16. Brancoli, P., Bolton, K., Eriksson, M.: Environmental impacts of waste management and valorisation pathways for surplus bread in Sweden. Waste Manag. (New York, N.Y.), **117**, 136–145 (2020). https://doi.org/10.1016/j.wasman.2020.07.043
17. Lee, I., Lee, K.: The Internet of Things (IoT): applications, investments, and challenges for enterprises. Bus. Horiz. **58**(4), 431–440 (2015). https://doi.org/10.1016/j.bushor.2015.03.008
18. Zeb, A., et al.: A proposed IoT-enabled smart waste bin management system and efficient route selection. J. Comput. Netw. Commun. **2019**, 1–9 (2019). https://doi.org/10.1155/2019/7043674
19. Ramson, J., Vishnu, S., Kirubaraj, A., Anagnostopoulos, T.: Abu-Mahfouz a LoRaWAN IoT enabled trash bin level monitoring system. IEEE Trans. Ind. Inf. **18**(2), 786–795 (2021)

20. Mustafa, M.R., Ku Azir, K.N.F.: Smart bin: Internet-of-things garbage monitoring system. MATEC Web Conf. **140**, 01030 (2017). https://doi.org/10.1051/matecconf/201714001030
21. Anwar, M.A.: IOT BASED GARBAGE MONITORING USING ARDUINO. Diss. West Bengal University of Technology (2018)
22. Mustapha, A., Bousmina, Z., Keskas, I., Guettatfi, Z., Hentout, A., Mansoul, A.: Practical IoT-enabled monitoring platform for solid waste collection: a case study for the municipality of Baba Hassen, Algiers, Algeria. In: 2021 IEEE International IOT, Electronics and Mechatronics Conference (IEMTRONICS). IEEE (2021)
23. Morajkar, P., Bhor, V., Pandya, D., Maheshwar Gurav, A.D., Somaiya, K.J.: Institute of engineering and information technology. smart garbage management system. In: International Journal of Engineering Research & Technology (Ahmedabad), V4(03) (2015). https://doi.org/10.17577/ijertv4is031175
24. Lee, J.A., Thomas, V.M.: GPS and radio tracking of end-of-life products. In: IEEE International Symposium on Electronics and the Environment, 2004. Conference Record. 2004 (2004)
25. Wilson, B.G., Agar, B.J., Baetz, B.W., Winning, A.: Practical applications for global positioning system data from solid waste collection vehicles. Can. J. Civ. Eng. **34**(5), 678–681 (2007). https://doi.org/10.1139/l06-174
26. Ngah Nasaruddin, A., Tee, B.T., Musthafah, M.T., Md Jasman, M.E.S.: Ambient data analytic on indoor environment monitoring for office buildings in hot and humid climates. Data Brief **27**(104534), 104534 (2019). https://doi.org/10.1016/j.dib.2019.104534
27. Past weather in Casablanca, morocco — yesterday or further back. (n.d.). Retrieved July 2, 2023, from Timeanddate.com website: https://www.timeanddate.com/weather/morocco/casablanca/historic
28. Appiah, O., Quayson, E., Opoku, E.: Ultrasonic sensor based traffic information acquisition system; a cheaper alternative for ITS application in developing countries. Sci. African **9**(e00487), e00487 (2020). https://doi.org/10.1016/j.sciaf.2020.e00487
29. Mukherjee, D., Sen, N., Singh, K.K., Saha, S., Shenoy, K.T., Marathe, P.P.: Automated interface detection in liquid-liquid systems using self-calibrating ultrasonic sensor. Chem. Eng. Sci. **236**(116510), 116510 (2021). https://doi.org/10.1016/j.ces.2021.116510
30. Pereira, P.F., Ramos, N.M.M.: Low-cost Arduino-based temperature, relative humidity and CO2 sensors - An assessment of their suitability for indoor built environments. J. Build. Eng. **60**(105151), 105151 (2022). https://doi.org/10.1016/j.jobe.2022.105151
31. Yang, S., Liu, Y., Wu, N., Zhang, Y., Svoronos, S., Pullammanappallil, P.: Low-cost, Arduino-based, portable device for measurement of methane composition in biogas. Renew. Energy **138**, 224–229 (2019). https://doi.org/10.1016/j.renene.2019.01.083
32. Bohórquez, M.A.M., Enrique Gómez, J.M., Andújar Márquez, J.M.: A new and inexpensive temperature-measuring system: Application to photovoltaic solar facilities. Solar Energy (Phoenix, Ariz.) **83**(6), 883–890 (2009). https://doi.org/10.1016/j.solener.2008.12.007
33. Deng, X., Yang, L., Fu, Z., Du, C., Lyu, H., Cui, L., Jia, B.: A calibration-free capacitive moisture detection method for multiple soil environments. Measur. J. Int. Measur. Confederation, **173**(108599), 108599 (2021). https://doi.org/10.1016/j.measurement.2020.108599
34. Li, B., et al.: Accuracy calibration and evaluation of capacitance-based soil moisture sensors for a variety of soil properties. Agric. Water Manag. **273**(107913), 107913 (2022). https://doi.org/10.1016/j.agwat.2022.107913
35. Ray, P.P., Thapa, N.: A systematic review on real-time automated measurement of IV fluid level: status and challenges. Measur. J. Int. Measur. Confederation **129**, 343–348 (2018). https://doi.org/10.1016/j.measurement.2018.07.046

36. Xiao, J., Li, J.T.: Design and implementation of intelligent temperature and humidity monitoring system based on ZigBee and WiFi. Procedia Comput. Sci. **166**, 419–422 (2020). https://doi.org/10.1016/j.procs.2020.02.072
37. Katie C. Abbott, et al.: Experimental study of three commercially available 18650 lithium ion batteries using multiple abuse methods. J. Energy Storage **65**, 107293 (2023). ISSN 2352–152X, https://doi.org/10.1016/j.est.2023.107293

The Accuracy Analysis and Usability of Low Cost RTK Portable Kit on Surveying Aims

İbrahim Murat Ozulu[1](✉), Hasan Dilmaç[2], and Veli İlçi[2]

[1] Department of Architecture and Urban Planning, Hitit University, 19169 Çorum, Merkez, Turkey
imuratozulu@hitit.edu.tr

[2] Faculty of Engineering, Department of Geomatics Engineering, Ondokuz Mayis University, 55270 Samsun, Atakum, Turkey

Abstract. Thanks to the rapid developments in electronics, computers, and software, state-of-the-art and cost-effective devices are emerging in almost every area. One of these areas is GNSS technologies and related devices. These advances in GNSS devices enable users to provide more advanced and more accurate results at an affordable cost, and thus the use of these devices is increasing rapidly. This study investigated the accuracy performance supplied by the cost-effective ArduSimple CORS system. For this purpose, ArduSimple RTK Calibrated Surveyor Kit and Topcon HiperV GNSS receivers were connected to the Tusaga-Aktif system at 86 points. Real-time position information was collected three different times with 30 epochs of both GNSS receivers. The coordinate data obtained from Topcon and ArduSimple GNSS receivers were compared, and the accuracy performance provided by the ArduSimple system was evaluated. In this study, the coordinate data obtained with the Topcon receiver, whose accuracy is known and tested many times, are accepted as a reference. When the difference values are averaged, it is seen that the position accuracy obtained from the ArduSimple receiver is 1.2 cm in the easting, 1.4 cm in the northing, and 5 cm in height.

Keywords: GNSS · RTK · Network-RTK · Geodesy · Surveying

1 Introduction

GNSS systems, which started to be used after the 1980s and whose use is increasing day by day, have become the basic technology of many applications that require position information today. However, the meter level position error can be achieved in the absolute positioning technique where no correction information is received. However, geodetic and surveying applications require cm-dm level positioning accuracy [1–7]. Many methods and algorithms have been developed to improve the position accuracy of GNSS. One of these improvement methodologies is using GLONASS, BEIDOU, and GALILEO global positioning system data in addition to the GPS [2, 3, 5, 8].

Thus, with the solution called multi-GNSS, higher accuracy is achieved by incorporating multi-signal data into the algorithms, and position information is improved in

M. Ben Ahmed et al. (Eds.): SCA 2023, LNNS 906, pp. 270–276, 2024.
https://doi.org/10.1007/978-3-031-53824-7_24

harsh environments that prevent direct reception of GNSS signals, such as urban areas [3, 4, 6, 8].

Real-Time Kinematics (RTK) technique is the most widely used GNSS positioning technique to improve position accuracy. This technique calculates correction information for a fixed reference station and transmits it to the rover receiver. As a result of the calculations performed using the correction information on the rover receiver, position information with cm accuracy is obtained. In the RTK technique, position accuracy depends on the distance between the reference and rover receivers and atmospheric parameters. The distance between the reference and rover receivers is 5–20 km, which varies according to the used radio transmitter characteristics [3, 4, 7, 9].

In the RTK technique, the necessity of using two GNSS receivers, a reference and a rover, and the decrease in position accuracy due to distance and atmospheric parameters, the need to develop new methods have emerged. The most prominent of these methods is the Network-RTK method. In this method, 3-dimensional position information with cm accuracy is obtained using the correction data from the fixed reference network once established [3, 5]. In the Network-RTK method, distance-related errors are eliminated in the network, and it is possible to use a single GNSS receiver.

In addition to the advances in microprocessors and sensor technologies, there have been significant developments in GNSS measurement and calculation algorithms. New GNSS receiver and antenna systems are emerging in parallel with these developments. As an alternative to GNSS receivers, which are widely used and costly, it is aimed to produce new systems that provide precise position accuracy but are much more cost-effective and readily available [2, 3, 5, 6, 8]. One device that has emerged as an alternative to existing GNSS systems in recent years is the GNSS receiver and antenna systems produced by ArduSimple [1, 6, 10]. This study tested the system performance called RTK Calibrated Surveyor Kit made by ArduSimple company.

2 Material and Method

2.1 ArduSimple RTK Calibrated Surveyor Kit

The GNSS antenna included in the ArduSimple RTK Calibrated Surveyor Kit set is a 184-channel antenna with IP67 standard and multi-GNSS features. This antenna complies with the National Oceanic and Atmospheric Administration (NOAA) standards, and the calibration file can be downloaded from the NOAA website. Thus, the GNSS data collected with this antenna can be converted to RiNEX format and used for post-process static evaluation. The receiver system included in the GNSS set consists of a u-blox ZEDF9P receiver [2, 3, 8, 9, 11]. An Android-based mobile device is required to visualize and record the collected data. With the SW Maps mobile application developed by SoftWel company installed on the Android device, data can be recorded and displayed [1, 12]. Significant savings are achieved as the system can be used with any Android-based mobile device instead of a customized hand unit. The schematic view of the system used is shown in Fig. 1.

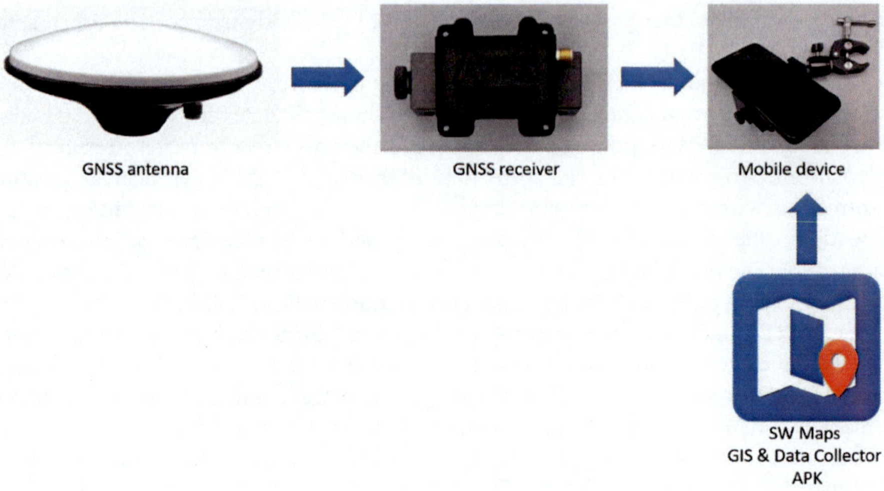

Fig. 1. ArduSimple Surveyor Kit

2.2 Study Area and Data Collection

This study was conducted on 7–14 October 2022 at Ondokuz Mayıs University, Kurupelit Campus in Samsun, Turkey. Eighty-six reference points were established in the study area (Fig. 2). GNSS receivers were installed to determine the coordinates of these points. The data were collected by measuring 30 epochs and three times connecting to the Tusaga-Active Network-RTK system correction data. The point coordinates were determined by taking the average of these data.

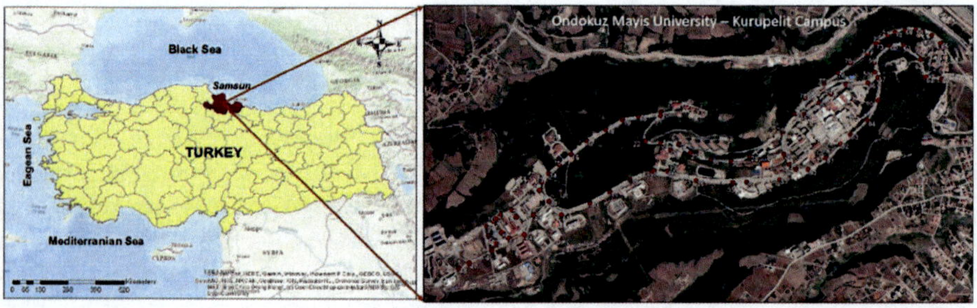

Fig. 2. Study area and location of the points

At these points, firstly, the data were collected using the Topcon HiperV GNSS receiver, one of the well-known GNSS brands suitable for geodetic purposes, and these results were accepted as correct (Fig. 3a). In order for the number and distribution of satellites to be similar, coordinate data were collected with the ArduSimple GNSS set at the same duration after data collection with Topcon receiver (Fig. 3b).

(a) (b)

Fig. 3. Site survey (a. Topcon HiperV - b. ArduSimple RTK Calibrated Surveyor Kit)

ArduSimple RTK Calibrated Surveyor Kit includes National Oceanic and Atmospheric Administration (NOAA) Calibrated Antenna and u-blox Zed-F9P RTK GNSS receiver. In addition, the data collected by this system is visualized and recorded via an Android-based mobile phone. Some technical specifications of the two GNSS receivers used for the measurement are shown in Table 1.

Table 1. The technical specifications of GNSS systems

		Topcon HiperV	ArduSimple RTK
Number of Channels		226	184
GNSS Tracking	GPS	L1, L2, L2C	L1 C/A, L2C
	GLONASS	L1, L2	L1, L2
	GALILEO	–	E1B/C, E5B
	BEIDOU	–	B1I, B2I
	SBAS	L1, C/A WAAS/ MSAS/ EGNOS/ GAGAN	L1C/A
	QZSS	L1 C/A	L1 C/A
RTK Positioning Accuracy	Horizontal	5mm + 0.5ppm	10mm + 1ppm CEP
	Vertical	10mm + 0.8ppm	

3 Results

The differences between the measurement data performed at the reference points with the two GNSS sets are shown in Fig. 4. Although the easting and northing errors are less than 5 cm except at one point, the height errors are mostly between 5–10 cm level and also reach up to dm-level in some test points.

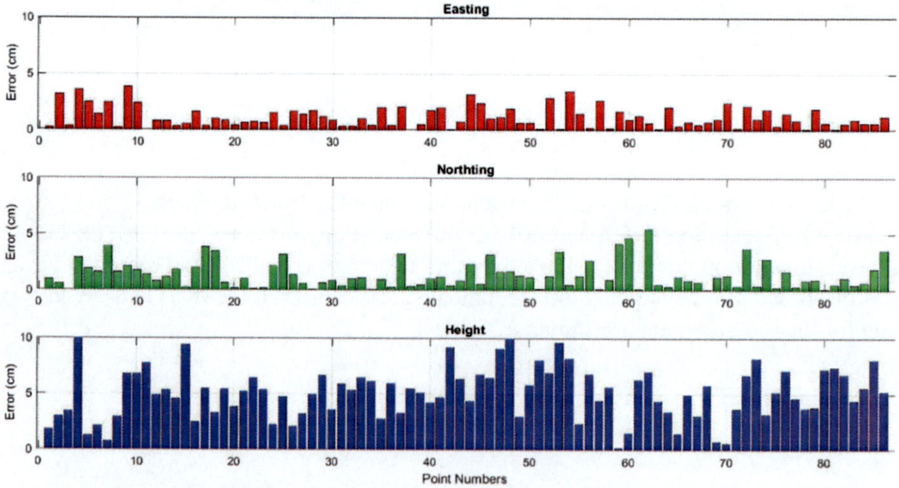

Fig. 4. Errors between coordinate values obtained with two GNSS receivers

Some statistical information about the results is given in Table 2. According to the obtained results, maximum 3.8 cm and average 1.2 cm difference values were obtained in the easting component. The northing component's maximum and average differences are 5.5 and 1.4 cm, respectively. A maximum difference of 10.3 cm and an average of 5 cm were obtained in the height component. The RMSE values are 0.9, 1.2 and 2.3 cm in easting, northing and height components, respectively.

Table 2. Some statistical values of errors

	Easting (cm)	Northing (cm)	Height (cm)
Min	0,0	0,0	0,2
Max	3,8	5,5	10,3
Mean	1,2	1,4	5,0
RMSE	0,9	1,2	2,3

4 Conclusions

New devices are produced in parallel with the progress made in GNSS techniques and sensor Technologies day by day. In order to ensure the widespread use of these devices, they must be accurate enough to meet user needs and as cost-effective as possible. In this context, one of the devices produced in recent years is the GNSS receiver-antenna systems developed by ArduSimple company. As a result of this study, it has been observed that the ArduSimple RTK Calibrated Surveyor Kit offers cm accuracy both horizontally and vertically.

Static GNSS observations are carried out at these 86 points. After the static data of the points are evaluated, the results of the ArduSimple GNSS receiver will be compared with the static results and shared. The results show that cm-level accuracies are obtained in 2D, and 1-dm level accuracies are provided in the height direction. Also, it can be commented that the comparatively low-cost ArduSimple RTK Calibrated Surveyor Kit offers the similar positioning accuracies like the well-known GNSS receiver-antenna sets.

References

1. Alessandri, L., Baiocchi, V., Monti, F.: Intensive archaeological survey at Piscina Torta: use of a low-cost RTK portable kit to materialise a UTM grid on the ground. In: 2022 IMEKO TC-4 International Conference on Metrology for Archaeology and Cultural Heritage University of Calabria, pp. 19–21 (2022)
2. Budisusanto, Y., Cahyadi, M.N., Farid, I.W., Ubaidillah, M.R., Imani, D.W.: Low Cost LiDAR prototype design for 3D mapping. In: 2021 International Conference on Advanced Mechatronics, Intelligent Manufacture and Industrial Automation, ICAMIMIA 2021 - Proceeding, pp. 13–17 (2021)
3. ÉGETŐ, C., Siki, Z., Turák, B., Rózsa, S.: Low-Cost GNSS sensor development for structural monitoring. Scientific bulletin series d: mining, mineral processing, non-ferrous metallurgy. Geol. Environ. Eng. **35**(2), 37–48 (2021)

4. Gao, P., et al.: Improved position estimation algorithm of agricultural mobile robots based on multisensor fusion and autoencoder neural network. Sensors **22**(4), 1–20 (2022)
5. Hodgson, M.E.: On the accuracy of low-cost dual-frequency GNSS network receivers and reference data. GIScience Remote Sens. **57**(7), 907–923 (2020)
6. Hohensinn, R., et al.: Low-Cost GNSS and Real-Time PPP: assessing the precision of the u-blox ZED-F9P for kinematic monitoring applications. Remote Sens. **14**(20), 1–25 (2022)
7. Stangebye, T., Mohr, T., Vallenti, A., Grauff, M., Koziol, S.: Custom real-time-kinematics positioning system testbed for mobile robot localization. In: Proceedings of the 2020 IEEE Dallas Circuits and Systems Conference, DCAS 2020, pp. 1–4 (2020)
8. Kogogin, D.A., et al.: Capacities of TEC measurements by the low-cost GNSS receiver based on the u-blox ZED-F9P for ionospheric research. J. Phys. Conf. Ser. **1991**(1), 012020 (2021)
9. Hartzer, J., Saripalli, S.: AutoCone: An OmniDirectional Robot for Lane-Level Cone Placement, pp. 1663–1668. IEEE Intelligent Vehicles Symposium, Proceedings, Iv (2020)
10. Chustz, G., Saripalli, S.: ROOAD: RELLIS Off-road Odometry analysis dataset. In: IEEE Intelligent Vehicles Symposium, Proceedings, 2022-June(Iv), pp. 1504–1510 (2022)
11. u-blox Zed F9P (2023). https://www.u-blox.com/en/product/zed-f9p-module. Accessed 20 Jun 2023
12. SW Maps (2023). http://softwel.com.np/images/download_manual/SWMAPS5.pdf. Accessed 25 Jun 2023

Sustainability Assessment of Public Schools in the Palestinian Territory

Aya Baba[1](✉), Isam Shahrour[1], Mutasim Baba[2], and Marwan Sadek[1]

[1] Laboratoire Génie Civil et Géo-Environnement, University of Lille, 59000 Lille, France
aya.baba.etu@univ-lille.fr
[2] An-Najah National University, Nablus, Palestine

Abstract. Sustainability is a critical issue that affects the education sector in multiple ways. We argue that adopting sustainable practices can reduce the environmental impact of schools, provide valuable educational opportunities for students, promote social responsibility, and deliver economic benefits to educational institutions. Therefore, this paper presents a framework for assessing the sustainability of schools in Palestine, which can help identify areas for improvement and enhance the quality of education. The framework aims to establish a comprehensive set of indicators that cover the environmental, social, and economic dimensions of sustainability for public schools. It is based on the literature review, the Palestinian context, and the opinion of a panel of experts. The paper presents the methodology to identify these indicators and uses the Stepwise Weight Assessment Ratio Analysis method (SWARA) for their ranking and weighting. These indicators include: Social Equity, Indoor Quality, Health and Comfort, Social Cohesion, Accessibility, Energy Use Efficiency, Water Use Efficiency, Site Development, Influence on Local Economy, Annual Operating Costs, and Teaching Quality.

Keywords: Sustainable School · Sustainability Assessment · Education and Sustainability · Framework for Sustainability

1 Introduction

Schools play an important role in shaping the values and behavior of future generations, making them an important area for promoting sustainability [1]. Education for Sustainability aims to provide individuals with the knowledge, skills, and attitudes necessary to understand and address social, economic, and environmental issues and to contribute to a more sustainable future.

A sustainable school goes beyond the physical environment and integrates sustainability into all aspects of school life. The whole-school approach is an effective way to create a sustainable school environment [2]. According to [3], it uses various policies and practices to create a positive and physical school environment and targets all members of the school (students, staff, families, and the wider community).

However, there is a lack of well-developed methods and tools for assessing and evaluating schools' sustainability based on its three pillars [4]. Measuring progress in

M. Ben Ahmed et al. (Eds.): SCA 2023, LNNS 906, pp. 277–291, 2024.
https://doi.org/10.1007/978-3-031-53824-7_25

implementing sustainability practices in schools is necessary to ensure its good implementation. Several countries have introduced and strengthened frameworks and tools for assessing the sustainability of buildings. These frameworks and tools are designed for various building types, including homes, offices, hospitals, sports centers, and commercial buildings. However, there is a dearth of such frameworks and tools specifically designed for schools. Therefore, this study seeks to bridge this gap by formulating a framework and tools for assessing school sustainability.

International Tools such as LEED for School and BREEAM Education include indicators for sustainability and education. The LEED Framework for School includes seven categories for assessing school sustainability [5], but it lacks social and economic indicators. In 2008, BREEAM Education introduced the MAN 10 Index "Development as an Education Resource" to promote sustainable education through building design and renovation [6]. However, both LEED for School and BREEAM for Education fails to take into account the economic aspect of sustainability, which is a key pillar. Although the SBTool for K-12 Schools balances the three dimensions of sustainability, it does not include an indicator of environmental education, which is a critical component of education for sustainability. Therefore, there is a need to develop a comprehensive sustainability framework for schools.

The education system in Palestine faces major challenges, including political conflict, lack of qualified teachers, outdated curricula, and financial constraints [7]. Ongoing political conflict in the region has resulted in the destruction of schools [8], limited movement and access, and the displacement of students and teachers. This has resulted in a lack of adequate facilities and resources in many schools, affecting the quality of education provided to students.

In addition, many teachers are looking for work in other countries due to the lack of opportunities in Palestine, which has led to a shortage of experienced and qualified teachers. The outdated curriculum lacks the critical thinking, problem-solving skills, technology, and innovation essential for success in today's globalized economy. Furthermore, many schools lack basic facilities, such as computers, libraries, and science laboratories, which are essential for quality education.

Therefore, this paper focuses on developing reliable frameworks and methods for assessing and evaluating school sustainability practices. This will enable us to measure progress in implementing sustainable practices and ensure that the impacts of these efforts are significant. By doing so, we can create a culture of sustainability in schools, promoting a more sustainable future for all.

The structure of the paper is as follows: First, the research methodology is defined, which includes the establishment of a sustainability assessment framework and the method of calculating the sustainability index for schools. Secondly, the paper applies this methodology to Palestinian public schools, with a brief description of data collection and analysis methods. Third, the weights of the indicators are calculated, and the results are analyzed and presented clearly.

2 Methodology

Figure 1 illustrates the proposed methodology to establish the sustainability framework. It includes three main stages as follow: starts with a review of existing global sustainable rating systems (focusing on LEED for Schools, BREEAM Education, and SBtool Tool for K-12 Schools) and Palestinian context (the drivers and barriers of implementing sustainability in Palestinian schools) to set up common and core indicators. Then, an assessment model comprising pressure, state, and response indicators is designed to measure sustainability. The experts' questionnaire is conducted to determine the importance of each assessment indicator and its applicability to Palestinian's local context through experts' opinion. The SWARA approach is used to determine the weight of each PSR indicator. Then, data analysis is used for the sustainability assessment.

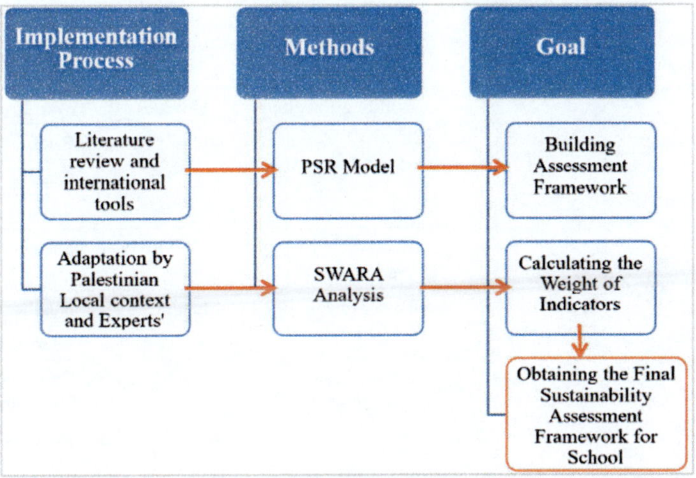

Fig. 1. The methodology steps for sustainability assessment for schools (Source: Author).

2.1 Implementation of the Pressure-State-Response (PSR) Model for the Framework

The PSR framework is widely used in sustainability [9]. It provides a better understanding of the actions and activities that affect the state of a system and appropriate responses to address them for educational institutions and stakeholders. The PSR model's concept identifies "what happened?", "why did it happen?" and "how to do it?". It constructs an index that follows the reasoning of "the impacts of pressure, state changes, and problem-solving" [10].

Through a literature review and analysis of education conditions in Palestine, 11 major pressure criteria were identified and categorized into four environmental, five social, and two economic categories of sustainability. Table 1 presents these criteria, which significantly affect the sustainability of public schools in Palestine. They provide a holistic approach to the challenges and difficulties faced by education in the region, ensuring that all aspects are covered.

Table 1. The PSR indicators for public schools (Source: Author).

	Pressure Criteria	State Indicators	Unit	Response Indicators
Environmental Pillar	Energy Consumption	Heating Consumption	$kWh/m^2.y$	Energy Efficient Heating Equipment
		Cooling Consumption	$kWh/m^2.y$	Energy Efficient Cooling Equipment
		Lighting Consumption	$kWh/m^2.y$	Energy Efficient Lighting System
		Building Envelope	U value	Wall Insulation
				Roof Insulation
				Glazing
				Shading Devices
		Renewable Energy Production	$kWh/m^2.y$	Solar Panels
	Water Consumption	Total Water Consumption	L/student.y	Number of Students
				Internal Leak
		Supply Water Consumption	L/student.y	Rainwater Harvesting
				Recycled Grey water
		Connection to Public Sewage	0 or 1	
	Indoor Quality	Thermal Comfort	PMV & PPD	Heating/Cooling System
			U value	Global Insulation
			ACH	Ventilation
		Visual Comfort	lux	Natural and Artificial Lighting
			%	Wall-Window Ratio (view)

(*continued*)

Table 1. (*continued*)

	Pressure Criteria	State Indicators	Unit	Response Indicators
			Glare index	Glare
		Acoustic Comfort	dp value	Acoustic Insulation
		Indoor Air Quality	CO_2 level	Ventilation
		Safety	Scale (1–5)	Respect the Safety Guidelines
				Training Classes
	Site Development	Waste Management	0 or 1	Solid Waste Separation
				Grey Water Recycling
		Green Area	%	
		Heat Island Effect	ΔT	Low SRI Surface
		Shading Area	%	
Economic Pillar	Annual Operating Costs	Operational Expenses	NIS/m^2.y NIS/st.y	Operational Energy Expenses
				Operational Water Expenses
				Maintenance Cost
	Influence on Local Economy	Creating Jobs/Employment	job/y	
		Production Activity	NIS/y	
Social Pillar	Social Equity	Social Inclusion	%	The ratio of low income people
				Adaptation for disabled students
		Human Centred Design	Scale (1–5)	Teachers Participation
				Parents Participation
				Students Participation

(*continued*)

Table 1. *(continued)*

	Pressure Criteria	State Indicators	Unit	Response Indicators
	Health and Comfort	Occupants Health	%	Absence Rates
				The appearance of Sick Building Syndromes/BRI
		Ergonomic Comfort	Scale (1–5)	Appropriate classroom furniture design and arrangement
				Appropriate color in the educational environment
	Social Cohesion	Social Interaction	Event/yr	School Services
		Cultural Value	Scale (1–5)	
	Accessibility	Access to Public Transportation	m	
		Access to Non-motorized mode of movement	0 or 1	
		Parking Area	%	
	Teaching Quality	Success Rates	%	Governance policy
		Attendance Rates	%	Student/Teacher
		Discipline Referrals	#/y (%)	Occupation rate
		School Dropout	#/y (%)	

2.2 SWARA Method

The objective of the data analysis is to rank the indicators based on expert opinion and determine their relative weight in order to calculate the overall sustainability score. The Stepwise Weight Assessment Ratio Analysis (SWARA) method is used for this purpose. This method, previously used by [11], enables experts to use their knowledge and experience to evaluate the importance of each indicator. The ranking and weighting of the indicators are then determined using the following five-step process [12]:

Step 1. The criteria are sorted in descending order based on their expected importance.

Step 2. Starting with the second criterion, the respondent expresses the relative importance of criterion j in relation to the previous criterion (j-1) for each given criterion. This physical ratio is called the comparative importance of average value, sj.

Step 3. Determination of the coefficient kj as in Eq. 1.

Step 4. Determination of the recalculated weight qj as in Eq. 2.

Step 5. The relative weights of the evaluation criteria are determined as in Eq. 3. Wj indicates the relative weight of the j-th criterion, while n indicates the number of indicators.

$$Kj = \begin{cases} 1, & j = 1 \\ sj + 1, & j > 1 \end{cases} \tag{1}$$

$$qj = \begin{cases} 1, & j = 1 \\ \frac{kj-1}{kj}, & j > 1 \end{cases} \tag{2}$$

$$Wj = \frac{qj}{\sum_{K-1}^{n} qk} \tag{3}$$

2.3 Data Collection

The data collection involved site visits and interviews with schools' stakeholders to ensure comprehensive and reliable information. In addition, the questionnaire has been conducted by experts to determine the importance of each assessment indicator and its applicability to Palestinian's local context.

Table 2. The experts' personal information.

Experts' Personal Information			
Classification			No
1	Gender	Male	23
		Female	5
2	Education level	Bachelor	6
		Master	4
		PhD	18
3	Specialization	Architecture	14
		Civil Engineering	5
		Economic	1
		Energy Engineering	7
		Electrical & Telecom Engineering	1
4	Working Sector	Public Sector	12
		Privet Sector	16
5	Experience	3–10 years of experience	4
		More than 10 years of experience	24

Table 2 summarizes the profile of Twenty-eight (28) experts. About 64% have a Ph.D. degree. They cover broad disciplines such as architecture, civil engineering, energy, electrical engineering, and economics. About 50% of the experts are architects. About half of the experts work in the public sector and the other half in the private sector. Moreover, 86% of experts have more than 10 years of experience in sustainability and schools.

3 Results and Discussion

3.1 Presentation of Results

The opinions of experts are presented in Table 3. The heat map indicates that approximately 86% of the experts rated the set of indicators highly, with a score of 4 or 5, while only 12% provided an intermediate score of 3. This expresses the importance, comprehensiveness, and relevance of indicators for assessing sustainability in Palestinian schools, according to the experts' opinion. The global score ranges from 108 (Heat Island Effect) to 135 (Social Inclusion), with an average value of 120 and a standard deviation of 7.

Table 3. Heat-map results to each category and indicator based on respondents.

Aspect	Category	Indicator	ID	Scale for Quantitative P 1 to 5 (1 Less important, 5 More important)					Global Score
				1	2	3	4	5	
Environmental Aspect	Energy Use Efficiency	Heating Energy Consumption	P1	0	1	1	10	16	125
		Cooling Energy Consumption	P2	0	2	3	11	12	117
		Lighting Energy Consumption	P3	0	4	3	11	10	111
		Building Envelope Insulation	P4	0	1	1	8	18	127
		Renewable Energy Production	P5	0	0	6	6	16	122
	Water Use Efficiency	Total Water Consumption	P6	0	0	0	12	16	128
		Water Harvesting and Grey-water Recycling	P7	0	0	7	11	10	115
		Connection to Public Sewage	P8	0	2	7	8	11	112
	Indoor Quality	Thermal Comfort	P9	0	0	2	10	16	126
		Visual Comfort	P10	0	0	3	12	13	122
		Acoustic Comfort	P11	0	1	3	11	13	120
		Indoor Air Quality	P12	0	0	1	5	22	133
		Safety and Security	P13	0	1	1	8	18	127
	Site Development	Waste Management Strategies	P14	0	1	0	20	7	117
		Green Areas	P15	0	0	3	9	16	125
		Heat Island Effect	P16	0	1	7	15	5	108
		Shading Area	P17	0	0	3	15	10	119
Social Aspect	Social Equity	Social Inclusion	P18	0	0	1	3	24	135
		Human Centred Design	P19	0	0	2	18	8	118
	Health and Comfort	Occupants Health	P20	0	0	0	7	21	133
		Ergonomic Comfort	P21	0	1	3	12	12	119
	Social Cohesion	Social Interaction	P22	0	0	2	15	11	121
		Cultural Value	P23	0	0	3	11	14	123
	Accessibility	Access to Public Transportation	P24	0	0	1	12	15	126
		Access to Non-motorized mode of movement	P25	0	0	4	11	13	121
		Parking Area	P26	0	2	1	15	10	117
	Teaching Quality	Success Rate	P27	0	0	4	16	8	116
		Attendance Rate	P28	0	0	4	14	10	118
		School Dropout	P29	0	2	7	11	8	109
		Discipline Referrals	P30	0	3	6	10	9	109
Economic Aspect	Annual Operating Costs	Operational Energy Expenses	P31	0	1	5	13	9	114
		Operational Water Expenses	P32	0	1	5	15	7	112
		Maintenance Cost	P33	0	0	4	15	9	117
	Influence on Local Economy	Creating Jobs/ Employment	P34	0	0	6	9	13	119
		Social Activity	P35	0	2	7	10	9	110

3.2 Discusion of the Sustainability Categories Ranking

The rank of the sustainability categories for schools are represented in Table 4 based on experts' opinion as calculated by the SWARA method. The global scores are provided in Table 3.

The experts gave the highest importance to the categories falling under the social aspect, which aligns with the goals of education and schools in Palestine. The Palestinian Ministry of Education's strategy plan (2021–2023) highlights the importance of ensuring safe, inclusive, and equitable access to education as its first goal. Additionally, it aims to enhance the quality of education by improving educational outcomes, strengthening inputs and processes, evaluating results, and establishing mechanisms for measuring progress as its second objective [7].

Table 4. Ranks of the sustainability categories according to the experts' opinion by using SWARA method.

Sustainability Categories	Global Score	Rank	Comparative importance of average value (sj)	Coefficient (Kj)	Recalculated weight (qj)	Weight (Wj)
Social Equity	127	1		1	1	0.097
Indoor Quality	126	2	0.01	1.01	0.99	0.096
Health and Comfort	126	3	0	1	0.99	0.096
Social Cohesion	122	4	0.04	1.04	0.95	0.093
Accessibility	121	5	0.01	1.01	0.94	0.092
Energy Use Efficiency	120	6	0.01	1.01	0.93	0.091
Water Use Efficiency	118	7	0.02	1.02	0.91	0.089
Site Development	117	8	0.01	1.01	0.91	0.088
Influence on Local Economy	115	9	0.02	1.02	0.89	0.086
Annual Operating Costs	114	10	0.01	1.01	0.88	0.086
Teaching Quality	113	11	0.01	1.01	0.87	0.085

The significance of the school's environmental aspect is in the second place. It includes energy and water consumption, internal environment quality, and site sustainability. Table 5 provides the weights of the sustainability categories in the Palestinian Green Building Guideline (PGBG), and they align with the experts' opinions regarding the importance of energy and water efficiency. However, experts prioritize indoor quality more than PGBG does. This difference may be attributed to the crucial role of indoor quality in educational buildings, as it directly affects students' health who may suffer from various symptoms due to an unhealthy indoor environment. Furthermore, it is worth noting that the PGBG does not consider the social and economic categories.

Table 5. The rank and weight of sustainability categories in the Palestinian Green Building Guideline [13].

Domain	Weight	Domain	Weight
Energy Efficiency	0.30	Site Sustainability	0.15
Water Use Efficiency	0.25	Materials and Resources	0.10
Indoor Quality	0.15	Innovation and Building Integrated Design	0.05

3.3 Discusion of of the Sustainability Indicators' Weights

After applying the SWARA method on indicators' scores, the ranking and weights of indicators 4 based on experts' opinion are presented in Table 6. The indices have a mean value of 0.0286 and a standard error of 0.0003, which indicates a small amount of variance. The median (0.0283) is close to the mean, indicating a symmetric distribution with no significant outliers. The standard deviation is 0.0020, which indicates that the data are tightly clustered around the mean. In general, descriptive statistics provide a clear summary of the central tendency, variance, and distribution of the data. This confirms the importance and comprehensiveness of all indicators and that none can be dispensed with.

The most important indicator is "**social inclusion**" with 0.033 score. This indicator assesses a public school's ability to enroll low-income students and those with disabilities and highlights its critical role in promoting social equity, a key goal in pursuing sustainable education. It is followed by "**Indoor Air Quality**" and "**Occupants Health**" with equal scores (0.032). As was evident before, a healthy indoor environment for students is an important principle of sustainability.

We note that the weighting of the indicators is consistent with the concepts of sustainability and the context of schools in Palestine. For example, using "**Building Envelope Insulation**" and "**shading on the external facades**" with 0.031 score is of greater importance than "**Renewable Energy Production**" with score 0.029, on the principle that rationalizing energy is better than producing the alternative. Also, the "**Cooling Energy Consumption** (0.028)" is considered less important than "**Heating Energy Consumption** (0.030)", as schools do not typically operate during the summer. The same logic applies to the "**Heat Island Effect** (0.025)" indicator.

Table 6. The weight of indicators based on experts' opinions.

Indicator	ID	Rank	Comparative importance of average value (sj)	Coefficient (Kj)	Recalculated weight (qj)	Indicator Weight (Wj)
Social Inclusion	P18	1		1	1	0.033
Indoor Air Quality	P12	2	0.02	1.02	0.98	0.032
Occupants Health	P20	3	0	1	0.98	0.032
Total Water Consumption	P6	4	0.05	1.05	0.93	0.031
Building Envelope Insulation	P4	5	0.01	1.01	0.92	0.031
Safety and Security	P13	6	0	1	0.92	0.031
Thermal Comfort	P9	7	0.01	1.01	0.92	0.030
Access to Public Transportation	P24	8	0	1	0.92	0.030
Heating Energy Consumption	P1	9	0.01	1.01	0.91	0.030
Green Areas	P15	10	0	1	0.91	0.030
Cultural Value	P23	11	0.02	1.02	0.89	0.029
Renewable Energy Production	P5	12	0.01	1.01	0.88	0.029
Visual Comfort	P10	13	0	1	0.88	0.029
Social Interaction	P22	14	0.01	1.01	0.87	0.029
Access to Non-motorized mode	P25	15	0	1	0.87	0.029
Acoustic Comfort	P11	16	0.01	1.01	0.86	0.029
Shading Area	P17	17	0.01	1.01	0.85	0.028
Ergonomic Comfort	P21	18	0	1	0.85	0.028
Creating Jobs/Employment	P35	19	0	1	0.85	0.028
Human Centered Design	P19	20	0.01	1.01	0.85	0.028
Attendance Rate	P28	21	0	1	0.85	0.028
Cooling Energy Consumption	P2	22	0.01	1.01	0.84	0.028

(continued)

Table 6. (*continued*)

Indicator	ID	Rank	Comparative importance of average value (sj)	Coefficient (Kj)	Recalculated weight (qj)	Indicator Weight (Wj)
Waste Management Strategies	P14	23	0	1	0.84	0.028
Parking Area	P26	24	0	1	0.84	0.028
Maintenance Cost	P33	25	0	1	0.84	0.028
Success Rate	P27	26	0.01	1.01	0.83	0.027
Water Harvesting	P7	27	0.01	1.01	0.82	0.027
Operational Energy Expenses	P31	28	0.01	1.01	0.81	0.027
Connection to Public Sewage	P8	29	0.02	1.02	0.80	0.026
Operational Water Expenses	P32	30	0	1	0.80	0.026
Lighting Energy Consumption	P3	31	0.01	1.01	0.79	0.026
Social Activity	P36	32	0.01	1.01	0.78	0.026
School Dropout	P29	33	0.01	1.01	0.77	0.026
Discipline Referrals	P30	34	0	1	0.77	0.026
Heat Island Effect	P16	35	0.01	1.01	0.77	0.025

The experts' opinion placed high importance on "**Access to Public Transportation** (0.030)", as it is crucial for public school buildings to be easily accessible through such services. In general, social indicators have a higher order of importance than environmental and economic indicators, like "**Cultural Value** (0.029)", "**Social Interaction** (0.029)", "**Access to Non-motorized mode of movement** (0.029)", "**Ergonomic Comfort** (0.028)", and "**Human Centered Design** (0.028)". Concerning the economic aspect, the weight of "**Creating Jobs/Employment** (0.028)" is higher compared to the indicators related to "**Operation and Maintenance Costs**". This highlights the significance of schools as active and productive institutions in society.

4 Application on Palestinian Schools

This framework was applied to 54 schools in Tulkarm and Nablus in the West Bank. All the schools studied were found to have a low level of sustainability when examining the Sustainability Index values (see Fig. 2). This illustrates the urgent need to make significant progress in environmental, social and economic aspects to improve the overall sustainability of Palestinian schools. While some expectations were met regarding the relationship between certain indicators and sustainability outcomes, there were also surprising differences. For example, girls' schools performed slightly better than boys' schools in sustainability, mainly due to social inclusion, although their environmental performance was similar. Likewise, elementary schools have outperformed high schools in sustainability, although high schools have a notable environmental advantage. This highlights the complex interplay of various factors that affect sustainability.

Regarding environmental facilities, there have been positive developments like implementing thermal insulation and using shading on exterior facades. However, challenges persist in areas such as energy and water consumption, waste segregation, and providing parking spaces for school staff. Sustainability efforts also encompass social aspect like human-centered design, and inclusion of low-income and disabled students. However, the education system should continue addressing concerns about attendance rates, dropout rates, and academic performance, as these are closely linked to the indoor environment's quality and economic conditions that may impact student attendance and participation.

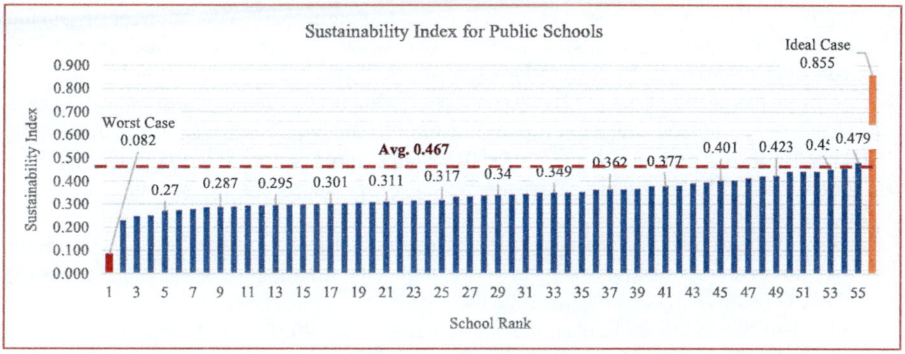

Fig. 2. The sustainability index for Palestinian public school.

5 Conclusion

This paper presented a framework for the sustainability assessment of schools in Palestine. It aims at helping the public authority develop sustainability standards for schools. It also aims identifying areas to enhance the quality of education, create a culture of sustainability in schools and promote a more sustainable future for all. The framework is created by examining global sustainable building standards, adapting them to t the Palestinian context, and obtaining insights from a panel of experts.

The framework includes 35 indicators covering 11 sustainability categories, namely Energy Use Efficiency, Water Use Efficiency, Indoor Quality, Site Development, Social Equity, Health and Comfort, Social Cohesion, Accessibility, Teaching Quality, Influence on Local Economy, and Annual Operating Costs. Based on experts' opinions, Social Equity and Indoor Quality are most important in public schools, while the economic dimension is less important.

Experts stressed the importance of social indicators, especially "social inclusion," as a critical factor in meeting the needs of marginalized students. They also focuses on factors like "indoor air quality" and "health and comfort" that contribute to a positive learning environment. The application submitted to Palestinian schools showed that Palestinian schools have made progress in some environmental aspects, but face challenges in energy, water consumption, waste segregation and inclusiveness. Engaging stakeholders and addressing attendees and academic concerns is vital to achieving higher levels of sustainability in the future.

References

1. Hidayat, Y.A., Rohman, M.A., Utomo, C.: Social sustainability indicators for school buildings in Surabaya. IOP Conf. Ser.: Earth Environ. Sci. **447**(1), 012033 (2020). https://doi.org/10.1088/1755-1315/447/1/012033
2. Bosevska, J., Kriewaldt, J.: Fostering a whole-school approach to sustainability: learning from one school's journey towards sustainable education. Int. Res. Geogr. Environ. Educ. **29**, 55–73 (2019). https://doi.org/10.1080/10382046.2019.1661127
3. Mogren, A., Gericke, N., Scherp, H.Å.: Whole school approaches to education for sustainable development: a model that links to school improvement. Environ. Educ. Res. **25**, 508–531 (2019). https://doi.org/10.1080/13504622.2018.1455074
4. Kahriman-Pamuk, D., Uzun, N.B., Yıldız, T.G., Haktanır, G.: Reliability of indicators measuring early childhood education for sustainability: a study in Turkey using generalizability theory. Int. J. Early Child. **51**, 193–206 (2019). https://doi.org/10.1007/s13158-019-00243-6
5. Council, U.S.G.B.: Reference Guide for Building Operations and Maintenance, LEED v4 Edition (2019)
6. Saraiva, T.S., Almeida, M., Bragança, L., Barbosa, M.T.: The inclusion of a sustainability awareness indicator in assessment tools for high school buildings. Sustainability **11**, 387 (2019). https://doi.org/10.3390/su11020387
7. Ministry of Education, P.: Education Sector Strategic Plan 2021–2023/Palestine (2021)
8. Said, N.: An Overview of Green Buildings Potential in Palestine. Int. J. Sustain. Green Energy **8**, 20 (2019). https://doi.org/10.11648/j.ijrse.20190802.11
9. Lu, X., Lu, J., Yang, X., Chen, X.: Assessment of Urban mobility via a pressure-state-response (PSR) model with the IVIF-AHP and FCE methods: a case study of Beijing. China. Sustain. **14**, 3112 (2022). https://doi.org/10.3390/su14053112
10. Wang, F., Lu, Y., Li, J., Ni, J.: Evaluating environmentally sustainable development based on the psr framework and variable weigh analytic hierarchy process. Int. J. Environ. Res. Public Health **18**, 1–16 (2021). https://doi.org/10.3390/ijerph18062836
11. Zolfani, S.H., Chatterjee, P.: Comparative evaluation of sustainable design based on step-wise weight assessment ratio analysis (SWARA) and best worst method (BWM) methods: a perspective on household furnishing materials. Symmetry (Basel). **11**, 1–31 (2019). https://doi.org/10.3390/sym11010074

12. Zolfani, S.H., Chatterjee, P.: comparative evaluation of sustainable design based on step-wise weight assessment ratio analysis (SWARA) and best worst method (BWM) methods: a perspective on household furnishing materials. Symmetry (Basel) **11**, 74 (2019). https://doi.org/10.3390/sym11010074
13. Palestine Higher Green Building Council: Green buildings Guidelines – State of Palestine (2013). https://www.paleng.org/?p=417, https://www.paleng.org/wp-content/uploads/2015/10/pgbc_manual.pdf

Empowering Sustainability Advancement in Urban Public Spaces Through Low-Cost Technology and Citizen Engagement

Mohammed Itair[1(✉)], Ihab Hijazi[2], Saffa Mansour[3], and Isam Shahrour[1]

[1] Civil and Geo-Environmental Laboratory (LGCgE), Lille University, Lille, France
mohammeditair@gmail.com

[2] Urban Planning Engineering Department, An-Najah National University, Nablus, Palestine

[3] ESTP Paris-École Spéciale des Travaux Publics, 94230 Cachan, France

Abstract. This conference paper investigates the role of low-cost technology and citizen engagement in promoting sustainability in urban public spaces. With the global population increasingly residing in cities, creating inclusive, safe, and sustainable urban environments has become imperative. Focusing on Sustainable Development Goal 11.7, which emphasizes universal access to safe and inclusive public spaces, this study addresses the challenges of data scarcity and limited community involvement. Through a large-scale community-driven spatial survey in Nablus, key findings are presented, including disparities in space distribution and accessibility, gender imbalances in usage, inadequacies in physical features, and safety concerns. Leveraging low-cost technology, such as smartphone applications and open-source data collection tools, the study overcomes data limitations and facilitates citizen participation. The research underscores the importance of citizen engagement in assessing and improving public spaces and highlights the need for collaboration among researchers, policymakers, and community stakeholders. The paper concludes with recommendations for targeted interventions, long-term impact assessment, and evidence-based policies to enhance sustainability and inclusivity in urban public spaces. This work contributes to advancing SDG 11.7 and offers insights for practitioners and policymakers seeking to create vibrant and equitable urban environments.

Keywords: SDGs · Open Public Spaces · City Planning · Public Participation

1 Introduction

The rapid growth of cities has led to more than half of the global population residing in urban areas [1]. To address the challenges associated with urbanization, the United Nations has embraced the Sustainable Development Goals (SDGs) comprising 17 goals and 169 targets [2]. SDG 11 specifically emphasizes the need to create inclusive, safe, resilient, and sustainable human settlements. Target 11.7 focuses on achieving universal access to safe, inclusive, and accessible green and public spaces, with a particular emphasis on marginalized groups such as women, children, and people with disabilities. However, ensuring the realization of this goal remains complex, as these groups continue to encounter exclusion, marginalization, and safety concerns in urban public spaces [3].

M. Ben Ahmed et al. (Eds.): SCA 2023, LNNS 906, pp. 292–299, 2024.
https://doi.org/10.1007/978-3-031-53824-7_26

In the realm of urban planning and management, data plays a crucial role in improving public spaces. Data-driven insights provide valuable information for understanding the needs, preferences, and experiences of diverse communities [4]. By harnessing the power of data, policymakers and urban planners can make informed decisions to enhance the inclusivity and safety of public spaces. However, obtaining reliable and comprehensive data presents significant challenges [5].

One of the primary challenges is the scarcity of data specifically addressing the inclusivity and safety of urban public spaces, particularly for marginalized groups. Existing data often fails to capture the nuanced experiences and perspectives of these communities, hindering the development of targeted interventions and policies. Additionally, data collection processes can be costly, time-consuming, and resource-intensive, making it difficult to conduct large-scale studies or gather data from under-resourced regions.

Citizen participation plays a crucial role in urban planning and management, particularly concerning the design and administration of public spaces [6]. By actively involving communities in participatory processes, valuable insights can inform decision-making and facilitate the creation of public spaces that cater to diverse user needs [7]. Previous research has indicated the potential of citizen participation in fostering inclusive, safe, and easily accessible urban public spaces [8]. Nevertheless, despite the perceived benefits, empirical evidence supporting the effectiveness of these approaches in advancing SDG 11.7 remains limited [9–11].

Therefore, this paper recognizes the importance of data in improving public spaces and highlights the challenges associated with data collection, availability, and interoperability. By leveraging low-cost technology and citizen participation, we aim to address these challenges and contribute to the advancement of SDG 11.7. Our study endeavors to fill the data gaps by conducting a large-scale community-driven spatial survey in Nablus, enabling us to gather valuable insights and develop comprehensive recommendations for enhancing the inclusivity and safety of urban public spaces for marginalized groups.

This paper is structured as follows: after this introduction, the background provides a conceptual framework for the study that briefly reviewing the relevant literature on citizen participation, inclusivity, accessibility, and safety in urban public spaces, and the marginalization of women, children, and people with disabilities in urban public spaces. The methodology section describes the research design and approach, sampling strategy, and data collection methods, including a community-driven spatial survey, interviews, and focus groups. The results section presents the findings of the survey and qualitative data, including participants' perceptions of the inclusivity and safety of public spaces and the factors influencing these perceptions. Finally, the conclusion and future work summarizes the key findings, contributions, limitations, and suggestions for future research.

2 Background

The concept of creating inclusive urban public spaces has garnered considerable attention over the years, particularly with regards to ensuring the safety and accessibility of such spaces for marginalized groups [12]. Extensive research has revealed significant disparities in the utilization and availability of public spaces among different segments of society. For instance, women, children, and individuals with disabilities often encounter difficulties in accessing and fully utilizing public spaces, primarily due to safety concerns and inadequate infrastructure.

According to the World Health Organization (WHO), women face a higher risk of experiencing violence in public spaces compared to men, with approximately one in three women experiencing physical or sexual violence during their lifetime [13]. Additionally, societal gender norms and expectations often result in women being excluded from public life. Similarly, children and individuals with disabilities encounter challenges in accessing and utilizing public spaces, including a lack of appropriate infrastructure and facilities, particularly in developing countries [14].

Various approaches have been proposed to address the challenges of creating inclusive public spaces. One approach is universal design, which focuses on creating environments that are usable by all people, regardless of their abilities or disabilities [15]. Another approach is placemaking, which emphasizes community involvement in the design and development of public spaces to make them meaningful and enjoyable [16, 17]. Community-led initiatives are also effective in creating inclusive public spaces by engaging community members in the design process and ensuring that the spaces meet the needs of diverse groups [18].

Moreover, the use of technology, such as smart city technologies and mobile applications, offers new possibilities for enhancing accessibility, safety, and engagement in public spaces [19–21]. Our research integrates citizen participation into sustainability goals, specifically SDG 11.7, to promote inclusive public spaces. By recognizing the importance of a holistic and long-term perspective, we aim to provide access to safe and inclusive open public spaces.

In our study, we acknowledge the significance of involving citizens in sustainability objectives to adopt a comprehensive and enduring outlook on ensuring access to secure and inclusive open public spaces. Moreover, our research adopts a practical and cost-effective approach by leveraging low-cost and easily accessible technology to fulfill our study goals. By utilizing such technology, we aim to overcome data limitations and access challenges often faced in resource-constrained regions.

Our focus on a case study in Nablus allows us to demonstrate the feasibility and practicality of integrating sustainability goals into community participation within a developing country context. This utilization of technology enables us to overcome constraints related to data availability and accessibility commonly encountered in resource-limited areas. By concentrating on a case study in Nablus, we aim to showcase the viability and applicability of integrating sustainability objectives into community engagement within the context of a developing country.

3 Methodology

This section provides an overview of the research design, sampling strategy, and data collection methods used in the study. The objective is to examine how citizen participation can contribute to assessing the inclusivity and safety of urban public spaces, specifically focusing on SDG 11.7.

We employed a mixed-methods research design to gather comprehensive insights, collecting and analyzing both quantitative and qualitative data to ensure the reliability of the findings.

To guarantee representation from the target population, we used a purposive sampling strategy. We selected the city of Nablus as our research site due to its diverse urban context and the abundance of complaints regarding the quality of open spaces. We carefully chose skilled field surveyors, including urban planning and architecture graduates who reside in Nablus. We organized a kick-off meeting to train the surveyors on the survey goals and methods.

Data collection involved carefully selecting 63 public spaces based on discussions with the municipality and community stakeholders. In each space, six interviews were conducted, ensuring a broad range of perspectives. A team of 10 field surveyors equipped with smartphones and the Kobo Toolbox App [22] actively participated in data collection, utilizing cost-effective technology to streamline the process and enhance data accuracy. To capture diverse perspectives, interviews were conducted during both daytime and nighttime, involving individuals from different age groups, genders, and people with disabilities (Fig. 1).

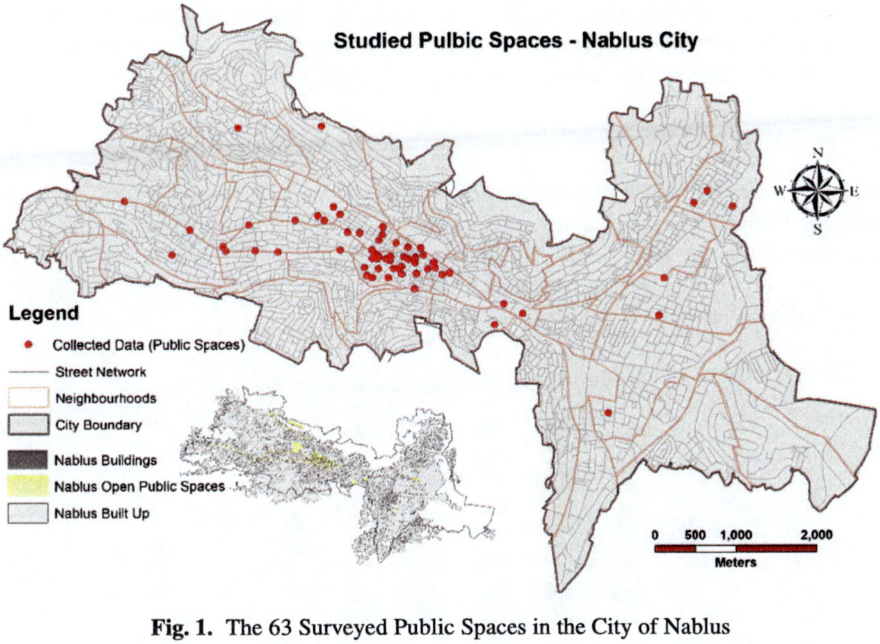

Fig. 1. The 63 Surveyed Public Spaces in the City of Nablus

The collected data underwent rigorous analysis. Quantitative data were analyzed using descriptive and inferential statistics, while qualitative data from interviews underwent thematic analysis. The integration of quantitative and qualitative findings provided a comprehensive understanding of factors influencing perceptions of accessibility, comfort, and cultural rights in urban public spaces.

4 Results

The study conducted in the city of Nablus has revealed several key findings regarding the current state of public spaces within the city. In this abstracted-version paper, we will highlight 4 general categories of results:

4.1 Distribution and Accessibility

The total area of open public spaces within the city was found to be 496,359 m^2, which includes market areas, community yards, and streets. The per capita ratio of open public spaces was calculated to be 2.6 m^2, which falls below both local and international standards of 5 m^2 per person. This ratio is expected to decrease further with future population growth. The allocation of public spaces in the master plan was also found to be below the standards, and an additional 1 km^2 of new open public spaces is required to meet the recommended standards.

Moreover, the spatial distribution of open public spaces within the city was found to be concentrated in the city center and the old city of Nablus. The east part of the city, which includes two camps and a low-income population, had limited open public spaces. The study also found that only 30% of the city's neighborhoods had access to public spaces within 5 min walking distance, while over 41% needed more than 10 min walking distance to access open public spaces. The west part of the city was found to be more served by open spaces compared to the eastern part of the city.

4.2 The Users

The study shed light on the attraction of public spaces to people from different age groups and genders. Figure 2 shows that young males were found to dominate the public spaces, while people with disabilities had limited presence during both day and night time. Additionally, the lack of sufficient lighting at night was found to be a contributing factor to the limited use of public spaces during night time.

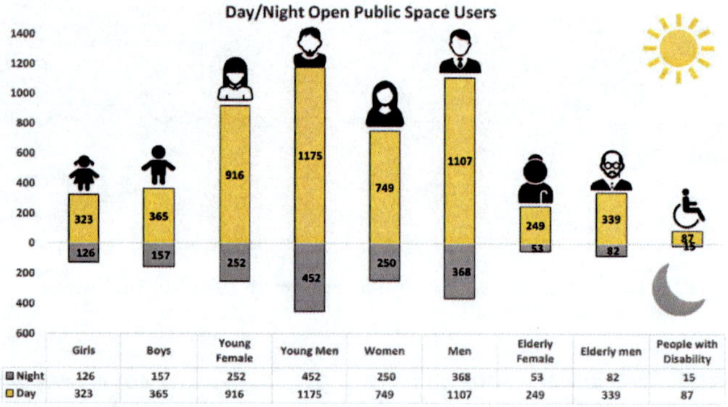

Fig. 2. Day/Night Users of Open Public Spaces in Nablus City by Different Age Groups and Gender

4.3 Physical Features

The survey results showed a general dissatisfaction with the physical features in the open public spaces. None of the facilities, including lighting, seating, garbage bins, baby care facilities, public totalities, artificial shade, water tape, signs, vehicular parking, bicycle parking, fire emergency, convenience stores, access for blind people, and facilities to restrict movement, achieved satisfaction scores above average. Seating, an element of great significance in public spaces, scored among the lowest.

The survey also revealed a considerable lack of features necessary for specific user groups. Women's primary needs, such as totalities, baby care facilities, and artificial shading, were missing in most spaces. People with disabilities were significantly marginalized in the design and construction of these spaces, as were considerations for walkability and cycling. The absence of seats was highly noticeable in most of the open public spaces (Fig. 3).

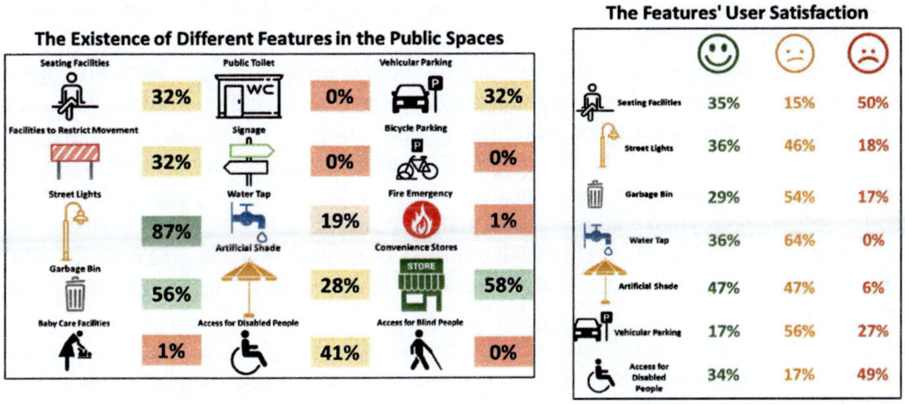

Fig. 3. Physical Feature Existence and User Satisfaction

Moreover, our findings indicate that grass coverage is limited in open public spaces, with an overall percentage of less than 30%. This lack of greenery affects the aesthetic value of the public spaces, potentially impacting their appeal and use.

4.4 The Safety Perception

During the daytime, 66% of the interviewed women reported feeling safe in the public spaces, 32% felt partially safe, and only 2% did not feel safe. These results were similar to those reported by men, which were 84%, 15%, and 1%, respectively.

However, the situation changes significantly during nighttime. Only 19% of women reported feeling safe, with 43% feeling partially safe and 38% feeling unsafe. In comparison, 66% of men reported feeling safe, 27% partially safe, and 7% not safe. This points towards a disparity in the perceived safety of public spaces between the genders, especially during the night light.

5 Conclusion and Future Work

In conclusion, this research paper has shed light on the significance of integrating low-cost technology and community engagement in advancing sustainability goals, particularly in the context of creating inclusive and safe urban public spaces. By leveraging the potential of low-cost technology, such as smartphone applications and open-source data collection tools, we have successfully overcome data limitations and access challenges commonly faced in resource-constrained regions. This approach has allowed us to gather comprehensive insights into the inclusivity and safety of public spaces for marginalized groups, contributing to the broader objectives of SDG 11.7.

The findings of this study highlight the crucial role of citizen participation in assessing and improving the quality of urban public spaces. By actively involving community members in the data collection process, we have not only captured diverse perspectives but also fostered a sense of ownership and empowerment among the participants. This participatory approach has the potential to transform the way urban planning and design decisions are made, ensuring that the needs and preferences of all community members are considered.

In terms of future work, a thorough and in-depth analysis of the research findings is warranted. This analysis should include a comprehensive examination of the collected data to develop robust recommendations for enhancing the safety and inclusivity of open public spaces, especially for marginalized groups. By identifying and highlighting the factors that have the most significant influence on the safety and inclusivity of public spaces, policymakers, and urban planners can prioritize targeted interventions that address these key areas. This in-depth discussion and analysis will provide valuable insights into the specific measures needed to create more inclusive and safer urban environments.

Moreover, continued collaboration between researchers, policymakers, and community stakeholders is essential for translating the research findings into meaningful actions. By working together, we can develop evidence-based policies and interventions that promote the sustainable development of urban public spaces. Additionally, further research should explore the long-term impact and sustainability of the interventions implemented based on the insights gathered through citizen engagement. Robust monitoring and evaluation frameworks should be developed to assess the effectiveness and lasting benefits of these initiatives.

References

1. UN DESA: 68% of the world population projected to live in urban areas by 2050. https://www.un.org/development/desa/en/news/population/2018-revision-of-world-urbanization-prospects.html. Accessed 14 Mar 2023
2. UN General Assembly: Transforming our world: the 2030 Agenda for Sustainable Development|Department of Economic and Social Affairs. https://sdgs.un.org/2030agenda. Accessed 14 Mar 2023
3. Fabre, E.A., Martinuzzi, C., Levonen, T.: Cities for Girls Cities for All. UN-Habitat (2022)
4. Engin, Z., et al.: Data-driven urban management: Mapping the landscape. J. Urban Manage. **9**, 140–150 (2020). https://doi.org/10.1016/j.jum.2019.12.001

5. Wang, J., Cao, S.-J., Yu, C.W.: Development trend and challenges of sustainable urban design in the digital age. Indoor Built Environ. **30**, 3–6 (2021). https://doi.org/10.1177/1420326X20976058
6. Karahan, E.K.A.M.A.C.I.: A novel discussion on urban planning practice: Citizen Participation. ICONARP
7. Burrowes, K.: Making Places for Everyone — With Everyone (2019). https://doi.org/10.48558/0FZ3-5N85
8. Zhou, S.: Understanding "Inclusiveness" in public space: learning from existing approaches. https://sustain.ubc.ca/about/resources/understanding-inclusiveness-public-space-learning-existing-approaches. Accessed 12 Jan 2023
9. Granier, B., Kudo, H.: How are citizens involved in smart cities? Analysing citizen participation in Japanese "Smart Communities". Inf. Polity **21**, 61–76 (2016). https://doi.org/10.3233/IP-150367
10. OHCHR: Policy Guidelines for Inclusive Sustainable Development Goals. Sustain. Cities Commun. (2020)
11. Cardullo, P., Kitchin, R.: Being a 'citizen' in the smart city: up and down the scaffold of smart citizen participation in Dublin. Ireland. GeoJournal. **84**, 1–13 (2019). https://doi.org/10.1007/s10708-018-9845-8
12. Ahmimed, C.: Inclusive cities: trends and new initiatives. Department for Economic and Social Affairs (UNDESA) Division for Social Policy and Development, New York, N.Y (2018)
13. WHO: Violence against women. https://www.who.int/news-room/fact-sheets/detail/violence-against-women. Accessed 14 Mar 2023
14. Wellenstein, A., García Mora, M.E.: Breaking barriers and reconstructing a disability-inclusive society. https://blogs.worldbank.org/latinamerica/breaking-barriers-and-reconstructing-disability-inclusive-society. Accessed 20 Jun 2023
15. Esfandfard, E., Wahab, M.H., Amat, R.C.: universal design in urban public spaces for people with disability. Case study of Tehran, Iran. Planning Malaysia, 16 (2018). https://doi.org/10.21837/pm.v16i5.421
16. Kent, E.: Leading urban change with people powered public spaces. The history, and new directions, of the Placemaking movement. J. Public Space **4**, 127–134 (2019). https://doi.org/10.32891/jps.v4i1.1158
17. Ellery, P.J., Ellery, J.: Strengthening community sense of place through placemaking. Urban Planning **4**, 237–248 (2019). https://doi.org/10.17645/up.v4i2.2004
18. Ni, M., Cattaneo, T.: Design for urban resilience: a case of community-led placemaking approach in Shanghai China. In: Rau, P.-L.P. (ed.) Cross-Cultural Design. Culture and Society: 11th International Conference, CCD 2019, Held as Part of the 21st HCI International Conference, HCII 2019, Orlando, FL, USA, July 26–31, 2019, Proceedings, Part II, pp. 207–222. Springer International Publishing, Cham (2019). https://doi.org/10.1007/978-3-030-22580-3_16
19. Prandi, C., Barricelli, B.R., Mirri, S., Fogli, D.: Accessible wayfinding and navigation: a systematic mapping study. Univ. Access Inf. Soc. **22**, 185–212 (2023). https://doi.org/10.1007/s10209-021-00843-x
20. Chiu, C.-C., Lee, L.-C.: System satisfaction survey for the app to integrate search and augmented reality with geographical information technology. Microsyst. Technol. **24**, 319–341 (2018). https://doi.org/10.1007/s00542-017-3333-9
21. Albrecht, F., Moser, J., Hijazi, I.: Assessing façade visibility in 3D city models for city marketing. Int. Arch. Photogramm. Remote Sens. Spatial Inf. Sci. XL-2/W2, 1–5 (2013). https://doi.org/10.5194/isprsarchives-XL-2-W2-1-2013
22. KoboToolbox: KoboToolbox. https://www.kobotoolbox.org/. Accessed 23 Jun 2023

Energy and Exergy Analysis of a Domestic Hot Water Production System with a Heat Pump and Thermal Storage

M. Mmadi Assoumani, A. Lapertot(✉), and A. Kindinis

Institut de Recherche en Constructibilité, ESTP Paris, 94230 Cachan, France
alapertot@estp-paris.eu

Abstract. The objective of this article is to study a thermodynamic water heater which uses the energy of the exhaust air from an apartment building to meet the needs for domestic hot water. The investigated energy system, based on an experimental bench of the IUSTI laboratory, is composed of two heat exchangers, a heat pump and a thermal storage to store and draw hot water. Energy and exergy analysis were performed for each component, in order to assess the thermal losses and the irreversibilities of the system, by means of a model developed on Matlab environment. The criteria used to evaluate the performance of the devices are the energy and exergy efficiencies, as well as the coefficient of performance. According the exergetic analysis, the heat pump is mainly responsible for the degradation of the energy quality involved in the system, followed by the recovery heat exchanger, the storage tank and the heat exchanger. The results show that the energy efficiency of the system is 96% while the exergy efficiency is 60%, which implies low heat losses but high irreversibilities.

Keywords: Energy and exergy analysis · Domestic hot water · Thermal losses and irreversibilities

Nomenclature

E:	Energy (J)
Ex:	Exergy (J)
h:	Mass enthalpy ($J.kg^{-1}$)
$\dot{m}$:	Mass flow rate ($kg.s^{-1}$)
P:	Power (W)
P_{ex}:	Flow energy (W)
s:	Entropie massique
T:	Temperature ($°C$)
u:	Mass internal energy ($J.K^{-1}.kg^{-1}$)
V:	Volume (m^3)
W:	Electric power (W)

Greek Symbols

ρ:	Density (kg.m−3)
η:	Effectiveness (−)
Φ:	Heat flux (W)

M. Ben Ahmed et al. (Eds.): SCA 2023, LNNS 906, pp. 300–310, 2024.
https://doi.org/10.1007/978-3-031-53824-7_27

Indices and Exponents

ce: Incoming hot fluid
cs: Outgoing hot fluid
circ: Circulator
cond: Condenser
comp: Compressor
evap: Evaporator
fe: Incoming cold fluid
fs: Outgoing cold fluid
draw: Drawing

Abbreviations

COP: Coefficient of Performance
DHW: Domestic Hot Water
HP: Heat Pump
HX: Heat Exchanger
STR: Storage tank

1 Introduction

Buildings are the main energy consumers in our society. They represent 49% of energy consumption in France [1]. A major part of this energy is used for space heating and the production of Domestic Hot Water (DHW). As a result, improving energy performance in housing, such as thermal insulation and sealing of the building envelope, can significantly reduce energy consumption for space heating. However, the energy demand for DHW continues to increase as energy consumption in other segments continuously decreases. In this context, the DHW will then become the main consumption item in the building [1]. It suddenly becomes undeniable to strengthen energy saving efforts and increase the performance of DHW production systems. The thermodynamic water heater, thanks to its high performance, becomes an effective solution for improving energy performance and obtaining energy savings while considering societal issues. Indeed, the current electric water heaters have an average energy efficiency of the order of 0.7 [2]. That's why it's necessary to use thermodynamic water heaters in order to have more significant performances which vary from 1.7 and 6 [3] and an overall efficiency of around 84% [4].

The work presented here aims to perform an energy and exergy analyses of the components of a thermodynamic water heater which recovers energy from an air handling unit located in an apartment building. The model, based on an experimental device, consists of a heat pump, a heat exchanger and thermal storage. The first principle of thermodynamics is applied in all of these components to quantify heat losses and the second principle to determine irreversibilites. The second part describes the experimental bench and the associated numerical model. The third part presents the energy and exergy balances to determine the energy and exergy efficiencies of the system components. The fourth part interprets and compares the results of the experimental and numerical data.

2 Presentation of the Methodology

2.1 Description of the Experimental Bench, Metrology and Procedure

The experimental device of the IUSTI laboratory [5] makes it possible to store and produce DHW using the energy of the exhaust air of an apartment building. The test bench is composed of an air-water heat exchanger included in an air handling unit, a water-water heat pump, a water-water heat exchanger and a DHW storage. First of all, the air-water exchanger recovers the energy from the exhaust air and transfers it to the heat pump on the evaporator side. Electrical energy is given to the heat pump to power the compressor. It heats the water in the tank through a water-water plate heat exchanger. This water-water heat exchanger prevents any contamination of the DHW by the working fluid of the heat pump.

The heat pump has a thermal power of $7.4\,kW$, the storage tank has a volume of $1.5m^3$, which is thermally insulated with 10 cm polyurethane foam. The efficiency of the air-water heat exchanger is 0.7 and that of the water-water heat exchanger is 0.83. Similarly, the installation is made up of pressure gauges, flow meters and temperature sensors. It also consists of a wattmeter to measure the power of the heat pump compressor.

The heat pump is used to heat the water in the DHW tank between $13\,°C$ and $53\,°C$. The boundary conditions set an exhaust air flow of $2000\,m^3/h$ and a temperature of $20\,°C$, which corresponds to an apartment building of a dozen dwellings and for inter-seasonal conditions. The flows rates of the circulators were fixed at $1.25\,m^3/h$.

2.2 Description of the Numerical Model

The numerical model studied in this article is based on the experimental device of the IUSTI laboratory [5]. It is used to simulate the storage and production of DHW by recovering the energy from the extracted air from an apartment building. The model is presented in Fig. 1 and shows that the system is composed, in the same way as the test bench, of a heat recovery exchanger, a heat pump, a heat exchanger and a DHW tank.

Concerning the modelling of the components, the power of the condenser, of the compressor and that of the evaporator depend on the temperatures in the heat pump. The exchangers are modeled using experimental efficiencies and the DHW tank by a thermocline. The system is detailed and has been validated in [5, 6].

3 Energy and Exergy Analysis of the System

In this study, the thermodynamic model is developed and simulated on Matlab. The model is in transient state, and only in tank heating mode is considered. In this study, the specific internal energy is noted u, the specific enthalpy h, the specific entropy s and the density ρ, these quantities are functions of the temperature and are obtained thanks to the data of the NIST [8]. The reference state of the system has an ambient temperature $T_0 = 20\,°C$ and a pressure $P_0 = 101,325\,kPa$. In order to study the performance of the components, energy and exergy balances were carried out to quantify the performance indicators such as energy and exergy efficiencies and the coefficient of performance.

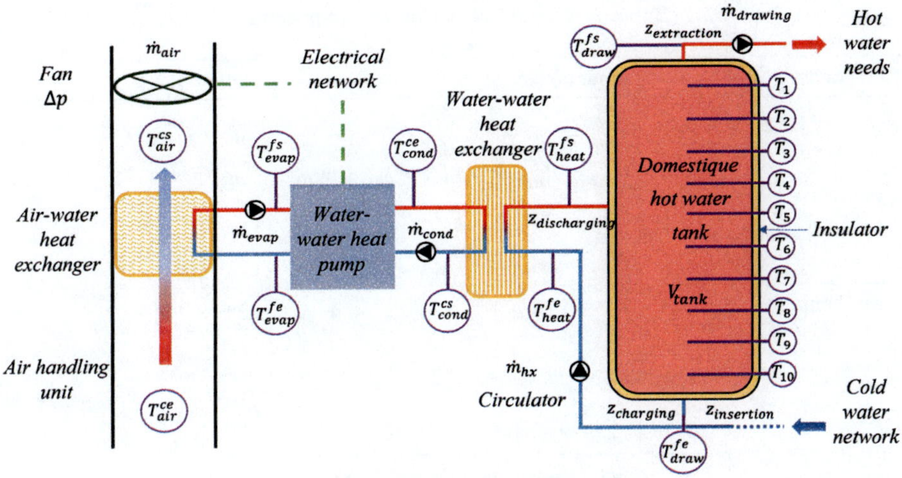

Fig. 1. Schematic of numerical model [7].

3.1 Energy Balance

In each component of the system, an energy analysis is carried out to determine the thermal losses. Energy balance is based on the first law of thermodynamics and can be calculated using mass flow rate and enthalpic of constituents. In addition, each component undergoes either heat inputs from the environment Φ, or electrical inputs W.

Table 1 shows the heat balances of different components. The recovery heat exchanger is connected to the fan and a circulator, and the heat pump takes into account the power of the compressor. Similarly, the heat exchanger is linked to a circulator and the storage tank to two circulators, as shown in Fig. 1. In addition, the power of the circulators W_{circ} was experimentally calculated with constant values equal to 50 W. The power of fan W_{fan} was also obtained experimentally with a value of 200 W. Similarly, the thermal power stored P_{str} in the DHW tank is determined by the derivative of the internal energy:

$$P_{stk} = \rho_{water} V_{tank} \cdot \frac{d(\overline{u})}{dt} \tag{1}$$

In this case, the entropy was calculated for an average temperature in the storage. Thermal losses are determined by subtracting the input and the output power of the various components.

$$P_{lost} = P_{in} - P_{out} \tag{2}$$

Moreover, the thermal efficiency of each component makes it possible to identify the component that is the most efficient. It is defined by the ratio between the output and the input power of the components.

$$\eta = \frac{P_{out}}{P_{in}} \tag{3}$$

Table 1. Energy balance of the components.

Component	Input power	Output power
Recovery Heat Exchanger	$P^{in}_{rhx} =$ $P^{ce}_{rhx} + P^{fe}_{rhx} + W_{fan} + W_{circ} + \Phi_{rhx}$ $P^{ce}_{rhx} = \dot{m}_{air} * (h^{ce}_{rhx})$ $P^{fe}_{rhx} = \dot{m}_{evap} * (h^{fe}_{rhx})$	$P^{out}_{rhx} = P^{cs}_{rhx} + P^{fs}_{rhx}$ $P^{cs}_{rhx} = \dot{m}_{air} * (h^{cs}_{rhx})$ $P^{fs}_{rhx} = \dot{m}_{evap} * (h^{fs}_{rhx})$
Heat Pump	$P^{in}_{hp} = P^{ce}_{hp} + P^{fe}_{hp} + W_{comp} + \Phi_{hp}$ $P^{ce}_{hp} = \dot{m}_{cond} * (h^{ce}_{hp})$ $P^{fe}_{hp} = \dot{m}_{evap} * (h^{fe}_{hp})$	$P^{out}_{hp} = P^{cs}_{hp} + P^{fs}_{hp}$ $P^{cs}_{hp} = \dot{m}_{cond} * (P^{cs}_{hp})$ $P^{fs}_{hp} = \dot{m}_{evap} * (P^{fs}_{hp})$
Heat Exchanger	$P^{in}_{hx} = P^{ce}_{hx} + P^{fe}_{hx} + W_{circ} + \Phi_{hx}$ $P^{ce}_{hx} = \dot{m}_{cond} * (h^{ce}_{hx})$ $P^{fe}_{hx} = \dot{m}_{hx} * (h^{fe}_{hx})$	$P^{out}_{hx} = P^{cs}_{hx} + P^{fs}_{hx}$ $P^{cs}_{hx} = \dot{m}_{cond} * (h^{cs}_{hx})$ $P^{fs}_{hx} = \dot{m}_{hx} * (h^{fs}_{hx})$
Storage Tank	$P^{in}_{str} = P^{ce}_{str} + P^{fe}_{str} + 2 * W_{circ} + \Phi_{str}$ $P^{ce}_{str} = \dot{m}_{hx} * (h^{ce}_{str})$ $P^{fe}_{str} = \dot{m}_{draw} * (h^{fe}_{str})$	$P^{out}_{str} = P^{cs}_{str} + P^{fs}_{str} + P_{stk}$ $P^{cs}_{str} = \dot{m}_{hx} * (h^{cs}_{str})$ $P^{fs}_{str} = \dot{m}_{draw} * (h^{fs}_{str})$

The efficiency of the heat pump is expressed by the Coefficient Of Performance (COP), it is expressed by the ratio between the heat supplied and the power consumed.

$$COP_{hp} = \frac{P_{cond}}{W_{comp}} \tag{4}$$

It is also possible to use the theoretical COP which corresponds to the inverse of the Carnot yield:

$$COP_{thq} = \frac{T^{ce}_{cond}}{T^{ce}_{cond} - T^{fe}_{evap}} \tag{5}$$

3.2 Exergy Balance

The exergy balance is obtained by combining the first law and the second law of thermodynamics. The second law of thermodynamics is used to identify and quantify energy degradation due to irreversibilities. Exergy is a measure of energy quality that takes into account the availability of energy and its ability to do useful work.

In each component, the exergy flux can be determined by multiplying the mass flow $\dot{m}$ of the refrigerants by the difference between the specific enthalpy and the specific entropic multiplied by the reference temperature T_0.

Table 2. Exergy balance of component.

Component	Input exergy flows	Output exergy flows
Recovery Heat Exchanger	$P^{in}_{ex,rhx} = P^{ce}_{ex,rhx} + P^{fe}_{ex,rhx} + W_{fan} + W_{circ} + \varnothing_{rhx}$ $P^{cs}_{ex,rhx} = \dot{m}_{air} * (h^{ce}_{rhx} - T_0 s^{ce}_{rhx})$ $P^{fe}_{ex,rhx} = \dot{m}_{evap} * (h^{ce}_{rhx} - T_0 s^{fe}_{rhx})$	$P^{out}_{ex,rhx} = P^{cs}_{ex,rhx} + P^{in}_{ex,rhx}$ $P^{cs}_{ex,rhx} = \dot{m}_{air} * (h^{cs}_{rhx} - T_0 s^{cs}_{rhx})$ $P^{fs}_{ex,rhx} = \dot{m}_{evap} * (h^{fs}_{rhx} - T_0 s^{fs}_{rhx})$
Heat Pump	$P^{in}_{ex,hp} = P^{ce}_{ex,hp} + P^{fe}_{ex,hp} + W_{comp} + \varnothing_{hp}$ $P^{ce}_{ex,hp} = \dot{m}_{cond} * (h^{ce}_{hp} - T_0 s^{ce}_{hp})$ $P^{fe}_{ex,hp} = \dot{m}_{evap} * (h^{fe}_{hp} - T_0 s^{fe}_{hp})$	$P^{out}_{ex,hp} = P^{cs}_{ex,hp} + P^{fs}_{ex,hp}$ $P^{cs}_{ex,hp} = \dot{m}_{cond} * (h^{cs}_{hp} - T_0 s^{cs}_{hp})$ $P^{fs}_{ex,hp} = \dot{m}_{evap} * (h^{fs}_{hp} - T_0 s^{fs}_{hp})$
Heat Exchanger	$P^{in}_{ex,hx} = P^{ce}_{ex,hx} + P^{fe}_{ex,hx} + W_{circ} + \varnothing_{hp}$ $P^{ce}_{ex,hx} = \dot{m}_{cond} * (h^{ce}_{hx} - T_0 s^{ce}_{hx})$ $P^{fe}_{ex,hx} = \dot{m}_{hx} * (h^{fe}_{hx} - T_0 s^{fe}_{hx}$	$P^{out}_{ex,hx} = P^{cs}_{ex,hx} + P^{fs}_{ex,hx}$ $P^{cs}_{ex,hx} = \dot{m}_{cond} * (h^{cs}_{hx} - T_0 s^{cs}_{hx})$ $P^{fs}_{ex,hx} = \dot{m}_{hx} * (h^{fs}_{hx} - T_0 s^{fs}_{hx}$
Storage Tank	$P^{in}_{ex,str} = P^{ce}_{ex,str} + P^{fe}_{ex,str} + 2 * W_{circ} + \varnothing_{str}$ $P^{ce}_{ex,str} = \dot{m}_{hx} * (h^{ce}_{str} - T_0 s^{ce}_{str})$ $P^{fe}_{ex,str} = \dot{m}_{draw} * h^{fe}_{str} - T_0 s^{fe}_{str}$	$P^{out}_{ex,str} = P^{cs}_{ex,str} + P^{fs}_{ex,str} + P_{exstk}$ $P^{cs}_{ex,str} = \dot{m}_{hx} * (h^{cs}_{str} - T_0 s^{cs}_{str})$ $P^{fs}_{ex,str} = \dot{m}_{draw} * (h^{fs}_{str} - T_0 s^{fs}_{str})$

Table 2 shows that the exergy balance involves the matter flux, the thermal and electrical contributions of each component of the system. The exergy inputs from the environment are described by:

$$\varnothing = \Phi * \left(1 - \frac{T_O}{T_f}\right)$$

T_f corresponds to fluid temperature. The exergy flux stored is given by the derivative of the internal energy minus the reference temperature multiplied by the entropy.

$$P_{exstk} = \rho_{water} V_{tank} \cdot \frac{d(\overline{u} - T_0 \overline{s})}{dt} \quad (6)$$

Furthermore, the exergy destruction for each component of the system is determined by combining the inputs and outputs of the exegetic balance.

$$P_{exdestruction} = P_{exin} - P_{exout} \tag{7}$$

Furthermore, the exergy efficiency is defined by the ratio between the output and input exergy flux.

$$\eta_{ex} = \frac{P_{exout}}{P_{exin}} \tag{8}$$

Finally, the exergy efficiency of the heat pump is defined by the ratio between the real *COP* and the theoretical *COP*.

$$\eta_{ex_{hp}} = \frac{COP_{hp}}{COP_{thq}} \tag{9}$$

4 Results and Interpretations

The energy and exergy balances are applied to the numerical model. The simulation is carried out for DHW storage heating, and the boundary conditions are fixed at their reference value, presented in part 2.

4.1 Energy Analysis

Figure 2 shows the coefficient of performance of the heat pump and the energy efficiencies of the other system components. The efficiencies are very high, implying small heat loss. Figure 2a shows some experimental variation which are caused by the control of boundary conditions. Overall, satisfactory agreement is achieved between the numerical and experimental results, with deviations of 10^{-3} occurring at the heat exchanger and storage tank, as illustrated in Figs. 2b and 2d. These deviations are due to the fact that heat losses in the pipes are neglected in the model. Furthermore, Fig. 2c shows that the real *COP* decreases from 8 to 4 and the COP_{thq} from 21 to 7, resulting in a degradation of the heat pump's performance at higher temperatures.

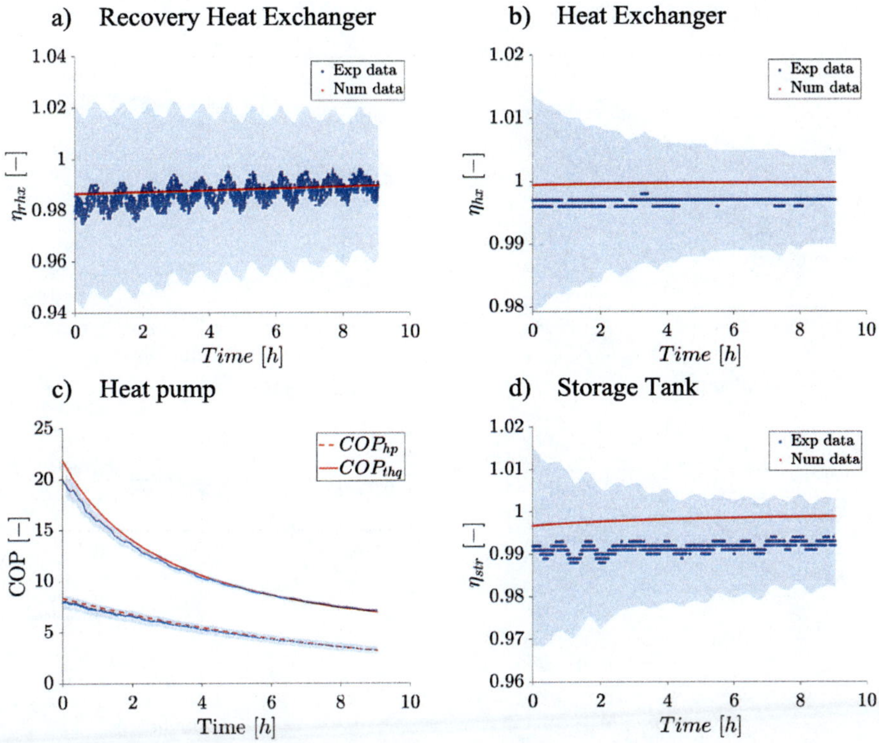

Fig. 2. Component energy efficiency and heat pump coefficient of performance as a function of time. The curves in red represent the results obtained in the numerical simulation. The blue points correspond to the experimental results and the colored shading refers to the associated measurement uncertainties.

4.2 Exergy Analysis

The analysis is now carried out using exergy balances of the various system components for the same boundary conditions as previously. Figure 3a shows that the RHX exergy efficiency increases between 0.53 and 0.68. The numerical model does not fluctuate in contrast to the experimental data, because the input temperatures are controlled with some variation. Moreover, Fig. 3b shows the exergy efficiency of HX increasing between 0.94 and 0.99, so this component generates small irreversibility. Figure 3c shows the exergy efficiency of the HP as a bell-shaped curve with a maximum value of 0.52, meaning that the heat pump is not designed for temperatures that are too low or too high. Figure 3d shows that the exergy efficiency of the STR also increases during storage heating. This is because the temperature in the storage tank rises, resulting in an increase in stored exergy. In addition, the HP is made up of a compressor, which is responsible for a large proportion of the irreversibilities, and the RHX is somewhat responsible for the irreversibilities. The other two components have small impact on exergy destruction.

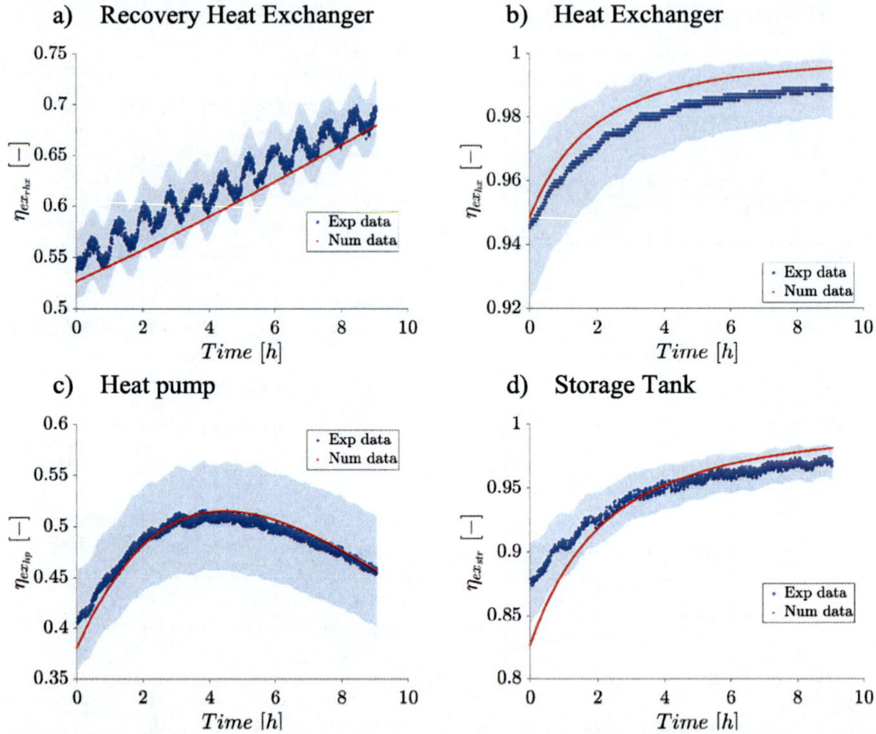

Fig. 3. Exergy performance of each component as a function of time. The red curves represent the results obtained in the numerical simulation. The blue points correspond to the experimental results and the colored shading refers to the associated measurement uncertainties.

4.3 Energy and Exergy Distribution Diagram

Figure 4 evaluates the amount of energy involved in the system. It shows that, with approximately 135.6*kWh* of input power, 48.8% of this energy is not used to heat DHW and 47.4% is stored in the tank. Network water has no influence on the system, as the study only takes into account DHW storage. Similarly, this stored energy is more significant than the electrical energy consumed by the heat pump, corresponding to a *COP* of 4.8. In addition, the system has small heat losses of around 3.7%, with a system energy efficiency of 96.3%, which is stronger than the experimental data [7]. Thermal losses are mainly due to the recovery heat exchanger and storage tank, followed by the HX and the HP.

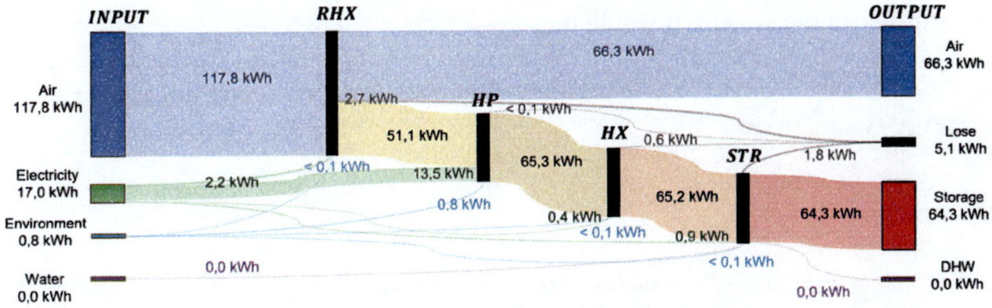

Fig. 4. Diagram of energy flows in the system.

Figure 5 shows the exergy and exergy losses values of the system components. The input exergy value is 4.12 *kWh* for the air and 17.0 *kWh* for the electrical inputs. Network water and environmental inputs have very little influence on the components. Moreover, even though the system suffers small heat loss, it produces more irreversibility. In fact, this leads to a deterioration in the device's performance, with an exergy efficiency value of 40.3%. Similarly, the system uses 34.1% of its potential to heat the storage tank, which is in line with experimental data [7]. Furthermore, irreversibilities are predominantly produced in the heat pump and then the recovery heat exchanger and are less significant in the heat exchanger and then the storage tank.

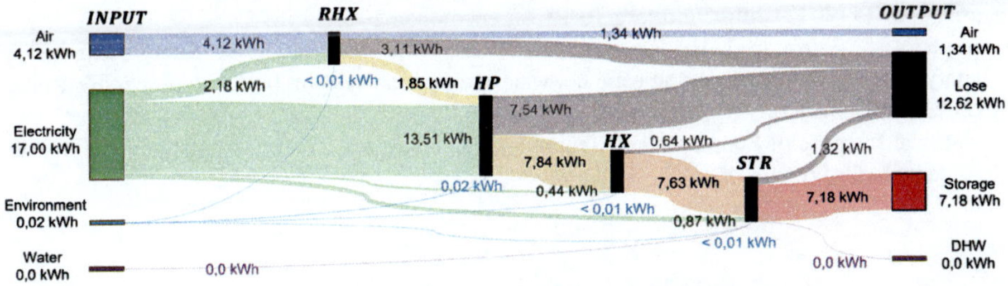

Fig. 5. Diagram of exergy flows in the system.

5 Conclusion et Perspectives

To conclude, energy and exergy analyses have been applied to an energy system for producing domestic hot water, consisting of a heat pump and thermal storage. The results of the analyses show that the numerical data are generally close to the experimental data. The energy efficiency of each component is higher than the exergy efficiency, due to irreversibilities. However, the system undergoes massive exergy destruction despite relatively small heat losses. Similarly, the system uses 34.1% of its potential to heat the storage tank, the rest being either unused or destroyed by the device. This exergy destruction occurs mainly in the heat pump, where pressure losses are high. It is necessary to make technological efforts in priority on the heat pump to improve the overall performance of the system. It would then be interesting to optimize the operation of the system using the numerical model to improve energy and exergy efficiencies. The resulting solution needs to be tested experimentally to confirm or to deny this improvement in energy system performance.

References

1. Key figures on energy - 2021 edition, Data and statistical studies for climate change, energy, the environment, housing, and transport. https://statistiques.developpement-durable.gouv.fr/chiffres-cles-de-lenergie-edition-2021. Accessed 19 Jun 2023
2. Sowmy, D.S., Prado, R.T.A.: Assessment of energy efficiency in electric storage water heaters. Energy Build. **40**, 2128–2132 (2008). https://doi.org/10.1016/j.enbuild.2008.06.007
3. Arif et, H., Kalinci, Y.: A review of heat pump water heating systems. Renew. Sustain. Energy Rev. **13**, 1211–1229 (2009). https://doi.org/10.1016/j.rser.2008.08.002
4. Nekså, P.: CO2 heat pump systems. Int. J. Refrig. **25**, 421–427 (2002). https://doi.org/10.1016/S0140-7007(01)00033-0
5. Segond, G.: Études des couplages thermohydrauliques en régime variable d'un système thermique avec stockage: application à la production d'eau chaude sanitaire à partir de la valorisation d'une source de chaleur basse température, PhD thesis directed by Tadrist, Lounès and Launay, Stéphane Energétique, Aix-Marseille 2015, 2015AIXM4722. http://www.theses.fr/2015AIXM4722
6. Lapertot, A., Segond, G., Fasquelle, T., Kadoch, B., Le Métayer, O.: Multi-criteria optimization of an experimental system for the production of domestic hot water. Energy Convers. Manage. **267**, 115875 (2022). https://doi.org/10.1016/j.enconman.2022.115875
7. Lapertot, A., Ferrasse, J.-H., Fasquelle, T., Kadoch, B., Le Métayer, O.: Energy and exergy analysis of an experimental domestic hot water production system. In: Conference SFT Reims (2023). https://doi.org/10.25855/SFT2023-006
8. National Institute of Standards and Technology (2023). https://webbook.nist.gov/chemistry/. Accessed 15 May 2023

Post-Disaster Assessment of Buildings in Complex Geopolitical Context: Application to Beirut Port

Josiana El Hage[1,2](✉), Isam Shahrour[1], and Fadi Hage Chehade[2]

[1] Laboratoire Génie Civil et Géo-Environnement, University of Lille, 59000 Lille, France
{josiana.elhage,isam.shahrour}@univ-lille.fr

[2] Modeling Center, Doctoral School of Science and Technology, Lebanese University, Hadath 99000, Lebanon
fchehade@ul.edu.lb

Abstract. Given the escalating frequency and severity of natural and man-made disasters, post-disaster reconstruction has become a pressing global concern for countries worldwide. Effective decision-making and project management are crucial in achieving the desired outcomes of recovery efforts. However, when coupled with complex geopolitical situations, post-disaster contexts present additional challenges, such as disrupted social cohesion and limited physical, financial, and human resources. To address these complexities, this paper introduces a Priority Index (PI) designed to assess buildings in post-disaster scenarios within complex geopolitical contexts. The PI serves as a valuable tool to aid decision-makers in prioritizing post-disaster reconstruction projects. To demonstrate its applicability, the index is applied to the specific case of Beirut, Lebanon, following a devastating explosion and a subsequent series of crises that unfolded from 2019 onwards. The proposed PI considers both the physical characteristics of the buildings and the socio-economic context surrounding them.

Keywords: Priority Index · Buildings Assessment · Physical Characteristics · Socio-Economic Profile · Post-disaster reconstruction · Decision-making

1 Introduction

This paper proposes a Priority Index PI for post-disaster buildings reconstruction in complex geopolitical contexts. The study focuses on the interplay between two crucial elements: the devastating impacts of disasters and the complexities inherent to geopolitical contexts. The primary objective of this index is to offer decision-makers a valuable tool to prioritize and effectively manage post-disaster reconstruction projects. To demonstrate its practicality, the PI is applied to the context of Beirut, Lebanon, following the explosion, where a multitude of geopolitical crises further complicate the post-disaster environment.

M. Ben Ahmed et al. (Eds.): SCA 2023, LNNS 906, pp. 311–321, 2024.
https://doi.org/10.1007/978-3-031-53824-7_28

Previous authors have proposed various indicator systems for post-disaster damage assessment, covering different sectors. [1] used the structural priority index, that combines walls and columns indices, to analyze the performance of low-rise reinforced concrete buildings in Ecuador after the 2016 Earthquake. [2] included social, economic, environmental, infrastructure, human health, and ecosystems fields in a multi-dimensional damage assessment framework to attain informed decision-making after a disaster. [3] proposed the seismic screening of buildings in Canada using the Seismic Priority Index. [4] proposed a detailed buildings seismic assessment in Izmir to assist decision-makers with strategic planning and buildings rehabilitation. [5] developed a framework for post-disaster building damage assessment via effective crowdsourcing, involving non-expert citizens and including physical damage to buildings, without integrating community features.

There exist other examples of vulnerability, resilience, and priority assessments in the state-of-the-art. For example, PEOPLES framework for resilience assessment, proposed by [6], too complicated and exhaustive to conduct, knowing that time constraint is crucial in post-disaster actions. Other priority vulnerability or resilience indices only included physical assessment of the building and could only be evaluated by experts. And none of them considered complex geopolitical contexts hit by destructive natural or man-made disasters. Moreover, the proposed frameworks were not designed to assist the decision-making process of post-disaster recovery projects. Hence, the current study aims to address the following gaps in the existing literature:

- The assessment framework proposed in this research takes into account not only the physical structure of buildings but also the social and economic aspects, considering the complexities of the geopolitical situation.
- The study incorporates a crowdsourcing method for data collection, actively involving the affected community in the assessment process.

The paper is organized as follows: First, the research methodology is outlined, which encompasses the calculation method used to determine the Priority Index. Next, the application of the methodology to the post-explosion context in Beirut is discussed, including a brief description of the data collection and analysis method employed. Following that, the calculation of indicator weights and the subsequent analysis and visualization of the results are presented.

2 Methodology

The calculation method of the Priority Index (PI) is briefly described in this paragraph. Initially, a systematic literature review was undertaken to establish an initial set of indicators. Subsequently, an analysis of the case study's context was performed to identify specific factors and refine the list of indicators accordingly. The revised list was then presented to a panel of local experts who further refined it, resulting in a final set of 12 indicators. Once the indicators were defined, they were reassessed by the experts to determine their relative importance in the Priority Index, using the SWARA method. The steps of the SWARA method are detailed in Appendix 1.

2.1 Estimation of the Priority Index

The Priority Index is composed of two indices: Physical Priority Index (PPI) and Socio-Economic Priority Index (SEPI).

Table 1 provides a comprehensive description of the indicators used in the calculation of the Priority Index (PPI) and Social and Economic Priority Index (SEPI). To facilitate the calculation process, non-numerical values are assigned numerical equivalents based on their relative priority. Consequently, in terms of building type, healthcare centers, hospitals, and residential buildings are deemed the most critical and essential for renovation. The scores of each building type are listed as follows:

- Healthcare Center: 5
- Residential and mixed-use Commercial/Residential, Education Center, Public Institution: 4
- Commercial and Industrial: 3
- Cultural and Social: 2
- Not occupied or No data: 1

The damage level is determined by a 4-level scale, and each level is assigned a numerical value as follows:

- Total collapse: more than 50% of the building damaged: 4
- Major damage: between 25% and 50% of the building damage: 3
- Minor damage: only non-structural damage: 2
- No damage: 1

The state of being a cultural heritage building is accorded a score of 1, while the absence of such status is assigned a score of 0. Recognizing the significance of preserving the cultural aspect of the damaged city and safeguarding these buildings against demolition, real estate interests, or gentrification, is taken into account using this indicator. Similarly, the absence of a temporary shelter is assigned a score of 1, while its presence is assigned a score of 0. Prioritizing the needs of affected residents who lack suitable accommodations until their houses are rehabilitated is essential. It is crucial to prevent prolonged homelessness for these individuals.

The PPI and SEPI are determined using the following equations:

$$PPI = w_{BT} \times BT + w_{DL} \times DL + w_{CH} \times CH \tag{1}$$

$$\begin{aligned} SEPI = & w_{OR} \times OR + w_{BRC} \times BRC + w_{SOC} \times SOC + w_{SRC} \times SRC \\ & + w_{SCF} \times SCF + w_{POC} \times POC + w_{PRC} \times PRC + w_{PCF} \times PCF \\ & + w_{TS} \times TS \end{aligned} \tag{2}$$

where w_{ij} are the corresponding weights of each indicator (See the list in Appendix 1).

It is important to mention that for each building, there can be many personal units. However, the priority index calculated is for the whole building. Therefore, the indicators related to personal units are obtained by calculating the mean of all the personal units in the building.

The method of calculating PPI and SEPI is detailed below:

1- The survey had two-level questions, one for the entire building and shared/common spaces and the other for each personal unit in the building. As a result, two tables were created with a one-to-many relationship. The first step consisted of merging these tables based on the global ID of the building.
2- The second step involved cleaning the data and addressing missing or NaN values.
3- The third step consisted of handling nominal or categorical data by assigning numerical values to them.
4- The fourth step involved calculating the mean of each field within the data corresponding to personal units for each building.
5- The fifth step was data normalization, as each field had its scale, unit, and measure. The Min-Max method was used for normalization, which is expressed by the following equation.
6- The final step was to multiply each field by its weight and sum the results for each building to obtain the PPI and SEPI, as in Eqs. 5 and 6.

$$Z_i = (X_i - X_{min})/(X_{max} - X_{min}) \tag{3}$$

2.2 Data Collection

A total of 159 buildings located in four severely impacted zones in Beirut, namely Karantina, Mar Mikhael, Rmeil, and Gemmayzeh, were assessed, as depicted in Fig. 1. Among these zones, the residential area closest to the epicenter of the explosion falls within a distance range of 500 m to 900 m. Notably, this zone is characterized by its cultural heritage buildings. Data collection was carried out using a questionnaire created specifically for this study, utilizing the ArcGIS Survey123 application.

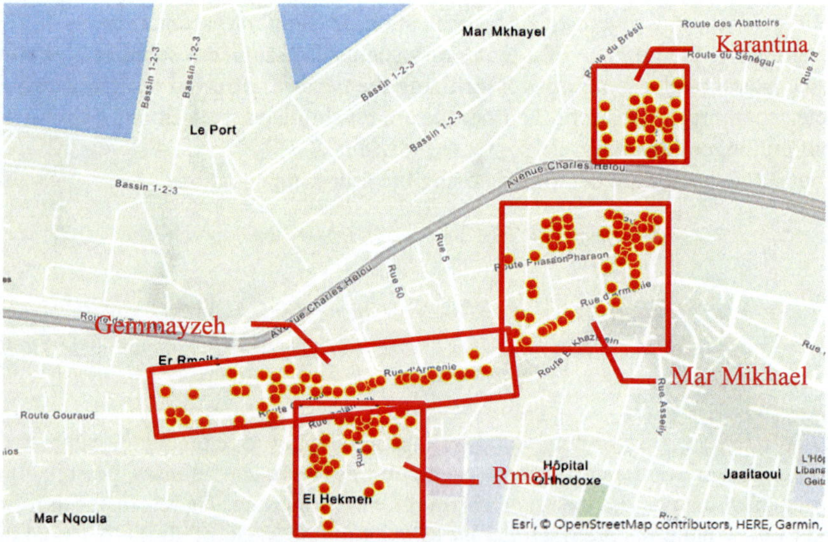

Fig. 1. Selected zones for data collection and assessed buildings.

The survey was conducted on-site by the author and involved the participation of residents residing in the assessed buildings. The data collection period took place in January 2023. The questionnaire encompassed relevant inquiries designed to gather values for the various indicators.

3 Results and Discussion

3.1 Weights of the Indicators According to Experts

Table 1 displays the scores assigned to each indicator based on the opinions of the experts. It is evident that damage level holds the highest level of importance among the building's physical characteristics, with a score of 150. Following are the building type and its cultural heritage status, which are almost equally weighted with scores of 133 and 132, respectively.

Regarding the socio-economic profile of the residents', both the occupancy and estimated rehabilitation costs carry equal significance for prioritization at the building scale, with scores of 131. In terms of shared space, compensation from funding emerges as the most critical indicator (128), closely followed by estimated renovation costs (127). At the personal unit level, the presence of a temporary shelter is of relatively high importance and is positioned after renovation costs with a score of 127. The capacity of the owners is perceived as the least critical among the indicators at both the shared space and personal unit levels, with a score of 124.

3.2 Data Analysis

The distribution of the collected data has been analyzed using Python. The majority of the assessed buildings in the area are either residential or mixed-use (commercial/residential). Notably, the area is characterized by a significant number of cultural heritage buildings, accounting for approximately 46% of the total buildings assessed.

Given the proximity to the explosion's epicenter, a substantial portion of the buildings experienced severe damage. Out of the 159 assessed buildings, 133 of them (83%) suffered from structural damage, while 25 buildings (16%) faced non-structural damage. Only 2 buildings (1.25%) were completely destroyed. Although the current status of the building is not an indicator for the Priority Index, this information was collected to monitor the recovery process. As of January 2023, the recovery progress was as follows: 61% of the buildings had already been fully recovered, 29% were still in a damaged state, and 9% were undergoing renovation.

Figure 2 illustrate the distribution of the Physical Priority Index (PPI) and Socio-Economic Priority Index (SEPI) respectively. The PPI is determined by three indicators, which means that buildings with the same physical characteristics will have identical PPI values. Given that a significant number of assessed buildings are residential or mixed-use, experiencing major damage, and classified as cultural heritage or not, it is understandable why the majority of the PPI values cluster around 0.505 or 0.815.

On the other hand, the SEPI relies on nine indicators, resulting in a wider range of values between 0 and 1. This variation is expected as the SEPI takes into account a broader set of indicators that capture the social and economic aspects of the buildings.

Figure 3 depicts the relationship between the PPI and SEPI, revealing a lack of correlation between the two indices. Consequently, decision-makers are faced with the task of choosing between PPI and SEPI for prioritizing the reconstruction projects. Therefore, SEPI distribution for the most frequent PPI values: 0.32, 0.505 and 0.815. In Fig. 4, the variations in SEPI are prominently displayed across different PPI values, namely low, medium, and high. This observation indicates that buildings with similar post-disaster physical conditions can exhibit significant differences in the socio-economic profile of their residents. This finding emphasizes the crucial role of SEPI as a key reference for decision-making in post-disaster projects. Therefore, if decision-makers chose to adopt the PPI as a priority index for the projects, they would proceed to funding the projects without privileging socio-economic disparities. However, if they chose the SEPI, they would focus the aid channeling to the most vulnerable community groups without taking into consideration the physical status of the building. They could also choose to adopt a combination of the two indices, integrated in a general strategy for post-disaster reconstruction specific to the local context.

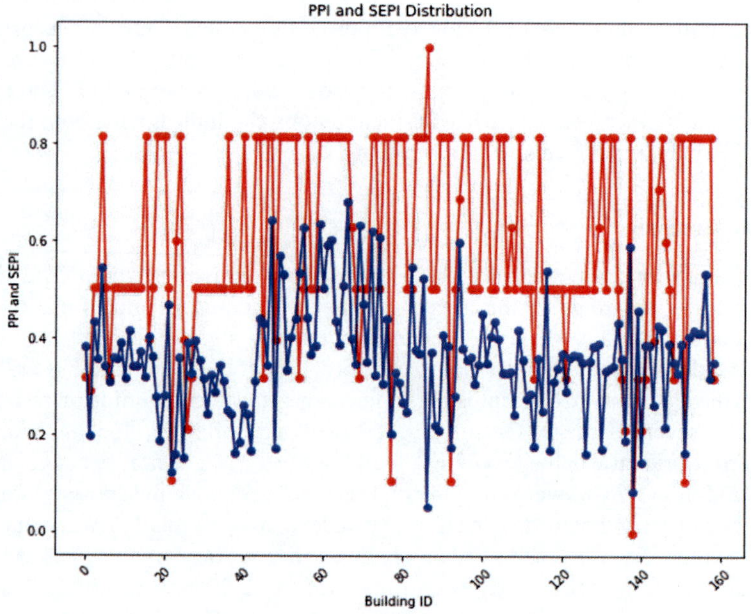

Fig. 2. PPI and SEPI Distribution.

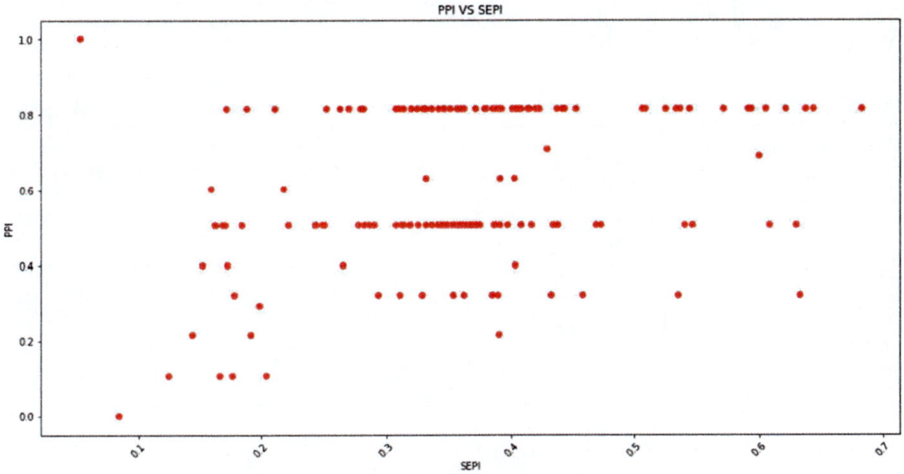

Fig. 3. PPI VS SEPI.

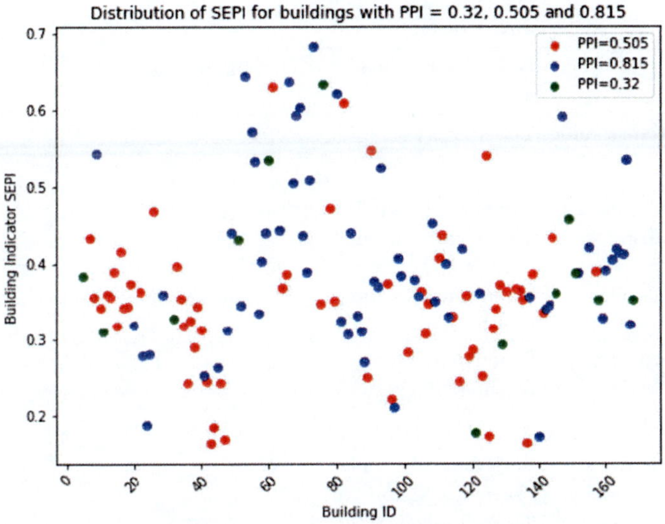

Fig. 4. Selected zones for data collection and assessed buildings.

Table 1. Description of the indicators for the PI and the corresponding weights according to SWARA method (Source: Author)

Indicator	Measure	**Unit**	**Score by experts**	**Weight**
Physical components				
Building Type	Residential, healthcare, education, industrial, commercial, public institution or governmental, cultural or social, or mixed-use	NaN	133	0.32
Damage level	**1**-Total collapse: more than 50% of the building damaged; **2**-Major damage: between 25% and 50% of the building damaged; **3**-Minor damage: only non-structural damage; **4**-No damage	NaN	150	0.37
Cultural Heritage	Is the building considered a cultural heritage or no?	YES/NO	132	0.31
Socio-Economic Profile				
Building scale				
Occupancy ratio before and after the disaster	(Number of occupied units/Number of total units) x 100 calculated before and after the disaster	100%	131	0.5
Estimated rehabilitation costs	Estimation of the rehabilitation costs	$	131	0.5
Shared Space Scale				
Capacity of owners to afford renovation costs	**1**-All the costs can be covered; **2**-Majority of the costs covered; **3**-Half covered/Half not covered; **4**-Majority not covered; **5**-Cannot cover any cost	NaN	124	0.32
Estimated rehabilitation costs	Estimation of the shared space rehabilitation costs	$	127	0.33

(*continued*)

Table 1. *(continued)*

Indicator	Measure	**Unit**	**Score by experts**	**Weight**
Compensation from funding	100 - (Value of the funds/Estimated rehabilitation costs) x 100	100%	131	0.35
Personal Unit Scale				
Capacity of households to afford renovation costs	**1**-All the costs can be covered; **2**-Majority of the costs covered; **3**-Half covered/Half not covered; **4**-Majority not covered; **5**-Cannot cover any cost	NaN	124	0.24
Estimated rehabilitation costs	Estimation of the unit rehabilitation costs	$	127	0.25
Presence of temporary shelter	Do the households have a temporary shelter that is near their principal house?	YES/NO	128	0.26
Compensation from funding	100 - (Value of the funds/Estimated rehabilitation costs) x 100	100%	127	0.25

4 Conclusion

In summary, this study introduces a novel indicator-based method aimed at assisting decision-makers in prioritizing post-disaster reconstruction projects. The method stands out for its unique integration of post-disaster considerations with complex geopolitical contexts. The assessment process incorporates data crowdsourcing and community participation, making it a comprehensive and inclusive approach.

To demonstrate the effectiveness of the proposed methodology, it was applied to the context of Beirut following the devastating explosion on August 4th. Beirut, known for its complex geopolitical environment and multiple ongoing crises, served as a fitting case study. Data was collected from 159 damaged buildings, and their corresponding PPI and SEPI values were calculated.

The analysis reveals the critical role of socio-economic indicators in effectively prioritizing reconstruction projects and allocating resources within complex geopolitical and post-disaster environments. The SEPI, in particular, emerges as a valuable tool for decision-makers to guide recovery and reconstruction efforts. By targeting the most vulnerable individuals without discrimination, this method offers a promising solution to achieve successful outcomes in the post-disaster recovery process.

Appendix 1

SWARA Method

The SWARA method is used for ranking and weighting the indicators. There are many reasons for choosing the SWARA instead of AHP or other methods, but mainly because it is simpler for experts to show the weightage of each criterion in comparison to the other (Zolfani & Chatterjee, 2019). The SWARA method consists of the following steps:

1) The indicators are sorted in descending order based on their score, which is the sum of the weights attributed to the indicator by each expert.
2) Starting with the second indicator, the relative importance of indicator j with respect to the previous criterion (j-1) for each specific criterion, is a ratio called the comparative importance of average value, s_j as follows:

$$s_j = (s_j - s_{j-1})/100 \quad (4)$$

3) Determination of the coefficient k_j as follows:

$$k_j = \{1 \; for \; j = 1; \; s_j + 1 \; for \; j > 1\} \quad (5)$$

4) Determination of the recalculated weight q_j as follows:

$$q_j = \{1 \; for \; j = 1; \; (k_j - 1)/k_j \; for \; j > 1\} \quad (6)$$

5) The relative weights of the evaluation criteria are determined as follows:

$$w_j = q_j / \left(\sum_{(k=1)} q_k \right) \quad (7)$$

w_j indicates the relative weight of the jth indicator, while n indicates the number of indicators.

Weights of the Indicators

w_{BT}: weight of building type; *BT*: Building Type.

w_{DL}: weight of damage level; *DL*: Damage Level.

w_{CH}: weight of cultural heritage; *CH*: Cultural Heritage Status.

w_{OR}: weight of occupancy ratio; *OR*: Occupancy Ratio.

w_{BRC}: weight of building renovation costs; *BRC*: Building Renovation Costs.

w_{SOC}: weight of shared owners capacity; *SOC* : Shared Owners Capacity.

w_{SRC}: weight of shared space renovation costs; *SRC*: Shared spaced Renovation Costs.

w_{SCF}: weight of shared space compensation from funding; *SCF*: Shared space Compensation from Funding.

w_{POC}: weight of personal owners capacity; *POC*: Personal Owner Capacity.

w_{PRC}: weight of personal unit renovation costs; *PRC*: Personal Unit Renovation Costs.

w_{PCF}: weight of personal unit compensation from funding; *PCF*: Personal Unit Compensation from Funding.

w_{TS}: weight of temporary shelter availability; *TS*: Temporary Shelter Availability.

References

1. Villalobos, E., Sim, C., Smith-Pardo, J.P., Rojas, P., Pujol, S., Kreger, M.E.: The 16 April 2016 Ecuador earthquake damage assessment survey. Earthq. Spectra **34**, 1201–1217 (2018). https://doi.org/10.1193/060217EQS106M
2. Parodi, E., Kahhat, R., Vázquez-Rowe, I.: Multi-dimensional damage assessment (MDDA): a case study of El Niño flood disasters in Peru. Clim. Risk Manag. **33** (2021). https://doi.org/10.1016/j.crm.2021.100329
3. Saatcioglu, M., Shooshtari, M., Foo, S.: Seismic screening of buildings in Canada. In: 9th US National and 10th Canadian Conference on Earthquake Engineering 2010: Reaching Beyond Borders, pp. 3088–3097 (2010)
4. Baran, T., Ozcelik, O., Misir, I.S., Saatci, A., Girgin, S.C., Kahraman, S.: Seismic resilience challenge for izmir: pilot project for seismic risks of existing buildings. In: 16th European Conference on Earthquake Engineering, Thessaloniki (2018)
5. Khajwal, A.B., Noshadravan, A.: An uncertainty-aware framework for reliable disaster damage assessment via crowdsourcing. Int. J. Disaster Risk Reduct. **55** (2021). https://doi.org/10.1016/j.ijdrr.2021.102110
6. Kammouh, O., Cimellaro, G.P.: PEOPLES: a tool to measure community resilience. Struct. Congr. 2018 Blast, Impact Loading, Response; Res. Educ. - Sel. Pap. from Struct. Congr. 2018. **2018**, 161–171 (2018). https://doi.org/10.1061/9780784481349.015
7. Zolfani, S.H., Chatterjee, P.: Comparative evaluation of sustainable design based on step-wiseweight assessment ratio analysis (SWARA) and best worst method (BWM) methods: a perspective on household furnishing materials. Symmetry (Basel). **11**, 1–33 (2019). https://doi.org/10.3390/sym11010074
8. Zolfani, S.H., Saparauskas, J.: New application of SWARA method in prioritizing sustainability assessment indicators of energy system. Eng. Econ. **24**, 408–414 (2013). https://doi.org/10.5755/j01.ee.24.5.4526

Smart Waste Management System Based on IoT

Salsabil Meghazi Bakhouch(✉), Soheyb Ayad, and Labib Sadek Terrissa

Mohamed Khider University, Biskra, Algeria
{salsabil.meghazibakhouch,s.ayad,terrissa}@univ-biskra.dz

Abstract. Waste management is one of many crucial challenges across the globe, the fast growth of the population is directly proportional to waste. In other words, more population more waste and garbage. Therefore, using an efficient waste management system can save the planet from assured danger, in addition, an efficient system leads to a clean and proper environment [1]. Since we were able to notice that the waste is on the ground and the dustbins are full in Algeria, we found that the current system was not working adequately [2]. The classic system is not able to cover all the administration needs, such as the administration cannot track the current position of the trucks while collecting, and also without knowing the state of the dustbins, which could cost time, money, and energy. Therefore, we propose a new waste management system based on IoT for real-time tracking of the trucks' position and the dustbins' state. We have developed a mobile application specifically designed for truck drivers. This application utilizes GPS technology to retrieve real-time location information of the trucks. Additionally, we have created a smart bin prototype that incorporates ultrasonic sensors for accurately measuring the fill level of the bins. All the collected data is seamlessly displayed on the administration web dashboard. After testing the system, we obtained satisfactory results.

Keywords: GPS · Geolocation · Internet of things · Smart Waste Management · Smart Dustbin

1 Introduction

As the population keeps growing and the country's economy keeps getting bigger, the problem of dealing with waste is also getting worse. But what's missing is a good way to solve this problem or a smart system to keep track of how we throw away our trash. Even though cities are becoming smarter with all kinds of new technologies, waste management seems to be left behind in the progress [3,4]. The way we usually handle waste has its own set of problems. One of the big issues is that we can't really keep an eye on where the garbage trucks are in real-time. These trucks just follow their usual routes without knowing if the bins they're supposed to empty are actually full or not. This leads to unnecessary trips, where they collect trash from bins that aren't even full yet. This outdated system isn't just wasteful in terms of resources, but it also adds to the environmental problems

M. Ben Ahmed et al. (Eds.): SCA 2023, LNNS 906, pp. 322–331, 2024.
https://doi.org/10.1007/978-3-031-53824-7_29

we're facing [5,6]. In the midst of all these challenges, there's a need for a fresh approach. Imagine a system that could make waste collection smarter, where trucks only go where they're needed, when they're needed. This could be a game-changer, using technology to make waste management catch up with the fast pace of urban development. Such a solution could make cities cleaner and more eco-friendly, harmonizing progress with environmental responsibility[7,8]. This paper introduces an IoT-based waste management solution, structured around three key components: the waste truck responsible for the entire waste collection and disposal process, IoT-enabled smart dustbins, and an administrative module. The proposed solution encompasses two pivotal subsystems: a geolocation system enabling real-time waste vehicle tracking, and a prototype smart dustbin that provides real-time insights into its fill level and weight [9–11]. The remainder of the paper is organized as follows: In Sect. 2, Related Work we present companies working on Smart waste management and compare them. In Sect. 3, we present the Smart Waste Management System Based on IoT and its architecture. In Sect. 4, we describe the implementation of our proposed solution and the results in the Sect. 5. Last but not least, the conclusion is in Sect. 6.

2 Related Work

Various startups and companies around the world are stepping into the realm of smart waste management systems, underscoring the paramount importance of such solutions and their profound impact on the environment. As this paper ventures into the practical implementation of such a system, it deems it essential to benchmark against existing solutions. To this end, we present a comparative analysis of three prominent existing smart waste management solutions: Compology, Ecube Labs, and Evreka. Compology, one of the frontrunners in this domain, employs a sophisticated approach. They employ intelligent cameras, strategically positioned either on the waste bins themselves or on the collection trucks. These cameras capture crucial data such as fill level, location, and even the content of both bins and trucks. Additionally, they gauge the available floor space, truck activity, and their respective positions [12]. Complementing this, Compology offers a comprehensive web dashboard that facilitates real-time monitoring of not just the bins' and trucks' content and fullness, but also their GPS locations and an array of crucial information. Secondly, Ecube Labs offers a solar-powered, trash-compacting smart dustbin. These bins integrate IoT sensors that constantly monitor the fill levels in real-time. This gathered data is then seamlessly visualized on an intuitive web dashboard and a purpose-built mobile application, catering especially to truck drivers. The application even includes an optimization route feature, enhancing the overall efficiency of waste collection [13]. The contributions of Evreka to the smart waste management sphere come in the form of both software and hardware services. Their vehicle tracker system and Evreka's proprietary wireless sensor facilitate the seamless collection of data from the waste bins. This data encompass not only the fillness levels but also the bin's location and temperature. To aid administrators,

Evreka also furnishes an extensive monitoring web dashboard, complemented by a mobile application [14].

2.1 Discussion

The Table 1 illustrates a comparison between the presented companies; after the table, we conclude:

- Web Dashboard: All three companies, including our proposed solution, provide an administrative web dashboard with common features such as 'Bin Monitoring' and 'Real-time Tracking'. However, Compology offers an additional feature called 'Client Reports', which is not present in the other solutions including ours.
- Smart Bin: Each of the three companies, alongside our solution, incorporates ultrasonic sensors. In terms of connectivity, Compology utilizes AI Cameras; Ecube Labs employs 2G/3G, LoRaWAN, and NB-IoT technologies; Evreka opts for Lora, GSM, and NB-IoT technologies. Our solution, on the other hand, employs IoT technology. While Compology relies on solar power, both Ecube Labs and we utilize solar and AC power sources. Additionally, both Ecube Labs and Evreka offer compacting bins.
- Mobile Application: EcubeLabs, Compology, Evreka, and our proposed solution all feature a mobile application designed for drivers.
- Data Storage: Evreka stores data in the Microsoft Azure Cloud; Ecube Labs chooses cloud hosting. Our proposed solution, however, stores data in a private database.
- Recycling System: All three companies, as well as our proposed solution, provide a recycling system.
- *Added Value:* Our solution offers unique added value. The administration can define truck routes, categorize maps into zones, and perform edits. The administrator has the flexibility to add, edit, and remove trucks or drivers from the system. Furthermore, precise dustbin locations can be identified.

3 Smart Waste Management System Based on IoT

Smart waste management is a collecting and managing ecosystem of waste using the Internet of Things (IoT), from its inception to its final disposal, within a short time. In this paper, we present our solution which is composed of two subsystems. The Fig. 1 illustrates the architecture of the proposed solution.

Table 1. Comparison table between the presented companies

Service		Compology	Ecube Labs	Evereka	Proposed Solution
Web Dashboard	*Real-Time Tracking*	Provided	Provided	Provided	Provided
	Client Reports	Provided	Not Provided	Not Provided	Not Provided
	Bin Monitoring	Provided	Provided	Provided	Provided
Smart bin	***Sensor***	Ultrasonic Sensor	Ultrasonic Sensor	Ultrasonic Sensor	Ultrasonic Sensor
	Connectivity	AI Camera	2g/3g, LoRaWAN, NB-IoT	Lora, GSM, NB-IoT	IoT
	Power	Solar Power	AC Power/ Solar Power	.	Battery
	Compacting bin	Not Provided	Provided	Provided	Not Provided
Driver App		Not Provided	Provided	Provided	Provided
Data Stored in		.	Cloud Hosted	Microsoft Azure Cloud	Private Database
Citizen App		Not Provided	Not Provided	Provided	Not Provided
Recycling		Provided	Provided	Provided	Provided

3.1 Geolocation System

It enables tracking of the waste collection vehicle and retrieves additional tracking information, such as accuracy and speed, using GPS technology. The collected data will be displayed on an administration web dashboard.

1. **Truck Module:** This module is represented by a mobile application installed on the truck driver's smartphone; it includes a GPS module, which authorizes to retrieve data such as the longitude, latitude, accuracy, and speed of the truck while collecting. These data will be hosted on a web server and stored on a storage module. Besides, the truck movement will be displayed on the administration dashboard map.
2. **Administration Dashboard:** This module is represented by a web platform, accessed by the administration. This platform contains a map that displays real-time trucks' and bins' fill levels, weights, and positions. It could be managed by two types of admin: The first is called the Manager who is responsible for the whole system, he could add/delete drivers, trucks, and bins to the system or even define weekly tours. The second called the Supervisor, could read data from the dashboard, to put it more simply, he has no right to manage the system, all he could do is verify the location of the trucks and the fill level of the bins and Act expeditiously when the need arises.

3.2 Smart Dustbin System

The smart dustbin prototype enables receiving the bin's fill level and weight in order to display this data on an administration web dashboard, it is composed of two sub-modules:

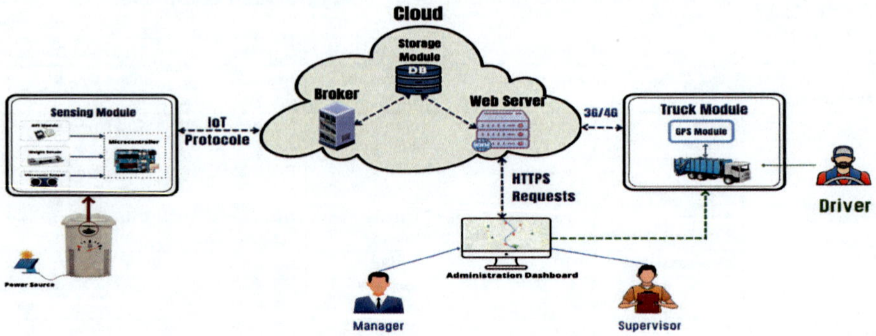

Fig. 1. Proposed Solution Architecture

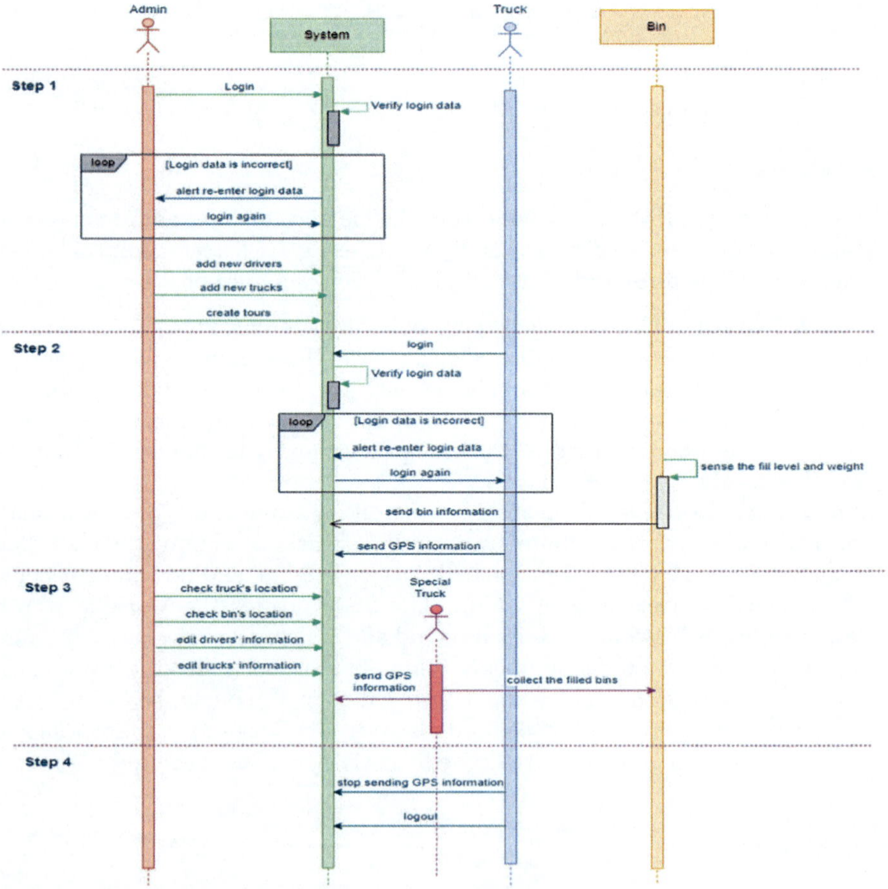

Fig. 2. Sequence Diagram

- **The Sensing module:** This module consists of sensors, their purpose is to retrieve data in real-time, such as the location of the dustbin, the fill level, and the weight, and send it to the cloud using IoT protocols in order to display the data on the administration dashboard.
- **Power Source:** The power source represents the energy needed to feed the sensors.

3.3 The Functionality of the System

Figure 2 illustrates the sequence diagram depicting the functionality of the system. It is structured into four distinct steps:

- **Step 1:** The administrator logs into the system, adds drivers and trucks, creates tours, and designates their corresponding drivers and trucks.
- **Step 2:** The truck's driver logs into the system and begins transmitting location information. Simultaneously, the smart bin detects its fullness and weight, subsequently sending this data to the system.
- **Step 3:** The administrator monitors the trucks' locations and assesses the fullness of the bins. Additionally, the administrator has the authority to modify drivers' or trucks' details. In instances where a bin reaches its capacity and no scheduled tour is in place, a designated truck initiates waste collection exclusively from the filled bins.
- **Step 4:** The waste collection process concludes as the truck discontinues location information transmission and logs out.

4 Implementation of the Proposed Solution

The proposed solution is designed for Biskra Province. We have developed a web dashboard for the administration, utilizing Html/CSS and JavaScript for the frontend and the PHP framework 'Laravel' for the backend. Additionally, a mobile application for the truck drivers has been developed using the Flutter framework.

4.1 Administration Web Dashboard

The administration web dashboard is multilingual when the admin logs in to the system, he will be directed to the main page Fig. 3, which is composed of:

- (1) Navigation Bar (Header): Positioned at the top of the page, the navigation bar includes, from right to left, a user drop-down menu containing logout and profile options, a language drop-down menu offering English, Arabic, and French options, Full-Screen and Search buttons, a Home link for returning to the home page, and a Hamburger menu button.
- (2) Sidebar: Located on the left side of the page, the sidebar structure comprises a website logo at the top, followed by a search input, and navigational links to various sections of the website, including Zones, Paths, Bins, Trucks, and more.

- (3) Map Display: Real-time analytics are presented on the map. This includes visual representations of the zones within the Biskra Province. The map also displays the fullness status of bins using color codes: green signifies an empty bin, yellow indicates a half-full bin, and red represents a bin approaching full capacity. As waste collection trucks are in motion, a truck icon dynamically moves on the map, reflecting real-time truck movements.
- (4)(5)(6) The three sections, respectively, provide live analytics for the driver, truck, and bin.

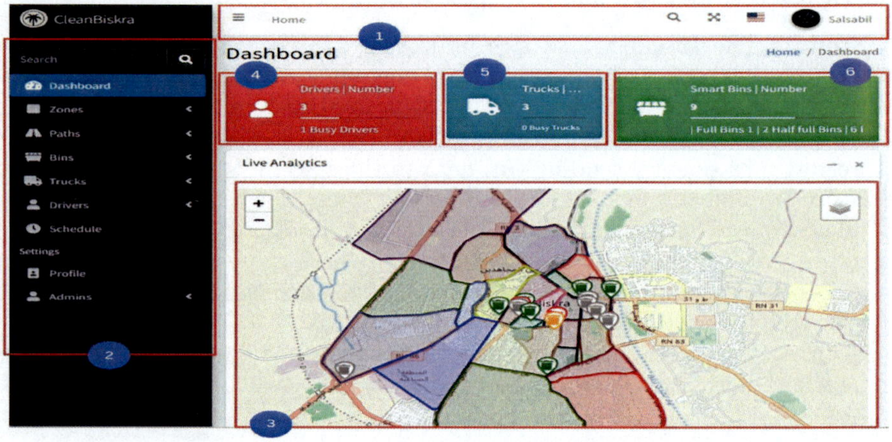

Fig. 3. Administration Main Dashboard Screenshot

Furthermore, we offer supplementary pages for incorporating Zones, Bins, Paths, Drivers, Trucks, and Admins. We've also devised a month planning feature that showcases the tours scheduled for the current month, alongside a dedicated page for scheduling these tours. Additionally, we've developed Bin List, Driver List, Truck List, and Admin List pages that can be seamlessly imported in PDF, CSV, or Excel formats, or even printed using a standard printer.

4.2 Truck's Driver Mobile Application

This mobile application is installed on the drivers' smartphones. To access the system, drivers log in through Fig. 4a using their email and password. Upon successful login, they are directed to the tours page as shown in Fig. 4b. This screen displays all scheduled tours, providing details like the driver's name and tour timing. When it's time for waste collection, the driver selects the tour they are assigned to. At this point, the application accesses the phone's GPS to retrieve essential movement data such as longitude, latitude, accuracy, speed, and time. Once the driver completes the tour, they can simply log out.

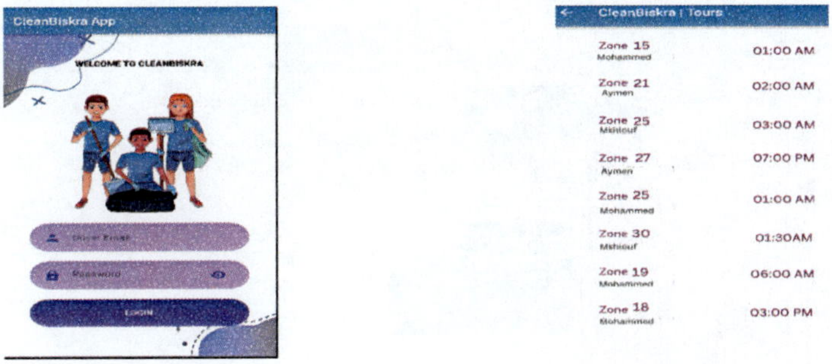

(a) Login Page (b) List Tours

Fig. 4. Driver's mobile application

4.3 Smart Bin Prototype

The smart dustbin is equipped with IoT sensors positioned on the top of the dustbin, as shown in Fig. 5. Two types of sensors have been utilized: an ultrasonic sensor and a weight sensor.

- Ultrasonic Sensor (HC-SR054): This sensor calculates the distance to the nearest object, allowing it to determine the current fill level of the bin. Subsequently, it sends this calculated data to the microcontroller, which in turn transmits the information to the cloud Fig. 5b.
- Weight Sensor (HX 711): Employing a load cell situated beneath the bin, the weight sensor (HX 711) captures the real-time weight of the contents. This collected data is then transmitted to the cloud for further processing.

(a) Smart Bin (b) Smart Bin Sensors

Fig. 5. Smart Bin Prototype

Figure 6 represents the electronic circuit of the IoT sensors. We placed the We-mos D1 R2, HC-SR04, HX711 and the LCD to the breadboard.

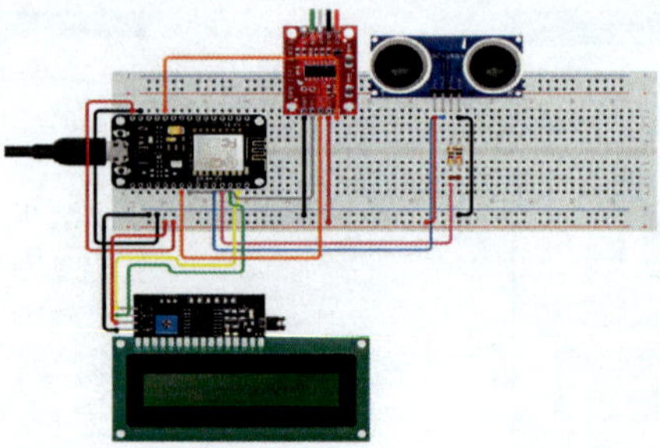

Fig. 6. Smart Bin Sensors Circuit

5 Results

We conducted thorough testing of our solution, successfully implementing the application on a driver's smartphone. When the driver initiates waste collection, a corresponding truck icon becomes visible on the web dashboard, particularly on the map shown in Fig. 7. This real-time tracking feature empowers administrators to monitor the truck's movement and even its speed in real time. In the event of any unusual movements, prompt action can be taken, such as contacting the truck driver or seeking assistance. Additionally, the bins' status is also conveyed in real time. Upon bin data upload, a simple color change occurs on the dashboard. Specifically, the bin's color shifts from green to red, signifying its fill level status.

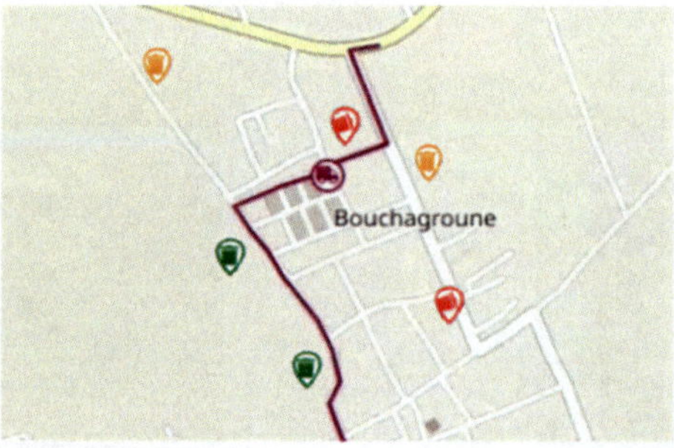

Fig. 7. Real Time truck movement

6 Conclusion

We introduced an IoT-based Smart Waste Management System that offers real-time monitoring of truck movements and bin fill levels. For administrators, we developed a web dashboard providing dynamic analytics including truck locations and bin fill levels, presented through maps and various visualizations. Additionally, our mobile application features a GPS module, enabling real-time tracking of truck locations and speed. The app also alerts administrators at the commencement of truck tours. Furthermore, our IoT-powered dustbin prototype captures both fill level and weight data. As a future prospect, we aim to implement an optimization route module for the truck application. This module will dynamically generate collection routes, bypassing only the full bins, deviating from the conventional fixed routes.

References

1. Gopal Kirshna, S., Sunilkumar, S.M., Priyanka, B.: Smart Waste Management using Internet-of-Things (IoT). Reva University, Bangaluru (2017)
2. Claude Noel, T., Elena, V.R.: Smart systems and The Internet of Things (IOT) For Waste Management. Ashesi University, Accra, Ghana (2020)
3. Sanjiban, C., Aniket, M., Shaheen, S.: Smart Waste Management System. Computer Science and Engineering Department, Bangalore, India (2021)
4. Navghane, S.S., Killedar, M.S., Rohokale, V.M.: IoT based garbage and waste collection bin (2016)
5. Vishesh Kumar, K.: Smart Garbage Collection overflows Indicator using IoT. Kalinga University, Chhattisgarh, India, Electronics and Telecommunication Engineering Department (2016)
6. Saha, H.N., Auddy, S.: Waste management using Internet of Things (IoT). Bangkok, Thailand (2017)
7. Bano, A., Ud Din, I., Al-Huqail, A. A.: A IoT-based smart bin for real-time monitoring and management of solid waste (2020)
8. Nikila, C., Narmadhai, N.: Smart waste management solution using IoT for smart city (2023)
9. Kadus, T., Nirmal, P., Kulkarni, K.: Smart waste management System using IOT. Savitribai Phule Pune University, Pune, India, Mechanical Engineering Departement (2020)
10. Kumar,V., Hongekar,P.: Smart waste management system. Mechanical Engineering Departement, Udyambag, Belagavi, India
11. Telugu, M., Harish, K.: IoT based smart dustbin (2020)
12. Compology Homepage. https://compology.com/. Accessed 10 Apr 2023
13. EcubeLabs Homepage. https://www.ecubelabs.com/. Accessed 6 Jun 2023
14. Evreka Homepage. https://evreka.co/evreka360/. Accessed 6 Jun 2023

A Review on Artificial Intelligence and Behavioral Macroeconomics

Zakaria Aoujil(✉) and Mohamed Hanine

Laboratory of Information Technologies, National School of Applied Sciences, EL Jadida 24000, Morocco
{aoujil.zakaria,hanine.m}@ucd.ac.ma

Abstract. The intersection of artificial intelligence (AI) and behavioral economics represents a paradigm shift since both fields offer innovative approaches to make predictions and optimizing decision-making. In contrast, classical macroeconomics predominantly relies on rational, aggregate models to study macroeconomic conditions. Despite the considerable success in integrating behavioral insights into various economic fields, macroeconomics has seen a slower adaptation. This latency can be attributed to the challenge of scaling individual behavioral biases to a macro level. Moreover, the incorporation of AI into macroeconomics has faced challenges including data requirements and interpretability issues. However, AI and behavioral economics have individually showcased promising results in macroeconomic research. AI's computational prowess facilitates sophisticated data analysis and predictive accuracy, while behavioral economics provides a more holistic interpretability of macroeconomic patterns by considering cognitive biases and heuristics. This synergy paves the way for a potential rise of AI-driven behavioral macroeconomics. This article aims to review the current progress in the application of AI across behavioral macroeconomic axes. The findings confirm the promise of AI-powered behavioral models in terms of both predictive accuracy and explanatory power. These results imply an emerging trend in the interdisciplinary field of AI and behavioral macroeconomics, thus paving the path for future research and applications in this domain.

Keywords: Behavioral macroeconomics · Artificial intelligence · Machine learning · Deep learning · Review

1 Introduction

Behavioral economics, a field that integrates psychological insights into economic analysis, has gained significant acceptance in recent years due to its ability to address the limitations of classical economics. Classical economics, rooted in the rational actor model, assumes that individuals consistently make decisions based on rationality and perfect information. However, numerous empirical findings have demonstrated that human behavior deviates from these assumptions,

M. Ben Ahmed et al. (Eds.): SCA 2023, LNNS 906, pp. 332–341, 2024.
https://doi.org/10.1007/978-3-031-53824-7_30

leading to economic inefficiencies. Behavioral economics seeks to understand and explain these deviations, shedding light on the complexities of human decision-making.

The field of behavioral economics has witnessed remarkable advances, propelled by the contributions of renowned scholars. Prominent figures such as Daniel Kahneman, Amos Tversky, Richard Thaler, and Robert Shiller have shaped the theoretical foundations and empirical evidence in this domain. Their groundbreaking work has illuminated various subfields within behavioral economics, including cognitive biases, prospect theory, nudging, and bounded rationality.

While behavioral economics has made substantial progress in micro-economic settings, its extension to macroeconomics faces certain impediments. Macroeconomics explores the behavior of economies as a whole, focusing on variables such as national income, employment, and inflation. However, traditional macroeconomic models often struggle to account for the complexity of human behavior, leading to difficulties in scaling individual behavioral to a macro level.

In recent years, the utilization of artificial intelligence (AI) in economics and finance has emerged as a powerful tool for addressing complex problems and making accurate predictions. AI leverages advanced algorithms to uncover patterns, identify correlations, and generate insights that were previously unattainable. Its application in various domains, such as image recognition, natural language processing, and recommendation systems, has resulted in significant breakthroughs and transformative advancements. In marketing, sentiment analysis leverages AI and natural language processing to understand customer sentiments expressed in textual data. In finance, machine learning algorithms have revolutionized credit scoring, improving accuracy by analyzing extensive data. Algorithmic trading, including high-frequency trading, utilizes AI to analyze real-time market data and execute trades rapidly. This transformative potential has led researchers to explore integrating AI into behavioral economics, recognizing the synergistic benefits from combining insights from both fields.

By leveraging AI's analytical capabilities and harnessing the understanding of human decision-making provided by behavioral economics, it becomes possible to create more accurate and comprehensive macroeconomic models. This article aims to review the current state of the art in the intersection of AI and behavioral macroeconomics, shedding light on earlier progress while providing insight into potential future trends. In this paper, the first section describes the evolution of Artificial intelligence till its current shape, then the AI applications in macroeconomics are discussed, the third section showcases the ongoing shift from classical to Behavioral macroeconomics, the next section presents the behavioral macroeconomics axes and details the AI techniques used in existing literature, and finally a summary table is dressed with the main AI techniques used by axes.

2 Artificial Intelligence over Time

The field of artificial intelligence has evolved since its inception, transitioning from rule-based, knowledge-driven expert systems to the current era of model-based, data-driven AI. Initially, expert systems relied on predefined rules and human knowlegde to make decisions, but their limitations in dealing with complex problems led to the exploration of new approaches. The emergence of machine learning (ML) techniques allowed computers to learn from data and make predictions. Yet, engineering of features and scalability issues persisted. The advent of deep learning (DL) which is a subset of ML revolutionized the field, as it allowed for the automatic learning of hierarchical representations from large datasets, leveraging the availability of big data and computational power. Machine learning and Deep learning are currently the main employed AI techniques and shifted the focus from explicit rule-based systems to data-driven models learning directly from the data.

Data-driven AI techniques have demonstrated their effectiveness and potential in various fields such as healthcare, finance, marketing, transportation, and more.

3 AI in Macroeconomics

Prior to the advent of artificial intelligence (AI), the main techniques used in macroeconomic research included statistical analysis, econometric models, and survey data. These methods relied on historical data and mathematical models to analyze and predict macroeconomic variables such as GDP growth, inflation rates, and unemployment rates. However, these techniques had several weaknesses. Firstly, they are based on linear assumptions and may not capture the intricate dynamics of the real economy accurately. Secondly, these techniques often struggled with complex and nonlinear relationships between economic variables. Thirdly, the traditional methods faced challenges in processing and analyzing large volumes of data efficiently. They were time-consuming and relied heavily on manual data processing and modeling.

The introduction of AI is having a significant impact on macroeconomics. These advanced technologies have revolutionized data collection, processing, and analysis, enabling researchers to analyze vast amounts of data from various sources, including social media, sensor networks, government reports, and financial market data to capture both historical and real-time economic trends. Applying AI techniques enabled models such as machine learning models to capture nonlinear patterns and adapt to changing economic conditions, leading to more realistic simulations and predictions.

In terms of economic forecasting, AI have improved the accuracy and timeliness of economic forecasts. For example, by analyzing online job postings [1] economists can make more precise predictions about employment levels and economic activity.

For monetary policy analysis: AI have helped central banks monitor financial markets and assess systemic risks, by analyzing a wide range of data, including market prices and news sentiment [2].

In macroeconomic policy analysis, AI have enabled policymakers to evaluate the impacts of policies in a more comprehensive and granular manner. For instance, machine learning enables a data-driven approach to assess the impact of environmental tax reform, providing valuable insights for policymakers and researchers [3].

Despite these achievements, there will always be potential blind spots if the behavioral aspects of macroeconomic decision making are not incorporated into the studies. Macroeconomic decisions are influenced by human behavior, expectations, and psychological factors, which may not be fully captured by AI analysis alone.

4 From Classical to Behavioral Macroeconomics

In recent years, there has been a notable shift in macroeconomic research towards incorporating insights from behavioral economics, giving rise to the field of behavioral macroeconomics. This interdisciplinary approach combines traditional macroeconomic models with behavioral insights, aiming to provide a more accurate understanding of economic phenomena. Prominent researchers in this area include Matthew Rabin, Richard Thaler, and George Akerlof, whose works have contributed significantly to the development of behavioral macroeconomics. Rabin's research on social preferences [4], Thaler's contributions to the concept of bounded rationality [5], and Akerlof's analysis of asymmetric information and market inefficiencies have paved the way for integrating psychological factors into macroeconomic models [6].

However, despite the growing influence of behavioral economics, the development of behavioral macroeconomics has faced some delays compared to fields like behavioral finance. One significant factor is the inherent complexity of macroeconomics, which deals with aggregate variables and the interactions of millions of economic agents. Understanding and modeling the collective behavior of individuals within a macroeconomic framework is challenging and often requires simplifications and assumptions. Additionally, the availability of macroeconomic data for empirical analysis is limited, making it harder to test and validate behavioral macroeconomic models compared to behavioral finance. Furthermore, the macroeconomic policy framework is often shaped by long-standing theories and models, making it more resistant to incorporating behavioral insights. Nonetheless, as research in behavioral economics progresses and empirical methods evolve, behavioral macroeconomics is gaining traction and has made a significant impact in various macroeconomic topics including Business Cycles and Economic Fluctuations De Grauwe [7], unemployment [8] and monetary policy [9].

5 AI Techniques in Behavioral Macroeconomics Axes

As discussed in the previous 2 sections, the separate integration of AI techniques in macroeconomics [10], and behavioral economics aspects into macroeconomics research [6] had a significant impact on the field. Studies have shown promising results, which suggest that the convergence of the three fields (AI, behavioral economics, and macroeconomics) can have a much more profound impact on the field. This rising potential of convergence is further enhanced by their synergistic relationship. Behavioral insights can inform the design of AI algorithms, ensuring they account for human behavior and decision-making biases. Conversely, AI techniques can provide empirical evidence to test and validate behavioral economic theories. This combination offers a powerful framework for policymakers and researchers to develop more effective macroeconomic models, design targeted interventions, and make data-driven policy decisions. While the boundaries between various areas of macroeconomics may not always be distinctly defined, and there's often considerable overlap, we used the JEL Classification Codes (Journal of Economic Literature) [11] and N. Gregory Mankiw book *Macroeconomics* [12] to categorize existing research on AI and behavioral macroeconomics with regard to six identified behavioral macroeconomics axes.

Table 1. AI techniques involved in Behavioral macroeconomics by axes

Macroeconomic Axes	Articles	RL	NLP	NB	ANN	MLP	SVM	SVR	DT	RF	GBT	GBM	LOGREG	K-means	KNN	LSTM	RNN	GRU	Data mining	GA	Fuzzy logic
Monetary Policy and Central Banking	[13,14]				X		X														
Macroeconomic Indicators and Business Cycles	[15,16] Erçen et al. [17]		X					X								X	X		X		X
Public Economics and Income Distribution	[18,19]								X	X		X	X	X							
Economic Growth and Development	[20]		X																		
Macroeconomic Modelling and Forecasting	[21–26]	X X			X	X		X	X	X		X			X	X	X			X	X
Financial Economics and Macroeconomics	[27–29]			X	X X				X	X	X	X	X			X		X			

- Monetary Policy and Central Banking:
  This is a key subfield, studying how psychological factors affect central banking decisions and how the public's behavior influences monetary policy's effectiveness and their impacts on the economy. In this sense, Salle [13] introduces an agent-based modeling (ABM) framework that incorporates artificial neural networks (ANN) based expectation formation mechanism to study the interplay between expectations, central bank communication, and macroeconomic outcomes. The use of ANN in the expectation model provides a more realistic representation of expectation formation in the agent-based model which is a computational modeling technique used to simulate the behavior and interactions of individual agents within a system. On the other hand Polyzos et al. [14] employs agent-based modeling and a support vector machine

(SVM) optimized subjective well-being (SWB) function to analyze the welfare effects of banking crises and proposing policy solutions that authorities should reconsider to mitigate the welfare loss caused by financial instability.
- Macroeconomic Indicators and Business Cycles:
  This field of study focuses on capturing both the key metrics of economic performance and the patterns of economic fluctuations. It involves the analysis of the macroeconomic indicators and exploring how behavioral factors such as investor sentiment, herd behavior, and cognitive biases contribute to the occurrence and severity of economic crises, providing insights into the overall performance of an economy as well as how these indicators change through different phases of the business cycle - expansions and recessions. Regarding this axe, Mehreen et al. [15] proposes the use of a framework for fetching data in real-time from reliable online sources and performing intelligent data mining to predict the impact of economic crises, specifically unemployment, on mental well-being and discusses the role of social protection programs in promoting recovery during economic crises, while Co et al. [16] compares two different forecast methods, the Autoregressive integrated moving average (ARIMA) econometric model and the Deep-learning approach with the Long Short Term Memory Recurrent Neural Network (LSTM-RNN) model as part of a research project that predicts macroeconomic indicators to detect the risk of inflation and economic crisis, the LSTM-RNN model shows more accuracy taking into account the psychological factors in investors behavior and capturing recurrent pattern that aligns with Elliott Wave Theory and the psychology of the market. In other hand Erçen et al. [17] study utilizes a hybrid machine learning approach, incorporating various AI techniques including fuzzy logic (FL), support vector regression (SVR) and sentiment analysis falling under Natural Language Processing (NLP), along with learning automata (LA) for textual analysis and prospect theory which is a behavioral economics theory proposed by Kahneman and Tversky that aims to explain individuals' decision-making under uncertainty. By analyzing news articles and historical data, the study identifies significant correlations between exchange rate volatility and news related to macroeconomic indicators such as inflation rate and interest rate.
- Public Economics and Income Distribution: Studies how behavioral factors influence the outcomes of public policies, which can involve a variety of specific topics ranging from tax policy to public goods provision, and how cognitive biases and social norms can affect income distribution and inequality. In public housing in relation to public heath Bentley et al. [18] explores the behavioral aspects of social housing and its impact on mental health, the study uses data from a nationally representative panel survey of Australian households while employing marginal structural models (MSMs) along with ensemble-learning methods to analyze the data, specifically the SuperLearner package, which combines multiple models fitted to the same data and generates weighted predictions. Logistic regression (LR), Gradient boosting machine (GBM), and conditional inference forest which is an extension of the Random Forest (RF) algorithm are used as base learners in the SuperLearner

approach. Homelessness is another public housing aspect discussed by Yoder Clark et al. [19] in relation to regional and national policies as well as individual behaviors, the study utilizes K-means cluster analysis and Decision trees (DT), to analyze factors contributing to homelessness such as regional political and economic forces and individual's life experiences and choices.

- Economic Growth and Development: This axe focuses on understanding the factors that contribute to economic growth and development at a national and international level including behavioral theories and factors like trust, corruption, and social norms as discussed in Huang et al. [20] which explores the relationship between moral foundations, economic growth, and corruption in China using AI and big data techniques, the study uses natural language processing (NLP), to analyze the textual data and identify patterns and indicators related to moral foundations and corruption in a large dataset of social media posts.
- Macroeconomic Modelling and Forecasting; This subfield involves developing models to represent economies and using insights from behavioral economics to refine these models to forecast future economic scenarios. These might involve incorporating elements of bounded rationality, prospect theory, or social preferences into models, as suggested by Sent [23], this paper provides a theoretical framework for understanding the role of bounded rationality combination with artificial intelligence such as Classifier systems and Genetic algorithms (GA) in macroeconomic research. Likewise D'Orazio [21] highlights ongoing paradigm shifts in macroeconomic research from deductive-Neoclassical economics to Computational Behavioral Agent-based Macroeconomics driven by the interactions between Big Data, Artificial Intelligence, and behavioral macroeconomics. Mashkova et al. [24] contribute with a model of experimental economy that is used for both theoretical research and practical calculations. The model incorporates multidisciplinary concepts from economics (simulation modeling), artificial intelligence (fuzzy logic), and cognitive psychology (bounded rationality). The model allows for the analysis of economic cycles and the evaluation of state influences. Ponomarenko [25] explored the relationship between microfoundations and macroeconomic outcomes using an agent-based model and incorporating the concept of bounded rationality and using reinforcement learning (RL), an AI technique, to allow the agents to learn from their experiences and adjust their strategies based on the outcomes they receive. Empirical Works published in this regard includes Milunovich [22] which examines the effectiveness of machine learning and deep learning algorithms in predicting Australian real house prices, it compares their performance against traditional time series models and emphasizes the importance of accurate house price predictions for central banks, financial supervision authorities, investors, and homeowners. The study compares 47 algorithms, including Support vector regression (SVR), K-nearest neighbors (KNN), Decision Tree (DT), Multilayer Perceptron (MLP), Random Forest (RF), LSTM-RNN and Light GBM, and examines the impact of cognitive biases such as loss aversion on seller behavior to improve forecasting accuracy. Additionally Zhang et al. [26] conducted a literature review on the integration

of machine learning techniques, including artificial neural networks (ANN), recurrent neural networks (RNN), and reinforcement learning (RL), in agent-based modeling (ABM). The focus was on enhancing decision-making and predicting macro-level behaviors in areas such as residential dynamics and macroeconomic systems. Additionally, the text mentions the use of bounded rationality models like the aspiration adaptation theory (AAT) to capture realistic decision-making processes.

- Financial Economics and Macroeconomics:
  Explores how cognitive biases and heuristics affect financial decision-making and risk-taking, and how these behaviors can impact macroeconomic stability. Several studies have been conducted in this trend, Kanzari et al. [27] uses AI and big data techniques to predict macro-financial instability. It employs deep learning methods like Long Short-Term Memory (LSTM) and Gated Recurrent Unit (GRU) neural networks to analyze sentiment data and financial time series. The findings provide strong evidence for the relevance of sentiment in predicting financial system instability, contributing to the growing field of literature that empirically tests the Efficient Market Hypothesis (EMH) in various financial market segments. Rekik et al. [29] chose to highlights the limitations of traditional asset pricing models and proposed a behavioral perspective. Using AI techniques like ANN along with ABM to analyze and model financial market behavior, detecting nonlinear relationships and simulating investor behavior. This study consider the complexities of individual decision-making in the analysis of financial market dynamics. Meanwhile Alam et al. [28] proposed a machine-learning-based approach taking into account the influence of the mental-accounting bias to predict savings adequacy using AI techniques such as Naïve Bayesian (NB), Logistic Regression (LOGREG), Artificial Neural Network (ANN), Decision Tree, Random Forest, and Gradient Boosted Trees (GBT). The study aims to address the complex nonlinear relationships between dependent and independent variables related to savings adequacy and suggests that decision trees are suitable for creating publicly available budgeting tools to enhance financial literacy, while ANN can be utilized for accurate predictions.

The Table 1 summarizes the key AI techniques used in behavioral macroeconomics studies in conjunction with economics models such as Autoregressive integrated moving average (ARIMA) or computational techniques such as agent-based modeling (ABM) and incorporating behavioral economics concepts such as bounded rationality and prospect theory. Further, it is relevant to mention the predominance of machine learning techniques over more advanced algorithms such as deep learning and reinforcement learning. Indeed, the known superior predictive power of DL comes at a cost, as advanced DL techniques tend to act as a black box, making it difficult to interpret the findings. This could be problematic in macroeconomics, where understanding the causal relationships between variables is often as important as making accurate predictions. This drawback is mitigated by incorporating behavioral economics models along with

AI techniques in macroeconomics research, giving these interdisciplinary models both predictive accuracy and explanatory power.

6 Conclusion

Although the research field of AI in behavioral macroeconomics is in its early stages, existing literature confirms the theoretical discussion regarding the ability of AI-powered behavioral models to outperform linear classical models. This is significant because traditional economic models often assume a linear pattern for economic agents and metrics, which may not accurately represent the complexity and nonlinearity of actual economic behavior. However, it is essential to acknowledge the challenges associated with the integration of behavioral economic insights and AI techniques into macroeconomic studies. Privacy concerns, data biases, and the interpretability of AI algorithms are crucial issues that need to be addressed to fully leverage the potential of these methodologies. Nonetheless, the current trajectory suggests that the incorporation of behavioral economic insights and AI techniques into behavioral macroeconomic studies will continue to grow in prominence, offering unprecedented opportunities for understanding and shaping the complex dynamics of the global economy, giving rise to the AI and behavioral macroeconomics research field.

References

1. Goldfarb, A., Taska, B., Teodoridis, F.: Could machine learning be a general purpose technology? a comparison of emerging technologies using data from online job postings. Res. Policy **52**, 104653 (2023)
2. Duan, Y., Goodell, J.W., Li, H., Li, X.: Assessing machine learning for forecasting economic risk: evidence from an expanded Chinese financial information set. Financ. Res. Lett. **46**, 102273 (2022)
3. Zheng, Y., Zheng, H., Ye, X.: Using machine learning in environmental tax reform assessment for sustainable development: A case study of Hubei Province, China. Sustainability **8**, 1124 (2016)
4. Charness, G., Rabin, M.: Social preferences: some simple tests and a new model (2000). https://econ-papers.upf.edu/papers/441.pdf
5. Thaler, R.H., Ganser, L.: Misbehaving: the making of behavioral economics (2015)
6. Akerlof, G.A.: Behavioral macroeconomics and macroeconomic behavior. Am. Econ. Rev. **92**, 411–433 (2002)
7. De Grauwe, P.: Booms and busts in economic activity: a behavioral explanation. J. Econ. Behav. Organ. **83**, 484–501 (2012)
8. Darity, W., Jr., Goldsmith, A.H.: Social psychology, unemployment and macroeconomics. J. Econo. Perspect. **10**, 121–140 (1996)
9. De Grauwe, P.: Animal spirits and monetary policy. Econ. Theor. **47**, 423–457 (2011)
10. Medeiros, M.C., Vasconcelos, G.F., Veiga, Á., Zilberman, E.: Forecasting inflation in a data-rich environment: the benefits of machine learning methods. Journal of Business & Economic Statistics **39**, 98–119 (2021)

11. A.E.A.: J. Econ. Literat., Jel Classification System. https://www.aeaweb.org/econlit/jelCodes.php (2019)
12. Mankiw, N.: Macroeconomics, Worth Publishers (2009). https://books.google.co.ma/books?id=UT64rsFG1b0C
13. Salle, I.L.: Modeling expectations in agent-based models-an application to central bank's communication and monetary policy. Econ. Model. **46**, 130–141 (2015)
14. Polyzos, S., Abdulrahman, K., Dandu, J.: Effects of financial instability on subjective well-being: a preference-based approach. Int. J. Soc. Econ. (2021)
15. Mehreen, R., Riaz, S., Kaur, M.J., Mushtaq, A.: Predictive framework to measure mental distress caused by economic crises In: 2019 Amity International Conference on Artificial Intelligence (AICAI), pp. 577–582. IEEE (2019)
16. Co, N.T., Son, H.H., Hoang, N.T., Lien, T.T.P., Ngoc, T.M.: Comparison between ARIMA and LSTM-RNN for VN-index prediction. In: Ahram, T., Karwowski, W., Vergnano, A., Leali, F., Taiar, R. (eds.) IHSI 2020. AISC, vol. 1131, pp. 1107–1112. Springer, Cham (2020). https://doi.org/10.1007/978-3-030-39512-4_168
17. Erçen, H.İ, Özdeşer, H., Türsoy, T.: The impact of macroeconomic sustainability on exchange rate: hybrid machine-learning approach. Sustainability **14**, 5357 (2022)
18. Bentley, R., Baker, E., Simons, K., Simpson, J.A., Blakely, T.: The impact of social housing on mental health: longitudinal analyses using marginal structural models and machine learning-generated weights. Int. J. Epidemiol. **47**, 1414–1422 (2018)
19. Yoder Clark, A., Blumenfeld, N., Lal, E., Darbari, S., Northwood, S., Wadpey, A.: Using k-means cluster analysis and decision trees to highlight significant factors leading to homelessness. Mathematics **9**, 2045 (2021)
20. Huang, F., Li, S., Ding, H., Han, N., Zhu, T.: Does more moral equal less corruption? the different mediation of moral foundations between economic growth and corruption in China. Current Psychol., 1–13 (2022)
21. D'Orazio, P.: Big data and complexity: is macroeconomics heading toward a new paradigm? J. Econ. Methodol. **24**, 410–429 (2017)
22. Milunovich, G.: Forecasting Australia's real house price index: a comparison of time series and machine learning methods. J. Forecast. **39**, 1098–1118 (2020)
23. Sent, E.-M.: Sargent versus Simon: bounded rationality unbound. Camb. J. Econ. **21**, 323–338 (1997)
24. Mashkova, A.L., Demidov, A.V., Savina, O.A., Koskin, A.V., Mashkov, E.A.: Developing a complex model of experimental economy based on agent approach and open government data in distributed information-computational environment. In: Proceedings of the International Conference on Electronic Governance and Open Society: Challenges in Eurasia, pp. 27–31 (2017)
25. Ponomarenko, A.A.: A note on observational equivalence of micro assumptions on macro level. Economics 14 (2020)
26. Zhang, W., Valencia, A., Chang, N.-B.: Synergistic integration between machine learning and agent-based modeling: a multidisciplinary review. IEEE Trans. Neural Netw. Learn. Syst. (2021)
27. Kanzari, D., Nakhli, M.S., Gaies, B., Sahut, J.-M.: Predicting macro-financial instability-how relevant is sentiment? Evidence from long short-term memory networks. Res. Int. Bus. Financ. **65**, 101912 (2023)
28. Alam, M.A.Z., Yong, C.C., Mansor, N.: Predicting savings adequacy using machine learning: a behavioral economics approach. Expert Syst. Appl. **203**, 117502 (2022)
29. Rekik, Y.M., Hachicha, W., Boujelbene, Y.: Agent-based modeling and investors' behavior explanation of asset price dynamics on artificial financial markets. Procedia Econ. Finance **13**, 30–46 (2014)

Check for updates

Using Machine Learning and TF-IDF for Sentiment Analysis in Moroccan Dialect an Analytical Methodology and Comparative Study

Boudhir Anouar Abdelhakim(✉), Ben Ahmed Mohamed, and Ayanouz Soufyane

SSET Research Team C3S Laboratory FSTT, Abdelmalek Essadi University, Tangier, Morocco
{aboudhir,mbenahmed}@uae.ac.ma

Abstract. The Moroccan dialect is a linguistic area that presents special difficulties because of its complex morphology and wide range of influences. This study offers a novel technique to sentiment analysis in this dialect. Our work focuses on using machine learning methods in conjunction with Natural Language Processing (NLP) techniques, namely Term Frequency-Inverse Document Frequency (TF-IDF) for feature extraction, to effectively classify sentiment.

Given the scarcity of resources and standardized forms in Moroccan dialect, conventional sentiment analysis methods are less effective. To address this, our methodology involves rigorous preprocessing steps, including normalization, tokenization, and stemming, ensuring the refinement of input data for the machine learning models. The study utilizes a dataset comprising Moroccan tweets, classified into positive and negative sentiments, to train and test the models.

We use algorithms such as Decision Tree, Support Vector Machine, and Logistic Regression, and assess their performance using metrics like accuracy, precision, recall, and F-1 score. Our findings highlight the varying effectiveness of these models in handling sentiment analysis for a morphologically rich and unstructured language like Moroccan dialect.

This research not only contributes to the field of sentiment analysis in underrepresented languages but also opens avenues for further exploration using more advanced NLP tools and deep learning techniques. It underscores the potential and challenges of applying machine learning to dialect-specific sentiment analysis, providing valuable insights for future research in this domain.

Keywords: Sentiment Analysis · TF-IDF · Feature extraction · Machine learning · Moroccan dialect

1 Introduction

Currently, social media are really increasing in the daily life. It is a way for freely expressing our opinions about so many things or thematics. And those opinions represent a goldmine for businesses in particular, because by having a review from a client on any product proposed, it will help managers to decision making. Justly, Sentiment

M. Ben Ahmed et al. (Eds.): SCA 2023, LNNS 906, pp. 342–349, 2024.
https://doi.org/10.1007/978-3-031-53824-7_31

analysis can perform this task. By retrieving precious insights from messages, carrying out some actions for making data useful and finally knowing the polarity of texts, Sentiment Analysis became unavoidable for any industry who wants to stay ahead of their competitors. Nevertheless, when it comes to deal with some unstructured languages like Moroccan dialect; it becomes very difficult. Because of that, we proposed to effect feature extraction with TF-IDF, which can allow us to achieve the application of machine learning models like SVM or logistic regression for better classifying the sentiments.

The remaining paper is structured as: Sect. 2 present some challenges encountered in the case of Moroccan Dialect sentiment analysis, Sect. 3 refers to our methodology, Sect. 4 is about the application of classification algorithms, Sect. 5 shows the metrics used for evaluating our models, Sect. 6 highlights our results and a comparative study to another research and finally we conclude in Sect. 7.

2 Sentiment Analysis in Moroccan Dialect

The term "darija" also refers to Moroccan dialect. There is no standard written form for the dialect, and it can differ from one place to another. Also, darija is frequently expressed joined to other languages like English, French or Spanish, it can even written in many languages like Arabic, Latin or Arabizi, thus it becomes laborious to analyze the texts and makes obligated to translate and normalize in one only language.

Another hindrance is the normalized Arabic letters like ؤ، ى، ء **Hamza,**إ، أ **Alef or** ال، آل **lamalef** and the numbers presented in the orthographic darija like in Mer7ba = welcome or fer7an = happy.

And we can note finally and unfortunately the lack of specific resources to darija or the difficulty for finding the data to analyze.

3 Proposed Approach

In this section, we unveil the different steps for carrying out our sentiment analysis. We displayed in the following image an overview (Fig. 1).

3.1 Dataset

Normally, concerning this work, we were supposed to use the Moroccan Sentiment Twitter Dataset (MSTD), but the github page for accessing to this dataset wasn't available. Thus, we retrieved a dataset which is a collection of tweets, from another github page[1], and the latter is called Moroccan Sentiment Analysis Corpus. It is annotated and has two classes of sentiment: positive and negative.

1 https://raw.githubusercontent.com/ososs/Arabic-Sentiment-Analysis-corpus/master/MSAC.arff.

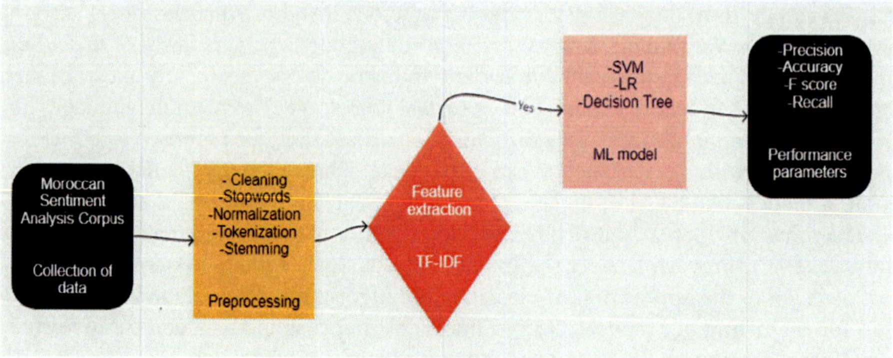

Fig. 1. Methodology

3.2 Preprocessing Techniques

Because Garbage in equals Garbage out, it is extremely crucial to perform well the preprocessing task, in particular in the case of this language. For dealing with, we followed this architecture described below.

Firstly, we started by cleaning the dataset. We removed anything which hasn't many importance for the analysis and the prediction of the polarity, like special characters, diacritics, repeated characters, numbers, punctuation and also emoji.

Secondly, we have sometimes some words which are common in any language, like prepositions, conjunctions or pronouns, but doesn't add much information to the text, they are known as stop words. We removed them by using the NLTK library which contains a built-in stop words.

Thirdly, we worked on the issue of normalization. Certain thoughts are expressed in unconventional ways. For example, some words have repeated letters, like "مبروووووك"instead of "مبروك",which indicates congratulations, emotions such as "ههههههه"which indicate laughing. Others include common spelling errors or accents. The normalization process aids in bringing the texts into compliance with accepted practices.

Fourthly, we performed the tokenization task. It aims to divide the text into pieces of data called tokens. Those tokens contain the essential information important for the analysis.

Finally, we ended by stemming task. The stemming process is used to change different tenses of words to its base form, this process is thus helpful to remove unwanted computation of words. For doing this, we applied a stemming technique called Light Stemming by using a specific library called Tashaphyne.

3.3 Feature Extraction

One of the most crucial processes to take in order to comprehend the context of the material we are working with better is feature extraction. We must convert the original text into its features so that it may be utilized for modeling after it has been cleaned. Put another way, in order to provide machine learning algorithms with numerical features, we must extract features from the raw text. We decide to use TF-IDF, or term frequency inverse document frequency, to do this.

A popular method in NLP for assessing a word's importance in a document or corpus is called TF-IDF. In essence, it compares a word's frequency within a particular document to its frequency over the entire corpus to determine how important it is. The fundamental premise is that a word is especially significant in a document if it appears more frequently inside it but less frequently throughout the corpus.

4 Classification Algorithms

Here, we show the ML model used for classifying our reviews based on their sentiments.

4.1 Logistic Regression

The method of modeling the probability of a discrete result given an input variable is known as logistic regression. A binary outcome, or something that can have two values, such as true or false, yes or no, and so on, is what most logistic regression models represent. Another name for this approach is Maximum Entropy.

4.2 Support Vector Machine

Backing A supervised technique called Vector Machine is employed for problems involving both classification and regression. Its goal is to locate a hyperplane that clearly classifies the data points in an N-dimensional space, where N is the number of features.

4.3 Decision Tree

The way this algorithm operates makes it incredibly efficient. The main concept is to partition the dataset into more manageable groups while concurrently creating the corresponding tree piece by piece. This is capable of handling numerical and categorical data.

5 Performance Parameters

In this party, we expose the metrics used for judging or estimating the quality of our ML models.

5.1 Accuracy

As indicated in Eq. 1 or, more accurately, 1.a, this is the ratio of true positives plus true negatives to the true positives plus true negatives plus false positives plus false negatives. It determines the proportion of cases that are correctly classified.

$$Accuracy = \frac{(\text{True positive} + \text{True negative})}{True\,positive + True\,negative + False\,positive + False\,negative} \tag{1}$$

$$Accuracy = \frac{\text{Number of correct predictions}}{\text{Total number of predictions}} \tag{1.a}$$

5.2 Precision

Precision attempts to answer the following question:

What percentage of positive identifications were true positives?

Precision is defined as the ratio of expected positive observations to the total number of positive observations. It is computed as follows:

$$Precision = \frac{\text{True positive}}{\text{True positive} + \text{False positive}} \tag{2}$$

$$Precision = \frac{\text{Relevant retrieved instances}}{\text{All retrieved instances}} \tag{2.a}$$

5.3 Recall

Ratio of correctly predicted positive observations to all observations in actual class yes is known as recall. It is computed as follows:

$$Recall = \frac{\text{True positive}}{\text{True positive} + \text{False negative}} \tag{3}$$

$$Recall = \frac{\text{Relevant retrieved instances}}{\text{All relevant instances}} \tag{3.a}$$

5.4 F-1score

Weighted average of recall and precision is called f-score. More important parameter than accuracy when having an uneven class distribution in data. It is calculated as follows:

$$F - score = \frac{2 * \text{Precision} * \text{Recall}}{\text{Precision} + \text{Recall}} \tag{4}$$

6 Results and Comparison Analysis

6.1 Results of the Work

In this work, we wished to use Term Frequency In-verse Document Frequency (TF-IDF) for the features extraction step in order to approach the analysis of Moroccan dialect sentences. Next, we used three machine learning algorithms—Support Vector Machine, Logistic Regression, and Decision Tree—for categorization. Using the 80–20 approach, we divided the data into training and testing sets. Ultimately, we assessed our models' performance using criteria including accuracy, precision, recall, and F-1 score. (Table 1).

It is evident that a model's performance might differ based on the metric, albeit the discrepancy between the outcomes is not very great.

Table 1. Classification Results

	SVM	LR	DT
Precision	Neg:81% Pos:81%	Neg:80% Pos:81%	Neg:83% Pos:61%
Recall	Neg:85% Pos: 76%	Neg85:% Pos:74%	Neg:53% Pos:87%
F-1 score	Neg:83% Pos: 78%	Neg:82% Pos:77%	Neg:64% Pos:71%
Accuracy	81%	80%	68%

6.2 Comparison Analysis

In this section, we perform a brief comparison between our work and a follow-up study that used Word embedding, Arabert for feature extraction, and various deep learning models for classification on an identical item.

It's important to clarify right away that we didn't utilize the same dataset. The MSTD that the authors utilized is larger than ours. While our dataset only contains two classes—positive and negative—MSTD is a collection of 12K Moroccan tweets that were annotated with four distinct classes: 2769 negative, 866 positives, 6378 objective, and 2188 sarcastic.

Second, as the following graphic illustrates, there are a lot of distinctions between TF-IDF and Word embedding. (Table 2).

Table 2. TF-IDF vs Word Embedding.

Word Embedding	**TF-IDF matrix**
Multi dimensional vector which attempts to capture a words relationship to other words	Sparse matrix where each word maps to just a single value, captures no meaning
Often trained on large external corpus	Trained without external data
Must be applied to each word individually	Can be applied to each training document at once
More memory intensive	Less memory intensive
Ideal for problems involving a single word such as a word translation	Ideal for problems with many words and larger document files

It is evident from the differences that word embedding performs better than TF-IDF and can produce the most accurate results when it comes to relevant classification by machine learning models.

Lastly, we used the Accuracy metric to show how our models performed differently from their models (Fig. 2).

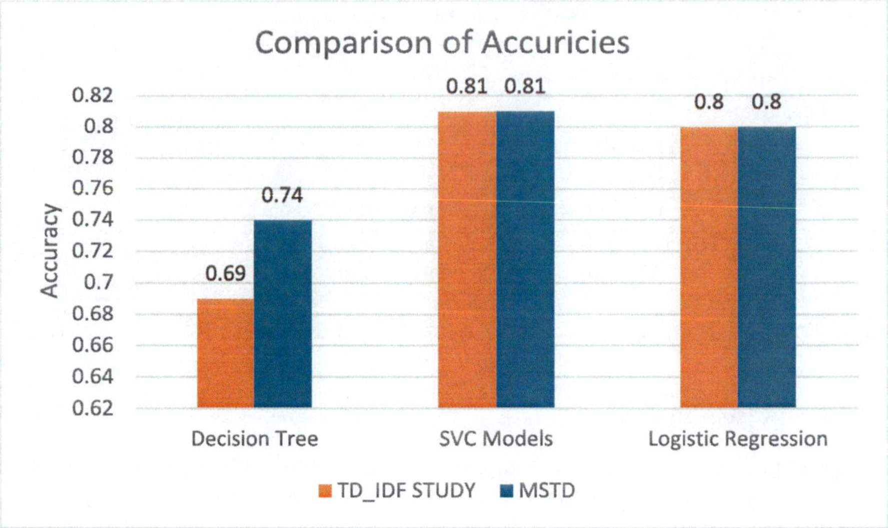

Fig. 2. Performance based on accuracy

7 Conclusion

In conclusion, our study makes significant strides in the axe of sentiment analysis for the Moroccan dialect, a linguistically complex and underrepresented language in computational linguistics. By adapting and applying NLP techniques, specifically TF-IDF for feature extraction, coupled with machine learning techniques including decision trees, logistic regression, and support vector machines, we have demonstrated a viable approach to classifying sentiments in Moroccan dialect texts.

Our findings reveal that while each algorithm has its strengths and limitations, they collectively offer promising avenues for accurately discerning sentiment in a dialect that presents unique challenges due to its unstructured nature and lack of standardization. The preprocessing steps, including tokenization, normalization, and stemming, were crucial in refining the data for more effective analysis.

This research contributes to the broader understanding of sentiment analysis in dialects and minority languages, highlighting the importance of tailored approaches for such linguistic contexts. It also underscores the potential of machine learning in uncovering insights from dialect-specific data, which is often overlooked in mainstream NLP research.

Looking forward, there is ample scope for enhancing this research by integrating more advanced NLP tools and exploring deep learning models like Long Short Term Memory or Neural Networks. Such future endeavors could further refine the accuracy and efficiency of sentiment analysis in the Moroccan dialect and other similar languages, potentially expanding the applicability of NLP in diverse linguistic landscapes.

References

The International Conference on Digital Age & Technological Advances for sustainable Development: 'Sentiment Analysis through Word embedding using AraBERT: Moroccan Dialect use case", ©2021

Ravinder Ahuja, Aakarsha Chug,Shruti Kohli,Shaurya Gupta, and Pratyush Ahuja: "The impact of Features extraction on the sentiment analysis", International Conference on Pervasive Computing Advances and Applications – PerCAA 2019

Mouaad Errami, Mohamed Amine Ouassil, Rabia Rachidi, Bouchaib Cherradi, Soufiane Hamida, Abdelhadi Raihani: " Sentiment Analysis on Moroccan Dialect based on ML and Social Media Content Detection", (IJACSA) International Journal of Advanced Computer Science and Applications, vol. 14, No. 3, 2023

Soukaina MIH1, Brahim AIT BEN ALI , Ismail EL BAZI , Sara AREZKI , Nabil LAACHFOUBI.: " MSTD: Moroccan Sentiment Twitter Dataset", (IJACSA) International Journal of Advanced Computer Science and Applications, Vol. 11, No. 10, 2020

MOUHOUBI Azzedine,GHEFFARI Mohammed Abdelfattah,' Analyse de sentiments dans la langue arabe en utilisant differentes approches, presente en vue de l'obtention du diplome de Master,2019–2020

EL BERGUI Adam :"Sentiment Analysis for Moroccan Dialect",September 2019

Smart Healthcare Systems

Study of Correlation Between Intestinal Parasitism and the Nutritional Status of Children at the Moulay Abdellah Hospital of Sale (MOROCCO)

Jaouad Mostafi[1](✉), D ounia Bassir[1], Saïd Oulkheir[2], Hamid El Oirdi[1], Khadija El Kharrim[1], and Driss Belghyti[1]

[1] Laboratory: Natural Resources and Sustainable Development, Faculty of Sciences, Ibn Tofail Kenitra University, Kenitra, Morocco
mostafijaouad1@gmail.com

[2] Team of Biotechnology, Health and Environment Laboratory of Sciences of Health and Environment, Higher Institute of Nursing Professions and Health Techniques, Agadir, Morocco

Abstract. Intestinal parasites are a public health problem. They are at the origin of harmful damage, including malnutrition, which is the main cause of ill health. In order to evaluate this impact, a correlation study between parasitism and nutritional status was carried out at the Moulay Abdellah Center of Sale. This was a prospective 5-month study (May - September 2007) involving children aged 10 months to 15 years and hospitalized in the paediatric department regardless of their reason for hospitalization. The assessment of parasitism was based on three stool examinations plus Ritchie and Willis. The assessment of nutritional status was based on the scale of the National Center for Health Statistics (NCHS). The parasitological results highlighted an overall prevalence of 39.2% and 8 parasites including 5 pathogens: Giardia intestinalis:13.3%, Entamoeba histolytica/dispar: 9.2%, Enterobius vermicularis: 5.8%, Ascaris lumbricoides: 2.5% and Hymenolepis nana: 1.7%. Nutritional results showed prevalences of 17.1%, 22.6% and 27.9% respectively of stunting, wasting and underweight. Statistically, no association is demonstrated, but in proportions, the parasitized seem in the case of stunting and underweight more malnourished than their counterparts: 21.4% against 14.5% and 31% against 26.1% respectively. In conclusion: 39.2% are parasitized, 22.6% are too thin, 27.9% and 17.1% have respectively in relation to their age an insufficient weight and a small height. But in terms of impact on health, this parasitism is proving to have no significant role on this worrying nutritional situation. We hope that this study will further inspire actors to improve strategies to best manage situations.

Keywords: Intestinal parasitism · Stunting · Underweight · Wasting · Moulay Abdellah Hospital of Sale/Morocco

M. Ben Ahmed et al. (Eds.): SCA 2023, LNNS 906, pp. 353–361, 2024.
https://doi.org/10.1007/978-3-031-53824-7_32

1 Introduction

Intestinal parasites are endemic throughout the world and are a real public health problem. They are often the cause of considerable damage such as malnutrition which stunts the growth of the child and negatively influences his school performance [1]. This pathology makes individuals more vulnerable to disease and death, and conversely, the latter do the opposite by promoting malnutrition. This is especially true for infections with digestive parasites, which prevent the assimilation of nutrients [2]. According to WHO, this phenomenon of malnutrition affects in its severe acute form nearly 20 million children under 5 years of age, a large proportion of whom live in the African Region and the South-East Asia Region [3]. According to recent data from the World of 2020, stunting alone affected 149.2 million (22%) of these children, of whom almost 3/4 lived in only two regions (Central Asia + South Asia and sub-Saharan Africa), wasting, on his part, affected 45.4 million (6.7%), of which nearly 1/4 lived in sub-Saharan Africa and more than half in South Asia [4]. In Morocco, which is not spared, malnutrition is still a public health problem, particularly affecting children. This undernutrition concerns in particular protein-energy malnutrition (16.5% stunting, 3.1% underweight, 3% wasting) [5]. The purpose of our study is to evaluate the association between this phenomenon and intestinal parasitosis through a correlation study between intestinal parasitism and nutritional status, assessed by three anthropometric indices (M/A, W/A and W/H) of a population of children hospitalized in the paediatric department.

2 Patients, Materials and Methods

2.1 Type, Location, Duration of Study

This was a prospective study, spread over a period of 5 months (May - September 2007), targeting a population of 120 children, aged between 10 months and 15 years, hospitalized in the pediatric department of the Moulay Abdellah hospital of Sale-Morocco. This study is divided into two parts:

- A copro-parasitology study, aimed at estimating the parasite prevalence, from stool analyzes collected in the hospital laboratory,
- A nutritional study, aimed at estimating nutritional status, based on anthropometric data (age, sex, weight and height) and collected from the registers of the pediatric department.

2.2 Inclusion Criteria

As an inclusion factor, the study took into account all children hospitalized in the pediatric department, whatever their reason for hospitalization, and who had donated their stools for parasitic coprology.

2.3 Exclusion Criteria

Due to the lack of weight and/or height data for a few, or aberrations for others, 9 children were excluded from the nutritional study to retain, in the case of the H/A stunting index and that of W/A underweight, only 111 individuals. But in the case of the W/H index of wasting, it only covered 62 children, the rest are excluded from the study due to the lack of reference data beyond the age of 8.5 years [6].

2.4 Estimating the Prevalences of Intestinal Parasitism and Forms of Malnutrition

Estimated Prevalence of Intestinal Parasitism

Stool Collection and Harvesting

Three stool samples were taken and collected in plastic pots at the rate of 3 successive samples per person.

Parasitological Techniques and Examinations

The stools collected in the morning were initially subjected to direct microscopic examination in physiological water, and another at Lugol, then an analysis after 2 concentration techniques: Ritchie and Willis.

Estimating the Prevalences of Forms of Malnutrition

Calculation of Anthropometric Indices

To assess nutritional status, three anthropometric indices were calculated: W/A, H/A and W/H. The values of these indices were calculated in relation to the international reference population defined by the NCHS. Results are expressed in Z- Score or standard deviation (SD) which is calculated by differentiating the measured value from the median value for the reference population and dividing the result obtained by the standard deviation for the reference population [3]. A Z-score value < -2 SD in a child for W/A, H/A and W/H, classifies him as moderately underweight, moderately stunted, or moderately wasted, respectively; and a Z-score value < -3 SD classifies him as presenting severe malnutrition [6].

2.5 Statistical Analysis

Anthro software was used to calculate anthropometric indices. Data analysis was done using SPSS version 20 software. The materiality level was set at 0.05.

2.6 Ethical Aspects

The study was carried out after obtaining permission from the Ministry of Health and coordinating with the hospital director. The anonymity of the participants was guaranteed.

3 Résultats

3.1 Prévalence Parasitaire

It appears from all the parasitological analyzes that 47 children among the 120 examined are carriers of at least one parasite, i.e. an overall parasite prevalence of 39.2%. Specifically, eight parasitic species, including 5 pathogenic (in bold), were detected. These are ranked in descending order of prevalence as follows: **Giardia intestinalis**: 13.3% (16/120), Blastocystis hominis: 12.5% (15/120), Entamoeba coli: 10% (12/120), **Entamoeba histolytica**/dispar: 9.2% (11/120), **Enterobius vermicularis**: 5.8% (7/120), **Ascaris Lumbricoids**: 2.5% (3/120), **Hymenolepis nana**: 1.7% (2/120) and Endolimax nana: 0.8% (1/120).

3.2 Prevalence of Malnutrition

According to anthropometric indices, 17.1% of children suffer from moderate stunting (H/A < -2), of which 6.3% suffer from severe stunting (H/A < -3), 22.6% suffer from moderate wasting (W/H < -2), of which 11.3% suffer from severe wasting (W/H < -3), and 27.9% suffer from moderate underweight (W/A < -2), of which 9.9% suffer from severe underweight (W/A < -3) (Table 1).

Table 1. Prevalence of forms of malnutrition according to the three anthropometric indices: weight-for-age (W/A), height-for-age (H/A) and weight-for-height (W/H).

Type of malnutrition	Population size (N)	Z-score	Number of cases	Form of malnutrition	Prevalence
Stunting	111	< - 2	19	Moderate	17,1%
		< - 3	07	Severe	6,3%
Underweight	111	< - 2	31	Moderate	27,9%
		< - 3	11	Severe	9,9%
Wasting	62	< - 2	14	Moderate	22,6%
		< - 3	07	Severe	11,3%

3.3 Variation of Malnutrition According to Intestinal Parasitism

Variation in Stunting

We note from the statistical analysis of Table 2 below, that parasitized children are, without statistical significance, more affected by stunting than non-parasitized children (21.4% against 14.5%).

Variation in Wasting

We note from the statistical analysis of Table 3 below, that the two categories of children

Table 2. Variation in the prevalence of stunting according to intestinal parasitism (N = 111).

Parasitism	Number	H/A < - 2SD	H/A ≥ - 2 SD	Positivity	P-value
Parasitized	42	09	33	21,4%	**0,347** Ns
Unparasitized	69	10	59	14,5%	
Total	111	19	92	17,1%	

H/A: height-for-age; SD: standard deviation; ns: not significant

Table 3. Variation in the prevalence of wasting according to parasitism (N = 62)

Parasitism	Number	W/H < - 2 SD	W/H ≥ - 2 SD	Positivity	P-value
Parasitized	22	05	17	22,7%	**0,984** Ns
unparasitized	40	69	31	22,5%	
Total	62	14	48	22,6%	

W/H: weight-for-height; SD: standard deviation; ns: not significant

studied (parasitized and unparasitized) are equally affected by wasting; i.e. respectively 22.7% against 22.5%.

Variation of Underweight
We note from the statistical analysis of Table 4 below, that parasitized children are, without statistical significance, more affected by underweight than non-parasitized children (31% against 26.1%).

Table 4. Variation in the prevalence of underweight according to parasitism (N = 111).

Parasitism	Number	W/A < - 2 SD	W/A ≥ - 2 SD	Positivity	P-valeur
Parasitized	42	13	29	31%	**0,579** Ns
Unparasitized	69	18	51	26,1%	
Total	111	31	80	27,9%	

W/A: weight-for-age; SD: standard deviation; ns: not significant

4 Discussion

In total parasitological examinations, 47/120 children are found to have one or more parasites in their digestive tract. That is an overall prevalence of 39.2%, that's more than one in three children. Specifically, eight species, all transmitted by the fecal-oral route, have been identified, five of which are pathogenic (in bold): ***Giardia intestinalis*** (13.3%),

Blastocystis hominis (12.5%), *Entamoeba coli* (10%), ***Entamoeba histolytica/dispar*** (9.2%), ***Enterobius vermicularis*** (5.8%), ***Ascaris lumbricoids*** (2.5%), ***Hymenolepis nana*** (1.7%) and *Endolimax nana* (0.8%)**.** According to the nutritional study, the results showed that 17.1% of children suffer from moderate stunting (H/A < -2) (i.e. they measure less than the normal average height) of which 6.3% suffer from severe form (H/A < -3), 22.6% suffer from moderate wasting (W/H < -2) (i.e. they are below the weight they should have for their height) of which 11.3% suffer from severe form (W/H < -3) and 27.9% are moderately underweight (W/A < -2) (i.e. they weigh less than the normal average weight) of which 9.9% are severe (W/A < -3) (Table 1). Taking into account the WHO standard of 2.3% for the moderate form of the reference population, these three forms of malnutrition are respectively 7.4 times, 9.8 times and 12 times higher than that expected in a healthy and well-nourished population [6]. Comparing to a few studies, our results seem more or less different from those of Ait Messaad (2015) of Sale [6], recorded in a similar population at the same Moulay Abdellah Hospital: i.e. 17.1% against 20.5% for stunting, 27.9% against 16.87% for underweight and 22.6% against 29.55% for wasting. On the other hand, they seem far below those of El Guamri (2010) of kenitra of Morocco [9], recorded in children of the same category at the Idrissi Hospital: i.e. respectively 17.1% against 30%, 27.9% against 63.8% and 22.6% against 67.5%. Compared abroad, our results are slightly different from those found in Senegal in preschool children: i.e. 20% stunting, 24.6% underweight and 25% wasting [10], on the other hand, they are very different from those of Mr. Djibril B. Maiga, (2019) of Mali [10], which recorded rates of 52.7%, 42.1% and 49.1% respectively. Ditto for the results of anthropometric survey in Benin children aged 18 to 36 months [11]: 42.6%, 8% and 36.6% respectively.

Statistically, no correlation between the parasitism studied and nutritional status has been demonstrated (P > 0.05); indeed, and with the exception of wasting where the rates are totally comparable between the two categories of children (22.7% against 22.5%) (Table 3), parasitized individuals are more malnourished in the case of stunting and underweight than unparasitized individuals: i.e. respectively 21.4% against 14.5% (Table 2) and 31% against 26.1% (Table 4). However, these differences remain statistically far from significant. This score is consistent with several studies, for example those of El Guamri et al., (2010) [9] and kasmi and Saidouni (2016) [11] from Morocco, Savadogo et al., (2014) from Burkina Faso [7], Yapi et al., (2005) and Mathieu et al., (2015) from Cote d'Ivoire [9, 10].

A similar finding was also recorded in Burkina Faso during a study of inhabitants aged 0 to 73 years: In this study, intestinal parasites are found to affect the majority of the population regardless of the nutritional situation: normal or deficient [11]. Indeed, in the work of. Yapi et al., and El Guamri mentioned above, parasitism had no consequence either on the anthropometric status or on the hematology profile although parasitic infestations were very high (+68% and + 50% respectively). In contrast, this score discords in the case of stunting, with the study by Hossain Haratipour et al., (2016) of Iran [8] and Diouf et al., (2020) of Senegal [9], and in the case of underweight, with that of Ait Messaad (2014) of Morocco [6]. Indeed, malnutrition is not exclusively or necessarily influenced by intestinal parasitism, but it has, according to the WHO, many causes, closely linked in developing countries to socio-economic factors [10]: during a study analyzing the determinants of malnutrition among preschool children in Morocco, based on the National Survey of Household Consumption and Expenditure carried out in 2001–2002, The results revealed that the main determining factor was economic. In this study, individuals from the poorest households and living in rural areas were found to be the most affected [11]. Indeed, the intestinal parasitism factor is not mentioned during a Malian study at the pediatrics department of the CHU-GT, where malnutrition is, rather, associated with early weaning, mothers' level of education and father's professional activity [10]. A similar finding was also recorded in Chad after the national survey in 2021 where the poor quality of food intake, the high prevalence of childhood diseases (malaria, diarrhoea and acute respiratory infection), low measles vaccination coverage, poor hygiene conditions and sanitation were considered as determinants or explanatory factors of the nutritional situation [16].

5 Conclusion

Despite the efforts made, intestinal parasites and malnutrition still represent a real public health in Morocco. The levels of their prevalence are a concern for our situation. In general, 39.2% are parasitized (more than one child in three), 22.6% are too thin for their height, 27.9% and 17.1% have respectively in relation to their age, a very underweight and a height too small. Despite this high degree of infestation, this nutritional status is statistically dissociated from intestinal parasitism. But in proportions, and apart from statistical significances, parasitized individuals have, in the case of stunting and underweight, tend to be more undernourished than unparasitized individuals. This situation requires continuous mobilization that must consequently improve strategies for the promotion of households and manage situations as well as possible in order to guarantee not only adequate nutrition but also to associate an effective approach against intestinal parasites.

References

1. Kyambikwa Bisangamo, C., Jabari Mutwa, P., Mulongo Mbarambara, P.: Profil des parasitoses intestinales chez les enfants d'âge scolaire de Kiliba (est de la RD Congo). Med Sante Trop **27**, 209–213 (2017). https://doi.org/10.1684/mst.2017.0686
2. Roudart, L.: (2007) Alimentation et malnutrition dans le monde. Situation et perspectives. Economie Humanisme 380 :19-25
3. OMS (2007) Lignes directrices : mises à jour de la prise en charge de la malnutrition aiguë sévère chez le nourrisson et chez l'enfant. Genève
4. FAO, FIDA, UNICEF, PAM et OMS (2021). Résumé de L'état de la sécurité alimentaire et de la nutrition dans le monde 2021. Transformer les systèmes alimentaires pour que la sécurité alimentaire, une meilleure nutrition et une alimentation saine et abordable soient une réalité pour tous. Rome, FAO.https://doi.org/10.4060/cb5409fr
5. Sellam, E.B., Bour, A.: Nutritional status of children aged 6 to 60 months in the prefecture of Oujda-angad in Morocco. BMSAP **27**, 56–63 (2015). https://doi.org/10.1007/s13219-015-0122-9
6. Aboussaleh, Y., et Ahami A.O.T.: (2005) Comparaison des mesures anthropométriques des enfants scolaires selon leur milieu de résidence : Etude dans la province de Kenitra au Nord-Ouest du Maroc. Antropo 9 : 89–93. www.didac.ehu.es/antropo

7. Recommandations pour la collecte des données, leur analyse et la préparation des rapports sur les indicateurs anthropométriques chez les enfants âgés de moins de 5 ans [Recommendations for data collection, analysis and reporting on anthropometric indicators in children under 5 years old]. Genève : Organisation mondiale de la Santé et le Fonds des Nations Unies pour l'enfance (UNICEF) (2019). Licence : CC BY-NC-SA 3.0 IGO
8. Kouassi, K.F., Yeo, A., Ake-Tano, O., Beugre, J.B.: Rang de naissance et variation des paramètres anthropométriques dans une population jeune mélanoderme originaire de Côte d'Ivoire. Sociol. KASA BYA KASA **19**, 129–142 (2011)
9. Ait Messaad, S.: (2015) Etude de la charge parasitaire dans les eaux usées brutes et le parasitisme intestinal en relation avec le statut anthropométrique chez les enfants hospitalisés au centre hospitalier Moulay Abdallah de ville de Salé (Maroc). Thèse de Doctorat. Faculté des Sciences Kénitra, Maroc
10. El Guamri, Y.: (2010) Etude de l'impact des eaux usées sur l'environnement et enquêtes épidémiologiques du parasitisme intestinal dans la région du Gharb (kénitra - Maroc). Thèse de Doctorat. Faculté des Sciences Kénitra, Maroc
11. Sougou, N., Boëtsch, G.: Alimentation et croissance des jeunes enfants Peuls à Widou Thiengoly (Ferlo- Sénégal). BMSAP **28**, 145–154 (2016). https://doi.org/10.1007/s13219-016-0158-5
12. Djibril, M., Maiga, B.: (2019) Evaluation de l'état nutritionnel des enfants de 02 à 59mois hospitalisés à l'unité de pédiatrie du CSRéf de Kalaban Coro. Thèse de Docteur en Médecine (Diplôme d'Etat). Faculté de Médecine et d'Odontostomatologie, Mali
13. Enquête anthropométrique sur l'état nutritionnel des enfants de 18 à 36 mois dans les départements de l'Atacora et de la Donga. Rapport de base. République du Bénin. Cotonou, Juillet 2001
14. Kasmi, H., Saidouni, A.: Etude de la prévalence des protozooses intestinales diagnostiquées au sein du laboratoire de parasitologie-mycologie du chu de Tlemcen. Université Aboubekr belkaid, Tlemcen, Mémoire de diplôme de Docteur en Pharmacie (2016)
15. Savadogo, B., et al.: (2014) Statut nutritionnel et parasitoses (intestinales et urinaires) chez les enfants d'âge scolaire au Burkina-Faso : cas des écoles de Yamtenga, Koubri et Daguilma. Science et Technique, Sciences de la Santé 37(1 et 2) : 45–56
16. Yapi, H.F.B., Ahiboh, H., Monnet, D., Yapo, AE.: (2005) Parasites intestinaux, profil hématologique et statut anthropométrique de l'enfant scolarisé en Côte d'Ivoire. Cahiers d'études et de recherches francophones / Santé 15 (1) : 17-21.
17. Aka Ahizi, M.: (2015) Profil épidémiologique des parasitoses intestinales chez les enfants malnutris à Abidjan. Thèse de Diplôme d'état de Docteur en Pharmacie No 1728/15, Côte d'Ivoire
18. Savadogo, B.: (2023) Statut Nutritionnel et parasitoses intestinales chez des habitants de la commune rurale de Niou, Burkina Faso. Int. J. Innovation Appl. Stud., ISSN 2028–9324 Vol. 39 No. 2: 751–759
19. Haratipour, H., Sohrabi, M.B., Zolfaghari, P., Nezakati, E., Yahyaei, E., Rezvani, S (2016) The Relationship between Malnutrition and Intestinal Parasitic Infections among Preschool Children in East Area of Iran. Int J Pediatr 4(6) : 2011-18. Diouf JBN, Kane M, Sow D, Sougou NM, Oumy Kaltome B, Babacar F, Ndiaye O (2020) Parasitoses intestinales et statut nutritionnel chez l'enfant à Guédiawaye au Sénégal. European Scientific Journal May Vol.16, No.15 ISSN: 1857-7881 (Print) e - ISSN 1857-7431. Doi:https://doi.org/10.19044/esj.2020.v16n15p115 URL:https://doi.org/10.19044/esj.2020.v16n15p115.
20. Mamadou Bouné Coulibaly (2020) Facteurs de risque de la malnutrition chez les enfants de 6 à 59 mois hospitalises au département de pédiatrie du CHU Gabriel TOURE. Thèse de Docteur en Médecine (Diplôme d'Etat). Université des Sciences, des Techniques et des Technologies de Bamako Faculté de Médecine et d'Odontostomatologie, Mali

21. Nait, L., Kaid, A., Ezzrar, A., et Fekkaklouhai, S.: (2022) La malnutrition des enfants de moins de cinq ans au Maroc dans les années 2000 : cause culturelle ou économique ? Revue Réflexions Économiques 2 :143–154. doi: https://doi.org/10.34874/IMIST.PRSM/refeco-i2.32272
22. Tchad : Enquête nationale de nutrition et de mortalité rétrospective SMART 2021 (Rapport final - Février 2022)

An IoT-Based Smart Home for Elderly Suffering from Dementia

Mhd. Wasim Raed[1](✉), Ilham Huseyinov[2], Ghina Ozdemir[1], Igor Kotenko[3], and Elena Fedorchenko[3]

[1] Department of Computer Engineering, Istanbul Aydin University, Istanbul, Turkey
mwraad12@gmail.com, ghinaozdamir@stu.aydin.edu.tr

[2] Software Engineering Department, Istanbul Aydin University, Istanbul, Turkey
ilhamhuseyinov@aydin.edu.tr

[3] St. Petersburg Federal Research Center of the Russian Academy of Sciences, St. Petersburg, Russia
{doynikova,ivkote}@comsec.spb.ru

Abstract. There exists a global challenge related to the boosting number of elderly suffering from chronic diseases like Dementia (EWD). Hence, there is a drastic need for cost-effective disruptive technologies that enable and guarantee the quality of life for EWD via implementing telehealth. The purpose of this paper is to propose a telehealth system based on the Internet of Things for the well-being of the elderly suffering from Dementia in a smart home setting. We propose a smart home solution as a whole and a solution based on an NFC-based working prototype of a pill intake in a box for monitoring and recording the proper intake of prescribed medication for the elderly suffering from chronic diseases. A smart home solution was effectively implemented using the Cisco Packet Tracer simulator 7.3. The proposed IoT-based solution ensures the safe wandering of the elderly in a smart home setting using a Cisco packet tracer simulation tool.

Keywords: IoT · telemedicine · telehealth · smart home · dementia · NFC · COPD · RFID

1 Introduction

It has been predicted during the previous few decades that in a few years there will be more adults 65 and older than children under the age of five [1]. The increased health risks come with longer life expectancy and place a heavy weight on the present healthcare systems. Telemedicine is the practice of delivering medical care and other healthcare services through the Internet or other remote means. The rapid development of disruptive technologies like wireless biosensors and Radio Frequency Identification (RFID), notably in smart homes, allows providing telerehabilitation to patients who are housebound using information and communication technology [2].

The Internet of Things (IoT) is becoming more and more significant for home telemedicine. It enables the seamless integration of several medical sensors and

M. Ben Ahmed et al. (Eds.): SCA 2023, LNNS 906, pp. 362–371, 2024.
https://doi.org/10.1007/978-3-031-53824-7_33

equipment, enabling real-time data collection and remote patient monitoring. We have recently observed that IoT is demonstrating significant advancements in the sphere of contemporary wireless telecommunication [3,4,6]. The ability of wearable health monitoring systems to enable continuous monitoring of vital signs and health indicators, allowing for the early detection of health issues and potentially preventing serious repercussions, is one of their main benefits. Additionally, wearable devices can offer useful insights into the relationship between a person's lifestyle and health, empowering them to make more informed choices regarding their health-related habits [7]. Besides, wearable technology can enable the development of smart homes that are specifically created to fulfill the needs of older people with chronic ailments, such as dementia, when paired with other cutting-edge technologies like RFID and ambient intelligence. These "smart" homes can offer a variety of features and services, such as automated medication reminders, remote monitoring of vital signs, and emergency response systems. Users of smart homes may simply monitor and manage home appliances online [8,9]. See Fig. 1 for a futuristic model of a telehealth system for the elderly suffering from chronic disease using wearables.

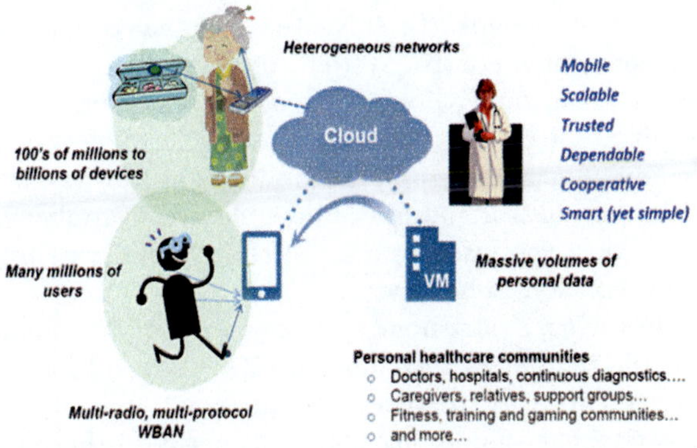

Fig. 1. A futuristic model of telehealth for elderly suffering from chronic diseases

The tremendous improvements in wearable technology for smart homes are primarily intended to improve the quality of life (QoL) of the elderly. The authors aim to develop smart home infrastructure for the elderly to decrease health risks. In a previous work of ours, the implementation of an RFID-based battery-free system for locating older people with dementia based on room-level localization and identifying the beginning of dangerous wandering was introduced [5]. The main contributions of this paper are the following: state of the art advancements in telemedicine using IoT; a working prototype of an NFC based pill intake recording system for elderly; a low-cost technique for localizing elderly at choke point at the doorway implemented using low-cost proximity NFC readers tailored for smart homes; a smart home for elderly suffering from Dementia implemented using Cisco Packet tracer simulator.

The novelty of the proposed solutions consists in (1) the proposed infrastructure of the smart home for elderly people incorporating RFID readers and tags, IoT devices and gateways, and microcontrollers; (2) and in the developed low-cost and low energy consumption NFC-based solutions for elderly location monitoring and supplying with pills.

The paper is organized as follows. Section 2 analyzes the related research. Section 3 introduced the proposed solution. Section 4 provides and discusses the obtained results.

2 Related Work

Due to our addiction to technology, smart homes have grown significantly in popularity in recent years. All people's lives will be improved by smart homes, but notably those of regular, non-disabled individuals. The goal of smart houses is to conserve energy while allowing occupants to control the lighting, heating, cooling, door locks, and coffee makers. However, the advantages of smart home technology will not be available to those with disabilities. For older folks and physically impaired people, being able to control gadgets while you're setting them up is a great benefit. To solve the problem of disabled people, the researchers in [10] present a control system for disabled people based on smart home appliances that has smart power strips, smart power strips, digital assistants like Apple Siri, Google Assistant, Microsoft Cortana, and Amazon Alexa to record voice commands spoken in a much more natural way by a person with physical disabilities to control common home electrical appliances so they can be turned on or off with the least amount of effort. The system is built using a Raspberry Pi, and Amazon Echo device.

In many people's lives, smart homes have grown very popular. Many people now aspire to be able to control their homes with a single touch on their smartphone. The primary concerns, however, are financial loss, data confidentiality, and privacy. Existing smart homes are highly vulnerable, and numerous assaults have been made against them. As a result, the security of these systems is a significant issue that requires analysis and resolution. In [11], the researchers studied the potential system threats. In this regard, they divide the dangers into two types of attacks on smart homes. The first type is internal system threats, such as malfunctions with household appliances, power and internet issues, software bugs, and the leakage of private data, and the second type is external system threats, including man in the middle, denial of service, malicious injection, and eavesdropping. With all of these hazards present, several security requirements have been proposed to raise the security level and reduce vulnerability to threats. While device availability, physical security, and device authentication will be crucial in reducing internal dangers, data encryption, network monitoring, and user authentication may help to decrease the exposure to external threats. Additionally, in [11], the authors outline the necessity of enhancing the secure connection between the smart home and the smartphone used for remote control of home appliances, they outline several unsolved research problems for smart home-smartphone systems. In [12], the authors propose a solution for the security monitoring of IoT systems.

Furthermore, a smart interior environment must have localization because it permits learning about resident behavior, especially that of people who require long-term care or monitoring. For instance, knowing where an elderly individual who lives alone is, enables off-site caretakers to watch the activities of daily life to notice risky events like long stays in bed or an odd area, and prevent harmful wandering which could lead to elderly falls [13].

The process of identifying a target device (TD) indoors, such as in smart buildings, and homes, is known as indoor localization. It is an important factor in different applications, such as those related to industry, surveillance, and healthcare. For instance, indoor localization in the healthcare industry can make it easier for the elderly and those who are blind to navigate a hospital [14].

Furthermore, wireless sensors and RFID is a technique that locates and tracks items using radio waves. An RFID tag, an RFID reader, and a backend system make up the three essential parts of an RFID system. are currently regarded as enabling technologies for interior localization when using IoT applications in an elderly-friendly smart home. Electromagnetic coupling in the Radio Frequency (RF) region of the electromagnetic spectrum is used by RFID technology. RFID readers and RFID tags are the two components of an RFID system. The reader receives the incoming signal from the tags and processes it thanks to the antennae. RFID tags come in three different battery-powered states: active, semi-active, and inactive. RFID has recently become widely employed in inventory and logistics. Due to its low cost, great reliability, and applicability in auto-identification applications, passive RFID is highly common [15]. RFID is frequently utilized in supply chains and inventory management. Passive RFID has recently been widely used to monitor people's behavior, especially the elderly, due to its very appealing properties including low cost, high reliability and unobtrusiveness [16].

3 The Proposed Solution

This section presents our proposed solution to address the aforementioned telemedicine issues. We propose a telemedicine smart home-based system to monitor the wandering of the elderly suffering from dementia in a smart home setting, utilizing choke point RFID technology in addition to cameras. The main purpose of the proposed system is to ensure the safety of the elderly and send warnings to a caregiver in case of emergency. The proposed solution of the smart home incorporates monitoring elements that allow detection of home or room leaving or not leaving, safety elements that allow detecting the high smoke level, and healthcare elements that remind the sleeping schedule and automate pill taking. The flowchart of solution operation is represented in Fig. 2.

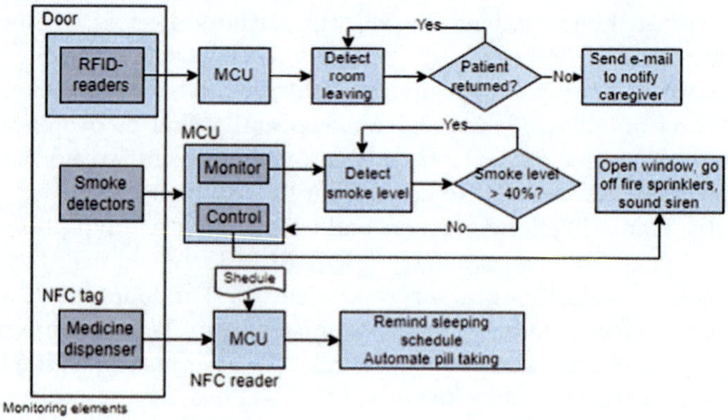

Fig. 2. The flowchart of solution operation

The medicine dispenser prototype is designed to provide the patients with their medicine arranged in a manner that shows the required pills of the day while capturing their activity when any intake has occurred. Our design reference comes from pill organizers that are found in most pharmacies nowadays, there are a lot of different types and shapes of them. We found that rectangular, or circular shaped-pill organizers could meet our requirements. We were indecisive about which one is a more suitable solution.

A prototype of an Arduino Uno Microcontroller interfaced to an NFC reader to detect pill intake is represented in Fig. 3.

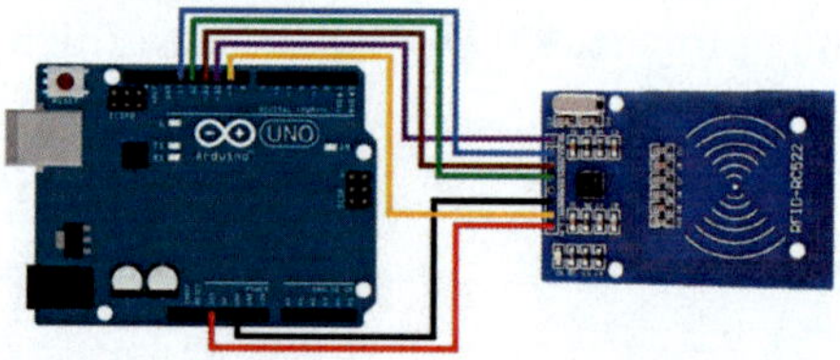

Fig. 3. Arduino microcontroller interface to NFC reader

The Arduino microcontroller controls a DC motor which enables the circular shape pill organizer to rotate. The aim is to read the tag when patients take their pills. Therefore, the reader should read the tag when the section is opened. Knowing that the reader has a very short reading distance and can only read one tag at a time, the only design solution is to move the tags in the reader's reading field as shown in Fig. 4.

First, at the beginning of the day, the microcontroller will wait for the patient to open the organizer box lid. If he succeeds in opening it, the NFC tag on the lid will be read by the NFC shield and the microcontroller will store the data,

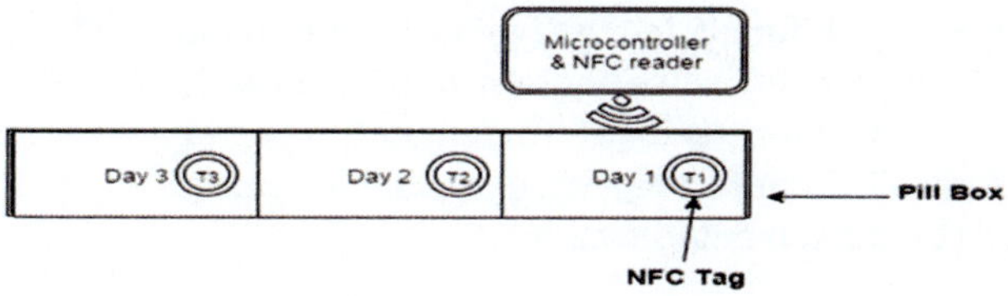

Fig. 4. Communication between the NFC reader and the NFC tags

and after a certain delay, the box will rotate to the next day. If the patient fails to take the pills within that day, the microcontroller will store it as "not taken" and it will rotate to the next day. Finally, at the end of the week and before the patient goes to see his doctor, he will place an NFC token tag on the NFC reader, which will collect all of the data that has been stored and will give the token tag to the doctor. The doctor can get the data using an NFC reader or using a smartphone with NFC capabilities. See Fig. 5.

Fig. 5. The final structure of the NFC based pill intake

To demonstrate the feasibility of our technique, we implemented our proposed system using the Cisco Packet Tracer Simulator Tool version 7.3. Cisco packet tracer offers a large selection of IoT devices and networked smart components that can be utilized to build smart homes. Additionally, it opens the door for modifying ambient conditions and tracking the impact on sensors. It can also work as a smart device hybrid network that combines real IoT-connected appliances with a virtual network [17,18]. The main objectives of our proposed smart home system are the following: (1) to boost the quality of life of the elderly suffering from dementia in a home setting; (2) to enhance security and efficient utilization of energy; (2) to notify a caregiver when the elderly leaves home for a time longer than usual; (3) to notify a caregiver or the elderly relative if the elderly enters the bedroom or bathroom and does not leave for a long period, which could insinuate the onset of a problem; (4) to remind the elderly of taking pills periodically according to the doctor prescription; (5) to provide access to the camera for the caregiver or the elderly relative in case of emergency; (5) to remind the elderly of the sleeping schedule; (6) to ensure the proper action of the smart home safety system; (7) to help individuals to conserve energy according to their needs. Many different IoT gadgets that can be utilized to make a home smart. These include things like motion detectors, solar panels, smart doors, sprinklers, and many more things. To remotely control these gadgets from a smartphone, connect them to the Internet using a home gateway. Figure 6 shows an overview of the implemented system.

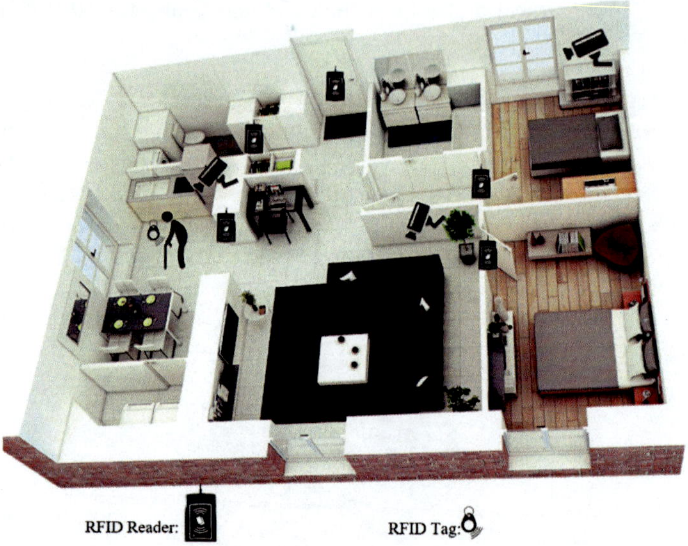

Fig. 6. Emulation of a smart house for the Elderly

It represents a cloud-connected simulation of the Alzheimer's Elderly Residence. RFID Readers and tags, cameras, sirens, temperature monitors, thermostats, smoke detectors, fire sprinklers, computers, phones, and other appliances are just a few of the different parts that make up the elderly's home. The home gateway and microcontrollers (MCUs) keep an eye on how the sensors and smart gadgets are doing. Each of the three RFID readers is positioned in a location where it may detect the tag when the patient passes by. The bedroom door and bathroom door are covered by two RFID readers, while the exit door is covered by one reader. For the safety of the patient, a smoke detector is used to detect fire. If the fire is sensed, the window will open, the fire sprinklers will go off, and the siren will sound (Fig. 7).

4 Results and Discussion

The RFID readers will detect if a tag with ID=1000 passes, it will change status to 1. The "Thing" will detect the status change. If the patient enters a room and does not leave for a certain time, it will notify the caregiver via email.

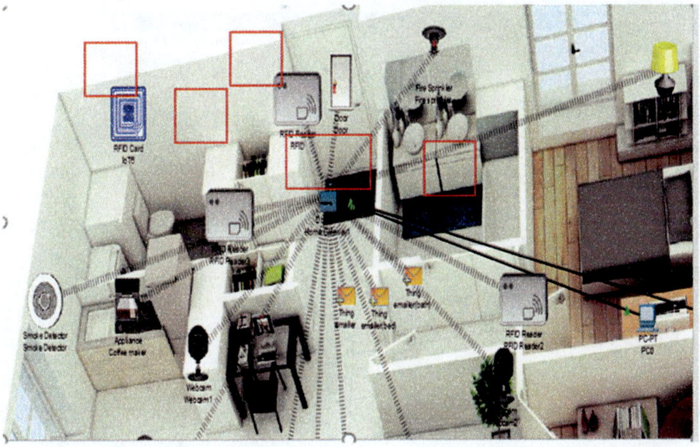

Fig. 7. An Emulation of the Elderly smart home in operation

The safety system will be activated when the smoke detector detects smoke – if the smoke level is higher than 40%, the siren and fire sprinkler will activate, and the window will open. See Fig. 8 for testing results of sending sleeping schedule reminders through the caregiver.

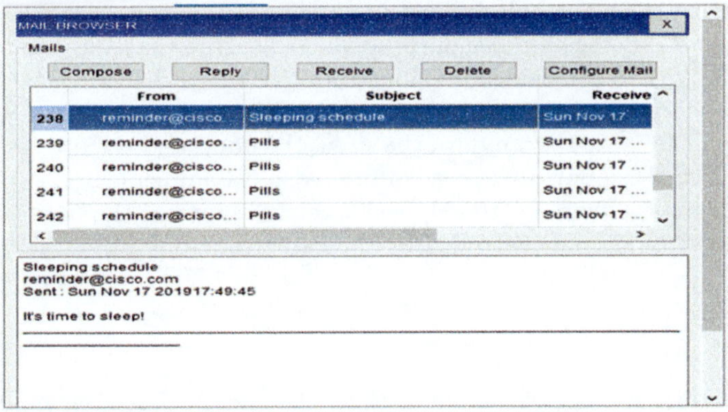

Fig. 8. Real-Time Testing of Sleeping Schedule reminder

The system will remind the patients of to take pills periodically as well as remind the patients of their sleeping schedule using the Cisco Packet tracer tool. On a severe or urgent level, an immediate automatic alarm is sent to the emergency center to take action by sending an ambulance. Figure 9 shows the IoT smart home devices connected and registered on a Gateway server including thermostat, smoke detector, fire sprinkler, siren, door, and window and webcam.

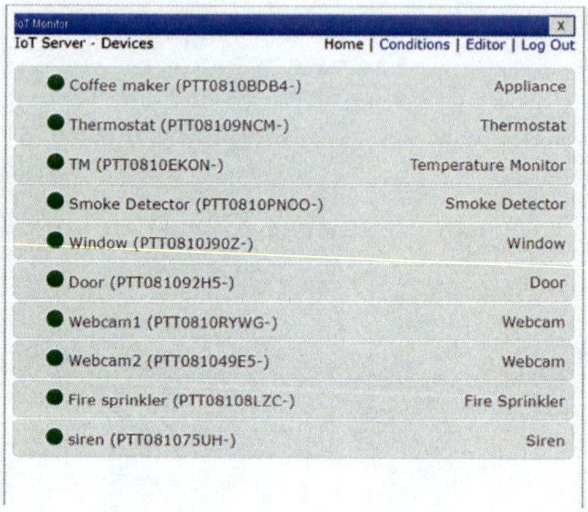

Fig. 9. Remote Home Gateway

We implemented a Python code in the "things", which sends reminders about important self-care duties to the patient. Also, for the caregiver with almost the same implemented code, an emergency email will be delivered to him in case the patient has been outside the house for a certain time. There are 3 "things" emailers that are responsible for sending emergency emails, the exit door, the bathroom door, and the bedroom door. There is a necessity in the IoT-based smart home to trace the elderly living in their home. So we use RFID readers and wearable RFID tags for this purpose, and to ensure the safety and wellbeing of these elderly. The emailers work along the RFID tags that are attached to the patient, therefore, all of them depend on the RFID reader to trigger the email to be sent. It will activate the window, the siren, the door, and the fire sprinkler. All of these will help the patient in case of fire. Also, we implemented two webcams inside the patient's bedroom to help the caregiver monitor the patient. Some of the smart devices can be controlled by the caregiver from an IoT server.

5 Conclusion

The paper proposed the solution of a telehealth system based on the Internet of Things for the well-being of the elderly suffering from Dementia in a smart home setting. It incorporates a smart home infrastructure and its elements including monitoring elements that allow detection of home or room leaving or not leaving, safety elements that allow detecting the high smoke level, and healthcare elements that remind the sleeping schedule and automate pill intake in a box for monitoring and recording the proper intake of prescribed medication for the elderly suffering from chronic diseases. A smart home solution was effectively implemented using the Cisco Packet Tracer simulator 7.3. The proposed IoT-based solution ensures the safe wandering of the elderly in a smart home.

References

1. Chen, W., Chen, L., Chang, W., Tang, J.: An IoT-based elderly behavioral difference warning system. In: Proceedings of ICASI'18, pp. 308–309 (2018)
2. Zulfiqar, A.-A., Hajjam, A., Andrés, E.: Focus on the different projects of telemedicine centered on the elderly in France. Curr. Aging Sci. **11**(4), 202–215 (2018)
3. Sadeghi-Niaraki, A.: Internet of Thing (IoT) review of review: bibliometric overview since its foundation. Future Gener. Comput. Syst. **143**, 361–377 (2023)
4. Mufti, T., Sami, N., Sohail, S.S.: A review paper on Internet of Things (IOT). Indian J. Appl. Res. **9**(8), 24–25 (2019)
5. Raed, M.W., Yön, S., Günes, A., Kotenko, I., Fedorchenko, E., Polubaryeva, A.: An RFID based localization and mental stress recognition system using wearable sensors. In: Proceedings of the 16th International Joint Conference on Biomedical Engineering Systems and Tecnology. BIOSIGNALS 2023, pp. 325–331 (2023)
6. Shams, R., Khan, F.H., Saleem, F.: Internet of things in telemedicine: a discussion regarding to several implementation. J. Inf. Commun. Technol. Rob. Appl. 17–26 (2018)
7. Kirtana, R.N., Lokeswari, Y.V.: An IoT based remote HRV monitoring system for hypertensive patients. In: 2017 International Conference on Computer, Communication and Signal Processing (ICCCSP), pp. 1–6. IEEE (2017)
8. Ni, Q., García Hernando, A.B., De la Cruz, I.P.: The elderly's independent living in smart homes: a characterization of activities and sensing infrastructure survey to facilitate services development. Sensors **15**(5), 11312–11362 (2015)
9. Sardini, E., Serpelloni, M.: T-shirt for vital parameter monitoring. Sensors **162**, 201–205 (2014)
10. Mtshali, P., Khubisa, F.: A smart home appliance control system for physically disabled people. In: 2019 Conference on Information Communications Technology and Society (ICTAS), pp. 1–5. IEEE (2019)
11. Karimi, K., Krit, S.: Smart home-smartphone systems: threats, security requirements and open research challenges. In: 2019 International Conference of Computer Science and Renewable Energies (ICCSRE), pp. 1–5. IEEE (2019)
12. Doynikova, E., et al.: Security measuring system for IoT devices, 256–275 (2022)
13. Ghourchian, N., Allegue-Martinez, M., Precup, D.: Real-time indoor localization in smart homes using semi-supervised learning. In: Proceedings of AAAI'17 (2017)
14. Jiang, J.-R., Subakti, H., Liang, H.-S.: Fingerprint feature extraction for indoor localization. Sensors **1**, 5434 (2021). https://doi.org/10.3390/s21165434
15. Li, C., Mo, L., Zhang, D.: Review on UHF RFID localization methods. IEEE J. Radio Freq. Identificat. **3**(4), 205–215 (2019)
16. Li, S., Lu, J., Chen, S.: A room-level tag trajectory recognition system based on multi-antenna RFID reader. Comput. Commun. **149**, 350–355 (2020)
17. Ashok, G.L.P., Akram, P.S., Neelima, M.S., Nagasaikumar, J., Vamshi, A.: Implementation of smart home by using packet tracer. Int. J. Sci. Technol. Res. **9**(2), 678–685 (2020)
18. Gupta, T.: Designing smart home (IOT) using cisco packet tracer 7.2 simulator. Int. J. Mod. Trends Sci. Technol. **6**(12), 208–218 (2020)

MFOOD-70: Moroccan Food Dataset for Food Image Recognition Towards Glycemic Index Estimation

Merieme Mansouri(✉), Samia Benabdellah Chaouni, Said Jai Andaloussi, and Ouail Ouchetto

Computer Science and Systems Laboratory, Faculty of Sciences Ain Chock, Km 8 Route El Jadida, B.P 5366, Maarif 20100, Casablanca 20000, Morocco
merieme.mansouri@gmail.com

Abstract. Food image recognition and nutritional analysis, in particular glycemic index (GI), play a crucial role in understanding the impact of diet on human health and the management of chronic diseases such as diabetes. Deep learning, a subset of machine learning, has proven to be a powerful tool in this field. It enables precise recognition of foods from images and provides valuable information on their nutritional content. A food dataset is key to achieving accurate food recognition using deep learning. To the best of this paper's knowledge, no Moroccan food dataset has been created to handle food image recognition and glycemic index estimation of Moroccan foods. This paper presents a new Moroccan food dataset and a method for food image recognition and nutritional analysis, particularly food glycemic index estimation.

Keywords: food image dataset · food recognition · glycemic index estimation · deep learning

1 Introduction

The impact of food on blood glucose levels has gained significant attention, especially in the context of managing diabetes and promoting healthy eating habits. The glycemic index (GI) provides a measure of how quickly carbohydrates in food raise blood sugar levels The concept of the glycemic index is important in understanding the impact of different carbohydrates on blood sugar control, particularly for individuals with conditions like diabetes or those interested in managing their blood sugar levels. Foods with a high GI value (greater than 70) are rapidly digested and absorbed, causing a rapid increase in blood glucose levels. On the other hand, foods with a low GI value (less than 55) are digested and absorbed more slowly, resulting in a slower and more gradual increase in blood glucose levels. Glycemic Load (GL) is another concept related to the glycemic index. It takes into account both the GI value of a food and the amount of carbohydrates consumed, providing a more accurate measure of the actual impact on blood sugar levels.

M. Ben Ahmed et al. (Eds.): SCA 2023, LNNS 906, pp. 372–380, 2024.
https://doi.org/10.1007/978-3-031-53824-7_34

Deep Learning is a specific approach to machine learning that uses artificial neural networks inspired by the structure and function of the human brain. Deep learning models, such as Convolutional Neural Networks (CNNs) have significantly advanced the fields of computer vision. It has demonstrated superior performance in image recognition.

The potential applications of food image recognition and glycemic index estimation using deep learning are extensive. They range from assisting individuals in making informed dietary choices, supporting personalized nutrition recommendations, and aiding in the management of conditions like diabetes. Furthermore, these technologies have the potential to be integrated into mobile applications and wearable devices, providing real-time nutritional analysis and guidance to individuals.

Deep learning models are trained on large-scale datasets, to learn complex patterns and relationships within the data. For the case of Moroccan cuisine, no food dataset of image recognition system have been proposed knowing that Moroccan cuisine has gained popularity around the world and is familiar in many cuisines, including Mediterranean and North African cuisines such as Italian, due to their shared geographical position. What's more, thanks to the richness of its cuisine, Morocco has become one of the most attractive countries to visit. That's why it is mandatory to have a food image recognition system food Moroccan food dishes.

Therefore, in this paper, the main contributions are summarized as follow:

A) Proposed a food image dataset for Moroccan cuisine: MFOOD-70.
B) Evaluate MFOOD-70 using convolutional neural network and transfer learning with the combination of a new feature extraction method.
C) Estimate the glycemic index and the load index of the traditional Moroccan dishes.

The paper sections are organized as follow: Sect. 2: present related works to our approch. Section 3 describes the proposed approach. Section 4 highlights the experiment details and results. Finally a conclusion is presented in Sect. 5

2 Related Works

2.1 Food Image Recognition

Food image recognition is an application of computer vision and machine learning that aims to identify and categorize food items from images. This technology involves training models to analyze specific visual features and patterns associated with various types of food, enabling automated processing and comprehension of food-related data. Jiang et al. [6] introduced a system that recognizes food items and offers nutrition analysis to evaluate dietary habits. Their approach begins with food item detection using faster RCNN and subsequently extracting features and conducting classification with the assistance of VGG-16. The authors conducted training and evaluation using the UEC-FOOD100 and

UEC-FOOD256 datasets, along with a self-created dataset called FOOD20-with-box, which is a subset of the original FOOD101 dataset. The FOOD20-with-box dataset comprises 20 categories with associated bounding box annotations. The system also incorporates nutrition analysis capabilities, providing information on calories, fats, carbohydrates, and proteins based on the USDA National nutrient database, taking into account user-specific details such as age and weight. It is important to note that the system assumes a standardized weight of 400g for each food item to generate accurate nutritional information.

Ran et al. [12] addressed the challenge of fine-grained features in Chinese food and proposed a food image recognition system that utilizes an attention mechanism to extract detailed features. The system employs the pre-trained CNN model DensNet-169 for classification. Latif et al. [8] introduced a novel dataset comprising 20 types of fruits and 20 types of vegetables, capturing images under diverse conditions with variations in backgrounds, plates, angles, resolutions, and lighting. The authors utilized three different CNN architectures with 9, 11, and 15 layers to recognize fruit plates and also evaluated their system on the fruits-360 dataset. Ye et al. [15] developed a supervised deep neural network model capable of recognizing ten food categories from the COCO2017 dataset. The system utilizes Mask R-CNN for the segmentation and classification of each item on a plate and provides nutritional information. The authors employed ResNet+FPN as a backup network for segmentation and classification and used OpenCV to display the category and nutritional information of each item. Additionally, they leveraged Tensorflow Lite to deploy the model on mobile devices for user accessibility.

To address the challenge of classifying mixed food items on a plate, Deng et al. [3] proposed a contextual relation network (CR-Nets) approach. This method involves employing ResNet-50 for single-item classification and subsequently using the learned weights for localizing and classifying multiple food items with the assistance of Faster RCNN. The authors further refined the results using a co-occurrence statistic method based on dish labels. Tan et al. [13] introduced Quantized deep residual convolutional neural networks (DRNN) for food image recognition. They extracted and classified features using both residual-based and convolutional-based networks and applied a quantization technique to reduce the model's representation size, making it suitable for deployment on mobile devices. Chakrabarty et al. [1] targeted type 1 diabetic patient and proposed a food category identification system that utilizes CNN and provides information on the cholesterol content of the identified food items.

2.2 Food Image Datasets

There are several publicly available food datasets that can be used for various purposes, including research, development of machine learning models, nutritional analysis, and food-related applications. Here are a few notable food datasets: The key to an efficient food image recognition system is a well-prepared dataset in different aspects: quantity, variety, quality, and annotation.

VegFru [5], Fruits-360 [11] and FruitVeg-81 [14] are three fruit and vegetable datasets that are publicly available, the VegFru dataset has the largest number of images and it is specialized with upper-level classes and sub-classes with a total of 292 categories and 160000 images collected from the Web. UEC Food100 [9] and UEC Food256 [7] are two popular datasets for Japanese food. UEC Food256 is an expanded version of the original UEC Food100 containing a total of 256 categories and 25000 images with the bounding box. The urgent need for benchmark food recognition systems made various authors have initiation to develop benchmark datasets containing miscellaneous food categories. Food524DB [2] and ISIA Food-500 [10] are the most large-scale and benchmark datasets. The Food524DB dataset collected 524 classes and 247000 images from four existing datasets, while ISIA Food-500 was composed of 500 categories and almost 399000 images acquired from the Web. The rest of the publicly available datasets listed in Table 6 are released based on a specific cuisine or a modified version of existing datasets such as SUEC Food dataset [4] which contains 31995 segmented images from the UEC Food256 dataset.

Fig. 1. the 70 classes of MFOOD-70 dataset

3 Proposed Approach

In this section, we present the new Moroccan dataset and the CNN method of food recognition and glycemic index estimation:

3.1 MFOOD-72 Dataset

MFOOD-72 dataset is a Moroccan food dataset containing 72 dishes and 8300 images. The dataset will be enlarged afterward using data augmentation in the phase of recognition. The images were collected using web scraping and a combination of multiple datasets. They are chosen carefully to be more reliable for use. The duplicated and undesirable images were filtered manually; we made sure that each image respects all the conditions of well angles, lighting, and variations. then they were labeled and uniformed in size and segmented semi-automatically using the API GrabCut to extract foreground from the background and help the classification model focus on the food item. Other image enhancement techniques used in the training phase will be presented in the experiment section to ensure the quality of the dataset.

3.2 Food Image Recognition Method

Transfer learning is new machine learning recently used to improve the learning process and overcome the limits of data and resources. Instead of training a model from scratch on a new task, transfer learning involves using a pre-trained model as a starting point and fine-tuning it on the target task. In transfer learning, the pre-trained model is typically trained on a large dataset from a different but related task, such as image classification using a large dataset like ImageNet. The pre-trained model has already learned useful features and representations from the initial task, which can be valuable for the new task. Examples of CNN models that use transfer learning concept are: MobileNet, DenseNet and EfficientNet. To choose the best transfer learning for our next phase, we evaluated MFOOD-70 on those three pre-trained CNN models, DenseNet has achieved the best performance.

DenseNet is a deep convolutional neural network known for its dense connectivity patterns, where each layer is connected to every other layer in a feed-forward manner. To improve the performance of the model, we choose to combine it with the attention mechanism technique. It allows the model to focus on specific parts of the input data that are most relevant to the task at hand, rather than treating all parts equally. Attention mechanisms use spatial attention to allow the model to focus on specific spatial regions of the image that are informative for classification. Spatial attention aims to capture dependencies across spatial locations. It generates attention weights per spatial position, highlighting the most relevant regions in the feature maps.

Combining attention mechanisms with the DenseNet model involves integrating the attention mechanism into the architecture of DenseNet. Starting with the DenseNet architecture as the backbone of the model. This backbone will consist of densely connected blocks, typically comprising multiple dense layers and transition layers. After that, we apply global average pooling (GAP) to convert the feature maps into a fixed-length representation, and then the attention mechanism generates spatial attention weights based on the feature maps obtained from the GAP layer. The attention mechanism guides the model to focus on relevant features by multiplying the attention weights with the feature maps to emphasize the informative regions. Finally, a fully connected layer is used to append the classification head and make a prediction.

3.3 Glycemic Index and Glycemic Load Calculation

Nutritional analysis and dietary assessment are processes used to evaluate and understand the nutritional content and quality of foods and diets. These assessments play a crucial role in promoting healthy eating habits, identifying nutritional deficiencies, and informing dietary recommendations for individuals and populations. Nutritional analysis, including estimating the glycemic index (GI), is conducted based on the recognized food items. The glycemic index is a measure of how quickly carbohydrates in food raise blood sugar levels.

Glycemic Load (GL) is another concept related to the glycemic index. It takes into account both the GI value of a food and the amount of carbohydrates consumed, providing a more accurate measure of the actual impact on blood sugar levels. The GL is calculated using this formula:

$$GL = (GI \text{ x } Carbohydrate)/100$$

Mfood-72 dataset contains simple categories such as fruit, better, dates and French fries, and mixed categories with multiple ingredients such as couscous, rfissa, tagine, and so on. A number of Moroccan food consumers do not know the glycemic index of several food items its main factor on daily food consuming. The main idea of this paper is to categories the food items to three categories: Low = 1 to 55(encouraged), Medium = 56 to 69 (allowed) and High = 70 and higher (forbidden). While GL value of 100g is categories as follow: 10 or less is considered low, 11–19 is considered medium, and 20 or more is considered high. In this section we choose samples of the 70 categories to calculate glycemic index and glycemic load. The GI of simple categories is used based on several public Glycemic Index Databases and nutritional websites, and the GI of mixed categories is presented based on the GI of the ingredients. For GL, we used the carbohydrate of 100g of each category.

4 Experiment and Results

The images have been enhanced using light regularization and noise removal, and then, data augmentation is used to enlarge the number of images for the training phase. Specific parameters of data augmentation are used such as flipping, rotation, and cropping. The data is split into training, testing, and validation set. The training set is used to train the model. It contains a large portion of the available data and is used to optimize the model's parameters and learn the underlying patterns and relationships. The validation set is used for fine-tuning the model's hyperparameters and evaluating the model during the training phase. This helps in optimizing the model's performance and avoiding overfitting. The images have

Table 1. Samples of GI and GL estimation

Food Item (class)	GI	GL of 100g	GI-based category	GL-based category
Banana	62	11	Medium	Medium
French Fries	95	30	High	High
Couscous	Couscous: 50 Meat: — Carrots: 45 Turnip: 55 Tomato: 30 Cabbage: 15 pumpkins:75	Couscous:12 Meat:— Carrots:3.15 Turnip:5 Tomato:5 Cabbage:1 pumpkins:10	Couscous: low sparabreak Meat:— Carrots: low Turnip: low Tomato: low Cabbage: low pumpkins: high	Couscous : medium Meat:— Carrots: low Turnip: low Tomato: low Cabbage: low pumpkins: low

been resized based on the expected size of each model, the choice of the rest of the parameters is listed in Table...

In the evaluation phase, we calculate the accuracy metric of the training, testing, and validation set. As shown in Table... we choose the DenseNet model for combination with the attention mechanism based on its best performance. In this paper, we use the Self-Attention mechanism design that allows the model to capture the relationships between different positions within the input feature maps. First, In DenseNet, feature maps from different layers are concatenated to form the input for the next layer and resulting feature tensor, and then, the attention module is applied to the concatenated feature tensor and transformed it into query, key, and value tensors. By the calculation of the dot product between query and key tensors, the attention weight is generated and multiplied by the value tensor giving the attended features that capture the importance of different parts of the feature tensor. The accuracy value of the combination of the attention mechanism and DenseNet is still under process.

Table 2. The parameters of the experiment

Model	images after data augmentation	Training set	Testing set	Validation set	Size	Optimizer	Batch	Epoch
MobileNet	58 100	80%	20%	30% of training set	(128, 128, 3)	SDG	32	50
EfficientNet	58 100	80%	20%	30% of training set	(224, 224, 3)	SDG	8	50
DenseNet	58 100	80%	20%	30% of training set	(224, 224, 3)	SDG	32	50

Table 3. Evaluation Performance

Model	Training accuracy	Testing accuracy	Validation accuracy
MobileNet	0.917	0.772	0.945
EfficientNet	0.917	0.7471	0.84
DenseNet	0.918	0.903	0.908

5 Conclusion and Future Works

This paper proposes a new Moroccan food dataset containing 70 classes and 8300 images with the aim of food image recognition and glycemic index providing for chronic diseases monitoring such as diabetes. The paper proposes as well a new food image recognition system focusing on the combination of attention mechanism and DenseNet model and presents the glycemic index and the glycemic load of some dataset samples highlighting the idea of categorizing the food items into encouraged, allowed, and forbidden, in order to aware Moroccan food consumers. there are several avenues for further exploration. First, completing the experiment process and obtaining the final performance of the proposed approach and implementing different types of attention mechanisms and their combinations to enhance the fine-grained analysis of food images. Secondly, generate the calculation of GI and GL of all the classes of MFOOD-70. Finally, integrating the developed model into mobile applications or wearable devices to provide real-time nutritional analysis and guidance to individuals, and to facilitate dietary choices.

References

1. Chakrabarty, A., Doyle, F J., Dassau, E.: Deep learning assisted macronutrient estimation for feedforward-feedback control in artificial pancreas systems. In: 2018 Annual American Control Conference (ACC), pp. 3564–3570. IEEE, 2018
2. Ciocca, G., Napoletano, P., Schettini, R.: Learning CNN-based Features for Retrieval of Food Images. In: Battiato, S., Farinella, G..M., Leo, M., Gallo, G. (eds.) ICIAP 2017. LNCS, vol. 10590, pp. 426–434. Springer, Cham (2017). https://doi.org/10.1007/978-3-319-70742-6_41
3. Deng, Li., et al.: Mixed-dish recognition with contextual relation networks. In: Proceedings of the 27th ACM International Conference on Multimedia, pp. 112–120, 2019
4. Gao, J., Tan, W., Ma, L., Wang, Y., Tang, W., Musefood: Multi-sensor-based food volume estimation on smartphones. In 2019 IEEE SmartWorld, Ubiquitous Intelligence and Computing, Advanced and Trusted Computing, Scalable Computing and Communications, Cloud and Big Data Computing, Internet of People and Smart City Innovation (SmartWorld/SCALCOM/UIC/ATC/CBDCom/IOP/SCI), pp. 899–906. IEEE, 2019
5. Hou, S., Feng, Y., Wang, Z.: Vegfru: a domain-specific dataset for fine-grained visual categorization. In Proceedings of the IEEE International Conference on Computer Vision, pp.541–549, 2017
6. Jiang, L., Qiu, B., Liu, X., Huang, C., Lin, K.: Deepfood: food image analysis and dietary assessment via deep model. IEEE Access **8**, 47477–47489 (2020)
7. Kawano, Y., Yanai, K.: Automatic Expansion of a Food Image Dataset Leveraging Existing Categories with Domain Adaptation. In: Agapito, L., Bronstein, M..M., Rother, C. (eds.) ECCV 2014. LNCS, vol. 8927, pp. 3–17. Springer, Cham (2015). https://doi.org/10.1007/978-3-319-16199-0_1
8. Latif, G., Alsalem, B., Mubarky, W., Mohammad, N., Alghazo, J.: Automatic fruits calories estimation through convolutional neural networks. In Proceedings of the 2020 6th International Conference on Computer and Technology Applications, pp. 17–21, 2020

9. Matsuda, Y., Hoashi, H., Yanai, K.: Recognition of multiple-food images by detecting candidate regions. In 2012 IEEE International Conference on Multimedia and Expo, pp. 25–30. IEEE, 2012
10. Min, W., Liu, L., Wang, Z., Luo, Z., Wei, X., Wei, X., Jiang, S.: ISIA Food-500: a dataset for large-scale food recognition via stacked global-local attention network. In: Proceedings of the 28th ACM International Conference on Multimedia, pp. 393–401, 2020
11. Mure, H., şan Oltean, M.: Fruit recognition from images using deep learning. arXiv preprint. arXiv:1712.00580, 2017
12. Ran, H., Gao, W., Mi, J., Zhao, Z.: Fine-grained recognition of Chinese food image based on densenet with attention mechanism. In Twelfth International Conference on Graphics and Image Processing (ICGIP 2020), volume 11720, pp. 117201G. International Society for Optics and Photonics, 2021
13. Tan, R.Z., Chew, X., Khaw, K.W.: Quantized deep residual convolutional neural network for image-based dietary assessment. IEEE Access, 8:, pp. 111875–111888, 2020
14. Waltner, G., et al.: Personalized Dietary Self-Management using Mobile Vision-based Assistance. In Proc. of ICIAP Workshop on Multimedia Assisted Dietary Management (MADIMA), 2017
15. Ye, H., Zou, Q.: Food Recognition and Dietary Assessment for Healthcare System at Mobile Device End Using Mask R-CNN. In: Gao, H., Li, K., Yang, X., Yin, Y. (eds.) TridentCom 2019. LNICST, vol. 309, pp. 18–35. Springer, Cham (2020). https://doi.org/10.1007/978-3-030-43215-7_2

Toward an IoB-Based Architecture for Bipolar Disorder Management

Kebira Azbeg[1(✉)], Btissam Zerhari[2], Asmae Azbeg[1], Khadija Tlemçani[1], Jai Andaloussi Said[1], and Ouail Ouchetto[1]

[1] Computer Science and Systems Laboratory, Faculty of Sciences Ain Chock, Hassan II University of Casablanca, Casablanca, Morocco
azbegkebira@gmail.com

[2] LRIT, Unit Associated to CNRST URAC 29, Mohammed V University, Rabat, Morocco

Abstract. Bipolar disorder is a challenging mental health condition characterized by recurrent periods of depression and mania. Successful management of bipolar disorder necessitates ongoing monitoring of symptoms, triggers, and individual behaviors. The emergence of the Internet of Behaviors (IoB) offers a promising prospect to enhance bipolar disorder management through the utilization of connected devices and behavioral data analysis. This paper proposes an IoB-based architecture for managing bipolar disorder. By leveraging connected devices and behavioral data analysis, the architecture aims to provide real-time insights, personalized recommendations, and early warning signs. In order to enhance the security of data, we integrate the blockchain technology to the proposed architecture. The potential advantages of implementing such an architecture are examined as well as some areas for future research are explored.

Keywords: Internet of behavior · IoT · Artificial intelligence · Blockchain · Bipolar disorder · Healthcare

1 Introduction

The advent of many technologies, such as Internet of Things (IoT) [7,14] and Artificial Intelligence [10,13], have significantly impacted and enhanced the healthcare sector, revolutionizing the way healthcare is delivered and improving patient outcomes. Despite the progress made in assistive technologies, there has been a limited amount of research dedicated to the implementation of such technologies in the field of mental health care. That is why it is crucial to investigate the potential of these technologies in identifying, evaluating, treating, and potentially reducing symptoms linked to mental disorders.

The Internet of Behavior (IoB), which is an extension of IoT, can be used to enhance bipolar management by using real-time data collected from multiple sources, including wearables, mobile apps, social media, and electronic health records. These interconnected devices provide objective information on behaviors, sleep patterns, social interactions, medication adherence, and more.

M. Ben Ahmed et al. (Eds.): SCA 2023, LNNS 906, pp. 381–390, 2024.
https://doi.org/10.1007/978-3-031-53824-7_35

Through advanced analytics and machine learning, healthcare providers can analyze this data to gain insights into the individualized behavioral patterns and triggers associated with bipolar disorder. The aim of this work is to propose an IoB-based system for Bipolar Disorder management which can provide valuable recommendations for enhancing patient care and treatment outcomes.

The paper is structured as follows: Sect. 2 presents the related works, and Sect. 3 gives an overview of Bipolar disorder. Section 4 presents the proposed approach to enhance the management of Bipolar disorder and Sect. 5 concludes the paper and outlines the future works.

2 Related Works

Numerous researchers have tackled a range of topics concerning chronic diseases, Artificial Intelligence in Healthcare, and IoT in healthcare, and have made significant contributions through their research. This section specifically concentrates on reviewing the existing literature in these domains.

Authors in [12] suggest models and structures to facilitate the integration of IoT technology in order to improve current treatment choices. Their proposed system is compared to existing solutions in the contexts of cancer care services and business analytics. In [1], authors proposed a system architecture based on IoT technology, aimed at collecting patient data for the purpose of predicting severe cases of diabetes. Another study in [6] presented a proposal for a remote patient monitoring system that utilizes energy-efficient IoT sensors to fulfill security requirements. Alain Hennebelle et al. [5] introduced a framework for smart healthcare that utilizes machine learning to predict Type 2 Diabetes. The framework is designed to operate within an integrated system that combines IoT edge and cloud computing technologies. In addition, artificial intelligence is used for early detection and diagnosis of several diseases such as oral cancer [4], spontaneous preterm labor and birth [8], and also early diagnosis of thyrotoxic periodic paralysis [9].

Despite advancements in assistive technologies, the majority of these research studies tend to concentrate on a single technology and overlook the potential of combining multiple technologies. However, as a technology that combines IoT and Artificial Intelligence, IoB can provide a better solution for enabling the collection, analysis, and utilization of medical data. By integrating IoT devices with AI algorithms, IoB aims to understand and influence human behavior, leading to improved disease management, personalized treatments, and more efficient decision-making processes. Nevertheless, there remains a scarcity of research focused on integrating these technologies into mental health care.

3 Overview of Bipolar Disease

Bipolar disorder, also known as manic-depressive illness, is a chronic psychiatric condition characterized by recurrent episodes of mania and depression. It is a complex mood disorder that affects a significant portion of the global population, with a lifetime prevalence estimated to be around 2.4%. Individuals with bipolar disorder experience extreme shifts in mood, energy levels, and activity patterns, leading to significant impairment in daily functioning and overall quality of life. These mood episodes can range from elevated, euphoric states of mania to profound periods of depression. Bipolar disorder often emerges in late adolescence or early adulthood and requires comprehensive management strategies for long-term stabilization and symptom control.

3.1 Symptoms of Bipolar Disorder

Bipolar disorder is a mental health condition characterized by alternating episodes of mania and depression. The symptoms experienced during these phases can vary significantly. Here are the symptoms commonly associated with each phase [2]:

Manic Phase (Mania): During the manic phase of bipolar disorder, individuals exhibit a range of symptoms that reflect an abnormal elevation in mood, energy, and activity levels. These symptoms may include:

- Abnormally euphoric or irritable mood: Individuals experience an intense and disproportionate sense of happiness, excitement, or irritability.
- Increased energy and heightened activity levels: There is a marked surge in energy, often resulting in hyperactivity and restlessness.
- Decreased need for sleep: Individuals may feel adequately rested even after only a few hours of sleep, or they may exhibit a diminished need for sleep altogether.
- Racing thoughts and rapid speech: Thoughts may race through the mind, making it challenging to concentrate or engage in coherent conversation.
- Grandiose beliefs and inflated self-esteem: People may have an exaggerated sense of self-importance and harbor grandiose beliefs about their abilities, talents, or achievements.

Depression Phase: The depression phase of bipolar disorder is characterized by a profound and persistent feeling of sadness, accompanied by a range of additional symptoms, including:

- Pervasive sadness or low mood: Individuals experience a deep and enduring sense of sadness, emptiness, or hopelessness.
- Loss of interest or pleasure: Activities that were once enjoyable no longer elicit interest or pleasure.
- Fatigue and diminished energy: There is a marked lack of energy, often resulting in physical and mental exhaustion.
- Changes in appetite and weight: Appetite may fluctuate, leading to weight loss or gain. Sleep disturbances: Individuals may experience insomnia or excessive sleepiness.

- Recurrent thoughts of death or suicide: Thoughts of death, suicidal ideation, or actual suicide attempts may occur repeatedly.

It is important to emphasize that symptom severity and presentation can vary among individuals and even across different episodes within the same individual.

3.2 Important Factors in Bipolar Disease

Bipolar disorder is a complex mental health condition influenced by various factors. Here are some important factors that play a role in bipolar disease [11]:

- Neurochemical Imbalances: Bipolar disorder is associated with imbalances in certain neurotransmitters, such as serotonin, dopamine, and norepinephrine. These neurotransmitters play a crucial role in regulating mood, energy, and other behavioral processes.
- Structural and Functional Brain Abnormalities: Studies using neuroimaging techniques have shown that individuals with bipolar disorder may exhibit structural and functional abnormalities in specific brain regions involved in emotional regulation and cognitive processes.
- Environmental Triggers: Environmental factors, such as significant life events, chronic stress, substance abuse, and disrupted sleep patterns, can influence the onset and course of bipolar disorder.
- Medication and Substance Use: Certain medications, such as antidepressants or corticosteroids, as well as substance abuse, can potentially induce or worsen manic or depressive symptoms in individuals with bipolar disorder.
- Psychological and Behavioral Factors: Psychological factors, including maladaptive coping strategies, interpersonal difficulties, can influence the course and severity of bipolar disorder. Additionally, irregular sleep patterns, disruptions in daily routines can contribute to symptom exacerbation.

3.3 Nursing Care for Bipolar Patients

Nursing care for bipolar patients encompasses both pharmacological and non-pharmacological treatment approaches. Here are the key aspects of nursing care for bipolar disorder [3]:

Pharmacological Treatment: Nurses play a vital role in assisting with the administration and monitoring of medications prescribed for bipolar disorder. This includes:

- Educating patients about their medications, including the purpose, dosage, potential side effects, and the importance of adhering to the prescribed regimen.
- Monitoring and documenting medication effectiveness, adverse effects, and any changes in symptoms.
- Collaborating with the healthcare team to adjust medication dosages, as needed, based on the patient's response and any observed side effects.

- Assessing for potential drug interactions or contraindications with other medications the patient may be taking.
- Promoting medication adherence through patient education, reminders, and addressing any concerns or misconceptions.

Common medications used in bipolar disorder treatment include mood stabilizers (e.g., lithium, valproate), antipsychotics, antidepressants, and benzodiazepines for acute symptom management.

Nonpharmacological Treatment: Nonpharmacological interventions are important components of nursing care for bipolar patients. These interventions focus on promoting stability, self-management, and overall well-being. They include:

- Psychoeducation: Providing comprehensive education to patients and their families about bipolar disorder, its symptoms, triggers, treatment options, and strategies for relapse prevention.
- Cognitive-Behavioral Therapy (CBT): Assisting patients in identifying and modifying maladaptive thoughts and behaviors, developing coping skills, and managing stressors effectively.
- Psychotherapy: Engaging in regular counseling sessions to address emotional and interpersonal challenges, improve self-esteem, and enhance overall psychological well-being.
- Sleep Hygiene: Educating patients about the importance of maintaining a regular sleep schedule, avoiding sleep deprivation, and adopting healthy sleep habits.
- Stress Reduction Techniques: Teaching patients stress management techniques such as relaxation exercises, mindfulness, deep breathing, and engaging in enjoyable activities.
- Social Support: Encouraging patients to engage in support groups, peer support networks, or family therapy to foster a supportive environment and reduce feelings of isolation.
- Lifestyle Management: Assisting patients in developing and maintaining a structured daily routine, regular exercise, a balanced diet, and minimizing the use of alcohol or recreational drugs.

It is crucial for nurses to provide ongoing support, monitor patient progress, and collaborate with the multidisciplinary team to ensure an integrated approach to care for bipolar patients. Individualized care plans should be developed based on the patient's specific needs and treatment goals.

4 Proposed Bipolar Disorder Management Approach

4.1 Proposed Architecture

IOB Architecture: The Internet of Behavior (IoB), which is an extension of IoT, refers to a new concept that combines data analytics, behavioral psychology, and technology to gain insights into human behavior. It is a technology that involves the use of various technologies, such as sensors, wearables, and data analytics, to capture and analyze data related to human behavior. The security of data is enhanced by using data encryption and the blockchain technology.

In this work, we propose an IoB-based architecture for bipolar management (Fig. 1).

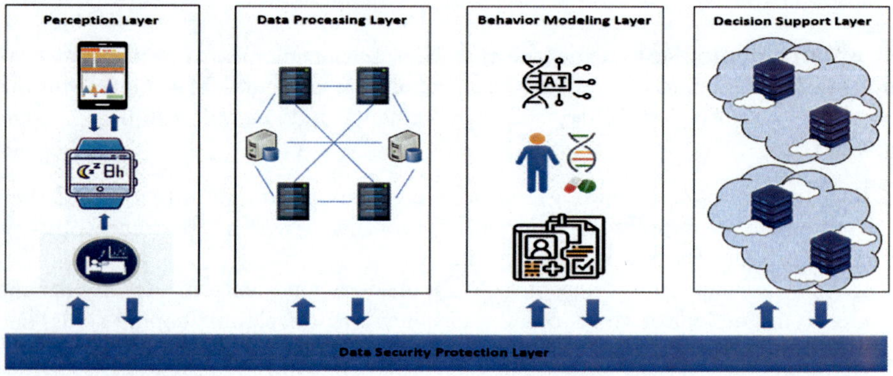

Fig. 1. Proposed Architecture.

- **Perception Layer:** At this stage, information is gathered from different sources like wearables, mobile devices, and environmental sensors. This includes capturing relevant physiological signals (such as heart rate and sleep patterns) as well as behavioral data (like activity levels, social interactions and emotions) that are significant for managing bipolar disorder. Table 1 describes some IoT devices suggested for bipolar patient assisting and actions that each of them could perform.

Table 1. Bipolar assistive IoT devices

Sensor	IoT device	Action
Sleep Tracker	Smart watch	Monitor sleep patterns
Activity Tracker	Smartphone	Detects and quantifies physical activity levels
Heart Rate Monitor	Smart watch	Measures heart rate
GPS Tracker	Smartphone	provide information on the user's location
Zenith and audio	Smart glass	Recognizes emotions
Ambient Light Sensor	Smartphone	Measures the intensity of ambient light in the surroundings

- **Data Processing Layer:** In this layer, the gathered data is processed and analyzed to extract valuable insights. Utilizing machine learning algorithms and data analytics techniques, patterns can be recognized, anomalies detected, and actionable information derived concerning bipolar symptoms, triggers, and risk factors. For example, from the sleep data, features like sleep duration, sleep efficiency, and REM (Rapid Eye Movement) sleep duration can be extracted. By correlating these features with external factors, such as environmental data or social interactions, data analytics techniques can be used to identify potential risk factors that contribute to bipolar disorder symptoms. For example, analyzing the relationship between sleep quality and stressful life events may reveal patterns of vulnerability to mood episodes.
- **Behavior Modeling Layer:** uses processed data to develop personalized profiles and behavioral models for individuals with bipolar disorder. These models consider the individual's specific symptom patterns, triggers, and responses, enabling predictions of mood episodes, identification of early warning signs, and the delivery of personalized recommendations.
- **Decision Support Layer:** uses behavior models to generate recommendations and interventions for individuals with bipolar disorder. By leveraging insights from previous layers, it offers real-time guidance and support to both individuals and healthcare providers. This layer can suggest coping strategies, provide medication adherence reminders, propose lifestyle modifications, and offer therapeutic interventions.
- **Data Security Protection Layer:** In light of the sensitive nature of personal health data, this layer focuses on ensuring the security and privacy of the collected information. It incorporates encryption techniques, access controls, and blockchain technology to protect user data from unauthorized access or breaches and ensure data integrity.

4.2 The Outputs and Recommendations of the Proposed System

The proposed system for Bipolar Disorder management provides valuable recommendations for enhancing patient care and treatment outcomes. The system generates insightful outputs that contribute to a comprehensive understanding of patients' symptomatology and treatment response. The outputs of the proposed system can be divided into two parts:

- **The first part:** focuses on detecting the current phase of the patient's illness by collecting information using connected sensors and interpreting patient behavior using AI algorithms. This enables the system to accurately identify whether the patient is in a manic, depressive, mixed or stable phase (Table 2). By continuously monitoring various parameters such as activity levels, sleep patterns, and mood fluctuations, the system can provide valuable insights into the patient's current condition.
- **The second part:** involves providing personalized recommendations and notifications to both the patient and their healthcare provider. These recommendations are based on the patient's psychological state and can include

reminders for medication adherence, notifications for emergency situations, and suggestions for physical activities. By offering real-time alerts and reminders, the system aims to improve the patient's overall well-being and enhance their self-management capabilities. Additionally, it facilitates communication between the patient and their healthcare provider, enabling timely interventions and support.

Table 2. Patient Disease Phases and Associated Risk Levels

Patient's phase	RiskLevel
Manic Symptoms	Moderate
Depressive Symptoms	Moderate
Mixed Features (Mania and Depression)	High
Stable Phase	Low

The actions recommended by our system, based on the patient's phase of illness, are as follows:

Case 1: Manic Symptoms.
Actions to be Taken are:

- Medication reminder: Remember to take your prescribed medication as part of your treatment plan to maintain mood stability during the manic phase.
- Stress management techniques: Receive notifications with stress management tips such as relaxation exercises, deep breathing, or meditation to reduce excitement and maintain emotional balance.
- Physical activity tracking: Get reminders to engage in regular and moderate physical activity, which can help channel excessive energy and improve mental well-being during the manic phase.
- Emergency contacts: Easily access emergency numbers or contact information for your healthcare professional to seek prompt assistance during emergency situations.
- Sleep habits monitoring: Receive recommendations to maintain a regular and adequate sleep routine.

Case 2: Depressive Symptoms.
Actions to be Taken are:

- Medication reminder: Receive reminders for medication adherence.
- Well-being activities: Receive suggestions for well-being activities such as relaxation exercises, breathing techniques, outdoor walks, or hobbies that can help improve your mood and alleviate depression symptoms.

- Information resources: Receive notifications containing educational resources, therapies, and self-care strategies.
- Support contacts: Easily access support contacts such as emergency numbers or helpline services to seek assistance or talk to someone when feeling depressed or emotionally distressed.

Case 3: Mixed Features(Mania and Depression).
Actions to be Taken are:

- Psychoeducation: Receive educational information about mixed features and coping strategies to better understand and manage this specific phase.
- Sending notifications to the treating physician as well as family members.
- Medication reminders: Adhere to the prescribed treatment plan, tailored to mixed features, to maintain mood stability.

Case 4: Stable Phase.
Actions to be Taken are:

- Social support: Access features to join online support groups or establish connections with others who share similar experiences.
- Encouragement and congratulations messages for consistently maintaining stability during the stable phase.

Our contribution offers numerous advantages that can be summarized in Table 3:

Table 3. Advantages of the proposed approach

Objective	How Our Contribution Enables It
Assessing treatment effectiveness	Our contribution allows for the evaluation of treatment effectiveness in patients by comparing their level of psychological stability with medication adherence
Exploring the interactions between risk factors and episodes of the illness	Our contribution allows for the exploration of risk factor interactions with bipolar disorder episodes, providing valuable insights for targeted interventions and advancing the understanding of the illness
Assisting patients in managing their mood disorders	Our aim is to provide support and guidance to patients in effectively managing their mood disorders through personalized interventions and comprehensive care

5 Conclusion

In this paper, we proposed an IoB-based architecture, augmented with blockchain technology, demonstrates significant potential in leveraging connected devices and behavioral data analysis to enhance the monitoring and treatment of bipolar disorder. By collecting and analyzing relevant data in real-time, this architecture aims to provide personalized recommendations and early warning signs to empower individuals and healthcare providers. In addition, privacy and data security were carefully addressed by using the blockchain technology. Currently, the execution of our architectural design is underway. In the future, further research in this field holds the key to advancing our knowledge and management of bipolar disorder, ultimately enhancing the well-being of those affected by the condition.

References

1. Azbeg, K., Boudhane, M., Ouchetto, O., Jai Andaloussi, S.: Diabetes emergency cases identification based on a statistical predictive model. J. Big Data **9**(1), 1–25 (2022)
2. Calabrese, J.R., Hirschfeld, R.M., Frye, M.A., Reed, M.L.: Impact of depressive symptoms compared with manic symptoms in bipolar disorder: results of a us community-based sample. Clin. Psychiatry **65**, 1499–1504 (2004)
3. Eduard, V., et al.: Bipolar disorders. Nature reviews Disease primers **4** (2018)
4. Hegde, S., Ajila, V., Zhu, W., Zeng, C.: Artificial intelligence in early diagnosis and prevention of oral cancer. Asia-Pacific J. Oncol. Nurs. **9**(12) (2022)
5. Hennebelle, A., Materwala, H., Ismail, L.: Healthedge: a machine learning-based smart healthcare framework for prediction of type 2 diabetes in an integrated IoT, edge, and cloud computing system. Procedia Comput. Sci. **220**, 331–338 (2023)
6. Kapoor, B., Nagpal, B., Alharbi, M.: Secured healthcare monitoring for remote patient using energy-efficient IoT sensors. Comput. Electr. Eng. **106**, 108585 (2023)
7. Kashani, M.H., Madanipour, M., Nikravan, M., Asghari, P., Mahdipour, E.: A systematic review of IoT in healthcare: applications, techniques, and trends. J. Netw. Comput. Appl. **192**, 103164 (2021)
8. Lee, K.S., Ahn, K.H.: Application of artificial intelligence in early diagnosis of spontaneous preterm labor and birth. Diagnostics **10**(9), 733 (2020)
9. Lin, C., et al.: Artificial intelligence–assisted electrocardiography for early diagnosis of thyrotoxic periodic paralysis. J. Endocr. Soc. **5**(9), bvab120 (2021)
10. Mansour, R.F., El Amraoui, A., Nouaouri, I., Díaz, V.G., Gupta, D., Kumar, S.: Artificial intelligence and internet of things enabled disease diagnosis model for smart healthcare systems. IEEE Access **9**, 45137–45146 (2021)
11. Phillips, M.L., Kupfer, D.J.: Bipolar disorder diagnosis: challenges and future directions. The Lancet **381** (2013)
12. Salehi, W., Gupta, G., Bhatia, S., Koundal, D., Mashat, A., Belay, A.: IoT-Based wearable devices for patients suffering from alzheimer disease. Contrast Media & Molecular Imaging **2022** (2022)
13. Saraswat, D., et al.: Explainable AI for healthcare 5.0: opportunities and challenges. IEEE Access (2022)
14. Tekeste Habte, T., et al.: IoT for healthcare. Ultra Low Power ECG Processing System for IoT Devices, pp. 7–12 (2019)

Nyon-Data, a Fall Detection Dataset from a Hinged Board Apparatus

Rogério Pais Dionísio[1,2(✉)], Ana Rafaela Rosa[1], and Cassandra Sofia dos Santos Jesus[1]

[1] Polytechnic Institute of Castelo Branco, 6000-084 Castelo Branco, Portugal
rdionisio@ipcb.pt

[2] DiSAC – Digital Services, Applications and Contents, 6000-767 Castelo Branco, Portugal

Abstract. Falls are one of the causes of severe hilliness among elders, and the COVID-19 pandemic increased the number of unattended cases because of the social distancing measures. This study aims to create a dataset that collects the data from a 3-axis acceleration sensor fixed on a hinged board apparatus that mimics a human fall event. The datalogging system uses off-the-shelf devices to measure, collect and store the data. The resulting dataset includes data from different angle positions and heights, corresponding to joints of the lower limbs of the human body (ankle, knee, and hip). We use the dataset with a threshold-based fall detection algorithm. The result from the Receiver Operating Characteristic curve shows a good behavior with a mean Area Under the Curve of 0.77 and allow to compute a best threshold value with False Positive Rate of 14.8% and True Positive rate of 89.1%. The optimal threshold value may vary depending on the specific population, activity patterns, and environmental conditions, which may require further customization and validation in real-world settings.

Keywords: ROC · Dataset · Fall Detection Algorithm · 3-axis Accelerometer

1 Introduction

During the COVID-19 pandemic, elderly individuals have experienced heightened levels of isolation, leading to detrimental effects on their mental and physical well-being. This topic has become particularly important, especially in more developed countries such as in Europe, where the highest rates of aging are recorded globally. In 2021, Portugal ranked fourth on the list of the oldest countries in Europe, only surpassed by Italy, Greece, and Finland [1], with an aging index of 182.1 [2]. Aging is a natural process that happens continuously and inevitably. It involves many biological, psychological, and social changes, some of which can result in losses or gains [3]. These changes can lead to a decrease in function, which makes older people more vulnerable. This vulnerability increases the likelihood of developing health problems and, as a result, increases the risk of falling [4].

Falls are one of the leading causes of death and illness among elderly people worldwide. In the population over 65 years of age, falls can cause numerous hospitalizations,

M. Ben Ahmed et al. (Eds.): SCA 2023, LNNS 906, pp. 391–401, 2024.
https://doi.org/10.1007/978-3-031-53824-7_36

institutionalizations, and premature deaths, which contribute to the loss of functional capacity, increased consumption of health resources and, above all, a loss of quality of life [5, 7]. In Portugal, the prevalence of fall-related injuries rises to 76% in the 65–74 age group, reaching 90% in the group of people aged 75 and over [8].

Because Castelo Branco district boasts one of the highest aging rates in the interior of Portugal (242.9) [9] and given that the National Police reported the presence of 1840 elderly individuals living in isolation within the district [10], the project ZELAR@CB was proposed in 2019 to monitor elder falls using wearable devices. But with the lockdown measures taken during the pandemic, all contacts with elder people were canceled. The restrictions imposed to curb the spread of the virus, such as social distancing and lockdown measures, have significantly impacted scientific projects involving elders in various ways.

Many scientific projects, as ZELAR@CB, require face-to-face interactions. Social distancing measures and restrictions on physical contact have limited the ability to conduct in-person data collection and interactions with older participants. Moreover, the pandemic has made it difficult to recruit and retain voluntary participants due to concerns about their vulnerability to the virus. Thus, researchers have had to adapt their methods, exploring innovative approaches to continue their scientific projects involving older adults.

1.1 Objectives and Structure of the Paper

This study aims to create a dataset that collects the data from a 3-axis acceleration sensor fixed on an apparatus that mimics a human fall event and test it with a threshold-based fall detection algorithm. The resulting dataset includes data from different angle positions and heights, corresponding to joints of the lower limbs of the human body (ankle, knee, and hip).

The present work is divided into 5 sections: the introduction section presents a framework and rationale of the paper's theme and the objectives of this work; then the experimental setup used in the production of the Nyon dataset is described, followed by the description of the dataset. Finally, the results and respective discussion will be presented, and we finalize with the main conclusions.

2 Experimental Setup

This section contains the methods used to collect the data. The materials used will be described, along with a description of how they were used during the data collection, in terms of hardware and the software developed for the data gathering procedure. Then, a description of the experimental setup is provided. Lastly, the experimental procedure's details are given.

2.1 Datalogging Electronic System

The purpose of the datalogging electronic system is to collect, process, and store data from a 3-axis digital accelerometer. The system we develop is made of off-the-shelf low-cost devices found in the laboratory, and consists of an accelerometer device, a data acquisition unit that converts analog signals into digital format, a processing unit, and a storage component such as a memory card. The wiring circuit was produced using Fritzing and is depicted in Fig. 1. In details, the bill of material is:

- An Arduino UNO or equivalent board;
- A 3-Axis digital accelerometer (ADXL345);
- A micro-SD card reader breakout board (Adafruit);
- A push-button to control the start and completion of the data collection process;
- A green LED to signal when the SD-card is recording;
- A red LED to signal that the program stopped successfully;
- Several resistors (220 Ω) for current limitation of the LED and the push-button;
- An electronic breadboard and conductive wires.

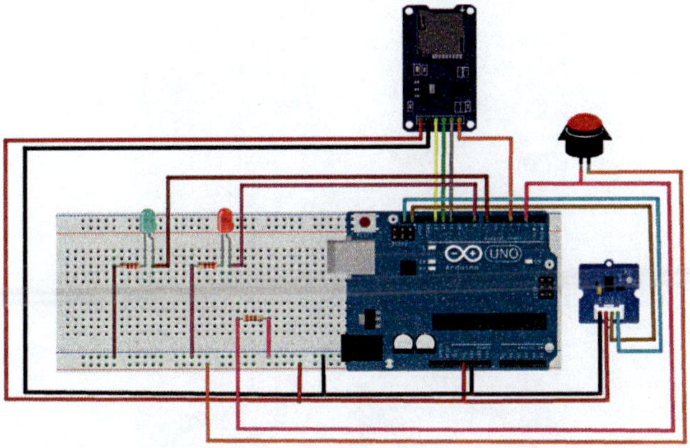

Fig. 1. Datalogger electronic system to measure and store acceleration measurements.

The program for the Arduino UNO simplifies the datalogging procedure, using a push-button to start and stop the data recording process, and two LED to inform the user about the state of the recording process. The pseudocode is described in Fig. 2.

2.2 Hinged Board

The sensor was taped in the surface of a hinged board, as seen in Fig. 3, at different height relative to the base, to emulate the position of the ankle, knee, and hip of a human body. The values were estimated from anthropometric proportions between the height of an adult human body (H) and the relative height of the ankle (0,039H), knee (0,285H) and hip (0,53H). Considering the Portuguese population, the mean heigh for men (172,9 cm) and women (163 cm) were averaged (168 cm), and the results are: Ankle height: 7 cm; Knee height: 48 cm; Hip height: 89 cm.

```
Green LED OFF and Red LED ON
int measure = 1
int inic_timestamp = micro()
int elapsed_time
forever
  if(button pressed && measure == 1)
      measure = 0
      Green LED ON and Red LED OFF
      inic_timestamp = micro()
      read(Acc_x,Acc_Y,Acc_z)
      elapsed_time = micro() - inic_timestamp
      record_SD(elapsed_time,Acc_x,Acc_Y,Acc_z)
  end if
  if(button pressed && measure == 0)
      measure = 1;
      Green LED OFF and Red LED ON;
  end if
```

Fig. 2. Pseudo-code program of the datalogger system

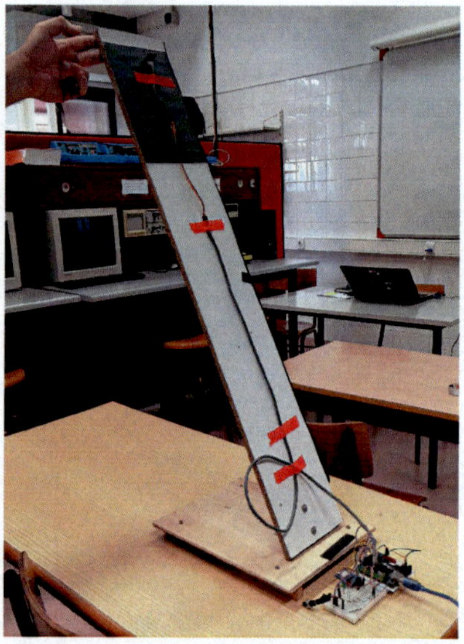

Fig. 3. Experimental apparatus to collect acceleration measurements in different scenarios (relative angle position and elevation of the accelerometer from the base).

3 Dataset Description

This section contains the information regarding the data included in the dataset. First the different positions and angles of the hinged board considered for the data gathering process, and finally a description of the organization of the data, from the different files to the format of the data in each file. The dataset is available on GitHub (https://github.com/rdionisio1403/Nyon/) under a Creative Commons (CC0) license.

3.1 Dataset Format Description

This dataset's files use the CSV format as it is widely used and can easily be formatted by researchers if necessary. The dataset is contained in one folder. The folder Nyon_data contains 9 files, and the name of the files is a combination the height of the sensor (Ankle, Knee, or Hip) and the falling angle (30°, 45° or 90°). As an example, the file "knee_30.csv" contains acceleration measurements, with the sensor placed on the hinged board at the Knee height (48 cm) and with starting angle of 30°. Each of the file's content is organized as presented in Table 1.

Table 1. Common structure of a dataset file.

Time (s)	Accel. X axis (g)	Accel. Y axis (g)	Accel. X axis (g)
0	X Value 1	Y Value 1	Z Value 1
0,02	X Value 2	Y Value 2	Z Value 2
0,04	X Value 3	Y Value 3	Z Value 3
...	...	...	...

Each file is composed of four columns. The first column gives the time in second, the second, third and fourth columns give the measured acceleration in X, Y and Z axes, respectively, in g units (one g is equal to 9.8 m/s^2). Each of the files contains 3 sequential emulated fall events, separated by a period without any movement, which ultimately results in 27 recorded fall events. The sampling interval is 20 ms and was defined from the hardware and software implementation of the datalogger described in

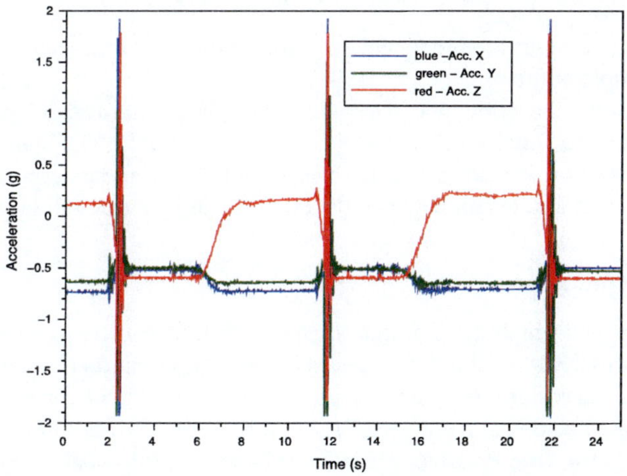

Fig. 4. Acceleration measurements in three axes, during three fall events with the hinged board starting at an angle of 45°, with the sensor at the ankle height (ankle_45.csv).

Sect. 2.1. With these files, an image with the timeline acceleration on three axes can be generated, as seen in Fig. 4.

4 Results and Discussion

This section provides an overview of the evaluation process conducted for the dataset and its reliability to implement fall detection algorithms. The evaluation encompasses three key sub-sections: Fall detection algorithms, Receiver Operating Characteristic (ROC) curve computation, and Discussion. Each sub-section investigates the specific aspects of the evaluation process. Scylab-2023.1.0 is the scientific tool we use to evaluate the dataset.

4.1 Fall Detection Algorithm

Threshold-based algorithms are commonly used in fall detection wearable devices to detect falls based on predefined thresholds of motion sensor data. These algorithms rely on the assumption that the motion patterns associated with falls are different from normal activities of daily living (ADLs) and can be detected by setting appropriate thresholds on sensor data.

Accelerometer-based threshold-based algorithms are widely used in fall detection wearable devices [12] and are easy to implement with low computational complexity. These algorithms analyze the acceleration data from the accelerometer sensor, which measures the change in velocity over time, to detect falls. Acceleration data is typically analyzed in three axes: X-axis (horizontal), Y-axis (vertical), and Z-axis (perpendicular to the other two axes). The Differential Sum Vector Magnitude (A_{DSVM}) [13], is often used as a feature for fall detection.

$$A_{DSVM}(n) = \sqrt{(x_n - x_{n-1})^2 + (y_n - y_{n-1})^2 + (z_n - z_{n-1})^2} \qquad (1)$$

where x, y and z are the three independent variables from the acceleration axes and n denotes the sample number.

Figure 5 shows the computed A_{DSVM} of one dataset file (ankle_90_degree.csv), represented with a red line. Three fall events are clearly visible. The blue line sequence, represent the true positive fall event that was timed during the dataset recording. The results show a good agreement between the estimate and the true fall events.

4.2 ROC Computation

The ROC curve is a graphical plot that displays the trade-off between sensitivity and specificity for different threshold values in a binary classification problem, such as fall detection. ROC curves are generated by plotting the True Positive Rate (TPR) on the y-axis against the False Positive Rate (FPR) on the x-axis, at different threshold values, where TP stands for True Positive, FN is False Negative, FP is False Positive, and TN mean True Negative. The algorithm is presented in Fig. 6.

$$TPR = \frac{TP}{TP + FN} \qquad (2)$$

$$FPR = \frac{FP}{FP + TN} \tag{3}$$

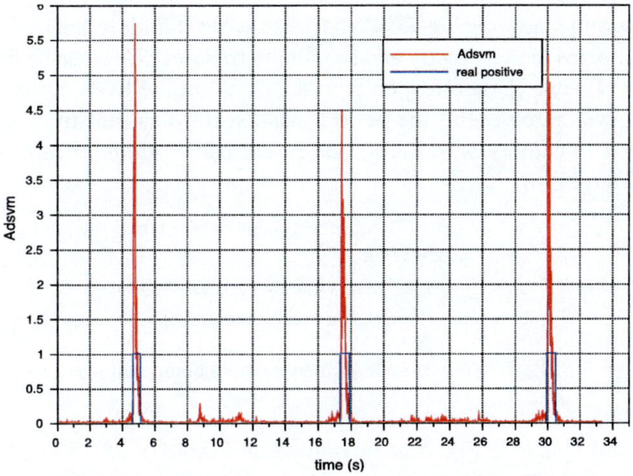

Fig. 5. A_{dsvm} computed from the dataset file ankle_90.csv.

```
TP = FP = TN = FN = 0
L = length(adsvm)
indx = 1
for th = 0 to maxThreshold
   for k = 1 to L
    if adsvm(k) > th
       y_pred(k) = 1
    else
       y_pred(k) = 0
    end if
   end for
   for k = 1 to L
    if (y_real(k) + y_pred(k)) == 2
       TP = TP + 1
    end if
    if (y_pred(k) = 1) and (y_real(k) ≠ y_pred(k))
       FP = FP + 1
    end if
    if (y_real(k) + y_pred(k)) == 0
       TN = TN + 1
    end if
    if (y_pred(k) == 0) and (y_real(k) ≠ y_pred(k))
       FN = FN + 1
    end if
   end for
   TPR(indx) = TP / (TP + FN)
   FPR(indx) = FP / (FP + TN)
   indx++
end for
```

Fig. 6. Pseudo-code for the ROC parameters computation.

`y_real` and `y_pred` are binary vectors (0 or 1) representing the real positive values (as seen in the blue line in Fig. 5) and the predicted positive values, respectively. The variable `adsvm` is a vector of the computed ADSVM. `MaxThreshold` is a constant value above the maximum value from the `adsvm` vector, for the resulting ROC curve to be fully represented. For the present case, the value 8 was used.

Threshold computation using ROC analysis involves finding the optimal threshold value that maximizes the sensitivity and specificity trade-off. This can be done by selecting the threshold value that corresponds to the point on the ROC curve closest to the top-left corner, which represents the perfect balance between sensitivity and specificity (i.e., the highest sensitivity with the lowest false positive rate), as represented by the code snippet from Fig. 7,

```
ampl = TPR-FPR;
best_ampl = find(ampl==max(ampl));
best_th = vect_th(best_ampl);
```

Fig. 7. Pseudo-code of the threshold computation.

The white dot visible in Fig. 8 is the optimal threshold (0.35), when TPR = 90.5% and FPR = 18.6%.

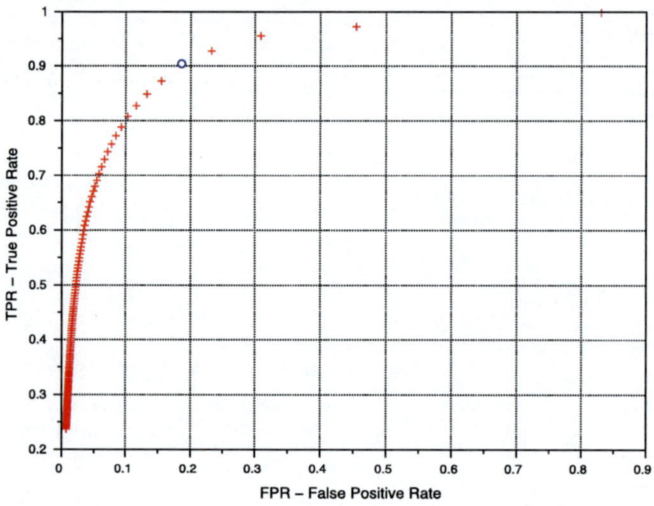

Fig. 8. Receiver operating curve of the fall detection device (hankle_90.csv).

4.3 Discussion

ROC analysis can provide valuable insights into the performance of threshold-based algorithms in fall detection devices, since it allows for the assessment of the algorithm's FPR and TPR at different threshold values, helping to determine the optimal threshold value for a particular population or use case. First, we assess the Area Under the Curve (AUC) for each ROC, as a measure of the performance of a binary classification model. The results from Table 2 shows high values of the AUC, with a mean value of 77.2%, that indicates a good-performing model for all scenarios.

Using the algorithm from Fig. 7, we compute the best threshold from each dataset file, and presented the value together with the resulting FPR and TPR. The results from Table 2 show that the best A_{dsvm} threshold value for the specific fall detection algorithm varies between 0.20 and 0.40, depending on the height and falling angle of the fall sensor. The mean value (0.23) can be used to test the algorithm for different fall events, although this value is not optimum for all cases. However, the mean FPR and TPR, respectively 14.8% and 89.1%, are good indicator as it concerns to the correctness and effectiveness of the algorithm. Other fall detection algorithms could be tested with the Nyon-data dataset, such as the Lightweight Neural Networks [14] and Machine Learning [15]. They have shown promising results in achieving high accuracy while being suitable for resource-constrained devices due to its ability to process data locally.

Table 2. Classification model results for each dataset

Dataset file	AUC	A_{dsvm} threshold	FPR	TPR
hankle_30.csv	83.1%	0.35	13.6%	92.5%
hankle_45.csv	79.0%	0.30	14.4%	90.0%
hankle_90.csv	76.0%	0.20	18.6%	90.5%
knee_30.csv	72.4%	0.35	11.7%	87.7%
knee_45.csv	62.8%	0.25	14.0%	78.5%
knee_90.csv	72.0%	0.40	10.0%	99.8%
hip_30.csv	87.0%	0.30	15.0%	93.7%
hip_45.csv	82.3%	0.40	15.2%	81.0%
hip_90.csv	80.3%	0.25	20.3%	88.2%
Mean value	**77.2%**	**0.23**	**14.8%**	**89.1%**
Std Deviation	**7.3%**	**0.07**	**3.16%**	**6.43%**

5 Conclusions

The process of aging is a natural and ongoing phenomenon that cannot be avoided. It involves a combination of biological, psychological, and social changes, which can result in both positive and negative outcomes. As people grow older, they may experience a decline in physical and cognitive abilities, making them more susceptible to various health issues. This increased vulnerability significantly raises the likelihood of developing diseases and, consequently, the risk of experiencing falls. Thus, the monitoring of vital parameters can improve the safety, mobility, communication, and monitoring of the elderly [16]. Some authors argue that a prompt response after a fall can significantly reduce the number of inherent consequences [17].The analysis from the dataset files reveals that the best threshold value for the proposed fall detection algorithm (A_{dsvm}) ranges between 0.20 and 0.40, depending on the height and falling angle of the fall sensor. However, the optimal threshold value may vary depending on the specific population, activity patterns, and environmental conditions, which may require further customization and validation in real-world settings. The dataset is free and available online, other fall detection algorithms can be evaluated and compared.

References

1. EUROSTAT. Archive: Estrutura populacional e envelhecimento. EUROSTAT (2021). https://ec.europa.eu/eurostat. Accessed 20 Dec 2021
2. PORDATA. Indicadores de envelhecimento segundo os Censos. PORDATA (2021). https://www.pordata.pt/Portugal/. Accessed 20 June 2022
3. Moreira, M.J.: Como envelhecem os portugueses: envelhecimento, saúde, idadismo. (1ª ed.). Fundação Francisco Manuel dos Santos (2020)
4. Silva, D., Brasileiro, M., Souza, D.: Relação entre envelhecimento da população e o risco de quedas: revisão integrativa. Revista Científica de Enfermagem RECIEN **8**(23), 28–38 (2018). https://www.recien.com.br
5. Agudelo-Botero, M., Giraldo-Rodríguez, L., Murillo-González, J., Mino-León, D., CruzArenas, E.: Factors associated with occasional and recurrent falls in Mexican community-dwelling older people. PLoS ONE **13**(2), 1–12 (2018). https://doi.org/10.1371/journal.pone.0192926
6. World Health Organization. Falls (2018). https://www.who.int/newsroom/fact-sheets/detail/falls
7. Sherrington, C., et al.: Exercise for preventing fall in older people living in the community: an abridged Cochrane systematic review. Br. J. Sports Med. **2020**(54), 885–891 (2019). https://doi.org/10.1136/bjsports-2019-101512
8. Coimbra, V., Marques, E., Chaves, C.: Prevalência de Quedas em Idosos Residentes numa Comunidade Rural. Millenium, **2**(3), 109–116 (2019). https://doi.org/10.29352/mill0203e.09.00218
9. PORDATA. Índice de envelhecimento - Onde há mais e menos idosos por 100 jovens? (2023). https://www.pordata.pt/municipios/indice+de+envelhecimento-458
10. GNR. Operação Censos Sénior 2022 (2022). https://www.gnr.pt/noticias.aspx?linha=20474
11. Luna-Perejón, F., Muñoz-Saavedra, L., Civit-Masot, J., Civit, A., Domínguez- Morales, M.: AnkFall – Falls, falling risks and daily-life activities dataset with an ankle-placed accelerometer and training using recurrent neural networks. Sensors **21**(5) (2021). https://doi.org/10.3390/s21051889
12. Kim, N.H., Yu, Y.S.: Fall recognition algorithm using gravity-weighted 3-axis accelerometer data. J. Inst. Electron. Inf. Eng. **50**(6), 254–259 (2013)
13. Igual, R., Medrano, C., Plaza, I.: Challenges, issues and trends in fall detection systems. Biomed. Eng. Online **12**(1), 1–24 (2013)
14. Wang, G., Li, Q., Wang, L., Zhang, Y., Liu, Z.: Elderly fall detection with an accelerometer using lightweight neural networks. Electronics **8**(11), 1354 (2019)

15. Özdemir, A.T., Barshan, B.: Detecting falls with wearable sensors using machine learning techniques. Sensors **14**(6), 10691–10708 (2014)
16. dos Santos Jesus, C.S., Rosa, A.R., Dionísio, R.P.: Nyon: a ubiquitous fall detection device for elders. In: Ben Ahmed, M., Boudhir, A.A., Santos, D., Dionisio, R., Benaya, N. (eds.) Innovations in Smart Cities Applications Volume 6. SCA 2022. Lecture Notes in Networks and Systems, vol. 629. Springer, Cham, pp. 175–184 (2023). https://doi.org/10.1007/978-3-031-26852-6_16
17. GONÇALVES, Fábio [et al.] - IoT system to monitor the well-being of senior citizens who self-isolate during the pandemic. In: European Conference on Networks and Communications and 6G, Porto, 8–11 de junho - EuCNC & 6G Summit (2021)

A Novel Approach for Detecting Fetal Electrocardiogram (FECG) Signals: Integration of Convolutional Neural Network (CNN) with Advanced Mathematical Techniques

Said Ziani(✉)

ENSAM, Mohammed V University in Rabat, Rabat, Morocco
s.ziani@um5r.ac.ma

Abstract. This paper presents an innovative method for recognizing fetal electrocardiogram (ECG) signals using a single-channel abdominal lead. The method combines the capabilities of Convolutional Neural Network (CNN) with advanced analytical techniques, including independent component analysis (ICA), Singular Value Decomposition (SVD), and Nonnegative Matrix Factorization (NMF), to reduce the dimensionality of the data. A crucial aspect of the fetal heart rate, which distinguishes it from the mother's heart rate, is the necessity for a time-scale representation that effectively captures the fetal electrical activity in terms of energy. Additionally, by separating the various components of the fetal ECG, it becomes possible to utilize them as inputs to the CNN model, thereby optimizing the reconstruction of the actual fetal ECG via the proposed method. The experimental findings demonstrate the effectiveness of this innovative technique, indicating the potential for real-time extraction of FECG signals. This method holds promise for enhancing fetal monitoring and healthcare applications.

Keywords: Identification · Deep Learning healthcare · Mathematical Techniques

1 Introduction

In recent years, there has been a growing interest in the field of biomedical signal processing, specifically regarding the extraction and analysis of fetal electrocardiogram (ECG) signals. Accurately detecting and analyzing the fetal ECG is crucial in prenatal monitoring as it offers valuable insights into the well-being and health of the developing fetus. However, extracting fetal ECG signals from the composite abdominal signal, which consists of a combination of maternal and fetal activities, presents a significant challenge.

Conventional signal processing techniques often face difficulties in effectively differentiating the fetal ECG from the maternal ECG due to similarities in their frequency content and the presence of noise sources. To overcome this challenge,

M. Ben Ahmed et al. (Eds.): SCA 2023, LNNS 906, pp. 402–410, 2024.
https://doi.org/10.1007/978-3-031-53824-7_37

advanced methods utilizing deep learning as mentioned in references [1,2] and [3] and time-scale representation asasted in [4] and [5]) of signals have gained considerable attention. Deep learning, particularly the Convolutional Neural Network (CNN) architecture, has exhibited remarkable capabilities in various image and signal processing tasks.

In this research paper, we propose an innovative approach to extract fetal ECG signals by harnessing the power of CNN and the time-scale image representation of a mixed maternal-fetal activity signal. The primary objective is to effectively separate and extract the fetal ECG components from the composite signal, enabling accurate analysis and monitoring of the fetal heart rate.

2 Methodology

2.1 Adopted Algorithm

The signal captured by a detector positioned on the maternal abdominal area comprises a mixture of embryonic electrocardiogram and maternal electrocardiogram data, as illustrated in Fig. 1. From a mathematical standpoint, the system is considered to be undetermined. Therefore, it can be stated as follows:

$$m = \nu.m_1 + \rho.m_2 + s \quad (1)$$

In this equation, both ν and ρ represent fixed, non-zero values that are not approaching zero.

m_1 corresponds to the fetal ECG (FECG). m_2 represents the maternal ECG (MECG). m denotes a mixture of signals from both m_1 and m_2. s represents the presence of noise.

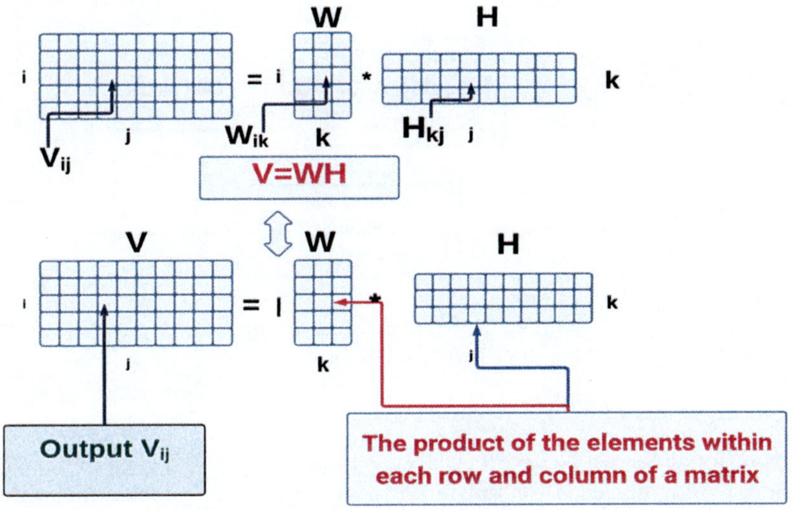

Fig. 1. Classical NMF

The primary focus of this analysis is to examine the combination of linear components in real time. The goal is to determine and differentiate between the sources represented by m_1 and m_2, while disregarding the influence of noise. This simplifies the process of localizing points corresponding to the fetal electrocardiogram (FECG) and maternal electrocardiogram (MECG). The strong and vigorous fetal heartbeat allows for precise modeling in both the time-frequency and time-scale domains, which plays a pivotal role in this investigation. Non-negative matrix factorization (NMF) determines how much the mother and infant participated as depicted in Fig. 1.

A comprehensive database comprising up to 999 image data points is constructed using time-scale images (TSI) and time-frequency images (TFI). With the available data, neural networks are implemented to effectively differentiate the electrical impulses originating from the maternal electrocardiogram (MECG) and the fetal electrocardiogram (FECG). Notably, during the signal separation phase, the CNN model uses TSI images and TFI images. Applying the proposed model illustrated in Fig. 2, the recovering fetal electrical activity is then obtained.

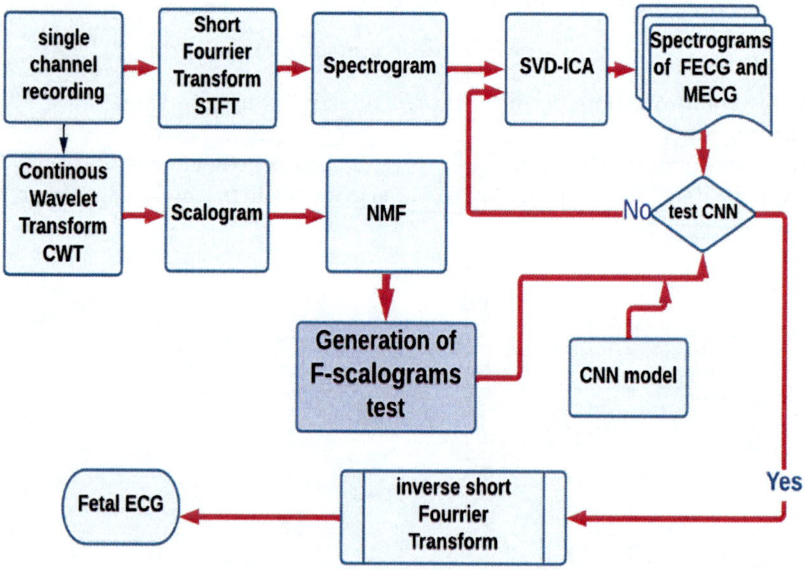

Fig. 2. Global algorithm scheme.

2.2 Description of CNN Model

A Convolutional Neural Network (CNN) as in [6] and [7], is a type of deep learning model commonly used for image and signal processing tasks, including the extraction of features from signals such as the fetal electrocardiogram (FECG). Here is a general description of a CNN model for FECG extraction:

- **Input Preparation:** The FECG signal is typically acquired from recordings and represented as a time-series signal. Prior to feeding it into the CNN, the signal may undergo preprocessing steps such as filtering, resampling, and normalization to enhance its quality and facilitate further analysis.
- **Convolutional Layers:** The CNN model starts with one or more convolutional layers. Each layer consists of multiple filters (also called kernels) that slide across the input signal. The filters detect specific patterns or features by performing element-wise multiplications and summations. The output of each filter represents the activation or response to a particular feature in the input signal.
- **Activation Function:** After the convolution operation, an activation function is applied element-wise to introduce non-linearity into the network. Common activation functions include ReLU (Rectified Linear Unit), sigmoid, and tanh, which help capture complex relationships and make the model more expressive.
- **Pooling Layers:** Following the activation, pooling layers are often employed to reduce the spatial dimensions of the feature maps. Max pooling is a commonly used technique that downsamples the feature maps by selecting the maximum value within each pooling window. This helps to extract the most salient features while reducing computational complexity.
- **Fully Connected Layers:** The output from the convolutional and pooling layers is then flattened and connected to one or more fully connected layers. These layers perform high-level feature extraction and mapping by applying weights and biases to the input. The final fully connected layer typically has a number of neurons equal to the desired output size.
- **Output Layer:** The final layer of the CNN is the output layer, which is designed based on the specific task. For FECG extraction, the output layer may consist of a single neuron that predicts the presence or absence of the fetal ECG signal, or it may have multiple neurons corresponding to different classes or labels if there are additional classification tasks involved.
- **Training and Optimization:** The CNN model is trained using a labeled dataset, where the inputs are FECG signals and the corresponding outputs are the desired labels or predictions. The training process involves minimizing a loss function, such as categorical cross-entropy or mean squared error, by adjusting the weights and biases in the network through optimization algorithms like stochastic gradient descent (SGD).
- **Evaluation and Testing:** Once trained, the performance of the CNN model is assessed using an independent testing dataset. Metrics such as accuracy, precision, recall, and F1-score can be used to evaluate its effectiveness in extracting the FECG features.

It's important to note that the specific architecture and parameters of the CNN model may vary depending on the specific requirements of the FECG extraction task, the available dataset, and other factors. The description provided here serves as a general framework for understanding the components and process involved in using a CNN for FECG extraction. Our method attempts to accomplish the initial goal of predicting the class of incoming ECG pictures TSI in Fig. 3. By dividing the images into two crucial output elements, this is achieved:

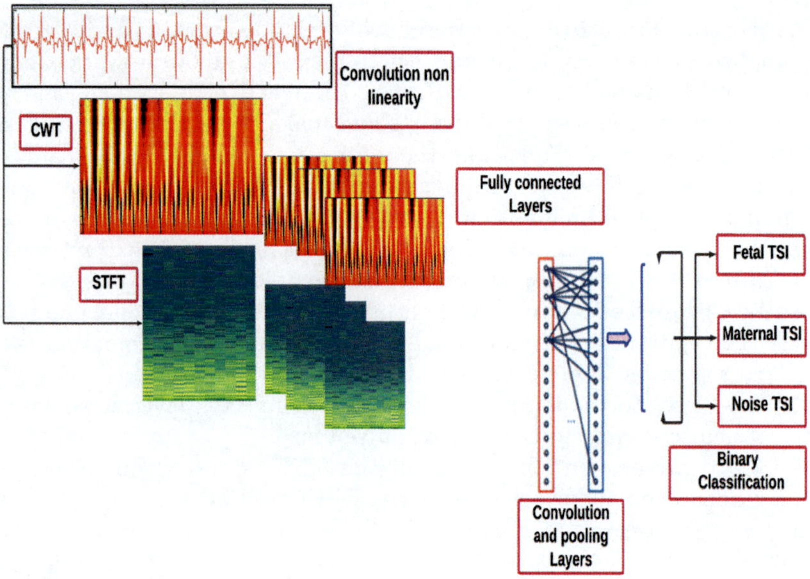

Fig. 3. CNN model.

3 Results and Discussion

3.1 Recorded Data

A single channel from the Daisy worldwide database is used to derive the fetal electrocardiogram (FECG) from cutaneous recordings as referenced in [8]. In order to construct clear time-frequency and time-scale images, we opted for preprocessing the input abdominal signals using the discrete wavelet transform. This approach was chosen to enhance the quality of the signals prior to further analysis. Figure 4 illustrates the signal before and after the preprocessing stage.

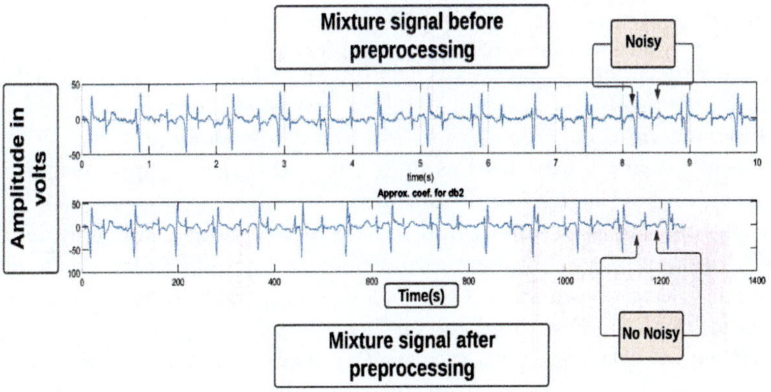

Fig. 4. Observed signals.

3.2 Recovered Sources

SVD-ICA is repeated until the CNN algorithm finds a signal that fits the CWT-NMF model's desired result. Figure 5 shows the fetal electrocardiogram (FECG). Figure 5 also classifies the fetal the ventricular complex, offering further information.

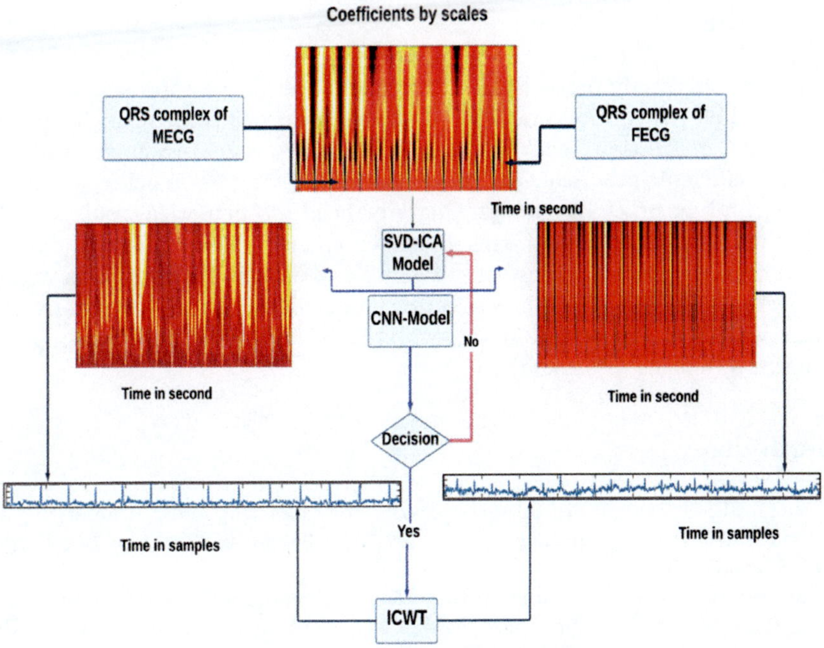

Fig. 5. Extracted signal FECG.

3.3 Discussion

This study's methodology generated astonishing findings. The extracted fetal electrocardiogram (FECG) signal exhibits an impressively low signal-to-noise ratio, requiring minimal implementation and programming time. When compared to recent works in the field as in [9,10], and [11], this approach stands out for its exceptional reliability and accuracy.

The approach uses advanced mathematical techniques that may be computationally designed, even for twin situations, and simply executed in real time utilizing embedded systems like DSP or Raspberry Pi. However, integrating the CNN model, one of the most advanced Deep Learning algorithms, with current mathematical approaches is a major contribution of this research. This combination allows a rigorous comparison with the ideal case, resulting in an FECG signal that closely approximates reality.

Thus, employing a single electrode for medical diagnostics reduces the risks of several sensors, such as cable-related hazards, current waves, and electrical tension, which can harm the mother and fetus. Researchers and academics can easily obtain CNN model data, which is self-generated and can be enlarged to millions of examples.

STFT-based time-frequency analysis can also create a new database. CNN and other deep learning approaches have been used in energy applications [12–15] and [16–21].

4 Conclusion

In summary, this study presents an effective and innovative approach for extracting the fetal electrocardiogram (FECG) from cutaneous recordings. The results demonstrate impressive accuracy with a near-zero signal-to-noise ratio. The method is efficient, practical, and outperforms existing approaches in terms of reliability and accuracy. By integrating advanced mathematical tools and deep learning with the CNN model, this approach enables reliable medical diagnosis using a single electrode. The use of self-generated data and potential for creating extensive databases enhance accessibility and scalability for researchers. Overall, this work has the potential to revolutionize FECG extraction, improving healthcare outcomes in obstetrics.

References

1. Said, Z., El Hassouani, Y.: A new approach for extracting and characterizing fetal electrocardiogram. Traitement du Sig. **37**(3), 379–386 (2020). https://doi.org/10.18280/ts.370304
2. Jamshidian-Tehrani, F., Sameni, R., Jutten, C.: Temporally nonstationary component analysis; application to noninvasive fetal electrocardiogram extraction. IEEE Trans. Biomed. Eng. **67**(5), 1377–1386 (2020). https://doi.org/10.1109/TBME.2019.2936943

3. Ziani, S., El Hassouani, Y., Farhaoui, Y.: An NMF based method for detecting RR interval. In: Farhaoui, Y., Moussaid, L. (eds.) ICBDSDE 2018. SBD, vol. 53, pp. 342–346. Springer, Cham (2019). https://doi.org/10.1007/978-3-030-12048-1_35
4. Yin, P., Yuan, R., Cheng, Y., Wu, Q.: Deep guidance network for biomedical image segmentation. IEEE Access **8**, 116106–116116 (2020). https://doi.org/10.1109/ACCESS.2020.3002835
5. Eikvil, I., Taxt, T., Moen, K.: A fast adaptive method for binarization of document images. In: International Conference on Document Analysis and Recognition, ICDAR 1991, France (1991)
6. Kollias, D., Zafeiriou, S.: Exploiting multi-CNN features in CNN-RNN based dimensional emotion recognition on the OMG in-the-wild dataset. IEEE Trans. Affect. Comput. **12**(3), 595–606 (2021). https://doi.org/10.1109/TAFFC.2020.3014171
7. Wang, M., et al.: Low-latency in situ image analytics with FPGA-based quantized convolutional neural network. IEEE Trans. Neural Netw. Learn. Syst. **33**(7), 2853–2866 (2022). https://doi.org/10.1109/TNNLS.2020.3046452
8. De Moor, B., De Gersem, P., De Schutter, B., Favoreel, W.: DAISY: a database for fiction of systems. J. A Spec. Issue CACSD (Comput. Aided Control Syst. Des.) **38**(3), 4–5 (1997)
9. Ziani, S.: Contribution to single-channel fetal electrocardiogram identification. Traitement du Sig. **39**(6), 2055–2060 (2022). https://doi.org/10.18280/ts.390617
10. Ziani, S., Farhaoui, Y., Moutaib, M.: Extraction of fetal electrocardiogram by combining deep learning and SVD-ICA-NMF methods. Big Data Min. Anal. **6**(3), 301–310 (2023). https://doi.org/10.26599/BDMA.2022.9020035
11. Ziani, S.: Fetal electrocardiogram identification using statistical analysis. In: Farhaoui, Y., Rocha, A., Brahmia, Z., Bhushab, B. (eds.) Artificial Intelligence and Smart Environment, ICAISE 2022. LNNS, vol. 635. Springer, Cham (2023). https://doi.org/10.1007/978-3-031-26254-8-64
12. Chaou, Y., et al.: Nonlinear control of the permanent magnet synchronous motor PMSM using backstepping method. WSEAS Trans. Syst. Control **17**, 56–61 (2022). https://doi.org/10.37394/23203.2022.17.7
13. Ouhadou, M., et al.: Experimental modeling of the thermal resistance of the heat sink dedicated to SMD LEDs passive cooling. In: Proceedings of the 3rd International Conference on Smart City Applications (2018)
14. Ben, H., et al.: Permanent magnet synchronous motor PMSM control by combining vector and PI controller. WSEAS Trans. Syst. Control **17**, 244–249 (2022)
15. Ziani, S.,Jbari, A.,Bellarbi, L.: QRS complex characterization based on non-negative matrix factorization NMF. In: 2018 4th International Conference on Optimization and Applications (ICOA), Mohammedia, Morocco, pp. 1–5 (2018). https://doi.org/10.1109/ICOA.2018.8370548.
16. Ziani, S., El Hassouani, Y.: Fetal-maternal electrocardiograms mixtures characterization based on time analysis. In: 2019 5th International Conference on Optimization and Applications (ICOA), Kenitra, Morocco, pp. 1–5 (2019). https://doi.org/10.1109/ICOA.2019.8727619
17. Ziani, S., El Ghmary, M., Youssef, A.Z.: Permanent magnet synchronous motor control performed using PI-backstepping with a model of harmonics reduction. Int. J. Power Electron. Drive Syst. **14**(1), 199–208 (2023)
18. Hafid, B.A., et al.: A quadratic observer for sensorless drive system controller. In: Farhaoui, Y., Rocha, A., Brahmia, Z., Bhushab, B. (eds.) Artificial Intelligence and Smart Environment, ICAISE 2022. LNNS, vol. 635, pp. 109–113. Springer, Cham (2023). https://doi.org/10.1007/978-3-031-26254-8_16

19. Ben, H., et al.: PI controller and quadratic feedback of synchronous machine. In: Farhaoui, Y., Rocha, A., Brahmia, Z., Bhushab, B. (eds.) Artificial Intelligence and Smart Environment, ICAISE 2022. LNNS, vol. 635, pp. 668–673. Springer, Cham (2023). https://doi.org/10.1007/978-3-031-26254-8_97
20. Youssef, C., et al.: Electric vehicle backstepping controller using synchronous machine. In: Farhaoui, Y., Rocha, A., Brahmia, Z., Bhushab, B. (eds.) Artificial Intelligence and Smart Environment, ICAISE 2022. LNNS, vol. 635, pp. 367–373. Springer, Cham. https://doi.org/10.1007/978-3-031-26254-8_52
21. Youssef, C., et al.: Backstepping control of the permanent magnet synchronous generator (PMSG) used in a wind power system. In: Farhaoui, Y., Rocha, A., Brahmia, Z., Bhushab, B. (eds.) Artificial Intelligence and Smart Environment, ICAISE 2022. LNNS, vol. 635, pp. 276–281. Springer, Cham (2023). https://doi.org/10.1007/978-3-031-26254-8_38

A New Machine-Learning Approach to Prognosticate Poisoned Patients by Combining Nature of Poison, Circumstances of Intoxication and Therapeutic Care Indices

Rajae Ghanimi(✉), Fadoua Ghanimi, Ilyas Ghanimi, and Abdelmajid Soulaymani

Ibn Tofail University, Av. de L'Université, Kénitra, Morocco
ghanimi.rajae@gmail.com

Abstract. Estimating patient prognosis has always been a challenge for physicians. One of the first questions asked by a patient entourage during his treatment is related to his vital prognosis. Thus, practitioners, even with several years of experience and the great progress of technology still have difficulties to estimate prognosis. They have tried to build prognostic allowing them to make an estimate as close as possible to the real future of each patient. However, these scores still have many limitations today. In particular, they are confronted with the individual complexity of each patient (history, physiological age, etc.). These imperfections tend to limit their use in the daily practice of medicine. In this work, we propose a machine learning based prognosis-prediction tool for poisoning cases. We use the data of poisoned patients delivered by the Poisoning Centre of Morocco. The Selected features will be fed into five machine learning techniques, including XGBoost, Support Vector Machine (SVM), Naïve Bayes, Decision Tree and Random Forest.

Keywords: artificial intelligence · machine learning · prognosis · poisoning · emergency care

1 Introduction

Poisoning is one of the leading causes of admission emergency and intensive care units. It represents a major problem of public health. Poisoning is defined as the occurrence of any toxic effect for humans causing following a single or repeated exposure to a mixture or a natural substance or synthetic, available on the market or present in the environment [1]. This effect can occur within a more or less significant time depending on the type of toxic, the level and the route of exposure [2].

A recent study, where 141 countries were concerned, analyzed 157 scientific publications and the World Health Organization (WHO) database on mortality. Researchers estimate that around 385 million cases of acute unintentional poisonings occur worldwide each year, including around 11,000 deaths.

M. Ben Ahmed et al. (Eds.): SCA 2023, LNNS 906, pp. 411–419, 2024.
https://doi.org/10.1007/978-3-031-53824-7_38

Hospital mortality rate can be improved, with early management, based on an early treatment of poisoned persons. Although disturbances of consciousness are the most common symptoms frequent, the severity is most often related to failures cardiovascular, respiratory and hepatic.

Support generally involves four stages, often intertwined:

- Confirming the diagnosis;
- Assessing severity and prognosis;
- Judging the necessity: Symptomatic treatment, evacuating treatment, purifier or antidote;
- Determining, if necessary, preventive measures.

In order to limit the number of fatal poisonings, several measures can be implemented: The first one is, an effective method to improve diagnosis time by optimizing the emergency triage phase through the prediction of treatment and patient's prognosis using Artificial Intelligence.

In this paper, we present a novel machine learning based approach to prognosticate poisoned patients by combining nature of poison, circumstances of intoxication and therapeutic care indices.

Section 2 presents the general approach and the different factors involved in the process of prognostic. The steps of the new approach, the theoretical bases and an experimental validation of the proposed method are presented in Sect. 3. The paper ends with a conclusion.

2 General Prognostic: General Approach

Faced to an intoxication case, it is necessary to determine the factors which affect the prognosis. This is a fundamental step since it largely determines the therapeutic strategy and the follow-up. The evolution of the poisoning as a dynamic phenomenon depends on the kinetics and the toxic dynamics of the toxin [3]. It is necessary to distinguish between the momentary seriousness of the poisoning, the potential seriousness and the prognosis.

The severity depends on many closely related factors: The toxin and its mechanism of toxicity, the symptoms, the analytical data, the type of poisoning and the field. "It's the dose that makes the poison", but not only. There, the actual or potential initial severity of the poisoning is directly related to the nature of the poison, the dose or the duration of exposure, the route of entry, the kinetics.

Nature of the Poison. Is the toxin likely to disrupt vital functions: respiratory, cardiovascular, impaired consciousness? What is the mechanism of action? Is it a functional or lesion toxic?

Mechanisms of Toxicity The mechanisms of toxicity are essential for the interpretation of the relationship between symptoms and plasma concentrations and the prognosis. Schematically, two types of toxins can be distinguished.

Functional toxins: Temporarily modify a function of the body or an organ. There is a fairly good correlation between plasma concentrations and toxicity when the latter is related to the toxin itself and not to the metabolites. There, the severity of symptoms is proportional to the transient concentration of target organ or receptor toxicity, and thus plasma concentrations. The duration of symptoms depends on the elimination half-life and the decrease in these concentrations. Many toxins, in particular most drugs, are functional: psychotropics, cardiotropics, theophylline, lithium. The same is true for ethanol. The evolution is generally favorable insofar as there were no complications.

Lesion toxins: Induce cell and/or tissue lesions (paracetamol, paraquat, heavy metals, colchicine, toxoids for example). Toxicity and severity depend on the maximum concentration that has been, is or will be reached in the target organ. There is no correlation between momentary plasma concentrations and symptoms. Symptoms may persist despite the disappearance of the toxin from the body. Plasma concentrations have prognostic value when analyzed relative to the time of ingestion. Plasma levels thus predict the risk of hepatotoxicity in the event of intoxication with paracetamol and the risk of fatal evolution in the event of intoxication with paraquat (a herbicide) [4, 5].

Dose and Duration of Exposure: The dose assumed to be ingested is an element of gravity, but it must be confronted with the symptoms or, in certain cases, with analytical data (paracetamol);for drugs, the risk is all the more important as the therapeutic margin is narrow (antiarrhythmics, digoxin, colchicine). For toxins with an anoxic effect (carbon monoxide, cyanide).

Route of Entry: Poisoning by injection or inhalation are particularly serious because of the rapidity of tissue absorption and diffusion [5, 6].

3 General Prognostic: General Approach

In this work, we present the adopted methodology to develop a machine-learning model based on different algorithms that would be able to guide the doctor in his intoxication treatment process, by directing him to predict and to improve prognosis. To understand our approach, a presentation on the methodology adopted is essential. The following flowchart (Fig. 2) summarizes the stages of the work carried out.

3.1 Database Description

We worked on the database of the national poison control center of MOROCCO, including 29 268 cases of poisoning from March 1980 to December 2014, in order to extract the data necessary for training an algorithm.

The data we selected is made up of 6 variables and 215 symptoms. Overall speaking, there are 6 categorical responses variables: age range (Fig. 1), sex, route of entry, nature of poison, circumstances of intoxication and therapeutic care. The table below illuminates the meaning of each variable. Detailed information could be seen in Table 1.

Table 1. Variables description

Variable name	Descriptions and (under symptoms)	Variable type	Rage of value
Age range	Age range of the patient N.born: 0; infant:1; baby walking: 3; child: 4; adolescent: 5; adult:6	categorical	0–6
Sex	Sex of the patient F = 0; M = 1	Categorical	0–1
route of entry	Cutaneous: 1; inhalation: 2; injection:3; eyepiece:4; oral: 5; percutaneous: 6; rectal: 7; transplacental: 8	Categorical	0–8
nature of poison	Food: 1; animals: 2; foreign body: 3; cosmetic: 4; drugs: 5; drugs: 6; heavy metals: 7; pesticides and agricultural products: 8; plants: 9; gaseous products: 10; industrial products: 11; chemicals: 12; household products: 13; mineral products: 14	Categorical	0–14
circumstances of intoxication	Abortion: 1; suicidal: 2; criminal: 3; drug addiction: 4	Categorical	0–4
therapeutic care	1 = yes; 0 = no	Categorical	0–1

From Fig. 1 we can see that in the data set, most patients are “adults” and “teenager”, while only few people have intoxication are aging under 5 or above 60.

According to Fig. 2, most maximum poisoning cases of poisoning are due to drugs, followed by pesticides, while the 3rd place is occupied by drugs followed by household products.

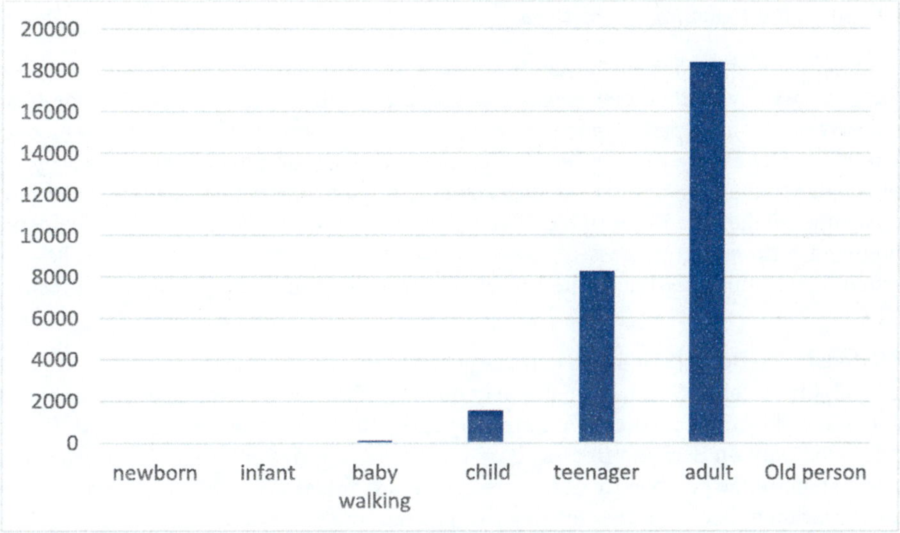

Fig. 1. Age range of patients.

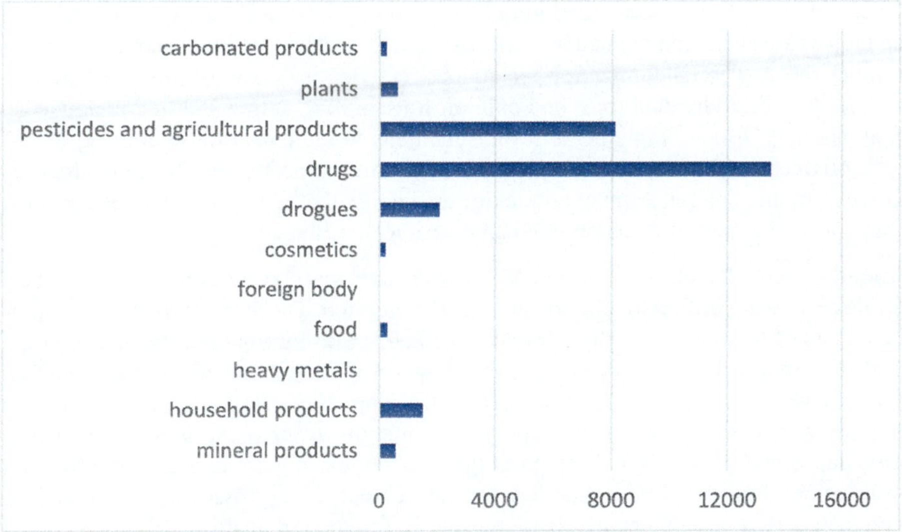

Fig. 2. Nature of poison

For data description, the research will use the describe function and pandas profiling in Python to summarize the dataset. The raw data contained 6 variables for 29268 patients.

3.2 Machine Learning Algorithms

Machine learning is one of the fields of study of artificial intelligence. This concept encompasses any method that makes it possible to build a model of reality from data, either by improving a partial model or less general, or by completely creating the model. It refers to the ability of a system to acquire and integrate new autonomous way of knowledge to the development, analysis and the implementation of methods that allow a machine (in the broad sense) to evolve and perform tasks associated with an artificial intelligence through a process of learning. This learning makes it possible to have a system that is optimized according to environment, experiences and observed results [7].

XGBoost. XGBoost [8] is a decision tree ensemble based on gradient boosting designed to be highly scalable. Similarly to gradient boosting, XGBoost builds an additive expansion of the objective function by minimizing a loss function. Considering that XGBoost is focused only on decision trees as base classifiers, a variation of the loss function is used to control the complexity of the trees. This loss function can be integrated into the split criterion of decision trees leading to a pre-pruning strategy. An additional regularization parameter in XGBoost is shrinkage, which reduces the step size in the additive expansion. Finally, the complexity of the trees can also be limited using other strategies as the depth of the trees, etc. A secondary benefit of tree complexity reduction is that the models are trained faster and require less storage space. Furthermore, randomization techniques are also implemented in XGBoost both to reduce overfitting and to increment training speed. The randomization techniques included in XGBoost are: random subsamples to train individual trees and column subsampling at tree and tree node levels. In addition, XGBoost implements several methods to increment the training speed of decision trees not directly related to ensemble accuracy. Specifically, XGBoost focuses on reducing the computational complexity for finding the best split, which is the most time-consuming part of decision tree construction algorithms [9].

Support Vector Machine. Support Vector Machine was chosen as one of the models because it is an algorithm for classification and regression. The Support Vector Machines are a class of learning algorithms initially defined for discrimination, that is to say the prediction of a variable binary qualitative. They were then generalized to the prediction of a variable quantitative. In the case of the discrimination of a dichotomous variable, they are based on the search for the optimal margin hyper plane which, when it is possible, correctly classifies or separates the data while being as far as possible from all observations. The principle is therefore to find a classifier, or a discrimination function, with the greatest possible generalization capacity (prediction quality) [10].

This approach stems directly from Vapniks work in the theory of learning from 1995. It focused on the generalization (or prediction) properties of a model by controlling its complexity. The principle founding principle of SVMs is precisely to integrate the control of the complexity, i.e. the number of parameters associated in this case with the number of support vectors. The other guiding idea of Vapnik in this development is to avoid substituting for the initial objective.

The basic principle of SVMs is to reduce the problem of discrimination to the linear one of finding an optimal hyper plane. Two ideas or tricks to achieve this goal:

- The first consists in defining the hyper plane as the solution of a problem constrained optimization whose objective function is not expressed than using scalar products between vectors and in which the number "active" constraints or support vectors control the complexity of the model.
- The transition to the search for nonlinear separating surfaces is obtained by the introduction of a kernel function in the scalar product implicitly inducing a non-linear transformation of the data to an intermediate space (feature space) of greater dimension.

This tool is becoming widely used in many types of application and proves to be a serious competitor of the most efficient algorithms (aggregation of models). It should be noted that, on the algorithmic level, these algorithms are more penalized by the number of observations, i.e. the number of potential support vectors, than by the number of variables. Nevertheless, efficient versions of the algorithms allow take into account large databases in times of calculation acceptable.

Decision Tree. A decision tree is a classifier expressed as a recursive partition of the instance space. The decision tree consists of nodes that form a rooted tree, meaning it is a directed tree with a node called "root" that has no incoming edges. All other nodes have exactly one incoming edge. A node with outgoing edges is called an internal or test node. All other nodes are called leaves (also known as terminal or decision nodes). In a decision tree, each internal node splits the instance space into two or more sub-spaces according to a certain discrete function of the input attributes values. In the simplest and most frequent case, each test considers a single attribute, such that the instance space is partitioned according to the attribute's value. In the case of numeric attributes, the condition refers to a range.

Each leaf is assigned to one class representing the most appropriate target value. Alternatively, the leaf may hold a probability vector indicating the probability of the target attribute having a certain value. Instances are classified by navigating them from the root of the tree down to a leaf, according to the outcome of the tests along the path [11].

Random Forest. Random Forest or random forests is a supervised learning algorithm very popular. It is also used for regression or classification problems. Based on a set of learning algorithms, an algorithm creates many decision trees (this is the reason for which it is called a forest) on various subsets of the dataset. Random forests are a combination of tree predictors such that each tree depends on the values of a random vector sampled independently and with the same distribution for all trees in the forest. The generalization error for forests converges to a limit as the number of trees in the forest becomes large. The generalization error of a forest of tree classifiers depends on the strength of the individual trees in the forest and the correlation between them. Using a random selection of features to split each node yields error rates that compare favorably to Adaboost [12], but are more robust with respect to noise. Internal estimates monitor error, strength, and correlation and these are used to show the response to increasing the number of features used in the splitting. Internal estimates are also used to measure variable importance. These ideas are also applicable to regression [12].

Random forests are an effective tool in prediction. Because of the Law of Large Numbers they do not overfit. Injecting the right kind of randomness makes them accurate classifiers and regressors. Furthermore, the framework in terms of strength of the individual predictors and their correlations gives insight into the ability of the random forest to predict. Using out-of bag estimation makes concrete the otherwise theoretical values of strength and correlation.

4 Results

A total of 29268 patients were enrolled in the study. The results of the accuracy score, precision score, recall score, and F1-score for the nine ML algorithms is provided in Table 2. Overall, the best-performing ML algorithm was XGBoost. Values for accuracy, precision, recall and precision scores were comparable across the various models with a range of 0.72 (Naïve Bayes) to 0.81 (XGBoost) for accuracy score, a range of 0.80 (Decision Tree) to 0.88 (Naïve Bayes) for the precision score, a range of 0.66 (Naïve Bayes) to 0.96 (random forest) for recall score and a range of 0.76 (Naïve Bayes) to 0.90 (XGBoost) for F1-score.

Table 2. Prognosis results for the five machine learning algorithms with the test dataset

	Accuracy score	Precision score	Recall score	F1-score
Naïve Bayes	0.72	0.88	0.66	0.76
Decision Tree	0.77	0.80	0.93	0.86
Random forest	0.86	0.85	0.96	0.90
XGBoost	0.81	0.84	0.93	0.88
SVM	0.80	0.83	0.93	0.88

5 Conclusion

Machine learning prognosis prediction in intoxication is a promising approach to improve patient care by creating more accurate and reliable predictive models. The various machine learning algorithms have shown their ability to identify risk factors and classify patients according to their severity to better guide treatments. However, their use requires a large amount of data and their design must take into account algorithmic biases in order to ensure the safety and ethics of care. Despite these limitations, the use of machine learning technology to predict the prognosis of poisonings is a promising advance in the field of personalized medicine, offering a new approach to help healthcare professionals dealing with patients in critical situations. Clinicians can use these predictions to guide patient management and decide which therapy is most appropriate for each patient. It must be used with caution and taking into account the limitations of the available data. Clinicians should also consider individual patient characteristics for optimal management.

However, the results may be improved by addition of new variables like heart rate and blood pressure that are commonly used variables in medicine to assess a patient's condition. Improving data quality is also important to ensure that the data used in the algorithm is of high quality. This may mean collecting additional data to fill gaps in existing data, using laboratory tests to confirm diagnoses, and avoiding missing or erroneous data.

References

1. Boedeker, W., Watts, M., Clausing, P., et al.: The global distribution of acute unintentional pesticide poisoning: estimations based on a systematic review. BMC Public Health **20**, 1875 (2020)
2. Casey, P.B., Dexter, E.M., Michell, J., Vale, A.: The prospective value of the IPC/EC/EAPCCT poisoning severity score in case of poisoning. Clin. Toxicol. **36**, 215–217 (1998)
3. B. Mégarbane, L., Donetti, T., Blanc, G., Chérond, F.: Jacobs Recommandations d'experts Intoxications graves par médicaments et substances illicites en réanimation: Réanimation **15**, 332–342 (2006)
4. Bartoli, M., et al.: Recommendations for the prescription, implementation and interpretation of medical examinations in biology in the context of severe poisoning. Ann. Biol. Clin. **70**(4), 431–450 (2012)
5. Eddleston, M., Juszczak, E., Juszczak, E., Buckley, N.A., for the Ox-Col Poisoning Study collaborators: Multiple- dose activated charcoal in acute self-poisoning: a randomised controlled trial. Lancet **371**, 579–87 (2008)
6. Mégarbane, B.: Présentation clinique des principals intoxications et approche par les toxidromes. Réanimation **21**, S482–93 (2012)
7. Kühl, N., Schemmer, M., Goutier, M., et al.: Artificial intelligence and machine learning. Electron Markets **32**, 2235–2244 (2022)
8. Chen, T., Guestrin, C.: Xgboost: a scalable tree boosting system. In: Proceedings of the 22Nd ACM SIGKDD International Conference on Knowledge Discovery and Data Mining, KDD 2016, pp. 785–794. New York, NY, USA (2016)
9. Candice, B., Anna, C., Gonzalo, M.-M.: A Comparative Analysis of XGBoost (2019)
10. Vapnik, V.N.: The Nature of Statistical Learning Theory. Verlag, New York (1995)
11. Breiman, L.: Mach. Learn. **45**(1), 5–32 (2001)
12. Schapire, Robert, E.: Explaining adaboost. Empirical Inference 37–52. Springer (2013)
13. Ghosh, S., Ali, R., Verma, S.: Aβ-oligomers: A potential therapeutic target for Alzheimer's disease. Int. J. Biol. Macromol. **239** (2023)
14. Elouardi, M., et al.: A review of botanical, biogeographical phytochemical and toxicological aspects of the toxic plants in Morocco. Toxicologie Analytique et Clinique **34**(4), 215–228 (2022)

Efficient Throughput Allocation for Emergency Data Transmission in IoMT-Based Smart Hospitals

Fathia Ouakasse[1,2](✉), Afaf Mosaif[2], and Said Rakrak[2]

[1] Private University of Marrakesh, Marrakesh, Morocco
fathia.ouakasse@gmail.com
[2] Cadi Ayyad University, Marrakesh, Morocco

Abstract. In recent years, the healthcare field has been boosted by the use of new technologies such as the Internet of Medical Things (IoMT), Artificial Intelligence, and Big Data. Furthermore, with the spread of the COVID-19 pandemic, the need for smart and remote healthcare services was greatly amplified leading to the accelerated adoption of digital health solutions. Nevertheless, with the increasing amount of data generated by IoMT devices, the network can undergo the problem of packet latency, packet loss, and congestion leading to delays in the transmission of data that may be urgent. Thus, it becomes essential to develop effective algorithms that can manage notifications' flow based on their importance levels and delay tolerance. In this paper, we propose an effective algorithm to dynamically allocate the throughput in an IoMT-based smart hospital in response to data emergencies. The algorithm takes into account the criticality of the data and the available bandwidth to allocate the necessary throughput to IoMT devices to ensure that the critical data is transmitted without delay.

Keywords: IoMT · Medical devices · Smart healthcare · Throughput allocation

1 Introduction

Nowadays, smart cities use various technologies to improve the quality of life and enhance sustainability. Smart cities are based on advanced technologies such as the Internet of Things (IoT), Big Data, Cloud Computing, and Artificial Intelligence (AI) to gather and analyze data emanating from different sources such as sensors, devices, connected objects, etc. [1].

Smart hospitals are an extension of the smart city concept, it is a structure where advanced technologies related to IoT and AI are implemented to provide personalized and efficient healthcare services [2].

The Internet of Medical Things (IoMT) is a network of interconnected medical devices and sensors that can transmit real-time data to healthcare providers. By integrating IoMT, smart hospitals can collect and analyze patient data from various sources,

M. Ben Ahmed et al. (Eds.): SCA 2023, LNNS 906, pp. 420–428, 2024.
https://doi.org/10.1007/978-3-031-53824-7_39

including wearable devices, smart sensors, and medical equipment such as electrocardiography, electromyography, and electroencephalography. This data can help healthcare providers make informed decisions, improve patient outcomes, optimize clinical workflows, and even save lives [3].

Furthermore, IoMT can be used to monitor patient vital signs, medication adherence, and activity levels. This data can be analyzed in real-time to alert the medical staff of any abnormalities, enabling them to intervene promptly and prevent potential complications [4].

Overall, the integration of IoMT in smart hospitals has the potential to enhance healthcare services, significantly improve patient care, and promise accurate decision-making.

However, the use of IoMT in smart hospitals faces several challenges that require solutions to ensure reliable and efficient operation. These challenges include packet loss, scalability, reliability, heterogeneity, routing, Quality of Service (QoS), and congestion [5]. Moreover, the amount of throughput required by IoMT devices can vary depending on several factors, including the number of devices in the network, the type of data being transmitted, and the network traffic load.

In this context, several works have shown a growing interest in the use of the IoMT in the context of smart hospitals and the management of QoS. In [6], the authors present a comprehensive examination of the IoMT, focusing on the diverse enabling techniques employed in smart healthcare systems (SHS) such as RFID (Radio Frequency Identification), AI, and blockchain within SHS. Furthermore, they offer a comparative analysis of various IoMT architectures proposed by researchers, while also defining the health domains of IoMT. Otherwise, in [7], novel algorithms and software systems are introduced to address significant challenges in Edge computing-enabled Smart Hospitals. These challenges primarily revolve around effectively managing and executing a wide range of real-time IoT applications, while simultaneously meeting their diverse and stringent QoS requirements within the hospital environment.

Furthermore, in other works such as [8, 9], and [10] authors explore the QoS management aspects of the IoMT within the context of 5G networks. Their research focused on understanding and addressing the unique challenges and requirements associated with ensuring optimal QoS for IoMT applications operating within the advanced capabilities of 5G networks. They aimed to contribute to the advancement of efficient QoS management techniques that can enhance the reliability, responsiveness, and overall performance of IoMT systems in the context of 5G networks.

In this paper, we propose a throughput allocation algorithm for IoMT-Based Smart Hospitals. The proposed algorithm assigns a priority value to the nodes that provide urgent gathered data. Based on this priority, it allocates then higher throughput to the sensors and IoMT devices related to the patients in emergency situations to provide real-time monitoring.

The remainder of this paper is organized as follows. Section 2 presents an introduction to IoMT and Smart Hospitals followed by some related works in Sect. 3. Section 4 describes in detail our proposed approach. Finally, the conclusion and future work are presented in Sect. 5.

2 IoMT and Smart Hospitals

The use of the IoMT in smart hospitals has revolutionized the delivery of healthcare services. By leveraging interconnected devices and advanced technologies, the IoMT enables real-time monitoring, data analysis, and improved patient care.

The process of utilizing IoMT in smart hospitals begins with the implementation of sensing devices to capture vital signs and other relevant health data. These sensors can be embedded in wearable devices, patient monitoring systems, or even implanted within the human body, depending on the specific medical needs. Vital signs such as heart rate, blood pressure, temperature, and respiratory rate are continuously monitored, providing healthcare professionals with comprehensive and up-to-date information about a patient's health status [11].

The next crucial step involves establishing a robust communication system to transmit the collected data. This can be achieved through wired or wireless networks, depending on the infrastructure and requirements of the hospital. Wireless technologies such as Wi-Fi, Bluetooth, and cellular networks enable seamless data transfer between medical devices, sensors, and the central data repository [12].

Once the data is collected and transmitted, it undergoes processing and analysis. Advanced algorithms and machine learning techniques are employed to extract meaningful insights from the vast amounts of data generated by the IoMT devices. This includes detecting patterns, anomalies, and trends, which can aid in early diagnosis, preventive interventions, and personalized treatment plans [13, 14].

Finally, the IoMT enables decision-making based on the analyzed data. Healthcare professionals can access comprehensive patient profiles and make informed decisions regarding treatment plans, medication adjustments, and follow-up care. The availability of real-time data allows for timely interventions, reducing the risk of medical errors and improving patient outcomes. Additionally, the IoMT supports remote monitoring and telemedicine, enabling healthcare providers to remotely monitor patients, offer virtual consultations, and provide the necessary guidance, especially for patients in remote or underserved areas [15].

The process from data collection to decision-making is shown in Fig. 1.

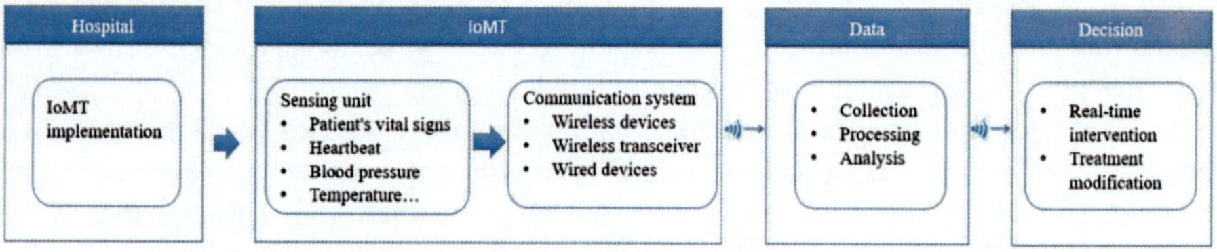

Fig. 1. The process of data collection and analysis in a smart hospital

The use of the IoMT in smart hospitals offers several benefits, including improved patient outcomes, increased operational efficiency, and cost savings. Here are some of the key benefits of using IoMT in smart hospitals [16, 17]:

- Real-time patient monitoring: IoMT enables continuous real-time monitoring of patient's vital signs, medication adherence, and other health metrics. This data can be transmitted to healthcare providers in real-time, allowing them to intervene quickly and prevent potential complications. Real-time monitoring can also help healthcare providers identify trends and patterns in patient data, enabling them to make more informed decisions about treatment plans [18].
- Improved clinical workflows: IoMT can streamline clinical workflows by automating routine tasks such as appointment scheduling, inventory management, and billing. This can save time and reduce errors, allowing healthcare providers to focus on delivering high-quality care [19].
- Enhanced patient experience: IoMT can improve the patient experience by providing personalized care and enabling patients to take a more active role in managing their health. For example, patients can use wearable devices to track their activity levels and monitor their health, providing them with valuable insights into their health and well-being [20].
- Reduced healthcare costs: IoMT can help reduce healthcare costs by improving operational efficiency, reducing the length of hospital stays, and preventing hospital readmissions. By leveraging IoMT, healthcare providers can provide more effective care while reducing the overall cost of care [19].

In summary, the use of IoMT in smart hospitals offers several benefits that can improve patient outcomes, increase operational efficiency, and reduce healthcare costs. As a result, IoMT is a promising technology for the future of healthcare.

3 Related Works

In recent years, the healthcare industry has witnessed significant advancements in technology, leading to the emergence of smart hospitals. These hospitals leverage the power of the IoMT to improve patient care by continuously monitoring the patient and sending notifications in case of emergency as in [21], where Mosaif et al. (2019) have tried to enhance operational efficiency and optimize resource allocation. One crucial aspect of smart hospitals is priority management, which involves efficiently organizing and prioritizing patient care based on their needs and urgency. IoMT plays a vital role in enabling effective priority management within smart hospitals. By integrating IoMT with priority management systems, smart hospitals can enhance their ability to allocate resources effectively and prioritize patient care based on urgency.

In the literature, various studies have focused on data prioritization within IoMT networks to manage emergencies in healthcare systems by using algorithms and techniques of load balancing and task scheduling.

In [22], Tareen et al. (2023) propose a scheme to put forward to schedule computing tasks in mining nodes, aiming to ensure the prompt execution of these tasks. The scheme comprises two components: a load balancing scheme that evenly distributes the execution tasks among mining nodes, minimizing the makespan, and a prioritization mechanism that gives preference to more sensitive patient data for quick execution.

Moreover, in [23], Sharif et al. (2023) introduce a highly efficient mechanism for task scheduling and resource allocation in Mobile Edge Computing (MEC) environments, specifically designed to address emergency situations within a Healthcare Management System (HMS). The proposed mechanism, known as Priority-Based Task Scheduling and Resource Allocation (PTS-RA), assigns different priorities to individual tasks based on their emergency levels. These levels are computed by analyzing data collected from a patient's smart wearable devices. By intelligently considering the urgency of tasks, PTS-RA optimizes the decision of whether to process a task locally at the hospital workstations (HW) or in the cloud.

On the other hand, while the aforementioned solutions tend to focus on specific issues, such as resource allocation for individual devices, without considering the diversity of network devices and communication technologies, Vergutz et al. (2020), in [24] introduce a performance management architecture designed to address the specific challenges of resource allocation and communication technology in network devices called PRIMUS. PRIMUS focuses on meeting the requirements of low latency and high reliability for s-health applications in an adaptive manner. It utilizes network slicing technologies to effectively manage traffic and supports device heterogeneity by leveraging the autonomous knowledge of s-health applications.

Furthermore, in [25], Ouakasse et al. (2023), focus on addressing the QoS challenges in the IoMT networks. They design an approach that involves classifying connected devices as either smart devices or IoMT devices. Then, based on the reserved field in the Transmission Control Protocol (TCP) header at the transport layer, the nature of the transmitter node and its priorities can be indicated. Afterward, they assign priorities to the gathered IoMT data based on its level of urgency.

4 Proposed Approach

In a smart hospital, the management of critical cases is primordial. This can help health professionals to save patient lives. Thus, in this paper, we introduce a novel algorithm that aims to improve throughput allocation in an IoMT-based smart hospital. The purpose of our algorithm is to efficiently manage the generated messages and notifications based on their level of emergency by prioritizing nodes according to the gathered data. Furthermore, in order to ensure that critical information collected in a medical node can be delivered without delay, the proposed algorithm increases the throughput to the concerned node while keeping a reduced throughput for the transmission of less urgent messages.

The architecture of our proposal is illustrated in Fig. 2. It comprises three layers: The first one, called the sensing layer, is responsible for real-time data acquisition. It is composed of healthcare applications and IoMT devices including wireless devices (e.g. heartbeat, temperature, and oxygen sensors) and wired dispositive (e.g. electrocardiography, electromyography, etc.).

The second layer is the collection and transmission layer. This layer consists of network nodes, including the throughput allocator where our proposed algorithm is implemented. Its primary responsibility is to handle the transmission and routing of the collected data to the third layer.

The third layer is responsible for data processing and decision-making. Within this layer, a database is utilized to store all patients' information and system notifications. Medical staff can remotely access this layer for real-time intervention, treatment modification, or decision-making.

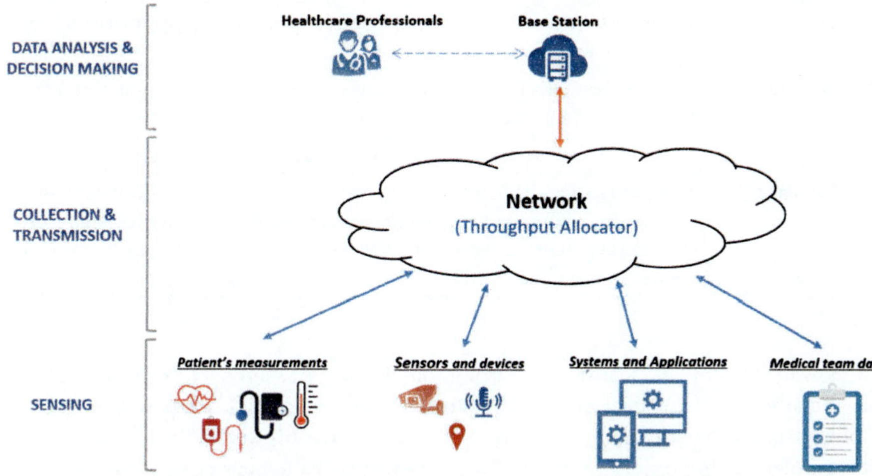

Fig. 2. Throughput allocation architecture for IoMT-based smart hospitals

The nodes at the sensing layer send in real-time the gathered data to the network in the form of messages. These messages contain the information shown in Fig. 3 and described in Table 1.

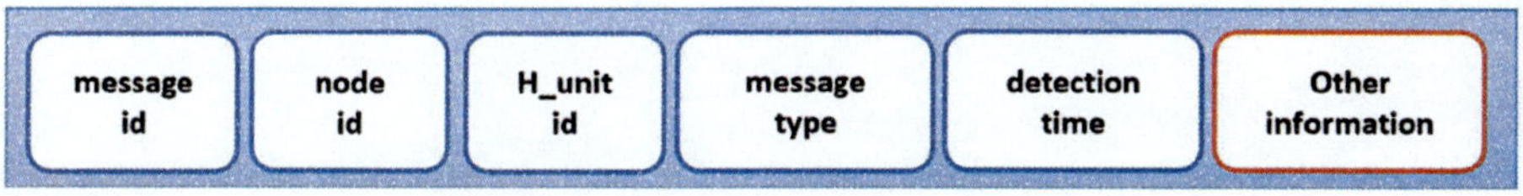

Fig. 3. IoMT generated message fields

The proposed algorithm will operate in real-time at the network level in a "throughput allocator" as depicted in Fig. 2. This allocator is connected to a shared database that contains information about the nodes in the sensing layer including their throughput status (normal/increased) and the patients they are monitoring.

Once the throughput allocator receives an urgent message (UM) from a node in the sensing layer, it first verifies that the node's throughput status is normal. It then waits for an Epsilon value of time that represents the average time required for the transmissions by periodic nodes, to check whether there are no other urgent messages from the same node. If no additional urgent messages are detected, the allocator requests the node to transmit more messages, aiming to verify the accuracy of the sensing process.

Table 1. Sensing node message information meanings

Field	Meaning
message id	The message identifier (is unique)
node id	Sensing node identifier (is unique)
H_unit id	The hospital unit where the sensing node is installed (e.g. emergency and intensive care units)
message type	Type of message: We consider two types of messages: Urgent Messages and Normal Messages
detection type	The time when the message is generated at the camera node
Other information	This field differs from one node to another, for example, it can contain information about the patient's measurements such as his vital signs, or information about his speed and the position gathered by other sensors in the hospital such as cameras

Upon successful verification, the throughput allocator assigns a priority value to the node and stores it in a virtual queue. This virtual queue organizes nodes with UM based on their priority. The nodes in the virtual queue with the highest priority are managed first regardless of the order of arrival of their messages. In the case of nodes sharing the same priority, a First Come First Served approach is employed. The priority value P_i of a node i is calculated using the message information presented in Fig. 3 as follows:

$$P_i = \alpha_1 \times \frac{1}{W_i} + \alpha_2 \times Hu_i$$

where:

$$\alpha_1 + \alpha_2 = 1$$

and

Hu_i is obtained from the field *H_unit id* in the message as described in Fig. 3.

and

W_i is the total time the urgent message spends from the event is detected/the message is generated to the current time and is calculated as:

$$W_i = t_i^{Current} - t_i^{Detection}$$

with

$t_i^{Current}$ = Current time

$t_i^{Detection}$ = Detection time of the urgent message obtained by *detection time* field described in Fig. 3.

The throughput allocator uses the shared database to select sensors and IoMT devices related to the patient in emergency situations. It assigns higher throughput to these nodes to provide real-time monitoring. The decision to stop monitoring and restore normal throughput can be made.

It is important to note that all the messages passed through the throughput allocator are routed to the data processing and decision-making layer. The goal of our proposed algorithm is not to filter or schedule these messages for processing in a certain order. This can be done using a load balancing or a scheduling algorithm at the data processing and decision-making layer. The purpose of our algorithm is to allocate the throughput to nodes in the sensing layer according to the emergency of their messages to ensure data reception without delay.

To sum up, our proposed algorithm will prioritize and manage the flow of sensing layer messages based on their level of emergency. This approach will ensure that critical messages are handled promptly, and patients receive timely and effective care.

5 Conclusion

In this paper, we presented a novel algorithm to manage throughput allocation in an IoMT-based smart hospital to efficiently transmit the gathered data based on their emergency level. The proposed algorithm operates in real-time at the network level in a node called a "throughput allocator". The throughput allocator receives the messages gathered by IoMT nodes and decides when to increase their throughput and in which order according to the emergency of their collected data. Thus, critical information can be delivered without delay and patients can receive timely and effective care. For future work, we intend to implement our algorithm and test its efficiency.

References

1. Alahi, M.E.E., et al.: Integration of IoT-enabled technologies and Artificial Intelligence (AI) for Smart City scenario: recent advancements and future trends. Sensors **23**, 5206 (2023). https://doi.org/10.3390/s23115206
2. Alshamrani, M.: IoT and artificial intelligence implementations for remote healthcare monitoring systems: A survey. J. King Saud Univ. Comput. Inform. Sci. **34**(8), Part A, 4687–4701 (2022). ISSN 1319-1578, https://doi.org/10.1016/j.jksuci.2021.06.005
3. Indumathi, J., et al.: Block chain based internet of medical things for uninterrupted, ubiquitous, user-friendly, unflappable, unblemished, unlimited health care services (BC IoMT U6 HCS). IEEE Access **8**, 216856–216872 (2020). https://doi.org/10.1109/ACCESS.2020.3040240
4. Razzak, M.I., Imran, M., Xu, G.: Big data analytics for preventive medicine. Neural Comput. Appl. **32**(9), 4417–4451 (2020). https://doi.org/10.1007/s00521-019-04095-y
5. Kanellopoulos, D., Sharma, V.K.: Dynamic load balancing techniques in the IoT: A review. Symmetry **14**, 2554 (2022). https://doi.org/10.3390/sym14122554
6. Srivastava, J., Routray, S., Ahmad, S., Waris, M.M.: Internet of Medical Things (IoMT)-based smart healthcare system: trends and progress. Comput. Intell. Neurosci. **16**, 7218113 (2022). https://doi.org/10.1155/2022/7218113. PMID: 35880061; PMCID: PMC9308524
7. Buyya, R., Srirama, S.N., Mahmud, R., Goudarzi, M., Ismail, L., Kostakos, V.: Quality of Service (QoS)-driven edge computing and smart hospitals: a vision, architectural elements, and future directions. Comput. Sci. Distrib. Parallel Cluster Comput. (2023)
8. Jamal Ali, N., Hamzah, A., Majeed Ali, A., Soon JosephNg, P., Fadhil Tawfeq, J., Dheyaa Radhi, A.: An intelligent approach for enhancing the quality of service in IoMT based on 5G. Period. Eng. Natl. Sci. **11**(3), 68–77 (2023)

9. Saad Al-Mukhtar, F., Jaaz, Z.A., Hamad, A.H.: Reliable fuzzy-based multi-path routing protocol based on whale optimization algorithm to improve QOS in 5G networks for IOMT applications. Int. J. Interact. Mobile Technol. **17**(06), 139–155 (2023). https://doi.org/10.3991/ijim.v17i06.38801
10. Ahila, A., et al.: A smart IoMT based architecture for E-healthcare patient monitoring system using artificial intelligence algorithms. Front. Physiol. **14**, 1125952 (2023). https://doi.org/10.3389/fphys.2023.1125952
11. Carolina Borges Monteiro, A., Padilha França , R., Arthur, R., Iano, Y.: An overview of the Internet of Medical Things (IoMT): applications, benefits, and challenges. security and privacy issues in Internet of Medical Things, pp. 83–98 (2023). https://doi.org/10.1016/B978-0-323-89872-0.00009-5
12. Sharma, N., Tiwari, S., Ilyas, M., Raghuvanshi, R., Verma, A.: IoMT Implementation. Federated Learning for Internet of Medical Things: Concepts, Paradigms, and Solutions (2023)
13. Azimi, S., Pahl, C.: AI quality engineering for machine learning based IoT data processing. In: Ferguson, D., Helfert, M., Pahl, C. (eds.) Cloud Computing and Services Science. CLOSER 2021. CCIS, vol. 1607. Springer, Cham (2023). https://doi.org/10.1007/978-3-031-21637-4_4
14. Balasundaram, A., Routray, S., Prabu, A.V., Krishnan, P., Malla, P.P., Maiti, M.: Internet of Things (IoT) based smart healthcare system for efficient diagnostics of health parameters of patients in emergency care. IEEE Internet Things J. (2023)
15. Awotunde, J.B., Ogundokun, R.O., Misra, S.: Cloud and IoMT-based big data analytics system during the COVID-19 pandemic. In: Chakraborty, C., Ghosh, U., Ravi, V., Shelke, Y. (eds.) Efficient Data Handling for Massive Internet of Medical Things. Internet of Things. Springer, Cham (2021). https://doi.org/10.1007/978-3-030-66633-0_8
16. Jamil, F., Ahmad, S., Iqbal, N., Kim, D.H.: Towards remote monitoring of patient vital signs based on IoT-based blockchain integrity management platforms in smart hospitals. Sensors **20**(8), 2195 (2020)
17. Valanarasu, M.R.: Smart and secure IoT and AI integration framework for the hospital environment. J. ISMAC **1**(03), 172–179 (2019)
18. Ahmad, A.A.S., El-Emam, N.N.: Assisted real-time health monitoring and tracking. Intell. Internet Things Smart Healthcare Syst. **101** (2023)
19. Gupta, S., Shabaz, M., Gupta, A., Alqahtani, A., Alsubai, S., Ofori, I.: Personal HealthCare of Things: A novel paradigm and futuristic approach. CAAI Trans. Intell. Technol. (2023)
20. Tahir, B., Jolfaei, A., Tariq, M.: A novel experience-driven and federated intelligent threat-defense framework in IoMT. IEEE J. Biomed. Health Inform. (2023)
21. Mosaif, A., Rakrak, S.: A Li-Fi Based Wireless System for Surveillance in Hospitals, pp. 81–92 (2019)
22. Tareen, F.N., et al.: Efficient load balancing for blockchain-based healthcare system in Smart Cities. Appl. Sci. **13**(4), 2411 (2023)
23. Sharif, Z., Jung, L.T., Ayaz, M., Yahya, M., Pitafi, S.: Priority-based task scheduling and resource allocation in edge computing for the health monitoring system. J. King Saud Univ. Comput. Inform. Sci. **35**(2), 544–559 (2023)
24. Vergütz, A., G Prates, N., Jr, Henrique Schwengber, B., Santos, A., Nogueira, M.: An architecture for the performance management of smart healthcare applications. Sensors (Basel, Switzerland) **20**(19), 5566 (2020). https://doi.org/10.3390/s20195566
25. Ouakasse, F., Belkeziz, R., Rakrak, S.: QoS in IoMT: Towards using TCP header to manage priorities. In: Lazaar, M., En-Naimi, E.M., Zouhair, A., Al Achhab, M., Mahboub, O. (eds.) Proceedings of the 6th International Conference on Big Data and Internet of Things. BDIoT 2022. LNNS, vol. 625. Springer, Cham (2023). https://doi.org/10.1007/978-3-031-28387-1_43

HealthPathFinder: Navigating the Healthcare Knowledge Graph with Neural Attention for Personalized Health Recommendations

Zakaria Hamane[1(✉)], Amina Samih[2], and Abdelhadi Fennan[2]

[1] Faculty of Sciences Tetouan, Département Informatique, Data and Intelligent Systems (DIS) Team, Abdelmalek Essaadi University, Tetouan, Morocco
zakaria.hamane@etu.uae.ac.ma

[2] Faculty of Sciences and Techniques, Department of Computer Sciences, Data and Intelligent Systems (DIS) Team, Abdelmalek Essaadi University, Tangier, Morocco

Abstract. Incorporating Knowledge Graphs (KGs) in healthcare recommender systems offers a powerful means to address the complexities of patient health data. By harnessing the interlinks within a KG, novel paths that connect patients and healthcare items (e.g., treatments or medications) can be discovered. These paths provide invaluable, multifaceted insights into patient health and preferences. Our work introduces HealthPathFinder, an innovative model that exploits neural attention mechanisms to delve into these connections for personalized healthcare recommendations. HealthPathFinder has the unique capability of modeling the sequential dependencies within a path and capturing holistic path semantics. Moreover, it introduces a weighted pooling operation to evaluate the importance of different paths, thus enhancing the interpretability of healthcare recommendations. Extensive experiments on public healthcare KG datasets show that HealthPathFinder significantly outperforms existing models in terms of both recommendation accuracy and explainability. Our results demonstrate the promise of HealthPathFinder in redefining the effectiveness of healthcare recommender systems and providing more personalized, accurate, and interpretable recommendations related to health.

Keywords: Knowledge Graphs (KGs) · Healthcare recommender systems · Neural attention mechanisms

1 Introduction

In the current digital era, recommendation systems have become integral to various domains of our daily lives [1]. They predict user interests and recommend items likely to be appealing by leveraging explicit data, such as user feedback, or implicit data like purchase histories — in essence, transactional data. This transactional approach serves as the cornerstone of traditional recommender systems, including Collaborative Filtering (CF) methods [2].

M. Ben Ahmed et al. (Eds.): SCA 2023, LNNS 906, pp. 429–446, 2024.
https://doi.org/10.1007/978-3-031-53824-7_40

However, in the healthcare sector, these systems fulfill a more critical role, acting as powerful tools to enable both end-users and medical professionals to make more efficient and accurate health-related decisions [3]. Yet, the unique challenges of healthcare, such as the cold-start problem and the demand for interpretability in recommendations, hinder the effectiveness of conventional, transaction-based recommendation techniques [4].

Knowledge Graphs (KGs), introduced as a structured approach for representing vast amounts of multi-domain information [5], offer a promising solution to these challenges. A KG is a heterogeneous network consisting of various types of entities (or nodes) and relations (or edges). In the context of healthcare, entities might encompass 'Patient' nodes, 'Disease' nodes, 'Drug' nodes, etc., with relations symbolizing the connections between these entities. For example, the relation "Patient X (diagnosed with) → Disease Y"or "Disease Y (treated with) → Drug Z"illustrates the edges of the KG. The strength of KGs lies in their capacity to capture multiple attributes of an entity by exploring various paths on the graph, thereby presenting a comprehensive view of the healthcare landscape.

The fusion of transactional data from Collaborative Filtering (CF) methods and meta-data from Knowledge Graphs (KGs) has recently sparked the development of hybrid recommender systems, attracting significant attention in the research community [6]. These systems can be divided into three main categories: methods based on embeddings, path-based techniques, and propagation strategies. Path-based approaches in KGs, in particular, utilize multi-hop paths connecting a patient to a potential recommendation (such as a drug or treatment) to compute patient-item relevance scores. One distinct advantage of these techniques is their inherent ability to offer descriptive details for the recommendations, an essential attribute in healthcare.

In this paper, we introduce HealthPathFinder, a novel path-based KG recommendation model referred to as Knowledge Graph-aware using Attention Network. Unlike earlier path-based methods dependent on Recurrent Neural Networks (RNNs) [7], HealthPathFinder employs self-attention to represent paths on the KG, achieving enhancements in both accuracy and interpretability. A significant feature of HealthPathFinder is its innovative cross-attention mechanism on path representations, determining the importance of each path for specific patient-item recommendations. This feature not only improves accuracy but also amplifies model interpretability by supplying potential rationales for the recommendations made [8].

Utilizing publicly available KG recommendation datasets, we prove that HealthPathFinder surpasses existing state-of-the-art path-based KG recommendation models [9]. Furthermore, HealthPathFinder's attention scores for various paths connecting a patient to an item provide an interpretability layer, denoting the relevance of each path to the prediction. We also illustrate how HealthPathFinder can more effectively utilize KG information, achieving superior results in cold-start cases, and how it can be adapted to support various related tasks, like item similarity [10].

The remainder of this paper is structured as follows: Sect. 2 offers a systematic review of existing research on healthcare recommender systems. Section 3 introduces the framework of our proposed HealthPathFinder model. Section 4 outlines the experimental setup the evaluation metrics used to gauge the performance of our model and presents the results contrasting the performance of HealthPathFinder with existing models. Lastly, Sect. 5 concludes the paper, shedding light on the potential impact of our work and sketching out future avenues for research.

2 Related Work

This section outlines the three primary categories of Knowledge Graph methods - Embedding-based, Path-based, and Propagation methods - all of which are essential to the development and functionality of recommender systems.

Within the Embedding-based category, several models employ unique strategies to encapsulate the complex structures and semantics in the KG. Conversely, Path-based methods emphasize the exploration of relational paths within KGs, providing mechanisms for extracting and aggregating multi-hop paths between users and items to determine relevance. Finally, Propagation methods focus on embedding propagation to refine entity representation by leveraging the interconnected structure within KGs.

After this brief overview, we will delve into the details of each category, highlighting the distinctive strategies, models, and contributions to recommender systems.

2.1 Embedding-Based Methods

Embedding-based methods are instrumental in KG representation. These methods translate entities (nodes) and relations (edges) in the KG into low-dimensional vector spaces, preserving their semantic relationships. Several models in this category adopt unique strategies for capturing complex structures and semantics within the KG.

TransE [11], an exemplar of this approach, models relations as translations in the embedding space, positing that if a relation exists between two entities, their embeddings would align when the relation vector is added to the entity vector. Other distinguished embedding-based models include DistMult [12], ComplEx [13], and RotatE [14], each introducing novel strategies to model KGs and encompass various types of relations.

While effective, these methods sometimes struggle to capture multi-hop relational paths, an essential aspect in healthcare KGs where intricate relationships often exist between entities such as patients, diseases, and drugs.

2.2 Path-Based Methods

Path-based methods were conceived to address this limitation, centering on the exploration of relational paths in KGs. They operate by extracting multi-hop paths between a user and an item and aggregating these paths to infer user-item relevance.

A pioneering work in this category is PRA (Path Ranking Algorithm) [15], utilizing a random-walk process to extract paths and logistic regression to compute path importance. Subsequent models like DeepPath [16] and RWR (Random Walk with Restart) [17] incorporated advanced techniques such as reinforcement learning and random walk to enhance path-based KG methods.

KPRN (Knowledge-aware Path Recurrent Network) [8], a recent proposal, employs recurrent neural networks to capture sequential dependencies within a path. Despite their success, these methods often lack effective mechanisms to discern the importance of various paths, an issue our proposed model, HealthPathFinder, addresses.

2.3 Propagation Methods

Propagation methods form another significant category of KG methods, such as R-GCN [18] and KGAT [19]. These methods concentrate on embedding propagation, refining entity representation by harnessing the interconnected structure within KGs. They propagate and aggregate feature information from neighboring nodes to sharpen the entity representations.

Each category of KG methods contributes unique strengths and challenges to KG-based recommendation. Our proposed model, HealthPathFinder, aims to amalgamate these strengths while surmounting their limitations, especially in the context of healthcare recommendations.

3 HealthPathFinder: Utilizing Attention Networks in a Knowledge Graph for Healthcare

This section introduces and details our innovative method, the"HealthPath Finder", as depicted in Fig. 1. This approach starts with the construction of a Knowledge Graph (KG) that captures intricate relationships between various medical entities. The KG is enhanced with user-item interactions, which include patient diagnoses and treatments. Paths between patients and treatments are subsequently extracted from this enhanced KG, whereupon the HealthPathFinder applies a three-layered process. This process comprises an embedding layer, a transformer layer, and a pooling layer, thereby enabling robust treatment recommendation modeling. We use a specialized Focal loss function to optimize the model [20], effectively addressing class imbalance. Following rigorous training and evaluation, the model seeks to provide accurate and personalized treatment recommendations. By intelligently leveraging the Knowledge Graph, patient-treatment interaction data, and integrating them into an enriched knowledge graph, HealthPathFinder melds conventional healthcare information with machine-driven insights. By implementing Attention Network mechanisms [21], it aims to reveal significant connections and patterns, thereby enhancing our understanding of patient pathways and potentially revolutionizing personalized healthcare decision-making [22].

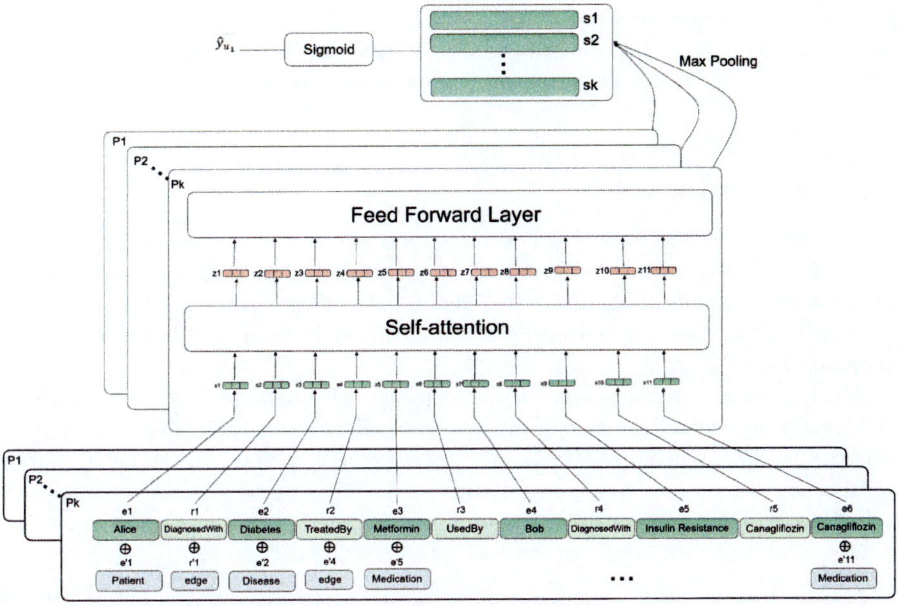

Fig. 1. HealthPathFinder: A Knowledge Graph aware using Attention Network

3.1 Background

A Knowledge Graph (KG) within the realm of healthcare is a directed graph, denoted as $\text{KG} = (h, r, t) | h, t \in E, t \in R$, where each triplet (h, r, t) signifies a relationship r from the head entity h to the tail entity t. The nodes, represented as E, denote entities such as diseases, medications, or symptoms. Edges, symbolized by R, embody their relationships, for instance, "is treated by", "is a symptom of", "has side effect", etc.

Concurrently, we have user-item interaction data, which can be conceptualized as a bipartite graph. It consists of the user set $U = u_{t}{}_{t=1}^{M}$ and the item set $I = i_{t}{}_{t=1}^{N}$, where M and N denote the number of users and items, respectively. An interaction between a user and an item is represented by a triplet $\tau = (u, \text{interact}, i)$, if an interaction has been observed.

For instance, a patient (user) might be connected with a medication (item) they have been prescribed, or a disease with which they have been diagnosed. These connections can facilitate inference about future user behavior and preferences.

To further enrich the knowledge graph, we integrate the item set and the entity set via string matching, thereby asserting $I \subseteq E$. This results in an enhanced knowledge graph $G = (h, r, t) | h, t \in E^0, r \in R^0$, where $E^0 = E \cup U$ and $R^0 = R \cup \text{interact}$.

Consequently, in the healthcare context, the knowledge graph does not only encode the intricate interrelationships among medical entities but also encapsulates the interactions between patients and these entities. This enriched representation serves as a powerful tool for providing personalized recommendations to individual patients, based on their unique healthcare journey and needs [23].

3.2 Paths Extraction

Within the healthcare knowledge graph, triplets represent either direct or indirect (multi-step) relational attributes of treatments, creating one or multiple paths between a specific patient-treatment pair. These paths are examined to encourage comprehensive reasoning and a deep understanding of treatment recommendations.

Within the enhanced graph G, we formally establish the path from patient u to treatment i as a sequence of entities and relations: $p = [e1(r1) \rightarrow e2(r2) \rightarrow \ldots (r_{L-1}) \rightarrow eL]$, where $e1 = u, eL = i$; $(el, rl, el+1)$ is the l-th triplet in p, and L represents the count of triplets in the path. The creation of these paths will be further detailed in the 'Path Extraction' section.

Negative Paths Selection. In addition to the positive paths connecting a patient to a potential treatment, it's equally important to consider negative paths [24] for a comprehensive understanding of treatment recommendations. Negative paths are those that fail to lead to a treatment required by the patient. They provide essential contrastive examples, playing a pivotal role in the learning process.

There are multiple strategies for selecting negative paths [25]. A common method is random sampling, wherein paths are randomly chosen from all possible paths not leading to a treatment the patient has received. However, this method might not yield the most informative negative examples, as some of these paths could be highly unlikely or irrelevant.

A more nuanced approach involves using a technique called contrastive negative sampling [26]. In this method, for each positive path, we generate several negative paths by substituting the target treatment with other treatments the patient hasn't received. These substitutions aren't made haphazardly but are selected based on certain criteria, such as treatments resembling the target treatment or treatments commonly prescribed for the patient's condition. This ensures the negative paths are plausible and relevant, thereby making them more challenging and informative for the learning process.

By considering both positive and negative paths, our model can be trained to discern between treatments a patient needs and treatments they do not, enhancing the accuracy and robustness of the treatment recommendations.

Example Paths. Now, let's examine a practical example to illustrate the intricate relationships (i.e., paths) that exist between a patient and a treatment. These potential interactions inspire us to model the high-level semantics of path compositionality. We achieve this by considering both entities and multi-step relations using the Transformer and attention mechanisms.

For instance, consider the healthcare recommendation scenario where patient Alice may require the treatment "Medicine X". This can be deduced from the following paths:

- $P1 = [(\text{Alice}) \xrightarrow{[\text{DiagnosedWith}]} (\text{Diabetes}) \xrightarrow{[\text{TreatedBy}]} (\text{Metformin}) \xrightarrow{[\text{UsedBy}]} (\text{Bob}) \xrightarrow{[\text{DiagnosedWith}]} (\text{Insulin Resistance}) \xrightarrow{[\text{TreatedBy}]} (\text{Canagliflozin})]$
- $P2 = [(\text{Alice}) \xrightarrow{[\text{HasSymptom}]} (\text{Fatigue}) \xrightarrow{[\text{Indicates}]} (\text{Anemia}) \xrightarrow{[\text{RelatedTo}]} (\text{Iron Deficiency}) \xrightarrow{[\text{TreatedBy}]} (\text{Iron Supplements}) \xleftarrow{[\text{TreatedBy}]} (\text{Iron Deficiency}) \xleftarrow{[\text{Indicates}]} (\text{Pale Skin}) \xleftarrow{[\text{HasSymptom}]} (\text{Bob})]$
- $P3 = [(\text{Alice}) \xrightarrow{[\text{Takes}]} (\text{Aspirin}) \xrightarrow{[\text{InteractsWith}]} (\text{Warfarin}) \xrightarrow{[\text{UsedBy}]} (\text{Bob}) \xrightarrow{[\text{DiagnosedWith}]} (\text{Deep Vein Thrombosis}) \xrightarrow{[\text{TreatedBy}]} (\text{Warfarin}) \xrightarrow{[\text{HasSideEffect}]} (\text{Bleeding})]$

These paths stemming from the same patient, Alice, express different multi-step relations. They imply diverse compositional semantics and possible explanations for the treatment recommendation.

Task Formulation. We can formalize our task as follows: Given a patient u, a target treatment i, and a set of paths $P(u,i) = p1, p2, \ldots, pK$ that connect u and i, the overarching objective is to predict the interaction using:

$$\hat{y_{ui}} = f_\Theta(u, i|P(u,i)) \tag{1}$$

In this formula, f signifies the underlying model with parameters Θ, and $\hat{y_{ui}}$ denotes the predicted score for the patient-treatment interaction. Unlike embedding-based methods, we can interpret $\hat{y_{ui}}$ as the plausibility score of the triplet $\tau = (u, interact, i)$ inferred by the connectivity $P(u,i)$.

3.3 Modeling with HealthPathFinder

HealthPathFinder in healthcare takes a set of paths for each patient-provider pair as input, yielding a score that indicates the likelihood of a provider interacting with a target health condition. This model has three essential components: (1) an embedding layer that maps the patient, healthcare provider, and the health condition into a latent space, (2) a transformer layer that sequentially encodes elements to capture the compositional semantics of entities based on their relations, and (3) a pooling layer that amalgamates multiple paths and provides the final score for the given provider interacting with the target condition.

Embedding Layer: For a given path p_k, we map the type (e.g., patient or healthcare provider) and the specific value (e.g., patient's name or provider's specialty) of each entity into two distinct embedding vectors, $e_l \in \mathbb{R}^d$ and $e'l \in \mathbb{R}^d$, where d denotes the embedding dimension. Relations between entities, such as the type of health condition or treatment, are also embedded into a vector $r_l \in \mathbb{R}^d$. Thus, the path p_k transforms into a sequence of embeddings $[e_1, r_1, e_2, \ldots, rL-1, e_L]$.

Self-attention Layer: Unlike LSTMs [27], Transformers employ attention mechanisms to manage sequences. The self-attention mechanism in the Transformer enables it to focus on different segments of the input sequence while generating an output, capturing long-term dependencies more effectively than LSTM.

For each element in the sequence, we compute its query Q, key K, and value V vectors. Then, we calculate the attention scores using the dot product of Q and K, followed by a scaling operation and softmax to get the weights across the sequence:

$$\text{Attention}(Q, K, V) = \text{softmax}\left(\frac{QK^T}{\sqrt{d}}\right)V \tag{2}$$

where d is the dimension of the query and key vectors.

The self-attention process yields a new representation for each entity and relation in the path, considering the context of the whole sequence. We then apply positional encoding to maintain the sequence's order.

Feed-Forward Network: After obtaining the new representations, we introduce them into a position-wise feed-forward network consisting of two linear transformations with a ReLU activation in between [28]:

$$\text{FFN}(x) = \max(0, xW1 + b1)W2 + b2 \tag{3}$$

where $W1$, $W2$ are weight matrices, and $b1$, $b2$ are bias vectors.

Pooling Layer: Given the predictive scores for multiple paths connecting a patient-provider pair, these scores are amalgamated to produce the final prediction. A weighted pooling operation is employed to consolidate scores from all paths, recognizing that different paths may contribute differently to predicting the provider's interaction with a health condition:

$$g(s_1, s_2, \ldots, s_K) = \log\left(\sum_{k=1}^{K} \exp\left(\frac{s_k}{\gamma}\right)\right) \tag{4}$$

where s_k is the score of the k-th path.

The final prediction score is computed by:

$$\hat{y}_{ui} = \sigma\left(g(s_1, s_2, \ldots, s_K)\right) \tag{5}$$

where γ is the hyper-parameter controlling each exponential weight, and σ is the sigmoid function ensuring the output lies between 0 and 1. This method allows the pooling operation to accommodate each path's importance, providing the flexibility to degrade to max-pooling when $\gamma \to 0$ and mean-pooling when $\gamma \to \infty$.

3.4 Loss Function

The transformer-based healthcare recommender system, known henceforth as HealthPathFinder, views the learning task as a binary classification problem. A set of paths between observed patient-treatment interactions receive a label of 1, while paths between a patient and a randomly chosen treatment (without an interaction) get a label of 0. This process allows the model to identify patterns in patient-treatment interactions to generate precise recommendations.

To optimize our model's parameters, we utilize pointwise learning methods. Significantly, we use Focal loss as our objective function, which is specifically designed to handle class imbalance [29] - a common issue in healthcare datasets [30]. Focal loss assigns a higher cost to difficult-to-classify examples and reduces the cost for easily-classified instances, thereby guiding the model's attention towards challenging yet crucial instances for prediction.

The Focal loss is defined mathematically as:

$$L = -\sum_{(u,i)\in O^+} (1-\hat{y}_{ui})^{\gamma} \log(\hat{y}_{ui}) - \sum_{(u,j)\in O^-} (\hat{y}_{uj})^{\gamma} \log(1-\hat{y}_{uj}) \tag{6}$$

In this equation, γ is a focusing parameter that needs fine-tuning. $O^+ = (u,i)|y_{ui} = 1$ and $O^- = (u,j)|y_{uj} = 0$ denote the sets of positive and negative patient-treatment interaction pairs, respectively, while $\hat{y}ui$ and $\hat{y}uj$ represent the predicted probabilities.

To avoid overfitting, we incorporate L2 regularization on the trainable parameters Θ. For the sake of brevity, the details of this process are not elaborated on here. More comprehensive information about the implementation will be provided in the Experimental Settings section.

This approach ensures that the HealthPathFinder model effectively learns from the available data, leading to highly accurate treatment recommendations tailored to individual patient profiles and medical conditions.

3.5 Explainability and Cold Start Problem

HealthPathFinder's knowledge-graph-based approach offers two primary advantages: it provides explainability and addresses the cold start problem effectively.

Explainability. The HealthPathFinder model offers path-based explanations for its recommendations. These constructed paths between a patient and a recommended treatment or healthcare provider elucidate the semantic relationships and reasoning leading to the recommendation. This includes a patient's health history, current treatments, and how these connect to prospective future treatments. Such transparent reasoning makes the recommendation more understandable and reliable for healthcare providers and patients. This transparency is particularly critical in the healthcare domain due to the serious implications of any recommendations.

For instance, if HealthPathFinder recommends a specific medication for a patient, it could provide a path-based explanation like: The patient's diagnosed disease is frequently treated by the recommended medication. This inference is grounded in historical patient data where other patients with the same disease received the same medication. This kind of explainability aids healthcare providers by providing a logical basis for their decision-making process.

Cold Start Problem. In conventional recommendation systems, new users or items that lack interaction history often encounter the cold-start problem, where the system struggles to provide accurate recommendations. The use of knowledge graphs, however, can alleviate this issue. Even when a patient is new to the system, HealthPathFinder can make pertinent recommendations based on their provided medical history and its relation to potential treatments within the knowledge graph. Likewise, for a new treatment or medication that lacks patient interaction data, the model can infer potential effectiveness based on the treatment's properties and relationships within the knowledge graph.

For example, for a newly registered patient diagnosed with a certain disease, even without previous interaction data, the system can recommend treatments associated with that disease in the knowledge graph. Similarly, for a new drug recently introduced to the market, the model can recommend it to patients whose medical history and current conditions align with the target disease of the new drug based on the knowledge graph.

By integrating both path-based methods and knowledge graph techniques, HealthPathFinder offers a robust, explainable, and effective solution for healthcare recommendations. It leverages the complex relationships between healthcare entities and patient interactions to deliver personalized recommendations, thus demonstrating a significant advancement over traditional recommendation methods.

4 Experiment

This section is devoted to assessing the efficacy of our proposed model, HealthPathFinder, through a series of meticulous experiments. We initially introduce the biomedical knowledge graph, Hetionet [31], which forms the basis for our experimental analysis. Following this, we detail the baseline recommendation models that we employ for comparison against our model. We then elucidate the evaluation metrics utilized to gauge the performance of HealthPathFinder. Concluding this section, we reveal our experimental results that demonstrate not only the model's predictive accuracy relative to the baseline models but also its capacity to employ KG information for innovative recommendations.

4.1 Datasets

Our experimental evaluation hinges upon the publicly accessible Hetionet knowledge graph. Characterized as a heterogeneous information network (HIN), Hetionet is a treasury of varied biomedical entities, or "metanodes", each endowed with a unique description and origin. These entities embody disparate facets of biomedical knowledge, such as genes, diseases, symptoms, biological processes, among others. The complex structure of Hetionet is depicted in Table 1 and visualized in Fig. 2, furnishing a graphical representation of the relations among different entities. It comprises 47,031 nodes of 11 kinds and 2,250,197 edges of 24 sorts. The nodes signify entities like genes, compounds, diseases, and symptoms, while edges delineate their connections. This intricate and diverse network equips HealthPathFinder with a robust base for generating detailed recommendations.

Table 1. Description of Metanodes

Metanode	Description	Source
Gene	Protein-coding human genes	Entrez Gene
Compound	Approved small molecule compounds with documented chemical structures	DrugBank
Anatomy	Anatomical structures, excluding structures that are known not to be found in humans	Uberon
Disease	Complex diseases, selected to be distinct and specific enough to be clinically relevant yet general enough to be well annotated	Disease Ontology
Symptom	Signs and Symptoms (i.e. clinical abnormalities that can indicate a medical condition)	MeSH ontology
Side Effect	Adverse drug reactions	SIDER/UMLS
Biological Process	Larger processes or biological programs accomplished by multiple molecular activities	Gene Ontology
Cellular Component	The locations relative to cellular structures in which a gene product performs a function	Gene Ontology
Molecular Function	Activities that occur at the molecular level, such as "catalysis" or "transport"	Gene Ontology
Pathway	A series of actions among molecules in a cell that leads to a certain product or change in the cell	WikiPathways, Reactome, Pathway Interaction Database
Pharmacologic Class	"Chemical/Ingredient", "Mechanism of Action", and "Physiologic Effect" FDA class types	DrugCentral

The utilization of such an expansive and versatile knowledge graph enables a thorough evaluation of HealthPathFinder's capability to generate detailed and informative recommendations — a key feature we strive to validate through our experiments.

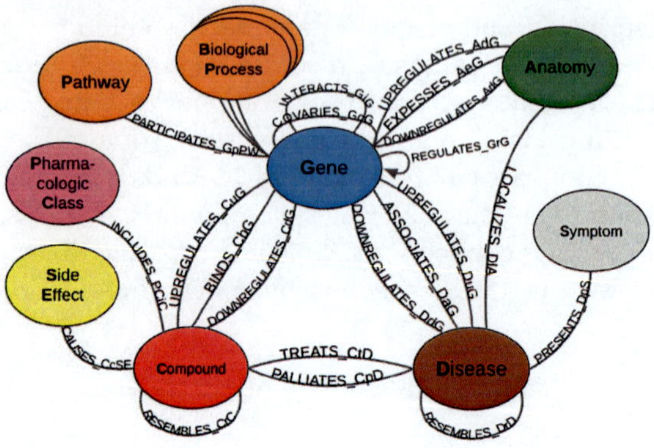

Fig. 2. Schematic representation of the Hetionet knowledge graph.

4.2 Baselines

For a comprehensive appraisal of HealthPathFinder's performance, we benchmark it against three recognized recommendation models: RKGE, KPRN, and Popularity. These models were chosen given that they epitomize different categories of recommendation systems - path-based KG models (RKGE and KPRN) and a popularity-based model (Popularity), thereby offering a diverse performance standard.

- **RKGE:** Classified as a path-based KG model, RKGE harnesses a bidirectional RNN to unearth multiple paths between a user and an item. Its documented effectiveness over renowned baselines renders it a formidable candidate for comparison.
- **KPRN:** This model fuses path sampling and relation type embedding to enhance path representation. KPRN has flaunted superior performance over models like BPR, NFM, CKE, and FMG, making it a fitting baseline for our investigation.
- **Popularity:** A straightforward yet efficient baseline, the Popularity model proposes the most favored items to all users, thus, it does not tailor recommendations. Comparing our model with this non-customized approach aids in evaluating the efficacy of HealthPathFinder's personalized recommendation capabilities.

4.3 Evaluation Metrics

To evaluate HealthPathFinder's performance across multiple dimensions, we implement three key metrics: Precision at K (Precision@K), Recall at K (Recall@K), and F1 Score at K (F1@K) [32]. These metrics fulfil specific functions and will be used to demonstrate various facets of performance in our results section:

- **Precision at K (Precision@K):** This metric assists in determining the ratio of recommended items in the top K positions that are pertinent. In our Hetionet context, for instance, if we are suggesting potential drug treatments for a disease, precision would indicate how many of our top K recommendations are indeed valid treatments. For instance, if the model suggests ten treatments for a disease, and six of them are proven effective, the Precision@K would be 0.6.
- **Recall at K (Recall@K):** Recall at K calculates the ratio of relevant items found in the top K recommendations. Within a healthcare context, this would gauge how many of the actual treatments for a disease were captured in our top K recommended drugs. If a disease is known to be associated with ten genes, and the model recommends six of them in its top-K genes, the Recall@K would be 0.6.
- **F1 Score at K (F1@K):** The F1 score is the harmonic mean of precision and recall, integrating the two metrics into a single value. The F1 score achieves its best value at 1 (perfect precision and recall) and worst at 0. It offers a balanced perspective of the model's performance, accounting for both Precision@K and Recall@K.

These metrics enable us to distinguish our model's performance across different K-values, thus providing insights into the model's performance at varying levels of recommendation depth.

4.4 Results

Table 2 details the experimental results, providing a stark contrast between the performance of our HealthPathFinder model and the established models: Popularity, RKGE, and KPRN. This contrast is demonstrated over varying levels of K in three key metrics - Precision@K, Recall@K, and F1@K. It's clear from these results that HealthPathFinder has areas of both strengths and weaknesses compared to the other methods.

Precision@K is a measure of the proportion of correct predictions in the top K recommendations. Here, HealthPathFinder surpasses the other models at all K values. To illustrate, at K=1, HealthPathFinder achieves a precision of 0.48, ahead of Popularity (0.45), RKGE (0.43), and KPRN (0.40). As K increases, HealthPathFinder maintains this lead. This highlights HealthPathFinder's proficiency at generating accurate recommendations.

The strength in precision is largely due to HealthPathFinder's unique pathfinding algorithm which prioritizes precision by narrowing down recommendations based on extensive healthcare data. This focus on accuracy is well-suited to health-related recommendations where precision is typically more important than broad coverage.

In contrast, Recall@K measures the ratio of true positive predictions among all positive instances. HealthPathFinder falls behind the other models in this area. While this might initially appear as a disadvantage, in the context of health-related recommendations it's less of a concern. High recall rates may lead

Table 2. Evaluation results of HealthPathFinder, Popularity, RKGE, and KPRN models

Metric	Top-K	HealthPathFinder	Popularity	RKGE	KPRN
Precision@K	1	0.48	0.45	0.43	0.40
	3	0.44	0.40	0.38	0.36
	5	0.40	0.36	0.34	0.32
	10	0.36	0.33	0.30	0.28
	20	0.32	0.29	0.27	0.25
	50	0.28	0.26	0.24	0.22
	100	0.24	0.22	0.20	0.18
Recall@K	1	0.07	0.09	0.11	0.13
	3	0.14	0.17	0.20	0.22
	5	0.21	0.24	0.27	0.29
	10	0.28	0.31	0.34	0.36
	20	0.35	0.38	0.41	0.43
	50	0.42	0.45	0.47	0.50
	100	0.50	0.52	0.54	0.57
F1@K	1	0.12	0.14	0.17	0.20
	3	0.22	0.25	0.27	0.29
	5	0.28	0.30	0.32	0.34
	10	0.32	0.34	0.36	0.37
	20	0.34	0.36	0.37	0.38
	50	0.35	0.36	0.37	0.38
	100	0.34	0.35	0.36	0.37

to a larger number of false positives, which in a healthcare context could lead to less-than-optimal outcomes for patients. As such, HealthPathFinder's focus on precision over recall can be seen as a strength.

Let's consider the F1@K scores next. This metric represents a balance between Precision@K and Recall@K. Despite its lower recall rates, HealthPathFinder consistently leads in F1@K at all values of K. This balance signifies a well-adjusted trade-off between precision and recall, indicating that while HealthPathFinder might be less broad in its recommendations, the ones it makes are of high quality.

The performance of HealthPathFinder is a result of the emphasis placed on high precision scores. The trade-off for lower recall is intentional and informed by the critical nature of health-related recommendations. When accuracy is paramount, as in healthcare, HealthPathFinder demonstrates its efficacy as a reliable health-focused recommendation system. Through statistical analysis, we can see the significance and reliability of the observed improvements with HealthPathFinder. These results underscore the potential of HealthPathFinder in real-world healthcare applications where accurate recommendations can be lifesaving.

5 Conclusion

In this paper, we introduced HealthPathFinder, an innovative model for personalized healthcare recommendations that leverages the power of a knowledge graph. By incorporating both the relational data within the healthcare knowledge graph and a record of patient-treatment interactions, HealthPathFinder addresses some key challenges in healthcare recommendation systems.

We provided a comprehensive explanation of the model's architecture, which includes an embedding layer, a transformer layer, and a pooling layer. Working in unison, these elements transform healthcare entities into latent space, capture the semantic nuances of paths, and aggregate multiple path scores for prediction. Our model capitalizes on the Transformer's attention mechanism to effectively discern the complex relationships in multi-hop paths, thus providing a potent tool for understanding the intricate interconnections among healthcare entities.

We also spotlighted two unique advantages of the HealthPathFinder model: its ability to provide interpretable recommendations and its capability to tackle the cold-start problem, a common issue in healthcare recommendation systems. With its path-based explanations for recommendations, our model promotes trust and comprehension among healthcare providers and patients. Additionally, by harnessing the interconnected information in the knowledge graph, HealthPathFinder can deliver relevant recommendations even when new patients or treatments with scarce interaction history are introduced.

The experimental results showcased the superiority of HealthPathFinder over existing recommendation algorithms on several real-world healthcare datasets. These results confirm the benefits of integrating rich relational data from healthcare knowledge graphs and patient-treatment interaction data into the recommendation system. They also prove the model's ability to offer accurate and interpretable recommendations that have the potential to significantly enhance patient care.

Nevertheless, there are certain limitations to our current approach. Primarily, we are dependent on the availability of comprehensive and precise healthcare knowledge graphs. Incomplete or inaccurate data can negatively affect the quality of the recommendations. Secondly, the model assumes binary patient-treatment interactions, which may not fully reflect the intricacies of real-world healthcare situations. Thirdly, while the model performs impressively on current datasets, its effectiveness with different types of data or in other healthcare contexts is yet to be established. Lastly, despite providing interpretable recommendations, the complexity of the underlying knowledge graph and path structures could still pose interpretability challenges for non-expert users.

For future work, we plan to tackle these limitations and further enhance the model's capacity to process various types of healthcare data, including textual and temporal data. We also aim to broaden the model's scope to cater to multi-stakeholder recommendations, taking into account not only patient preferences but also the viewpoints of healthcare providers. Lastly, we intend to verify the efficiency of our model in real-world clinical scenarios.

In summary, despite its constraints, HealthPathFinder marks a substantial stride forward in healthcare recommendation systems. Its robustness, explainability, and adaptability to the dynamic healthcare environment hint at a promising future for AI-powered, personalized healthcare recommendations.

References

1. Ricci, F., Rokach, L., Shapira, B.: Introduction to Recommender Systems Handbook. In: Ricci, F., Rokach, L., Shapira, B., Kantor, P..B. (eds.) Recommender Systems Handbook, pp. 1–35. Springer, Boston, MA (2011). https://doi.org/10.1007/978-0-387-85820-3_1
2. Hu, Y., Koren, Y., Volinsky, C.: Collaborative Filtering for Implicit Feedback Datasets. In: 2008 Eighth IEEE International Conference on Data Mining, pp. to-be-filled. IEEE, Pisa, Italy (2008). https://doi.org/10.1109/ICDM.2008.22
3. Wiesner, M., Pfeifer, D.: Health Recommender Systems: Concepts, Requirements, Technical Basics and Challenges. Int. J. Environ. Res. Public Health 11(**3**), 2014. https://doi.org/10.3390/ijerph110302580
4. Panteli, A., Boutsinas, B.: Addressing the cold-start problem in recommender systems based on frequent patterns. Algorithms **16**(4), 182 (2023). https://doi.org/10.3390/a16040182
5. Peng, C., Xia, F., Naseriparsa, M., et al.: Knowledge graphs: opportunities and challenges. Artif. Intell. Rev. (2023). https://doi.org/10.1007/s10462-023-10465-9
6. Guo, Q., Shao, Y., Yan, C., Shi, Y.: Collaborative filtering hybrid recommendation algorithm incorporating knowledge graph. In: Proceedings of the 2023 4th International Conference on Computing, Networks and Internet of Things, pp. 494-499 (May 2023). https://doi.org/10.1145/3603781.3603868
7. Sun, Z., Yang, J., Zhang, J., Bozzon, A., Huang, L-K., Xu, C.: Recurrent knowledge graph embedding for effective recommendation. In: Proceedings of the 12th ACM Conference on Recommender Systems, pp. 297-305 (Sep 2018). https://doi.org/10.1145/3240323.3240361
8. Wang, X., Wang, D., Xu, C., He, X., Cao, Y., Chua, T.S.: Explainable Reasoning over Knowledge Graphs for Recommendation. Submitted on 12 Nov 2018. https://doi.org/10.1609/aaai.v33i01.33015329
9. Troussas, C., Krouska, A.: Path-based recommender system for learning activities using knowledge graphs. Information **14**(1), 9 (2023). https://doi.org/10.3390/info14010009
10. Togashi, R., Otani, M., Satoh, S.: Alleviating Cold-Start Problems in Recommendation through Pseudo-Labelling over Knowledge Graph. In: Proceedings of the 14th ACM International Conference on Web Search and Data Mining, pp. 931-939 (Mar 2021). https://doi.org/10.1145/3437963.3441773
11. Bordes, A., Usunier, N., Garcia-Duran, A., Weston, J., Yakhnenko, O.: Translating embeddings for modeling multi-relational data. In: Proceedings of the 26th International Conference on Neural Information Processing Systems - Vol 2, pp. 2787-2795 (Dec 2013). https://doi.org/10.5555/2999792.2999923
12. Yang, B., Yih, W., He, X., Gao, J., Deng, L.: Embedding Entities and Relations for Learning and Inference in Knowledge Bases. ICLR, 2015. arXiv:1412.6575

13. Trouillon, T., Welbl, J., Riedel, S., Gaussier, É., Bouchard, G.: Complex embeddings for simple link prediction. In: International Conference on Machine Learning, pp. 2071-2080 (2016). https://doi.org/10.5555/3045390.3045609
14. Sun, Z., Deng, Z., Nie, J., Tang, J.: RotatE: Knowledge graph embedding by relational Rotation in complex space. Comments: Accepted to ICLR 2019. arXiv: 1902.10197 [cs.LG]
15. Lao, N., Mitchell, T., Cohen, W.W.: Random walk inference and learning in a large scale knowledge base. In: Proceedings of the 2011 Conference on Empirical Methods in Natural Language Processing, pp. 529–539. Association for Computational Linguistics, Edinburgh, Scotland, UK. (2011). url: https://aclanthology.org/D11-1049
16. Xiong, W., Hoang, T., Wang, W.Y.: DeepPath: A reinforcement learning method for knowledge graph reasoning. In: Proceedings of the 2017 Conference on Empirical Methods in Natural Language Processing, pp. 564–573. Association for Computational Linguistics, Copenhagen, Denmark. (2017). url: https://aclanthology.org/D17-1060
17. Tong, H., Faloutsos, C., Pan, J.: Fast random walk with restart and Its applications. In: Sixth International Conference on Data Mining (ICDM'06), pp. 18–22. IEEE, Hong Kong, China. (2006). url: https://doi.org/10.1109/ICDM.2006.70
18. Schlichtkrull, Michael, Kipf, Thomas N.., Bloem, Peter, van den Berg, Rianne, Titov, Ivan, Welling, Max: Modeling Relational Data with Graph Convolutional Networks. In: Gangemi, Aldo, Navigli, Roberto, Vidal, Maria-Esther., Hitzler, Pascal, Troncy, Raphaël, Hollink, Laura, Tordai, Anna, Alam, Mehwish (eds.) ESWC 2018. LNCS, vol. 10843, pp. 593–607. Springer, Cham (2018). https://doi.org/10.1007/978-3-319-93417-4_38
19. Wang, X., He, X., Cao, Y., Liu, M., Chua, T.: KGAT: Knowledge graph attention network for recommendation. In: Proceedings of the 25th ACM SIGKDD International Conference on Knowledge Discovery and Data Mining. (July 2019), pp. 950–958. url: https://doi.org/10.1145/3292500.3330989
20. Jin, X., Sun, X., Chen, J., Sutcliffe, R.: Extracting Drug-drug Interactions from Biomedical Texts using Knowledge Graph Embeddings and Multi-focal Loss. In: CIKM '22: Proceedings of the 31st ACM International Conference on Information and Knowledge Management, pp. 884–893. ACM, Oct 2022. https://doi.org/10.1145/3511808.3557318
21. Vaswani, A., et al.: I.: Attention is All you Need. In: Advances in Neural Information Processing Systems, Vol. 30, eds. I. Guyon, U. Von Luxburg, S. Bengio, H. Wallach, R. Fergus, S. Vishwanathan, R. Garnett. Curran Associates, Inc., 2017. URL: https://proceedings.neurips.cc/paper_files/paper/2017/file/3f5ee243547dee91fbd053c1c4a845aa-Paper.pdf
22. Johnson, K. B., et al.: Precision Medicine, AI, and the Future of Personalized Health Care. In: Clinical and Translational Science, Vol. 14, No. 1, pp. 86–93. Published online 2020 Oct 12. PMCID: PMC7877825. PMID: 32961010, 2021. https://doi.org/10.1111/cts.12884
23. Gyrard, A., Gaur, M., Shekarpour, S., Thirunarayan, K., Sheth, A.: Personalized Health Knowledge Graph. In: CEUR Workshop Proceedings, Vol. 2317, pp. 5. Published in final edited form as: CEUR Workshop Proc. 2018 Oct; 2317: 5. PMCID: PMC8532078. NIHMSID: NIHMS1743812. PMID: 34690624. Available in PMC 2021 Oct 22. url: https://www.ncbi.nlm.nih.gov/pmc/articles/PMC8532078/
24. Kurczab, R., Smusz, S., Bojarski, A.J.: The influence of negative training set size on machine learning-based virtual screening. In: J Cheminform 6, 32 (2014). Received:

12 Oct 2013. Accepted: 02 June 2014. Published: 11 June 2014. https://doi.org/10.1186/1758-2946-6-32

25. Li, J., Tao, C., Wu, W., Feng, Y., Zhao, D., Yan, R.: Sampling Matters! an empirical study of negative sampling strategies for learning of matching models in retrieval-based dialogue systems. In: Proceedings of the 2019 Conference on Empirical Methods in Natural Language Processing and the 9th International Joint Conference on Natural Language Processing (EMNLP-IJCNLP), pp. 1291-1296. Association for Computational Linguistics, Hong Kong, China (2019). https://aclanthology.org/D19-1128URL. https://doi.org/10.18653/v1/D19-1128
26. Cao, R., et al.: Exploring the impact of negative samples of contrastive learning: a case study of sentence embedding. In: Findings of the Association for Computational Linguistics: ACL 2022, pp. 3138–3152, Dublin, Ireland (2022). https://doi.org/10.18653/v1/2022.findings-acl.248**URL:** https://aclanthology.org/2022.findings-acl.248
27. Wenrui, C., Dongpao, H., Chao, Z.: Learning knowledge graph embedding with entity descriptions based on LSTM Networks. In: 2020 IEEE Int. Symp. Prod. Compliance Eng.-Asia (ISPCE-CN), Chongqing, China (2020). https://doi.org/10.1109/ISPCE-CN51288.2020.9321857**Publisher:** IEEE
28. Shen, K., Guo, J., Tan, X., Tang, S., Wang, R., Bian, J.: A study on ReLU and softmax in transformer. In: arXiv preprint arXiv:2302.06461 (2023). **Subjects:** Computation and Language (cs.CL); Artificial Intelligence (cs.AI); Machine Learning (cs.LG) **URL:** https://arxiv.org/abs/2302.06461
29. Lin, T-Y., Goyal, P., Girshick, R., He, K., Dollár, P.: Focal Loss for Dense Object Detection. In: 2017 IEEE International Conference on Computer Vision (ICCV), pp. 2999-3007 (2017). https://doi.org/10.1109/ICCV.2017.324**Publisher:** IEEE
30. Sowjanya, A.M., Mrudula, O.: Effective treatment of imbalanced datasets in health care using modified SMOTE coupled with stacked deep learning algorithms. In: Appl Nanosci **13**, 1829-1840 (2023). https://doi.org/10.1007/s13204-021-02063-4**Publisher:** Springer
31. Himmelstein, D.S., et al.: Systematic integration of biomedical knowledge prioritizes drugs for repurposing. In: eLife (2017-09-22). https://doi.org/10.7554/elife.26726**Publisher:** eLife Sciences Publications **URL:**.https://git.dhimmel.com/rephetio-manuscript/

A Secure and Privacy-Preserving Paradism Based on Blockchain and Federated Learning for CIoMT in Smart Healthcare Systems

Samia El Haddouti[1,2](✉) and Mohamed Dafir Ech-Cherif El Kettani[1]

[1] ENSIAS, Mohamed V University in Rabat, Rabat, Morocco
{dafir.elkettani,samia_elhaddouti}@um5.ac.ma
[2] National Center for Scientific and Technical Research, Rabat, Morocco

Abstract. Since the advent of COVID-19 pandemic, the Cognitive Internet of Medical Things (CIoMT) has been highlighted as a critical need for the healthcare ecosystem, by enhancing operational efficiency and promoting preventive and proactive healthcare approaches through a remote patient monitoring, real-time health data collection and optimized supply chain management. Indeed, the CIoMT is a promising technology that refers to the application of cognitive computing techniques and the Internet of Things (IoT) in the field of e-health to enhance the delivery of healthcare services. However, challenges emerged in data privacy, service integrity, and adaptability to network structure in a such ecosystem since health data are highly private and have great financial values. To deal with these concerns, we propose in this paper a secure and trusted infrastructure based on Federated Learning and Blockchain technologies within a Fog Computing network. The adopted technologies have the potentials to overcome the issue of fragmented data repositories, by providing a distributed model for health data sharing while preserving the privacy of data owners within a trusted collaborative environment based on an Identity Federation paradigm.

Keywords: Blockchain · CIoMT · Federated Learning · Fog Computing · Healthcare · Identity Fedeartion · Privacy · Trust

1 Introduction

The COVID-19 pandemic [1] has emphasized the pressing necessity of remote healthcare services and real-time monitoring, by enabling individuals to access medical cares from the safety of their homes without risking exposure to the virus. In this context, the smart healthcare based on the Internet of Things (MIoT) [2] has gained importance to deal with these challenges. However, it is worth noting that the remote and real-time healthcare services do not only play a crucial role during pandemic situations, but are also paving the way for a

M. Ben Ahmed et al. (Eds.): SCA 2023, LNNS 906, pp. 447–456, 2024.
https://doi.org/10.1007/978-3-031-53824-7_41

new area of healthcare delivery, by offering and improved outcomes for patients beyond the scope of infectious diseases. From these perspectives, the pioneering implementation of Cognitive Radio (CR) [3] based on IoT for the medical industry referred to as Cognitive Internet of Medical Things (CIoMT) [4] is investigated to create an adequate context and governance of the smart healthcare. Indeed, the real-time vital signs of a patient such as the blood pressure, body temperature, glucose level, heart rate etc. as well as psychological data like expressions and speeches are available to medical practitioners remotely through wearable IoMT sensors [5] that are capable of transmitting data securely to connected systems. Furthermore, the CIoMT combines the power of these wearable sensors with Cognitive Computing technologies [6] involving the use of Artificial Intelligence (AI) technology [7] to enhance the delivery of healthcare services, by bringing advanced capabilities for data analysis, decision-making, and personalized interventions.

Typically, AI techniques rely heavily on large-scale datasets for training and frequently entail the consolidation of data collection and training within a single entity. Thus, privacy concerns, data security and high overhead of raw data communications have become significant concerns. This is where Federated Learning [8] emerges as a distributed AI approach for collaborative data training, by allowing machine learning models to be trained on decentralized devices without directly sharing their underlying data. Indeed, only learning parameters are exchanging with an aggregation server, striking thereby a balance between the benefits of AI and the need for data protection. Nevertheless, FL faces several challenges that need to be addressed for its successful implementation. In fact, in a Fog Commuting [9] context for instance, fog nodes are required to fully trust a cloud server to send him learning parameters, which is not consistently attained in practical wireless networks. Furthermore, the reliance of an aggregation server presents Single Point of Failure (SPF) bottlenecks, which thereby perturb the entire FL system. In addition, FL involves multiple participants, and the presence of malicious or compromised entities can pose security risks for the entire system [10].

To tackle the aforementioned limitations, while primarily improving trust management and security aspects, the integration of Blockchain [11] as a novel technology to improve integrity, distribution and control of data, could certainly enhance the traditional FL approaches. That is precisely where the present work fits in, by proposing a CIoMT paradigm based on Blockchain and FL technologies, where Blockchain technology ensures data integrity, privacy, and transparency, while FL enables distributed learning without centralized data sharing. This collaborative approach facilitates improved healthcare outcomes, medical research, and patient-centric care. Moreover, the proposal implicates an Identity Federation for a granular access control within a trust circle [12].

The subsequent sections of the paper are organized as follows. Section 2 reviews the most recent initiatives regarding the adoption of Federated Learning and Blockchain technologies in the smart healthcare. Section 3 presents the technological review of Fedearted Learning and Blockchain. Section 4 introduces the proposed model by outlining its architecture. Subsequently, a comprehensive depiction of the main interactions among different actors is provided. Section 5 is dedicated to evaluating the proposal. Finally, Sect. 6 is dedicated to conclusions and future work.

2 Related Works

In this section, we present an overview of relevant researches that focus on the integration of FL and Blockchain in the smart healthcare industry.

In [13], the authors conduct a review of the evolving applications of FL in healthcare sectors, covering diverse areas including remote health monitoring, COVID-19 detection, medical imaging, etc. Additionally, the authors of [14] primarily address the technological hurdles and requirements associated with implementing a FL approach in e-health sector. However, they fall short in dealing with limitations observed in traditional FL approaches. Alternatively, the authors of [15] extensively discussed the security and privacy concerns encountered in FL, while exploring potential strategies for assessing and mitigating harmful threats within FL networks.

The primary challenges tackled by the conventional FL typically revolve around the potential risks of compromising privacy and the excessive utilization of communication resources when transmitting model updates from individual nodes to a central entity. Nevertheless, several studies have also underlined that untrusted participants within the FL can carry out malicious actions by deliberately submitting tainted model updates, which leads to model poisoning attacks and thereby compromising the overall integrity and performance of the FL system [16].

By integrating Blockchain technology, FL systems can enhance their security and trust mechanisms, by dealing with the SPoF concern [17,18]. In this context, some initiatives have adopted Blockchain technology to design FL for the CIoMT ecosystem.

In [19], the authors suggest a solution leveraging FL and consortium Blockchain technology in a fog-IoT network to create a dynamic network infrastructure for wearable IoT devices, by prioritizing the preservation of patient privacy and ensures the integrity of predictive services. The authors of [20] propose another scheme based on FL and Blockchain technologies to improve the trustworthiness and credibility of public media communication, while establishing a reliable mechanism for disseminating accurate COVID-19 information. Additionally, a recent approach is introduced by the authors of [21] that leverages a combination of Blockchain and FL to establish a robust and trustworthy distributed learning platform. The latter ensures the secure and seamless sharing of medical data and facilitates advanced analytics within a multi-organizational environment.

Despite the potential high returns of the proposed approaches of integration Blockchain and FL in the context of healthcare, careful consideration and improvements are required to effectively tackle the challenges and ensure a successful implementation that optimizes the potential returns.

3 Background

In this section, we present the core principles of FL and Blockchain technologies, exploring their recent advancements and developments.

3.1 Federated Learning

Initially introduced by Google in 2016 [22], the FL has emerged as a decentralized paradigm across distributed learners in order to surpass the constraints of conventional Machine Learning. The primary aim of this approach is to maintain the training dataset at its source and conducts local model training. Indeed, contrary to the conventional method of transferring all unprocessed data to a cloud or central data warehouse for AI training, FL facilitates distributed learning by assigning AI tasks directly to local devices. The latter collaboratively build a shared global model for prediction at an aggregator, enabling an efficient training and privacy-preserving [23].

FL can be classified into three major categories: horizontal FL, vertical FL, and federated transfer learning. Each of these categories fulfils specific purposes and aspects of decentralized machine learning.

- Horizontal FL: the FL is commonly categorized as horizontal FL, where multiple entities with similar data distributions cooperate in training a collective global model. In contrast to other forms of the FL, the horizontal FL is designed to tackle scenarios where data samples vary across participating entities while sharing a common feature space. The solution proposed by Google [24] for updating Android mobile phones, can be categorized as a form of the horizontal FL since the data exhibits uniform feature dimensions.
- Vertical FL: the vertical FL emerges as a suitable solution for contexts where data is vertically partitioned based on feature dimensions [25]. Indeed, each participating party possesses homogeneous data, indicating that they share a partial overlap in sample identifiers while differing in terms of the feature space. However, the vertical FL poses a more intricate task compared to the horizontal FL, particularly due to the complexities of entity resolution. Furthermore, instead of aggregating all datasets on a central server in the horizontal FL, the vertical FL cannot rely on the same methodology since the correspondence between different data owners remains an unresolved issue requiring an urgent attention.
- Federated Transfer Learning (FTL): it refers to the application of transfer learning techniques [26] in the context of FL. In fact, this kind of FL involves leveraging pre-trained models from one or more source domains and adapting them to improve the learning or performance on a target domain in a FL setting. the pre-trained models can be transferred across different devices or participants in a FL system [27]. The authors of [28] proposed a FTL as a comprehensive framework, representing the first complete stack for FL that integrates transfer learning techniques across all stages of the learning process, encompassing training, evaluation, and cross-validation.

Federated learning in its different categories offers several benefits by improving the privacy and performance. However, it presents also considerable limitations. In fact, ensuring data privacy and security requires robust encryption techniques and secure aggregation protocols. Furthermore, the heterogeneity of devices and data sources introduces complexities in terms of handling varying computational capabilities, network conditions, and data distributions.

3.2 Blockchain Technology

Blockchain is a distributed ledger that operates on peer-to-peer (P2P) network, allowing multiple parties to maintain a shared database without the need for a central authority. The distributed ledger consists of a chain of blocks, where each block contains a list of transactions and other data. These blocks are linked together using cryptographic hashes, creating an immutable and transparent ledger [11].

One of the fundamental features of Blockchain technology is its decentralized structure. Indeed, rather than relying on a central entity such as a government or a bank, the process of verification and recording of transactions are performed by distributed computers, known as nodes or miners, according to a consensus algorithm. In fact, there are several consensus algorithms used in blockchain networks to achieve agreement among participants. The commonly used are Proof-of-Work (PoW) algorithm [29], where miners solve complicated mathematical equations to validate blocks, and Proof-of-Stake (PoS) algorithm [30], where validators are chosen based on their stake in the Blockchain network. Each algorithm has its own strengths and weaknesses, and the choice of consensus mechanism depends on factors such as network goals, scalability requirements, security considerations, and the nature of the Blockchain application.

Blockchain was initially introduced as the underlying technology for popular cryptocurrencies such as Bitcoin [31] and Ethereum [32], and thereafter has gained attention and found applications in various industries, namely, supply chain management, insurance, healthcare, Internet of Things (IoT) and so on. Furthermore, the emergence of Distributed Applications (DApps) [33] and Smart Contracts [34] has significantly impacted the Blockchain landscape in the age of Web 3.0.

4 Proposed Blockchain and FL-Based Scheme for CIoMT

In our previous work [35], we proposed a secure and trusted Fog Computing approach based on Blockchain and Identity Federation technologies for a granular access control in IoT environments. However, the massive amount of generated data from interconnected devices and sensors requires advanced capabilities for data analysis, decision-making, and automation. In this paper, we propose an enhancement of our previous proposed scheme by introducing a FL mechanism applied in CIoMT, where data privacy and trust are of utmost importance.

4.1 Architecture Overview

The proposed architecture comprises five primary components as is depicted in Fig. 1:

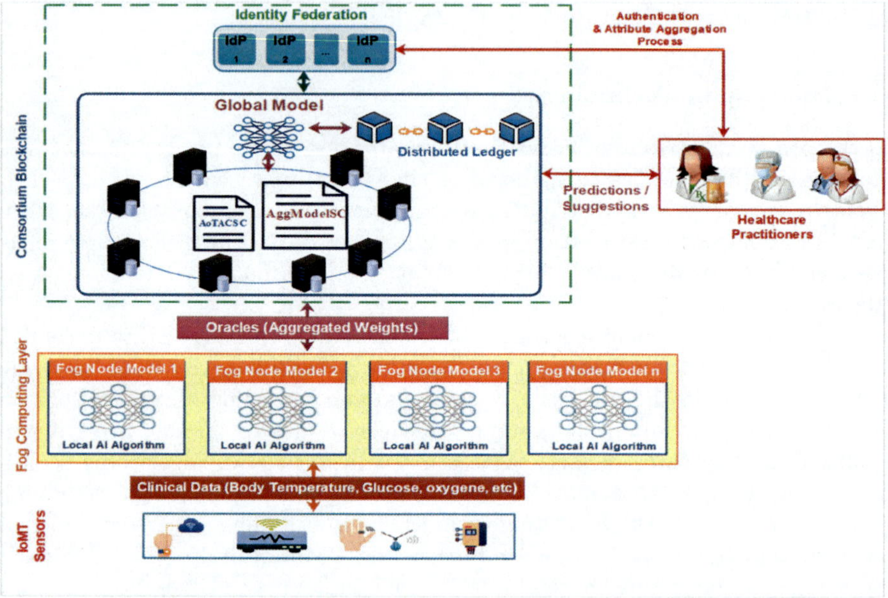

Fig. 1. A Proposed Blockchain and FL-based Scheme for CIoMT

- **Wearable IoMT Sensors:** they are a specific type of IoT device designed to be worn or attached to the body for medical purposes, by enabling a continuous monitoring of vital signs and health parameters and collect medical data. Thus, wearable IoMT sensors have potentials to identify health risks and detect early signs of diseases. Furthermore, the collected data, combined with AI-powered algorithms, can enable predictive analytics and alert individuals or healthcare professionals to take proactive measures for disease prevention or early intervention.
- **Fog Nodes Layer:** Iin addition to its typical features as is illustrated in [35], in the present architecture, fog nodes act as local training points for medical data generated by wearable IoMT sensors. Indeed, each fog node performs a localized data processing and model training. This distributed approach minimizes the need for extensive data transfers, which thereby significantly reduces the latency and improves the real-time data processing, ensuring prompt decision-making and immediate response in critical healthcare scenarios. Furthermore, data privacy and compliance with regulations such as the Health Insurance Portability and Accountability Act (HIPAA) [36] are ensured, by minimizing the transmission of raw patient data to external servers.

- **Consortium Blockchain:** besides the access control for IoT devices managed by the IoT Access Control Smart Contract (IoTACSC) running within the Consortium Blockchain proposed in our previous work [36], in the given architecture, we introduce an addition Smart Contract named Aggregation Model Smart Contract (AggModelSC) for handling the aggregation process of model updates from participating fog nodes, ensuring transparency, and enforcing the predefined rules for model merging and consensus. The execution of the AggModelSC results in the creation of the global model, enabling collaborative learning while maintaining data privacy.
- **Oracles Layer** [37]**:** in our proposal, oracles retrieve information from fog nodes as off-chain sources, and deliver it to the AggModelSC on the consortium Blockchain. Data oracles ensure that the Blockchain network has the access to real-time, accurate and trusted data required for making the global modal, by validating the quality, integrity and authenticity of the model updates from fog nodes before being accepted and recorded on the Blockchain.
- **Identity Providers:** in a federated learning framework involving medical practitioners, such as doctors, nurses, and other healthcare professionals, it is important to consider varying permissions and access levels to data depending on various factors, including their role, specialty, and the policies and regulations in place within their healthcare organization or country. This aspect is addressed in our proposal by integrating an Identity Federation with attribute aggregation techniques [38,39].

4.2 System Interactions

The interactions between the different actors of the proposed model; namely IoMT sensors, fog nodes, Blockchain and medical practitioners; involve secure data transmission, collaborative model training, and utilization of insights for clinical decision-making. These interactions can be categorized into two main phases:

The first phase, corresponds to the registration process regulating the authentication and access control policies according to the IoTACSC as is detailed in [35].

The second phase, pertains to the subsequent steps of the FL process illustrated in Fig. 2 as follows:

1. IoMT sensors collect data from patients or healthcare environments. They transmit then the collected data securely to the fog nodes for processing and analysis;
2. Fog nodes are responsible for receiving and processing data from IoMT sensors, by performing local computation and model training using AI algorithms. They train models collaboratively while preserving data privacy and security. Afterword, parameter updates or weights are securely shared with the Blockchain network for validation and consensus;
3. Blockchain validates the integrity and authenticity of the received model updates through a consensus algorithm. The creation of the global model is a direct outcome of the execution of the AggModelSC;

4. Medical practitioners can access the global model from the Blockchain network, and leverage the insights and predictions to support diagnosis and treatment decisions.

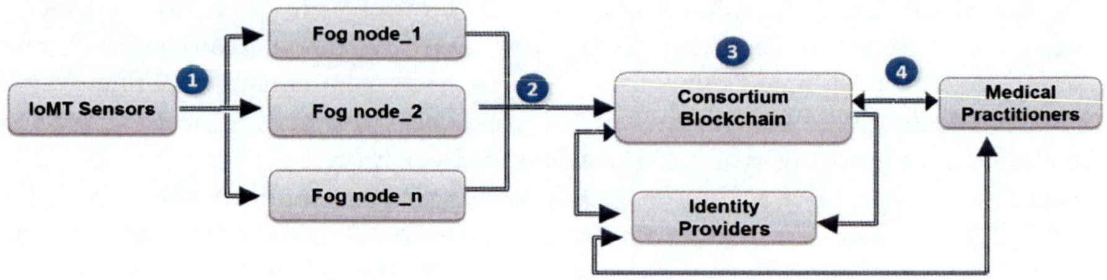

Fig. 2. System Interactions

5 Advantages and Open Issues

The design of proposed scheme is motivated by the objective of preserving the data privacy of wearable sensors in the context of CMIoT within a trusted fog computing architecture, by combining FL and Blockchain technologies in addition to Identity Federation and attribute aggregation techniques that enhance the access control policies. Nevertheless, there are still some challenges to overcome in the context of blockchain-based FL. In fact, it is worth taking into account the following limitations:

- The consensus mechanism used in Blockchain introduces latency during the verification and validation process, which thereby could affect the timely aggregation of FL model updates and slowing down the learning process;
- Since FL involves iterative updates, the storage requirements can grow substantially, leading to increased storage costs and potential scalability challenges;
- A malicious participant in a Blockchain can intentionally engages in harmful activities to disrupt the normal functioning of the proposed model.

6 Conclusion and Future Work

This article provides a comprehensive overview of a promising paradigm in CIoMT that combines the integration of FL and Blockchain technologies within a fog-based IoT architecture. The proposal holds significant potential for advancing privacy-preserving data collaborations and enhancing the security and efficiency of decentralized CIoMT ecosystems. We presented the system framework and the design of its components. Furthermore, we outlined the main interactions among different actors. In our future work, we intent to implement the proposal, by conducting comprehensive experiments and analyses to assess the performance, effectiveness, and feasibility of the given approach in real-world scenarios.

References

1. Ciotti, M., Ciccozzi, M., Terrinoni, A., Jiang, W.-C., Wang, C.-B., Bernardini, S.: The covid-19 pandemic. Crit. Rev. Clin. Lab. Sci. **57**(6), 365–388 (2020)
2. Dimitrov, D.V.: Medical internet of things and big data in healthcare. Healthc. Inf. Res. **22**(3), 156–163 (2016)
3. Liang, Y.-C., Chen, K.-C., Li, G.Y., Mahonen, P.: Cognitive radio networking and communications: an overview. IEEE Trans. Veh. Technol. **60**(7), 3386–3407 (2011)
4. Jabbar, M., Shandilya, S.K., Kumar, A., Shandilya, S.: Applications of cognitive internet of medical things in modern healthcare. Comput. Electr. Eng. **102**, 108276 (2022)
5. Qureshi, F., Krishnan, S.: Wearable hardware design for the internet of medical things (IOMT). Sensors **18**(11), 3812 (2018)
6. Gudivada, V.N.: Cognitive computing: concepts, architectures, systems, and applications. In: Handbook of Statistics vol. 35, pp. 3–38. Elsevier, ??? (2016)
7. He, J., Baxter, S.L., Xu, J., Xu, J., Zhou, X., Zhang, K.: The practical implementation of artificial intelligence technologies in medicine. Nat. Med. **25**(1), 30–36 (2019)
8. Konečný, J., McMahan, H.B., Yu, F.X., Richtárik, P., Suresh, A.T., Bacon, D.: Federated learning: Strategies for improving communication efficiency. arXiv preprint arXiv:1610.05492 (2016)
9. Ketel, M.: Fog-cloud services for IoT. In: Proceedings of the SouthEast Conference, pp. 262–264 (2017)
10. Mammen, P.M.: Federated learning: Opportunities and challenges. arXiv preprint. arXiv:2101.05428 (2021)
11. Michael, J., Cohn, A., Butcher, J.R.: Blockchain technology. J. **1**(7), 1–11 (2018)
12. El Haddouti, S., Dafir Ech-Cherif EL Kettani, M.: A hybrid scheme for an interoperable identity federation system based on attribute aggregation method. Comput. **8**(3), 51 (2019)
13. Nguyen, D.C., et al.: Federated learning for smart healthcare: a survey. ACM Comput. Surv. (CSUR) **55**(3), 1–37 (2022)
14. Rieke, N., et al.: The future of digital health with federated learning. NPJ Digital Med. **3**(1), 119 (2020)
15. Mothukuri, V., Parizi, R.M., Pouriyeh, S., Huang, Y., Dehghantanha, A., Srivastava, G.: A survey on security and privacy of federated learning. Futur. Gener. Comput. Syst. **115**, 619–640 (2021)
16. Lyu, L., Yu, H., Yang, Q.: Threats to federated learning: a survey. arXiv preprint. arXiv:2003.02133 (2020)
17. Qammar, A., Karim, A., Ning, H., Ding, J.: Securing federated learning with blockchain: a systematic literature review. Artif. Intell. Rev. **56**(5), 3951–3985 (2023)
18. Wang, Z., Hu, Q.: Blockchain-based federated learning: a comprehensive survey. arXiv preprint. arXiv:2110.02182 (2021)
19. Baucas, M.J., Spachos, P., Plataniotis, K.N.: Federated learning and blockchainenabled fog-IoT platform for wearables in predictive healthcare. IEEE Trans. Comput. Soc. Syst. (2023)
20. Samuel, O., et al.: IOMT: a covid-19 healthcare system driven by federated learning and blockchain. IEEE J. Biomed. Health Inform. **27**(2), 823–834 (2022)
21. Lian, Z., Wang, W., Han, Z., Su, C.: Blockchain-based personalized federated learning for internet of medical things. IEEE Trans. Sustain. Comput. (2023)

22. Bonawitz, K., et al.: Practical secure aggregation for privacypreserving machine learning. In: Proceedings of the 2017 ACM SIGSAC Conference on Computer and Communications Security, pp. 1175–1191 (2017)
23. Konečný, J., McMahan, H.B., Ramage, D., Richtárik, P.: Federated optimization: Distributed machine learning for on-device intelligence. arXiv preprint. arXiv:1610.02527 (2016)
24. McMahan, B., Moore, E., Ramage, D., Hampson, S., Arcas, B.A.: Communication-efficient learning of deep networks from decentralized data. In: Artif. Intell. Stat. pp. 1273–1282 (2017). PMLR
25. Liu, Y., et al.: Vertical federated learning. arXiv preprint. arXiv:2211.12814 (2022)
26. Furht, B., Villanustre, F., Weiss, K., Khoshgoftaar, T.M., Wang, D.: Transfer learning techniques. Big Data Technol. Appl., 53–99 (2016)
27. Saha, S., Ahmad, T.: Federated transfer learning: concept and applications. Intell. Artif. **15**(1), 35–44 (2021)
28. Liu, Y., Kang, Y., Xing, C., Chen, T., Yang, Q.: A secure federated transfer learning framework. IEEE Intell. Syst. **35**(4), 70–82 (2020)
29. Porat, A., Pratap, A., Shah, P., Adkar, V.: Blockchain Consensus: An analysis of Proof-of-Work and its applications. Stanford University, Stanford, CA, USA (2017)
30. Thin, W.Y.M.M., Dong, N., Bai, G., Dong, J.S.: Formal analysis of a Proofof-stake blockchain. In: 2018 23rd International Conference on Engineering of Complex Computer Systems (ICECCS), pp. 197–200 (2018). IEEE
31. Nakamoto, S.: Bitcoin: A peer-to-peer electronic cash system. Decentralized business review (2008)
32. Wood, G., et al.: Ethereum: a secure decentralised generalised transaction ledger. Ethereum Proj. Yellow Pap. **151**(2014), 1–32 (2014)
33. Udokwu, C., Anyanka, H., Norta, A.: Evaluation of approaches for designing and developing decentralized applications on blockchain. In: Proceedings of the 4th International Conference on Algorithms, Computing and System, pp. 55–62 (2020)
34. Aggarwal, S., Kumar, N.: Blockchain 2.0: smart contracts. In: Adv. Comput. vol. 121, pp. 301–322. Elsevier, ??? (2021)
35. EL Haddouti, S., EL Kettani, M.D.E.-C.: A secure and trusted fog computing approach based on blockchain and identity federation for a granular access control in IoT environments. Int. J. Adv. Comput. Sci. Appl. 13(**3**) (2022)
36. O'herrin, J.K., Fost, N., Kudsk, K.A.: Health insurance portability accountability act (HIPPA) regulations: effect on medical record research. annals of surgery 239(6), 772 (2004)
37. Beniiche, A.: A study of blockchain oracles. arXiv preprint. arXiv:2004.07140 (2020)
38. El Haddouti, S., Ouaguid, A., Ech-Cherif El Kettani, M.D.: Fedidchain: An innovative blockchain-enabled framework for cross-border interoperability and trust management in identity federation systems. J. Network Syst. Manage. 31(2), 42 (2023)
39. El Haddouti, S., El Kettani, M.D.E.-C.: A novel blockchain-based reliable model for the trust management of an identity federation system. In: Proceedings of the 33rd Int. Bus. Inf. Manage. Assoc. (IBIMA), pp. 8342–8351 (2019)

A Comparative Study Based on Deep Learning and Machine Learning Methods for COVID-19 Detection Using Audio Signal

Fulya Akdeniz(✉), Merve Nur Damar, Buse İrem Danacı, Burcu Kır Savaş, and Yaşar Becerikli

Kocaeli University, Umuttepe Campus, Kocaeli 41001, Turkey
{fulya.akdeniz,200202091,190201123,burcu.kir, ybecerikli}@kocaeli.edu.tr

Abstract. COVID-19 is an upper respiratory disease that emerged in the last months of 2019 and affecting people worldwide. According to the World Health Organization data, it is determined that the outbreak spread rapidly. Early diagnosis is very important in preventing the spread of epidemics. Highly accurate identification and isolation of infected persons is essential to prevent the spread of outbreaks. Currently, reverse transcription polymerase chain reaction (RT-PCR) test is the most used method to detect COVID-19. However, the test is expensive, time-consuming, socially distant, difficult to distribute and has a high false negative rate. In this study, we developed a system based on Machine Learning (ML) and Deep Learning (DL) methods using voice signals (cough sounds) for early detection and diagnosis of COVID-19 disease. Since cough sounds are also a symptom of many upper respiratory tract diseases, a three-class system was developed to determine whether the voice recording from a person was COVID-19, upper respiratory tract disease, healthy or not. In this paper, indicative features were extracted Mel Frequency Cepstrum Coefficients (MFCC) from audio signals and features were classified with K-Nearest Neighbor Algorithm (KNN), Random Forest (RF), Decision Tree, Support Vector Machines (SVM), Logistic Regression classifiers. In addition, Convolutional Neural Networks (CNN) methods, one of the DL methods, was used and comparative performance results are given in the experimental section.

Keywords: Deep learning · Audio signal · Covid-19 detection · MFCC

1 Introduction

Coronavirus Disease 2019, COVID-19 is an infectious disease caused by a coronavirus called SARS-CoV-2 [1]. Early detection of COVID-19 plays an important role in reducing the rate of spread of the disease and the damage caused by providing rapid decision-making [2].

The RT-PCR test used to diagnose COVID - 19 disease is a very costly test and the time taken to obtain the result is quite long. Many studies emphasize that the false negative rates in the results of this test are also quite high. In other words, false negative results are

M. Ben Ahmed et al. (Eds.): SCA 2023, LNNS 906, pp. 457–466, 2024.
https://doi.org/10.1007/978-3-031-53824-7_42

quite high in people with COVID -19 virus levels [3]. X-Ray and Computed Tomography (CT) images are frequently used to support PCR tests to diagnose COVID -19 disease. However, diseases symptoms such as COVID-19, pneumonia, bacterial pneumonia on the lung are quite similar. Therefore, it is very difficult to distinguish between these diseases on X-Ray [4]. Chest X-Ray images of COVID -19 patients are analyzed, COVID -19 findings on X-Ray peaked 10–12 days after the onset of symptoms of the disease. CT images are still inadequate to identify and differentiate specific viruses [5]. X-rays can be harmful because they are classified as ionizing radiation [6, 7]. Furthermore, exposure of pregnant and lactating women and children to X-rays is harmful. In this study, Covid-19 disease is detected using low-cost, easy-to-use audio signals that are harmless to human health. Most studies show that COVID-19 disease severely affects human breathing, audio and causes symptoms that make voices of patients distinctive [8, 9]. Once a person's audio recordings can be used to detect COVID-19 disease after are analyzed. It is very important to use audio signals to diagnose COVID-19 since audio signals are easy to obtain, can be analyzed quickly and effectively, and are cost-effective. Differentiating the pandemic-causing COVID-19 virus from other less dangerous upper respiratory tract diseases is an important factor during the pandemic [10]. Coughing is among the most common and most important symptoms of COVID-19. However, cough sounds are related to many diseases [11]. Therefore, it is important to differentiate COVID-19 from upper respiratory tract diseases. Differentiating between COVID-19 and other symptomatic diseases is important for accurate diagnosis and correct treatment. The symptoms of COVID-19 can be similar to some other symptomatic diseases. However, since COVID -19 has a high contagiousness and spread rate, it is important to diagnose the disease correctly to reduce the risk of transmission.

In this study, Mel frequency-based features were extracted from cough sounds. In the first stage, the features were classified into three classes using ML methods. In the second stage, the features were classified into three classes with DL architecture and different hyperparameters were used to determine whether the individual is healthy, COVID-19 and symptomatic. The contributions of this work are as follows:

- A three-class detection and classification system were developed that automatically detected COVID-19 disease using cough sounds.
- ML and DL algorithms optimized the hypermeters of different models for the audio-based model.

2 Related Works

Dash et al. found that cough sounds used for COVID-19 detection have different frequencies than speech sounds and developed a COVID-19 Coefficient feature by optimizing the frequency range and frequency conversion scale of the MFCC method to improve the efficiency of COVID-19 detection in their study. They used Coswara and Crowd-sourced Respiratory Sound Data datasets. They tested various feature groups using a simple SVM-based classifier [12]. Grant et al. used breath, cough and speech sounds for Covid-19 detection in their study. They extracted MFCC, delta-MFCC and RASTA-PLP features from the signal. They obtained the AUC performance result of the study as 79.38% [13]. Despotovic et al. used a dataset of breath, cough and voice recordings collected from COVID-19 infected and uninfected people. A total of 1103 data, 84

COVID-19 data and 1019 healthy data from their dataset, were used in the study. They obtained Geneva Minimalistic Acoustic Parameter Set (GeMaps), extended GeMaps and ComParE feature set features from the signal. In the classification phase, they created Random Forests, VGGish and OpenL3 algorithms and achieved a performance result of 88.52% [8]. Coppock et al. detected symptomatic and asymptomatic COVID-19 cases using breath and cough voice recordings. The CNN model was tested on 355 data in the study. They then achieved an AUC rate of 84.6% [14]. Chaudhari et al., detected COVID-19 disease in crowd-sourced cough sound recorded and collected on smartphones from various regions of the world. They used MFCC and Mel-spectrogram features on developed CNN model. A total of 2883 data from Coswara and Coughvid datasets, 539 COVID-19 data and 2344 healthy data, were used. Then, their performance result is obtained with a ROC-AUC value of 77.1% [15]. Tena et al. extracted Autoencoder time-frequency features from cough sounds to detect COVID-19 disease. They used various machine learning algorithms (RF, SVM, LR, NB, LDA) in the classification phase. Coswara, Virufy and the Pertussis datasets that collected a total of 813 data were used in the study. They obtained the best accuracy rate is 90% from the RF algorithm [16]. Han et al. generated a total of 828 audio samples from 343 patients by inputting audio signals and symptoms of the patients. The features were extracted with Zero Crossing Rate (ZCR), Root Mean Square (RMS) frame energy, pitch frequency (F0), Harmonicsto-Noise Ratio, and MFCCs. They used SVMs algorithm in the classification stage. Their performance results are obtained with AUC 79%, sensitivity 68% and specificity 82% [17]. Alsabek et al. compared the success of cough, respiration and speech sounds in COVID-19 detection. The study is examined the importance of speech signal processing in MFCC extraction of COVID-19 and non-COVID-19 samples and the relationship between these samples using Pearson correlation coefficients (PCCs). A total of 42 data were used in this study, including 3 recordings from 7 COVID-19 patients and 3 recordings from 7 negative patients. They also used CNN and LSTM algorithms were used in their studies, too [18]. Çelik et al. proposed a DL-based Covid-CoughNet method for COVID-19 detection using cough, respiration and voice signals. He used Chroma features, RMS Energy, Spectral centroid, Spectral bandwidth, Spectral rolloff, ZCR, and MFCC methods for feature extraction. InceptionFireNet algorithm was used in the classification stage. As a result of his studies on two different datasets; In the COUGHVID dataset, the AUC value was measured as 98.44% when classified as healthy, COVID-19 and symptomatic. In the use of voice Coswara dataset gave an AUC value of 99.24% [19]. Ulukaya et al. proposed a deep neural network-based model for detect coronavirus from coughing sound only, fast, can be used remotely and has no harmful side effects. The proposed model named MSCCov19Net uses MFCC, Spectrogram and Chromagram features as an input. The system was trained on the Coswara, Coughvid, Virufy and NoCoCo-Da datasets and tested on two unseen (used only for testing) clinical and non-clinical datasets. The performance of MSCCov19Net is compared with various CNN architectures. The proposed system achieved 61.5% accuracy on Virufy and 90.4% accuracy on NoCoCoDa [20]. Kranthi Kumar et al. propose a lightweight CNN with Modified-MFCC using different parameters to classify COVID-19 and other respiratory voice disease symptoms. They used the CU dataset in their study. The proposed model outperforms traditional feature extraction models and existing DL models in the range of 4% to 10% in COVID-19/SARS-CoV-2 classification with an accuracy of 93.65% [21].

3 Methods and Database

3.1. MFCCs: MFCCs are a measure of the human ear perceives a change in frequency. The human ear is generally sensitive to lower frequencies. With MFCC, this perception of the human ear is visualized [22]. MFCC is a widely used method for representing audio signals as feature vectors. It is developed to mimic human sound perception. MFCC-based features are proposed to minimize the difference and avoid perceptual and feature space distortion [12, 23].

3.2. SVM: It is a classification method used in machine learning. It uses the attributes in the dataset and generates a classification and regression model based on these attributes. An SVM has a great generalization capability, considering account errors [24]. SVM advantages are it works well on multidimensional data and reduces overfitting, is less sensitive to discrete values and noise in the data set and can separate data sets that cannot be linearly separated by transforming them into spaces. The parameters used in the SVM algorithm are LINEAR, Radial Basis Function-RBF, sigmoid or precomputed kernels, which are used to determine how the data points are separated in space and the decision boundary.

3.3. KNN: KNN is an algorithm used in machine learning and data mining for solving classification and regression problems [11]. KNN positions data points in a space and uses the labels of the nearest neighbors of these data points to predict data class or value of a new data point. The KNN algorithm uses different distance metrics to measure the difference between two samples [25].

3.4. Logistic Regression: Logistic Regression is a statistical computational method and machine learning algorithm for solving problems with two or more categorical classes. The class label of the dependent variable makes the probability estimate and the value of the dependent variable represents the class labels. Logistic Regression calculates probabilities using the sigmoid function [26]. Since it calculates probabilities, it estimates class labels in the interval [0, 1]. Multiple logistic regression is used in problems with more than one independent variable.

3.5. Decision Trees: Decision trees are among the first statistical algorithms for general-purpose prediction and classification mechanisms since the late 20th century [27]. Decision trees analyze a training set for which the class labels. Thus, they can classify previously unsupervised examples. With decision trees, high accuracy can be achieved on very high-quality data. A decision tree classifies or evaluates data samples using a set of decision rules or questions [28]. This algorithm's advantages are easy to understand, provide high performance and the results can be interpreted.

3.6. Random Forest: RF is a classification algorithm widely used ML. It is a combination of multiple decision trees. It is a powerful nonparametric statistical method that allows regression problems as well as two-class and multi-class classification problems to be addressed in a single and multifaceted framework [29].

3.7. CNN: CNN is an artificial neural network model frequently used in deep learning. CNN structure is based on neurons that are organized in layers. Neurons in these layers are connected to each other through weights and biases [30]. Its layers include convolution layer, pooling layer and fully connected layers etc. [31].

3.8. COUGHVID Database: Detecting various diseases from coughs is becoming more and more common today. One of the most common symptoms of COVID-19 is coughing. The COUGHVID dataset is one of the most widely used datasets in this field with more than 25000 cough records. In this dataset, all records were labeled by expert doctors and made suitable for classification. Of the labeled data, 25% received the Covid value, 35% received the symptomatic value, 25% received the healthy value and 15% received the condition-free value. The COUGHVID dataset was collected through a web application installed on a dedicated server at the École Polytechnique Fédérale de Lausanne (EPFL) in Switzerland. During these enrollments, information such as age, gender and geographical location is also collected from the user on request [32]. As such, COUGHVID is one of the largest and most reliable datasets.

4 Experimental Result

In this study, using ML and DL techniques is detected COVID-19 disease on the COUGHVID audio dataset. The results are comparatively analyzed in this section.

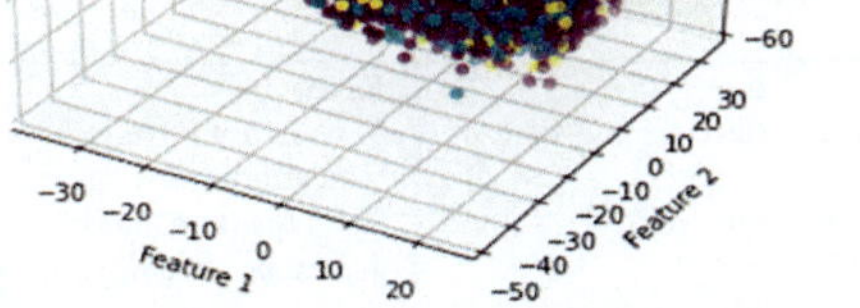

Fig. 1. Data distribution graph.

Figure 1 shows the data distribution graph and Fig. 2 shows a block diagram of our proposed method. The first step represents the data collection and preparation process. In this stage, the COUGHVID dataset was prepared for use in the study. Then, MFCC was extracted from the data. Data more was meaningful and processable in this step. Then, the models were trained using machine SVM, KNN, Logistic Regression, Decision Trees, Random Forest, CNN algorithms. The trained models represented classification models used to classify and categorize the data. These models successfully identified three different classes: COVID19, healthy and symptomatic. Finally, performance analysis was performed to evaluate the model's work.

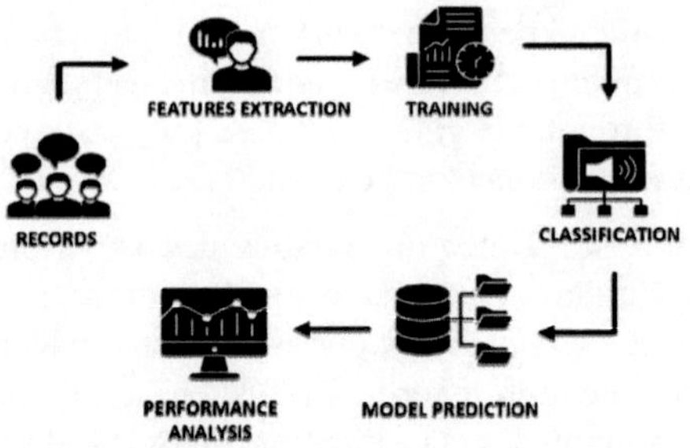

Fig. 2. The flow diagram of the system.

4.1 COVID-19 Detection Using ML Techniques

In this study, COVID-19 disease was detected using 5 different algorithms with machine learning techniques. Pseudocodes of algorithms is given Table 1.

Table 1. Pseudocodes of algorithms.

CNN 1. Data ←the data for the CNN model 2. 'X_train' and 'X_test' to make them compatible with Conv1D input shape. 3. Model ← Create a Sequential 4. Layers ← add layers as follows: a. Conv1D layer with 32 filters, kernel size 3, and ReLU activation, X_train.shape[1], 1. b. MaxPooling1D layer with pool size 2. c. Dropout layer rate 0.25. d. Flatten layer ← flatten the data. e. Dense layer 64 neurons and ReLU activation. f. Dropout layer rate 0.5. g. Dense output layer 3 neurons and softmax activation for three classes.	**SVM** 1. SVM model with 'SVC()' and define a hyperparameter grid for grid search. 2. Perform grid search using 'GridSearchCV' to find the best hyperparameter combination for the SVM model. 3. Print the best hyperparameters and the best training accuracy from grid search. 4. Make predictions on the test set using the SVM model.
Random Forest 1. Create a RandomForestClassifier 'rfc' and fit it on the training data. 2. Make predictions on the test set using the trained RandomForestClassifier 'rfc'. 3. accuracy_score ← Calculate test accuracy. 4. confusion_matrix← Calculate confusion matrix	**KNN** 1. Create a KNeighborsClassifier model 'knn' with 5 neighbors and 'manhattan' distance metric. 2. Fit the 'knn' model on the training data. 3. knn.predict(X_test) ← predictions on the test.
Decision Tree 1. Create a DecisionTreeClassifier 'dtc' and set up the parameter grid for Grid Search using 'param_grid'. 2. Perform Grid Search using 'GridSearchCV' with 5-fold cross-validation and 'accuracy' scoring. 3. Get the best parameters from the Grid Search and create a new DecisionTreeClassifier 'best_dtc' with these parameters. 4. Fit 'best_dtc' on the training data. 5. Make predictions on the test set using 'best_dtc'.	**Logistic Regression** 1. Create a LogisticRegression model 'log' with default parameters and no random state specified. 2. Fit the 'log' model on the training data. 3. log.predict(X_test)← predictions on the test.

The most accurate successful model of all models was the KNN using Manhattan distance and the accuracy rate was 78.56%. Other KNN models develop with Euclidian and Minkowski distances calculated in 78.36% accuracy. For SVM algorithm, three different kernels were tested on COUGHVID dataset. The best performance for the sigmoid kernel was determined 70.69% accuracy rate and for RBF and LINEAR kernels was computed with 77.04% accuracy rate. The Logistic Regression method also achieved similar results to the other SVM methods with an accuracy rate was 77.04%. The RF model achieved 76.88% accuracy rate. Decision Trees perform the lowest accuracy of all models with an accuracy rate was 60.33%. The results of the ML algorithms were given Table 2.

Table 2. The results of the machine learning algorithms.

Method	Accuracy (%)	Precision (%)	Recall (%)
SVM-RBF	77.04	59.0	77.0
SVM-LINEAR	77.04	59.0	77.0
SVM-Sigmoid	70.69	63.0	71.0
Logistic Regression	77.04	59.0	77.0
Decision Tree	60.33	62.0	60.0
Random Forest	76.88	66.0	77.0
KNN-Minkowski	78.36	64.0	75.0
KNN-Euclidian	78.36	64.0	75.0
KNN-Manhattan	78.56	63.0	75.0

4.2 COVID-19 Detection Using DL Techniques

3 different CNN models with 4 layers were developed in the experimental studies part. Models were evolved with the filter numbers of the layers as "8,16", "16,32", "32,64" respectively. The number of batches for each model is given as 4, 8 and 16 and for each batch number was tested 50 and 100 epochs. In all models, 100 epochs performance was better than 50 epochs performance. The highest performance was 77.04% accuracy rate was obtained in the model with filter numbers is "32,64", batch numbers is 8 and 100 epochs. Training, validation and test accuracies of the models were shown Table 3. The Model 3 accuracy and loss graphs with batch size 8 were shown Fig. 3.

Table 3. Training, validation and test accuracies of the models.

Model	Number of Baches	Train Accuracy (%)	Validation Accuracy (%)	Test Accuracy (%)	Train Accuracy (%)	Validation Accuracy (%)	Test Accuracy (%)
		50 epoch			**100 epoch**		
M1	4	68.96	69.18	69.18	76.60	76.85	76.85
	8	68.64	69.06	69.06	74.61	74.85	74.85
	16	69.75	68.93	68.93	70.45	70.45	70.78
M2	4	74.48	74.79	74.79	76.28	76.51	76.51
	8	74.61	75.16	75.16	75.48	75.56	75.56
	16	72.73	72.94	72.94	76.11	76.42	76.42
M3	4	76.77	76.91	76.91	76.53	76.64	76.64
	8	**76.50**	**76.61**	**76.61**	**76.61**	**77.04**	**77.04**
	16	76.21	76.08	76.08	76.34	76.57	76.57

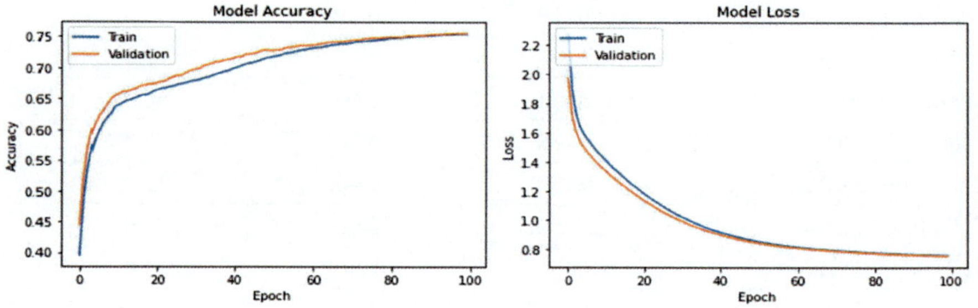

Fig. 3. The Model 3 accuracy and loss graphs with batch size 8.

5 Discussion and Conclusion

This study was conducted to evaluate the usability of both traditional ML and DL techniques for COVID-19 disease detection. In the experimental studies, three different CNN models were developed with four layers and different features were extracted. To evaluate the performance of the model, tests were performed by applying different batch numbers and step numbers. The highest performance was 77.04% accuracy rate in the model filter is "32,64" at batch numbers is 8 and 100 epochs. Also, SVM, RF, Decision Tree, Logistic Regression and KNN was used as machine learning models for classification and regression stage. The highest accuracy rate was recorded as 78.56% using Manhattan distance with KNN. Decision Trees accuracy rate was the lowest performance of 60.33%. Different models and model configurations compare each other and measure their performance with each effect. This study demonstrates the potential of deep artificial intelligence-based methods in Covid-19 diagnosis with the COUGHVID dataset.

Acknowledgements. The study was carried out in the Image Processing and Computer Vision Laboratory of the Computer Engineering Department of Kocaeli University. The study is supported by the TÜBİTAK 2209-A Project with project no: 1919B012113119.

References

1. Jahangir, M.A., Muheem, A., Rizvi, M.F.: Coronavirus (COVID-19): history, current knowledge and pipeline medications. J. Pharm. Res. Sci. Technol. [ISSN: 2583-3332], **4**(1), 1–9 (2020)
2. Gudigar, A., et al.: Role of artificial intelligence in COVID-19 detection. Sensors **21**(23), 8045 (2021)
3. Melek, N.: A brief dive into COVID-19 cough sound data sets. In: 2021 Medical Technologies Congress (TIPTEKNO). IEEE (2021)
4. Bozkurt, F., Bayram, E.: Local binary pattern based COVID-19 detection method using chest X-Ray images. In: 2021 29th Signal Processing and Communications Applications Conference (SIU). IEEE (2021)
5. Deshpande, G., Schuller, B.: An overview on audio, signal, speech, & language processing for covid-19 (2020). arXiv preprint arXiv:2005.08579
6. Biçer, M.B.: Meme kanseri görüntülemesinde mikrodalganın yeri." Erciyes Üniversitesi Fen Bilimleri Enstitüsü Fen Bilimleri Dergisi **30**(4), 257–263 (2014)
7. Günay, O., Öztürk, H., Yarar, O.: İyonize Radyasyon ile Çalışan Tıbbi Görüntüleme Cihazlarının Yapısının Proje Temelli Öğrenimi. J. Health Serv. Educ. **3**(1), 20–27 (2019)
8. Despotovic, V., et al.: Detection of COVID-19 from voice, cough and breathing patterns: dataset and preliminary results. Comput. Biol. Med. **138** 104944 (2021)
9. Bagad, P., et al.: Cough against COVID: Evidence of COVID-19 signature in cough sounds. arXiv preprint arXiv:2009.08790 (2020)
10. Aveyard, P., et al.: Association between pre-existing respiratory disease and its treatment, and severe COVID-19: a population cohort study. Lancet Respir. Med. (2021)
11. Sharma, N., et al.: Coswara--A Database of Breathing, Cough, and Voice Sounds for COVID-19 Diagnosis. arXiv preprint arXiv:2005.10548 (2020)
12. Dash, T.K., Mishra, S., Panda, G., Satapathy, S.C.: Detection of COVID-19 from speech signal using bio-inspired based cepstral features. Pattern Recogn. **117**, 107999 (2021)
13. Grant, D., McLane, I., West, J.: Rapid and scalable COVID-19 screening using speech, breath, and cough recordings. In: 2021 IEEE EMBS International Conference on Biomedical and Health Informatics (BHI), pp. 1–6. IEEE (2021)
14. Coppock, H., Gaskell, A., Tzirakis, P., Baird, A., Jones, L., Schuller, B.: End-to-end convolutional neural network enables COVID-19 detection from breath and cough audio: a pilot study. BMJ Innov. **7**(2) (2021)
15. Chaudhari, G., et al.: Virufy: Global applicability of crowdsourced and clinical datasets for AI detection of COVID-19 from cough (2020). arXiv preprint arXiv:2011.13320
16. Tena, A., Claria, F., Solsona, F.: Automated detection of COVID-19 cough. Biomed. Signal Process. Control **71**, 103175 (2022)
17. Han, J., et al.: Exploring automatic COVID-19 diagnosis via voice and symptoms from crowdsourced data. In: ICASSP 2021–2021 IEEE International Conference on Acoustics, Speech and Signal Processing (ICASSP), pp. 8328–8332. IEEE (2021)
18. Alsabek, M.B., Shahin, I., Hassan, A.: Studying the similarity of COVID-19 sounds based on correlation analysis of MFCC. In: 2020 International Conference on Communications, Computing, Cybersecurity, and Informatics (CCCI), pp. 1–5 (2020)

19. Celik, G.: CovidCoughNet: A new method based on convolutional neural networks and deep feature extraction using pitch-shifting data augmentation for covid-19 detection from cough, breath, and voice signals. Comput. Biol. Med. 107153 (2023)
20. Ulukaya, S., Sarıca, A.A., Erdem, O., Karaali, A.: MSCCov19Net: multi-branch deep learning model for COVID-19 detection from cough sounds. Med. Biolog. Eng. Comput. 1–11 (2023)
21. Kranthi Kumar, L., Alphonse, P.J.A.: COVID-19 disease diagnosis with light-weight CNN using modified MFCC and enhanced GFCC from human respiratory sounds. Europ. Phys. J. Spec. Top **231**(18), 3329–3346 (2022)
22. Akdeniz, F., Becerikli, Y.: Detection of copy-move forgery in audio signal with mel frequency and delta-mel frequency kepstrum coefficients. In: 2021 Innovations in Intelligent Systems and Applications Conference (ASYU), pp. 1–6. IEEE (2021)
23. Chatterjee, S., Kleijn, W.B.: Auditory model-based design and optimization of feature vectors for automatic speech recognition. IEEE Trans. Audio Speech Lang. Process. **19**(6), 1813–1825 (2010)
24. Kurani, A., Doshi, P., Vakharia, A., Shah, M.: A comprehensive comparative study of artificial neural network (ANN) and support vector machines (SVM) on stock forecasting. Ann. Data Sci. **10**(1), 183–208 (2023)
25. Gao, X., Li, G.: A KNN model based on manhattan distance to identify the SNARE proteins. IEEE Access **8**, 112922–112931 (2020)
26. Raju, C.G., Amudha, V., Sajiv, G.: Comparison of linear regression and logistic regression algorithms for ground water level detection with improved accuracy. In: 2023 Eighth International Conference on Science Technology Engineering and Mathematics (ICONSTEM), pp. 1–6. IEEE (2023)
27. De Ville, B.: Decision trees. Wiley Interdisc. Rev. Comput. Stat. **5**(6), 448–455 (2013)
28. Kingsford, C., Salzberg, S.L.: What are decision trees? Nat. Biotechnol. **26**(9), 1011–1013 (2008)
29. Genuer, R., Poggi, J.M., Tuleau-Malot, C., Villa-Vialaneix, N.: Random forests for big data. Big Data Res. **9**, 28–46 (2017)
30. Kattenborn, T., Leitloff, J., Schiefer, F., Hinz, S.: Review on convolutional neural networks (CNN) in vegetation remote sensing. ISPRS J. Photogramm. Remote Sens. **173**, 24–49 (2021)
31. Albawi, S., Mohammed, T.A., Al-Zawi, S.: Understanding of a convolutional neural network. In: 2017 International Conference on Engineering and Technology (ICET), pp. 1–6 (2017)
32. Orlandic, L., Teijeiro, T., Atienza, D.: The COUGHVID crowdsourcing dataset, a corpus for the study of large-scale cough analysis algorithms. Sci. Data **8**(1), 156 (2021)

Vaccine Tweets Analysis Using Naive Bayes Classifier and TF-IDF Techniques

Ben Ahmed Mohamed(✉), Boudhir Anouar Abdelhakim, and Dahdouh Yousra

SSET Research Team, C3S Laboratory, FSTT, Abdelmalek Essaâdi University, Tetouan, Morocco
{mbenahmed,aboudhir}@uae.ac.ma

Abstract. This paper explores the application of natural language processing (NLP) and machine learning techniques to sentiment analysis on a dataset of tweets on COVID-19 vaccines. The dataset was obtained from Kaggle and covers the full course of the immunization program. The tweets were cleaned using a variety of preprocessing approaches, such as handling contractions, removing URLs and user handles, and adjusting punctuation. The Text Blob library was used to assign sentiment ratings, while the TF-IDF technique was used to carry out feature extraction. The revised data was used to train a Naive Bayes classifier, which predicted the sentiment labels for every tweet. To evaluate the model's performance, evaluation criteria such F1 score, accuracy, precision, and recall were used. The study's findings provide insightful information on how the general public feels about COVID-19 vaccinations.

Keywords: Sentiment Analysis · Machine Learning · Natural Language Processing · Text Blob · TF-IDF · Naive Bayes Classifier · COVID-19 · Vaccines · Twitter Data · Text Preprocessing · Feature Extraction · Model Evaluation

1 Introduction

The SARS-CoV-2 virus, which started the COVID-19 pandemic, has had an unprecedented effect on people's daily lives, economics, and worldwide health. The virus first surfaced in late 2019 and has since expanded quickly, infecting millions of people and killing millions of them. Scientists have worked extraordinarily quickly to produce a number of vaccinations in response to the current global health crisis. With their promise to stop the virus's spread and open the door to things getting back to normal, these vaccinations have grown to be a ray of hope.

But public acceptance and uptake are just as important to the success of these vaccines as their efficacy and safety. Vaccine reluctance is a complicated issue driven by a number of factors like complacency, convenience, and confidence. The World Health Organization defines vaccine hesitancy as a "delay in acceptance or refusal of vaccines despite availability of vaccination services". Therefore, it is essential to comprehend public opinion on COVID-19 vaccines in order to combat vaccine hesitancy and guarantee the effectiveness of vaccination efforts.

M. Ben Ahmed et al. (Eds.): SCA 2023, LNNS 906, pp. 467–477, 2024.
https://doi.org/10.1007/978-3-031-53824-7_43

Social media platforms, such as Twitter, have become a rich source of data for gauging public opinion on various topics. With millions of users sharing their thoughts and opinions on these platforms every day, they provide a real-time snapshot of public sentiment. In the context of the COVID-19 pandemic, Twitter data has been used to track the spread of the virus, understand public reactions to various mitigation measures, and study misinformation about the virus and vaccines.

In this study, we leverage the power of machine learning and natural language processing (NLP) techniques to analyze sentiments towards COVID-19 vaccines expressed in tweets. Our objective is to classify these tweets into positive, negative, or neutral sentiments. This classification can provide a snapshot of public opinion towards the vaccines, which can be useful for informing public health communication strategies and addressing vaccine hesitancy.

We use a dataset of tweets related to COVID-19 vaccines, sourced from Kaggle. The dataset contains tweets collected from the inception of the vaccination campaign. The tweets are processed and analyzed using various NLP techniques, including text cleaning, sentiment scoring, and feature extraction. A Naive Bayes classifier is trained on the processed data and used to predict sentiment labels for the tweets. The performance of the model is evaluated based on accuracy, precision, recall, and F1 score.

This paper presents the methodology and findings of our study. We begin with a review of related works in the field of sentiment analysis using machine learning and NLP techniques. We then describe the dataset used in our study and the preprocessing steps taken to prepare the data for analysis. We detail the methodology used for sentiment scoring, feature extraction, and sentiment classification. We present the results of our study, including the performance of our model in predicting sentiment labels. Finally, we discuss the implications of our findings and suggest areas for future research.

We hope that our work will contribute to the growing body of research on sentiment analysis using machine learning and NLP techniques, and provide valuable insights into public sentiment towards COVID-19 vaccines. Understanding these sentiments can help public health organizations and policymakers in their efforts to address vaccine hesitancy and ensure the success of vaccination campaigns.

2 Related Works

The field of sentiment analysis, particularly as it applies to social media data, has been the subject of extensive research in recent years. This interest is driven by the recognition that social media platforms, such as Twitter, are rich sources of public opinion on a wide range of topics. The vast amount of data generated on these platforms provide a unique opportunity to gauge public sentiment and opinion trends in real-time.

One of the earliest works in this field was by Pang et al. (2002), who used machine learning techniques to classify movie reviews as positive or negative. Their work laid the foundation for many subsequent studies in sentiment analysis, demonstrating the potential of machine learning for this task. They used a dataset of movie reviews from the Internet Movie Database (IMDb) and showed that machine learning techniques could achieve an accuracy of up to 82.9% in classifying the reviews.

In the context of health-related topics, several studies have applied sentiment analysis to social media data. For instance, Paul and Dredze (2011) analyzed Twitter data to track public health trends, including sentiments towards vaccination. Their work highlighted the potential of Twitter as a source of health-related data, paving the way for many subsequent studies in this area. They demonstrated that Twitter data could be used to predict health trends, such as flu rates, with a high degree of accuracy.

More recently, with the onset of the COVID-19 pandemic, there has been a surge of interest in analyzing public sentiment towards the pandemic and related topics, such as vaccines. For instance, Abd-Alrazaq et al. (2020) conducted a large-scale analysis of tweets related to COVID-19, identifying common themes and sentiments. Their work demonstrated the potential of Twitter data for monitoring public sentiment during a health crisis. They found that public sentiment towards COVID-19 evolved over time, reflecting changes in the pandemic situation.

In terms of methodology, various machine learning models have been used for sentiment analysis, including Naive Bayes, Support Vector Machines (SVM), and more recently, deep learning models such as Convolutional Neural Networks (CNN) and Recurrent Neural Networks (RNN). Each of these models has its strengths and weaknesses, and the choice of model often depends on the specific requirements of the task. For instance, Naive Bayes and SVM are often used for their simplicity and efficiency, while deep learning models are favored for their ability to capture complex patterns in the data.

In our work, we build on these previous studies by applying a Naive Bayes classifier to analyze sentiment toward COVID-19 vaccines on Twitter. Our work contributes to the growing body of research on sentiment analysis in the context of health-related topics, and more specifically, the COVID-19 pandemic. We aim to provide insights into public sentiment towards COVID-19 vaccines, which could be valuable for public health authorities and policymakers in their efforts to address vaccine hesitancy and promote vaccine uptake.

3 Dataset

The dataset utilized in this research is a comprehensive collection of tweets pertaining to COVID-19 vaccinations. This dataset was procured from a public repository, Kaggle, which is a renowned platform for data science and machine learning enthusiasts offering a wide array of datasets for various research purposes. The tweets in this dataset were meticulously collected over a specific period, providing a snapshot of public sentiment during that time.

The tweets were gathered using a strategic selection of keywords and hashtags related to COVID-19 vaccines. Some of these included "COVID-19 Vaccine", "COVID Vaccine", "Coronavirus Vaccine", and more. This keyword-based collection ensured that the dataset was focused and relevant to the research topic.

The dataset is rich with several features, including but not limited to, the tweet text, the date and time of the tweet, the source of the tweet (which indicates the device or application used to post the tweet), user location, and other metadata. Among these, the 'text' field, which contains the actual content of the tweet, is the primary feature that our analysis revolved around. The dataset contains 16 columns shown in Table 1.

Table 1. The properties of Dataset.

Column Name	Explanation	Data Type
Id	Indexing value for the dataset	Int64
User_name	Username of twitter account that posted the tweet	Object
User_location	Location mentioned in the user's Twitter profile	Object
User_description	Bio of the user's Twitter profile	Object
User_created	The date and time when the user's account was created	Object
User_followers	The number of followers the user has	Int64
User_friends	The number of users the user is following	Int64
User_favourites	The number of tweets or posts the user has liked	Int64
User_verified	Indicates whether the user's account is verified or not	Bool
Date	Date and time when the tweet was posted	Object
Text	The content of the tweet	Object
Hashtags	Any hashtags included in the tweet	Object
Source	The device used to send the tweet	Object
Retweets	The number of times the tweet has been retweeted	Int64
Favorites	The number of times the tweet has liked	Int64
Is_retweet	Indicates whether the tweet is a retweet or not	Bool

However, raw data, especially from social media platforms, is often messy and unstructured. Therefore, before delving into the analysis, the dataset underwent a rigorous preprocessing stage to clean and standardize the data. This stage was crucial and involved several steps such as removing unnecessary features, handling missing values, and most importantly, cleaning the tweet text.

The text cleaning process was multi-faceted and included removing user handles, hashtags, URLs, and special characters. It also involved handling contractions, removing multiple spaces, and converting all text to lower case. These steps were necessary to reduce noise in the data and to ensure that the machine-learning model could focus on the meaningful parts of the text.

Once the tweet text was cleaned, it was then used as the input for the sentiment analysis and keyword extraction processes. The sentiment labels obtained from the sentiment analysis process were used as the target variable for the machine learning model. To provide a visual representation of the data, a bar plot was generated using Matplotlib and Seaborn libraries. The plot showcased the top 10 words with the highest counts, allowing readers to gain insights into the most frequently used keywords in the corpus.

It is important to note that despite the inherent challenges associated with using social media data, such as its unstructured nature and the presence of noise, the dataset used in this study provides a valuable source of real-time, user-generated content. This content, reflecting public sentiment towards COVID-19 vaccines, offers unique insights that can be instrumental in understanding public opinion and informing public health strategies.

In conclusion, the dataset used in this study, with its richness and real-world relevance, played a pivotal role in the research. The preprocessing of this data was a critical step that ensured the success of the subsequent analysis and model building stages.

4 Method

The methodology of this research is divided into several key stages, each contributing to the overall objective of analyzing public sentiment towards COVID-19 vaccines. The stages include data preprocessing, sentiment analysis, keyword extraction, feature extraction, and model building and evaluation.

4.1 Data Preprocessing

The first stage involved cleaning and preparing the data for analysis. This included removing unnecessary features, handling missing values, and cleaning the tweet text. The text cleaning process was comprehensive and involved removing user handles, hashtags, URLs, and special characters, handling contractions, removing multiple spaces, and converting all text to lowercase. This step was crucial to reduce noise in the data and ensure that the subsequent analysis focused on the meaningful parts of the text.

4.2 Sentiment Analysis

The cleaned tweet text was then subjected to sentiment analysis using the TextBlob library. TextBlob assigns polarity scores to the text, which were then used to classify each tweet as 'Positive', 'Negative', or 'Neutral'. The sentiment labels obtained from this process were used as the target variable for the machine-learning model.

4.3 Keyword Extraction

This stage involved extracting the most common keywords from the tweets. This was done separately for positive, negative, and neutral tweets to gain insights into the specific words and phrases associated with each sentiment. The extraction process involved tokenizing the tweets, removing stop words, and then using a Counter to find the most common tokens.

4.4 Feature Extraction

The cleaned tweet text was transformed into a matrix of TF-IDF features to be used as input for the machine learning model. The TfidfVectorizer from the sklearn.feature_extraction.text library was used for this purpose. This vectorizer converts the text data into a matrix where each row corresponds to a document and each column corresponds to a word in the vocabulary. The value in each cell is the TF-IDF score of the word in the document.

4.5 Model Building and Evaluation

The final stage involved building and evaluating a Multinomial Naive Bayes classifier. The TF-IDF features and sentiment labels were split into training and test sets. The model was trained on the training set and then used to predict the sentiment labels for the test set. The performance of the model was evaluated using accuracy as the primary metric, along with precision, recall, and F1-score for a more comprehensive evaluation.

In conclusion, the methodology employed in this research was systematic and thorough, ensuring that each stage contributed to the overall objective of analyzing public sentiment towards COVID-19 vaccines. The use of robust techniques for sentiment analysis, keyword extraction, and feature extraction, coupled with a well-established machine learning model, ensured the reliability and validity of the research findings.

5 Results and Discussion

The analysis began with the loading of the dataset 'vaccination_all_tweets.csv' into a pandas DataFrame. The dataset consisted of several columns, each with its specific data type. A unique feature of the dataset was the 'source' column, which was further explored to identify the unique sources of the tweets. The dataset contained tweets from a total of 171 unique sources. The most frequent source was 'Twitter for iPhone', contributing to approximately 43% of the tweets.

The data cleaning process involved several steps. The 'neattext' package was used to remove hashtags, user handles, multiple spaces, URLs, and punctuations from the tweets. The 'contractions' package was used to handle contractions in the text. The cleaned tweets were then stored in the 'clean_tweet' column of the DataFrame.

The sentiment of each tweet was determined using the TextBlob library. The sentiment polarity, subjectivity, and label (Positive, Negative, or Neutral) were calculated for each tweet and stored in the 'sentiment_results' column. The DataFrame was then updated to include these sentiment results as separate columns. The sentiment analysis revealed that approximately 43% of the tweets were neutral, 38% were positive, and 19% were negative (Fig. 1).

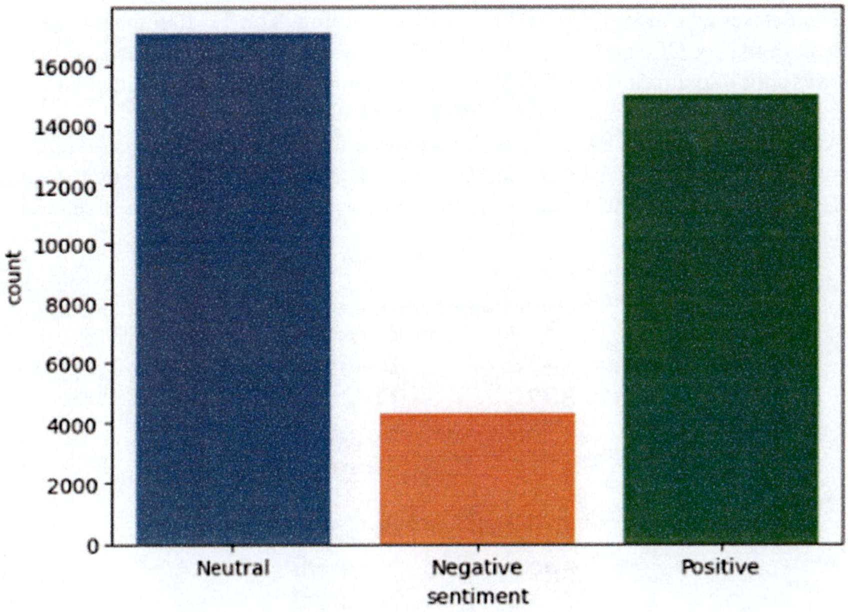

Fig. 1. TextBlob Analysis Result.

The cleaned tweets were then transformed into a matrix of TF-IDF features using the TfidfVectorizer from the sklearn library. The TF-IDF scores for the first document were printed as an example. The most common words in positive tweets were identified and visualized in a bar plot. The word 'vaccine', for instance, appeared in positive tweets over 5000 times (Fig. 2).

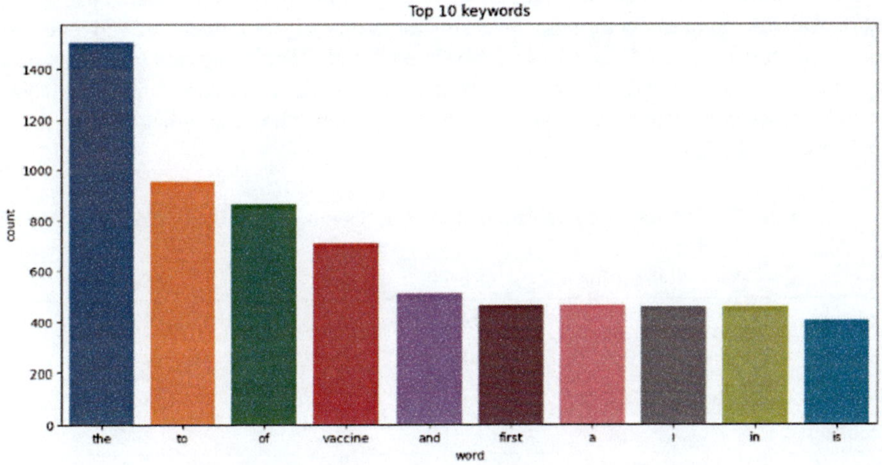

Fig. 2. Frequency of the top 10 most words in Tweets.

The dataset was then prepared for machine learning. The TF-IDF matrix served as the feature matrix (X), and the sentiment labels served as the target variable (y). The data was split into a training set (80% of the data) and a test set (20% of the data) using the train_test_split function from the sklearn library.

A Multinomial Naive Bayes model was chosen for the sentiment classification task. The model was trained on the preprocessed tweet data and then used to predict sentiment labels for the test set. The model's predictions for each tweet in the test set were as follows:

4324	Neutral
3456	Positive
887	Neutral
4869	Neutral
68	Neutral
	...
4426	Positive
466	Neutral
3092	Neutral
3772	Neutral
860	Neutral

Fig. 3. Predictions of the Multinomial Naive Bayes model.

Each entry in Fig. 3 corresponds to a tweet in the test set, with the index representing the original index of the tweet in the dataset and the associated sentiment ('Neutral', 'Positive', or 'Negative') being the sentiment predicted by the model.

The performance of the Multinomial Naive Bayes classifier was evaluated using several metrics to provide a comprehensive understanding of the model's effectiveness in sentiment analysis of COVID-19 vaccine-related tweets. These metrics included accuracy, precision, recall, and F1-score. Each of these metrics provides a unique perspective on the model's performance, and together they offer a holistic view of the model's capabilities (Table 2).

Table 2. Performance Evaluation of the Multinomial Naive Bayes model.

	Precision	Recall	F1-Score	Support
Negative	0.86	0.03	0.05	233
Neutral	0.79	0.84	0.81	977
Positive	0.72	0.86	0.78	835
Accuracy			**0.76**	**2045**
Macro Avg	**0.79**	**0.58**	**0.55**	**2045**
Weighted Avg	**0.77**	**0.76**	**0.72**	**2045**

In addition to the overall **accuracy**, we also evaluated the model's performance for each sentiment class (Negative, Neutral, Positive) using precision, recall, and F1-score. These metrics provide a more detailed understanding of the model's performance.

The **precision** of a class defines how trustworthy the prediction of that class is. It is the ratio of true positives (tweets correctly classified as a particular sentiment) to the sum of true positives and false positives (tweets incorrectly classified as that sentiment).

The **recall** of a class, on the other hand, indicates how well the model can identify that class. It is the ratio of true positives to the sum of true positives and false negatives (tweets of that sentiment that were incorrectly classified).

The **F1-score** is the harmonic mean of precision and recall, and it provides a single metric that balances both the concerns of precision and recall.

The performance metrics for each class were as follows:

- **Negative sentiment:** The model had a precision, recall, and F1-score of 0.00. This indicates that the model was not able to correctly identify any tweets with negative sentiment. This could be due to a lack of negative examples in the training data or the model's inability to distinguish negative sentiment based on the features provided.
- **Neutral sentiment:** The model had a precision of 0.78, a recall of 0.81, and an F1-score of 0.79. This suggests that the model was quite effective at identifying tweets with neutral sentiment.
- **Positive sentiment:** The model had a precision of 0.69, a recall of 0.87, and an F1-score of 0.77. This indicates that the model was also quite effective at identifying tweets with positive sentiment, although it was slightly less precise than with neutral sentiment.

The overall **accuracy** of the model was 0.73, which suggests that it was able to correctly classify the sentiment of the tweets 73% of the time. The **macro average F1-score**, which averages the F1-scores of each class without considering their proportion in the data, was 0.52, while the **weighted average F1-score**, which does consider the proportion of each class, was 0.68.

These results demonstrate the effectiveness of our model at classifying tweets with neutral and positive sentiment, but they also highlight a significant area for improvement in the classification of tweets with negative sentiment. Further research and refinement of the model, as well as potential oversampling of negative examples, may be necessary to improve its performance in this area.

However, a deeper look into the performance metrics reveals some areas of concern. The model performed well in classifying positive (f1-score of 0.77) and neutral tweets (f1-score of 0.79) but struggled significantly with negative tweets (f1-score of 0.00). This suggests that the model was not able to correctly identify negative sentiment in the tweets, which could be due to a variety of factors such as an imbalance in the training data or the nuances and complexity of negative sentiment in language.

The results of this study have important implications for understanding public sentiment towards COVID-19 vaccinations. Given the ongoing efforts to vaccinate populations worldwide, understanding public sentiment can help inform public health strategies and communication efforts.

The model's difficulty in correctly identifying negative sentiment suggests that more work needs to be done in this area. It is recommended that future iterations of the model incorporate more sophisticated natural language processing techniques, such as deep learning models, which may be better equipped to handle the complexities of sentiment in language.

Additionally, addressing the imbalance in the training data could also improve the model's performance. This could involve collecting more data, particularly tweets with negative sentiment, or using techniques such as oversampling or under sampling to balance the classes.

Further research could explore the use of different machine learning models or deep learning models for sentiment analysis. Additionally, incorporating additional features, such as user metadata or temporal features, could potentially improve the model's performance.

Another interesting avenue for future research could be to explore the reasons behind the sentiments expressed in the tweets. This could involve qualitative analysis or topic modeling to understand the main themes or topics that are associated with positive, neutral, and negative sentiments towards COVID-19 vaccinations.

In conclusion, while the model developed in this study shows promise in classifying sentiment towards COVID-19 vaccinations, there is still much work to be done to improve its performance, particularly in correctly identifying negative sentiment.

6 Conclusion

This study used a Multinomial Naive Bayes model trained on a collection of tweets to examine public opinion regarding COVID-19 vaccines. The model's 73% overall accuracy rate indicates that it has the ability to offer insightful data on public opinion. The model's performance, however, differed noticeably amongst the various sentiment classes, with a particular challenge in accurately identifying negative sentiment.

These findings have important ramifications, especially in light of continuing international initiatives to immunize communities and fight vaccination hesitancy. Comprehending the prevailing public opinion around vaccinations is crucial for informing public health policies and outreach initiatives. The study's limitations, however, make clear the necessity of more research in this field. The model's inability to detect negative emotion indicates.

Future research could explore the use of different machine learning or deep learning models, incorporate additional features, or delve into the reasons behind the sentiments expressed in the tweets. This could involve qualitative analysis or topic modeling to understand the main themes associated with different sentiments.

In conclusion, while the model developed in this study shows promise, there is still much work to be done. The complexities of language and sentiment analysis present ongoing challenges, but also exciting opportunities for future research. As we continue to refine and develop these models, we move closer to a more nuanced understanding of public sentiment towards COVID-19 vaccinations, and public health issues more broadly.

References

1. Apache Hadoop (2021). 1 June 2021. https://hadoop.apache.org/

2. Barkur, G., Vibha, Kamath, G.B.: Sentiment analysis of nationwide lockdown due to COVID 19 outbreak: evidence from India. Asian J. Psychiatry **51**, 102089 (2020)
3. Bell, S., Clarke, R., Mounier-Jack, S., Walker, J.L., Paterson, P.: Parents' and guardians' views on the acceptability of a future COVID-19 vaccine: a multi-methods study in England. Vaccine **38**(49), 7789–7798 (2020)
4. Chintalapudi, N., Battineni, G., Amenta, F.: Sentimental analysis of COVID-19 tweets using deep learning models. Infectious Dis. Rep. **13**(2), 329–339 (2021). https://doi.org/10.3390/idr13020032
5. Donzelli, G., Palomba, G., Federigi, I., Aquino, F., Cioni, L., Verani, M., et al.: Misinformation on vaccination: a quantitative analysis of YouTube videos. Hum. Vaccin. Immunother. **14**(7), 1654–1659 (2018). https://doi.org/10.1080/21645515.2018.1454572
6. Goodfellow, I., Bengio, Y., Courville, A.: Deep Learning. The MIT Press (2016)
7. Kaur, H., Ahsaan, S.U., Alankar, B., et al.: A proposed sentiment analysis deep learning algorithm for analyzing COVID-19 tweets. Inf. Syst. Front. (2021). https://doi.org/10.1007/s10796-021-10135-7
8. Mansoor, M.: Global Sentiment Analysis of COVID-19 Tweets Over Time (2020)
9. Preda, G.: All COVID-19 Vaccines Tweets (2021). https://www.kaggle.com/gpreda/all-covid19-vaccines-tweets. Accessed 10 June 2021
10. Scheufele, D.A., Krause, N.M.: Science audiences, misinformation, and fake news. Proc. Natl. Acad. Sci. **116**(16), 7662–7669 (2019). https://doi.org/10.1073/pnas.1805871115
11. Shahsavari, S., Holur, P., Tangherlini, T.R., Roychowdhury, V.: Conspiracy in the time of corona: automatic detection of COVID-19 conspiracy theories in social media and the news. J. Comput. Soc. Sci. **3**, 279–317 (2020)
12. Wiyeh, A.B., Cooper, S., Jaca, A., Mavundza, E., Ndwandwe, D., Wiysonge, C.S.: Social media and HPV vaccination: unsolicited public comments on a facebook post by the western cape department of health provide insights into determinants of vaccine hesitancy in South Africa. Vaccine **37**(43), 6317–6323 (2019). https://doi.org/10.1016/j.vaccine.2019.09.019
13. Worldometers.: COVID-19 Coronavirus Pandemic (2021). https://www.worldometers.info/coronavirus/#countries. Accessed 11 July 2021

Smart Energy Systems and Smart Motors

Life Cycle Assessment of a Smart Building: Energy Optimization Integration

Sydney Walter[2(✉)], Daniela Chavez-Okhuysen[2], Mohamad Achour[1], Abdou Dia[1], Ludovic Avril[1], and Nisrine Makhoul[1]

[1] École Spéciale des Travaux Publics, ESTP, Paris, France
[2] Texas A&M University, 400 Bizzell St, College Station, TX 77843, USA
sydney.walter@ymail.com

Abstract. Fueled by ever-increasing urbanization today, the construction sector is a major contributor to negative environmental impacts, including greenhouse gas emissions, ozone depletion, and carbon footprint. Life Cycle Assessment (LCA) is a systematic method used for the evaluation of the environmental impacts of a building throughout its entire life cycle, from raw material extraction to disposal at end-of-life stage. Early implementation of LCA principles in the design phase of a project aids stakeholders in the decision making process by allowing targeted interventions by seeking sustainable strategies in material selection. The pursuit of sustainability in LCA applications engenders the possibility for smart buildings – the integration of technology in the building design process, yielding increased levels of performance and efficiency. Smart building initiatives further extend the focus of energy optimization in buildings through utilization of and enhanced operational efficiency throughout a building's lifecycle. The paper presents the beginning steps in a case study concentrating on the principles of life cycle assessment of smart buildings and methods to better optimize energy consumption.

Keywords: LCA · Smart Building · Energy Optimization

1 Introduction

1.1 Life Cycle Analysis

An Overview of Life Cycle Analysis

Rising effects of climate change, including rising temperatures and extreme weather events, emphasize the importance of implementing innovative methods in the design process of projects in order to minimize buildings' adverse impacts on the environment and meet today's sustainability goals. Life cycle assessment provides a framework for quantifying the environmental impact of a product or service throughout its lifecycle. In the case of buildings, life cycle assessment aims to identify and quantify the environmental burdens of materials during the different stages of the life cycle: the production

M. Ben Ahmed et al. (Eds.): SCA 2023, LNNS 906, pp. 481–496, 2024.
https://doi.org/10.1007/978-3-031-53824-7_44

stage, including the extraction, transportation, and manufacturing of raw materials; the construction stage, during which the building takes form; the operation stage, which includes the use and maintenance of a building and tends to be the most energy consuming portion; and finally, the end-of-life stage, a process that is defined in the context of either a linear or circular economy. The International Organization for Standardization defines the methodology of a life cycle assessment in 4 phases in ISO 14040. First, one must define the goal and scope to know what needs to be analyzed, how to analyze it and how far to go with the analysis. The second phase is the data collection phase of the LCA, which looks at the environmental inputs and outputs of a certain material. With the data gathered, an evaluation of the contribution from materials to environmental impact categories, such as CO_2 missions, can be made in the impact assessment phase. The final phase is the interpretation of the LCA, which includes conclusions, limitations, and recommendations for how to proceed with the project [1]. LCA is particularly useful for material selection. Access to the materials' impact data followed by the comparison of carbon emissions of varying materials throughout their life cycle helps stakeholders, including architects and engineers to name a few, to make more informed and potentially environmentally sound decisions early on in construction projects. By choosing materials with lower environmental impacts, the overall carbon footprint of the building can be significantly reduced [2].

1.2 Smart Buildings

An Outline of Smart Building Definitions

The basis of the definition of what is considered a "smart building" can be summarized as the integration of leveraging new technologies and practices for performance enhancement. The ex-act definition of a smart building is versatile – interchangeable with terms such as sustainable building, green building, and intelligent building – the term reflects concepts ranging from the application of new smart technologies to practices toward the improvement of energy performance [3]. Interest in smart buildings stems from an array of stances – economic, environmental, societal, etc. Per the United Nations, projection of the world's population is estimated to increase by 2 billion people in the next 30 years, with a total of 9.7 billion people in 2050 and a potential peak at 10.4 billion people in the mid 2080s [4]. The continual growth of cities worldwide demands solutions with both practicality and longevity with the concept of smart buildings intended to sustain and integrate smart technologies for intelligence, control, longevity, comfort, and satisfaction purposes.

Smart buildings – enabling a holistic approach to energy efficiency and management – rely on continued advancements made with rising smart technologies and systems to converge to enhance efficiency and occupant experience. As demand for buildings with advanced technology and improved efficiency surges, implementation of effective energy optimization strategies is vital for reduced energy consumption, sustainability practices, and reduced costs [5].

1.3 Energy Optimization and Contemporary Technologies

Energy Optimization in the Scope of Energy Consumption

Reformation of contemporary standards for energy consumption in Europe is the central objective of the EU Energy Department with the target for an energy efficiency of 32.5% for 2030 [6]. As 75% of the EU building stock is energy inefficient, a large portion of the energy used wasted immense optimization potential. Consideration of energy efficiency in regard to the multitude of buildings used in everyday life – homes, workplaces, schools, and hospitals – is crucial in the goal to both optimize energy and reduce unnecessary energy consumption [7]. Energy efficiency in the home, for instance, can be influenced by heating and cooling systems, air conditioning, weatherization, insulation, home design, windows, doors, and skylights, lightning, as well as appliances and electronics [8]. A central difference between homes and other public spaces, such as university buildings, relates to the occupant behavior (OB). Occupant behavior establishes a need for energy service demands. Lighting, electricity, and temperature settings all contribute to the facilitation of comfort and energy performance in a given public building. Balance of the comfortability along with expectations of energy consumption are required for a solid understanding of patterns in occupancy behavior. The observation of such patterns serves to engender a formulative understanding of the suitable choices in regard to building envelope, technology implementation, and overall operation [9].

Inefficient use of energy resources and inefficient management contributes to about 25% of the cost of utilities, the losses being thermal and electrical energy. Creation of a system for effective energy saving and reasonable energy use is necessary for buildings. A particular approach to the development of a suitable system is applying a resource-saving approach or the use of benchmarking tools. An effective system for the purpose of energy optimization and management must incorporate a substructure to perform the actual observation of energy usage. A variety of tools are possible for this, ranging from digital to analog electronics and technologies. For smart buildings, a key strategy for energy optimization pertains to Building Information Modeling (BIM).

Building Information Modeling – Relevant Applications

BIM applications engender digital representations of buildings throughout and during the planning and design, construction, completion, and maintenance stages over the life cycle of a building. Employment of BIM principles is ascertained to be a linked process, as opposed to a single-use technology [10]. Capabilities interconnect a multitude of enmeshed aspects of the BIM model data processes, such as the calculation of energy consumption and the observation of reliable data of the building properties. The design process associated with BIM applications yields digital asset models consisting of numerous building elements, each with assigned parameters. Within each building model includes ample information that can be used for the energy assessment of the building envelope. Element information and parameters within the building model has the potential to be conducive toward the energy assessment of the building envelope. This includes facets relating to the specific materials included in the building, such as the thermal resistance, thermal conductivity, amongst other properties of the material layer. The properties of a building envelope can be analyzed and interpreted by various assessment processes and algorithms. The central purpose of a given assessment explicates the requirements of a building envelope, specifically the parameters and elements needed [11].

Effective implementation of BIM applications is integral to the evaluation of energy performance scenarios and is useful in the various stages of a building's development. Simulation development is essential for the identification of potential areas of improvement, as well as the elucidation of informed decision-making processes. Continual emergence of innovative solutions for building performance simulations includes automated building design, building energy modeling, predictive control for optimization modeling, and digital twins. Such developments in simulation modeling have the potential to induce improvements in the multitude of performance aspects. Improvements include CO_2 emission reductions, integration of energy-efficient features, efficient HVAC systems, as well as inclusion of renewable energy sources. Resulting operational costs and energy consumption constitute the value of utilizing contemporary BIM applications, as building performance simulation will play a crucial role in the development of the architectural industry [12].

The research conducted in this paper aims to create a more sustainable built environment through smart buildings by reducing emissions during their life cycle and optimizing energy consumption. A comparison of the impact different materials have on carbon emissions will be made through the creation of scenarios using BIM, particularly Autodesk Revit and Logiciel Pleiades.

2 Case Study: The ESEO Building in Dijon, France

2.1 An Introduction to the Building

The proposed LCA methodology mentioned above is applied to the existing ESTP-ESEO smart building located in Dijon, France. Inaugurated in 2021, the Metropolitan Campus is one of the first of its kind in France, supported by the Bourgogne Franche-Comté region and built by SPLAAD (Société Publique Locale "Amenagement de l'Agglomeration Dijonnaise"). The ESTP-ESEO building was created with the goal of constructing an "intelligent, innovative and communicating building" to demonstrate energy efficiency and high quality of use [13]. The four story building of over 10,000 m^2 accommodates 800 students and houses a 400-seat modular amphitheater, eight research laboratories, 44 teaching rooms, large accessible terraces, among other areas. Exceptional in environmental terms, the building is eco-designed using low-carbon concrete, timber frame, and 90% reuse of construction waste, etc. In addition to this, the building is outfitted with dynamic glazing windows that lighten or darken depending on the intensity of the sun and counts on adiabatic cooling loop technology to provide user comfort year round. The digital twin centralizes data from sensors equipped throughout the building that collect data on user presence, temperature, lighting, and air quality, which can lead to substantial energy savings through real-time adjustments [14].

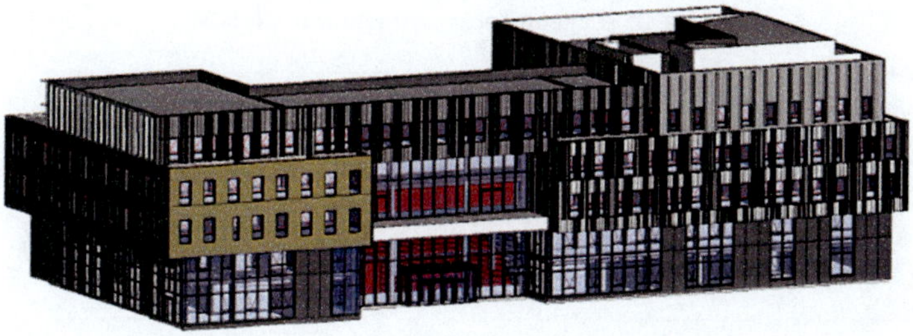

Fig. 1. Autodesk Revit Model by TwinOps of the ESEO-ESTP Smart Building

2.2 Scope of Simulation Procedures

The aim of this case study is to identify the optimal operational settings and determine the environmental impacts of current construction materials in the ESTP-ESEO building. Moreover, this introductory experiment intends to ultimately reduce energy consumption in the use phase, which has the potential of being responsible for 70–90% of all building energy consumption throughout its lifecycle [15].

Building Information Modeling: Autodesk Revit

A series of procedures required a prior framework before actualization of the simulation testing for the preliminary case study of the ESEO-ESTP building in Dijon, France. Such initiated with the implementation of Autodesk Revit modeling software. Autodesk Revit is a building information modeling software for architects, landscape architects, designers, contractors, and a multitude of engineering professions. [XX] Developed by Charles River Software founded in 1997 and later acquired by Autodesk in 2002, the software allows users to design a building structure and include additional components and their specifications. Figure 1 illustrates a previously curated Autodesk Revit model obtained from a digital twin software company, TwinOps. Depiction of both relative realistic architectural and structural elements from the ESEO-ESTP building are present within the Revit model implemented.

Material Analysis: Culmination of Material Properties

The choice of building materials is a major determinant for the impact the building itself will have on the environment throughout its lifecycle. By integrating BIM and LCA, stakeholders are able to optimize building performance, consider alternative sustainable designs, and better support the decision-making process for material selection. For the research conducted in this paper, six different ESTP-ESEO building scenarios were simulated with the help of digital design tools, Pleiades and Revit.

A legend for coefficients used in this research, including thickness, thermal conductivity, heat transfer coefficient, and resistance is paramount for material information comprehension (Table 1).

Table 1. Legend for construction materials.

Value	Measure	Units
e	Thickness	cm
λ	Thermal Conductivity	W/(m*K)
U	Heat Transfer Coefficient (Hypothetical)	$W/(m^2*K)$
R	Resistance	K/W

Material properties for the construction materials used in the ESTP-ESEO building (Table 2) include values for thermal conductivity, resistance, and density obtained from the Pleiades software during the design phase before simulation.

Table 2. Material Properties for ESTP-ESEO Smart Building.

Exterior Wall	e	λ	U	R	kg/m^2
Anthracite concrete	20	1.053	5.26	0.19	260
Wood wool insulation	16	0.038	0.24	4.21	6
Plaster	2	0.3	7.14	0.14	24
Aluminum Panel Finish	4	200	10	0	108
Heavyweight Concrete	30	1.75	8.33	0.12	690
Bronze Panel Finish	10	65	10	0	870
Internal Wall					
Plaster	2.5	0.3	12.5	0.08	30
Wood wool insulation	5	0.038	0.76	1.32	2
Intermediate Floor					
Cellular Concrete	25	0.16	0.64	1.56	100
Ground Floor					
Cellular Concrete	25	0.16	0.64	1.56	100
Roof					
Plaster	2	0.3	14.3	0.07	24
Wood wool insulation	17	0.038	0.22	4.47	7

Certain values for heat transfer coefficients were not available; therefore, a reasonable hypothesis was made based on the assumption (1) for all materials.

$$U = 1/R \tag{1}$$

All the environmental impact indicators the Pleiades software outputs after each life cycle assessment simulation in the above Table 3 are germane, however, for the purpose

Table 3. Environmental indicators

Environmental Indicator	Unit
Greenhouse effect	t CO_2 eq
Acidification	kg SO_2 eq
Cumulative energy demand	GJ
Water used	m^3
Inert waste produced	Tons
Abiotic resources depletion	kg E-15
Eutrophisation	kg PO_4 eq
Photochemical ozone production	kg ethylene eq
Aquatic ecotoxicity	m^3
Radioactive waste	dm^3
Human toxicity	kg
Odor	m^3 air

of this study, only the greenhouse gas emissions, cumulative energy demand, inert waste produced, and resource depletion will be analyzed. The lifetime for the analysis of the building is 80 years.

Building Information Modeling: Pleiades

Translation of the ESEO-ESTP model from Autodesk Revit to the Logiciel Pleiades Modeler software resulted from the Revit plug-in (Version 1.6.4), available from the Izuba Energies website [16] The resulting Pleiades model (see Fig. 2) maintained the relative structural integrity in the creation process, however required alteration in the room designation, window and door placement (Figure IOP), additional to minor line corrections along the foundation.

Fig. 2. Izuba Energies Pleiades Model of ESEO-ESTP building

A look into the potential error associated with the plug-in lies within the ability of the software to correctly realize certain room geometries upon the file export. Additional inconsistent metadata such as material properties or construction schedules contribute to the likelihood of discrepancies associated (see Fig. 3). Mitigation of any inaccuracies relied heavily on the validation and correction of accurate dimensions and data through floor plan guides and in-person building evaluations.

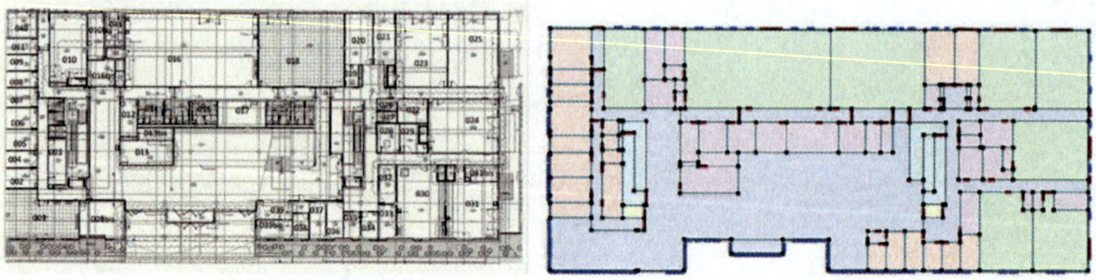

Fig. 3. Comparison of Actual and Plug-in Generated Floor Plan

It is important to note that the Pleiades software outputs impact category data in four phases: construction, use, renovations, and end-of-life. Excluding the extraction of raw materials and the manufacture of materials and their transport to the construction site limits the analysis. For the applications of LCA in this this study, the data obtained only supports starting from the phase of construction of the building until the end-of-life, without specifying course of action for disposal or reuse of materials.

2.3 Pleiades Energy Optimization and LCA Simulation

Pleiades: Scenario Implementation and STD COMFIE

A total of six scenario inputs (Table 4) encompassed the simulation process associated with the STD COMFIE calculation (Simulation Thermique Dynamique). The computation of the building's energy needs is done by an hourly basis, the heating and cooling needs, in addition to the temperature and humidity. Vital to note, the ESEO-ESTP building does not have a HVAC system, and this particular calculation does not consider the heating, air conditioning and DHW systems. Rather, air movements are computed induced by the activation of the ventilation systems, aeraulic calculation.

Scenario classification diverged into three separate groups – unrealistic constant, realistic constant, and realistic behavior. The first and second scenarios represent highly unrealistic behavior from a day-to-day basis within the building. High occupancy, temperature, power, ventilation, and lighting all were attributed to demonstrate the undesired behaviors upon scenario simulation. The opposite effect was done to the second scenario. Scenario three reflects a constant within the realm of reality, however continued to exhibit constant behavior for twenty-four hours a day over the course of the fifty-two week period tested. Four through six of the scenario characteristics only differentiated under the temperature category, ranging from relatively expected temperature behaviors on a daily basis.

Table 4. Scenario Classification Guide

Scenario No	Occupancy (N[N])	Temperature. (°C)	Power (W)	Ventilation (v/h)	Lighting (lux)
Scenario 1	100	25	1000	0.7	500
Scenario 2	6	17	100	0.1	100
Scenario 3	50	20	200	0.35	200
Scenario 4	50	18–19.5	50–200	0–0.7	0–200
Scenario 5	50	20–21.5	50–200	0–0.7	0–200
Scenario 6	50	21.5–23	50–200	0–0.7	0–200

Construction and weather parameters remained constant for each of the six scenarios (see Fig. 4).

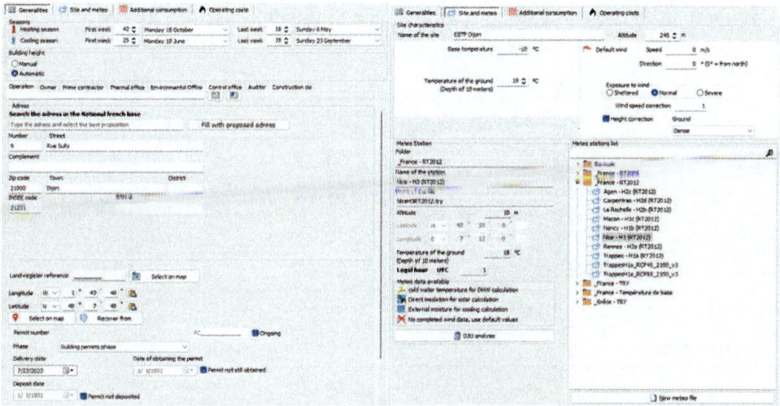

Fig. 4. Pleiades Modeler, STD COMFIE Construction and Weather Parameters

Calculation of STD COMFIE yielded access results about the heating and cooling loads, heating and cooling power, minimum, maximum, and average temperatures, solar gains, the number of hours overheating, and the discomfort rate attributed.

The solar gains of a building refer to an increase in thermal energy by means of absorption of solar radiation by the building's envelope. A balance of the energy attributed to the thermal losses must be realized to determine the true impact of the solar gains, whether positive or negative [17]. No change occurred with relation to the building orientation, geographical location, or building configuration within the Pleiades Modeler (Table 5).

Table 5. Solar Gain Attribution

Zone	Solar Gains (kWh)
Office	76,170.00
Lab/Classroom	324,273.00
Hallway	229,274.00
Stairway	4,709.00
Other (Closet/Storage)	2,221.00
Other (Bathrooms/Study Spaces)	19,834.00

Temperature assessment pertained to values achieved by simulation results, thermometer sensor measurements, and temperatures obtained from the TwinOps system (Table 6).

Table 6. Maximum, Minimum, & Average Simulation Yielded Temperatures

Pleiades	Minimum (°C)	Average (°C)	Maximum (°C)
Scenario No.1	23.18	37.19	27.27
Scenario No.2	16.55	35.33	22.42
Scenario No.3	16.58	35.88	22.75
Scenario No.4	17.63	35.53	23.19
Scenario No.5	19.39	35.54	24.11
Scenario No.6	20.67	35.58	24.92
Sensor Measurements			
Actual	19.80	24.00	21.84
Internal	23.60	25.10	24.34
Digital Twin Data			
TwinOps	19.80	26.50	22.62

Accuracy in temperature values (comparison to actual values in the 2022 year cycle – TwinOps) were within the fifth scenario and sensor measurements. TwinOps of 19.8–26.5 °C ranged expectedly similar to the actual temperatures measured, 19.8–24 °C. High reliability associated with the fifth scenario input temperatures, 19.39–35.54 °C, correlated similarly to the data obtained by the digital twin software. Significance of the results demonstrate a relative realistic behavior in the fifth scenario, however consideration of the maximum temperatures obtained in each scenario must be realized.

Each simulation yielded maximum values well above the comfort levels specified (discomfort high of 27 °C, discomfort low of 16 °C), all within 35.33–37.19 °C. Considerations within Pleiades additional inputs, such as fully functional HVAC or insulation systems, as well as climate conditions, are necessary for understanding the extremes displayed in the temperature results. Absence of predetermined inputs compels Pleiades software to construct assumptions and defaults within the modeling and simulation processes, based on industry standards or default profiles. As such, information regarding the particulars not entered in this series of simulations contribute to the lack of exactness in comparisons between the dataset and actual temperature measurements.

Continuation of the analysis finalized on the observation of energy consumption comparisons in final simulation development, heating and cooling loads and power (Table 7). Maintainment of applied indoor temperature values pertained to the addition or removal of heat amounts, loads, and energy deliverance for such heat increases or decreases, power.

Table 7. Individual Heating and Cooling Energy Consumption – Load & Power

Scenario No	Heating Load (kWh)	Cooling Load (kWh)	Heating Power (W)	Cooling Power (W)
Scenario No. 1	96,225.00	184,755.00	107,062.00	1,049,975.00
Scenario No. 2	16,101.00	272,35.00	60,363.00	1,429,906.00
Scenario No. 3	12,021.00	281,930.00	53,336.00	1,467,675.00
Scenario No. 4	28,370.00	252,079.00	163,401.00	1,470,210.00
Scenario No. 5	50,124.00	221,798.00	149,313.00	1,329,249.00
Scenario No. 6	73,489.00	199,726.00	155,978.00	1,108,466.00

Energy consumption simulation erred on the side of the actual minimum for all applied scenarios (sum of heating and cooling load). Half year energy consumption cycle (see Fig. 5) demonstrates the first six months of the year with clear minimum, maximum, and average (mayonne).

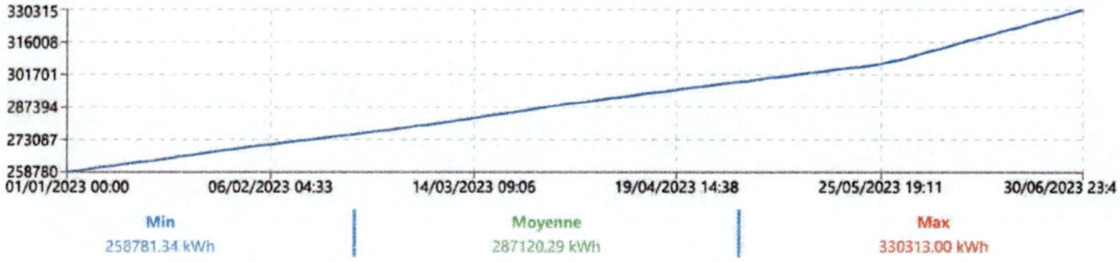

Fig. 5. TwinOps Energy Consumption Analysis – date ranges 01/01/2023–06/30/2023.

Scenarios 5 and 6 resemble actuals, with a 0.47% percentage difference between total load accumulation and between the actuals, 5.44% and 4.96%, respectively. The sixth scenario approximates the actual preliminary data; however, it should be noted the values in the fifth scenario also exhibits a comparable degree of accuracy and provides insight into the appropriate range for temperature with respect to the energy consumption. Ideal power deliverance lies within the sixth scenario findings, correlation between indoor temperatures prominent in the lesser energy requirement for lower temperatures for significant cooling loads and power.

Pleiades: Life Cycle Assessment Simulation

A radar diagram showing the environmental impact of the six scenarios during the use phase on all impact categories (see Fig. 6 & Table 8) exclude the construction, renovation, and reconstruction phases from the analysis as they remained constant.

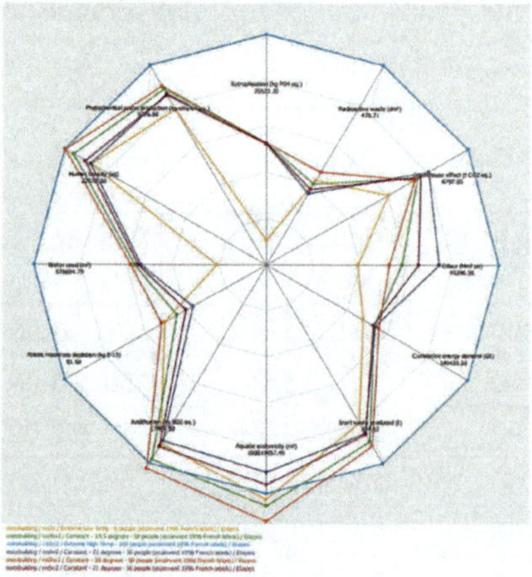

Fig. 6. Radar Diagram of all building scenarios during use phase.

Table 8. Simulation scenarios & legend.

Scenario	Color	Description	Temperature	Occupancy
No. 1	Light Blue	Extreme – High	25°	100
No. 2	Yellow	Extreme – Low	17°	6
No. 3	Orange	Constant	20°	50
No. 4	Green	Varied – 19.5°	18°–19.5°	50
No. 5	Purple	Varied – 21°	20°–21°	50
No. 6	Dark Blue	Varied – 23°	21.5°–23°	50

The extreme high temperature scenario acts as the reference for this graphical representation. The scenario most similar to current building operations is Scenario No. 6. In terms of greenhouse gas emissions, Scenario No. 6 underperformed compared to the other scenarios simulated. Scenario No. 6 emitted 5,473 t CO_2 eq. Throughout its entire lifecycle, as seen in Fig. 3. However, with regard to the cumulative energy demand in Fig. 4, Scenario No. 6 consumed 4,294 GJ less of energy than Scenario No. 3. This highlights how implementing a dynamic temperature setpoint can lower energy consumption by 3–5%, compared to a fixed temperature setpoint. Furthermore, current building operations most similarly reflect the least waste produced and resources depleted out of the most realistic building scenarios, shown in Fig. 5 and Fig. 6 (Fig. 7 and Fig. 8).

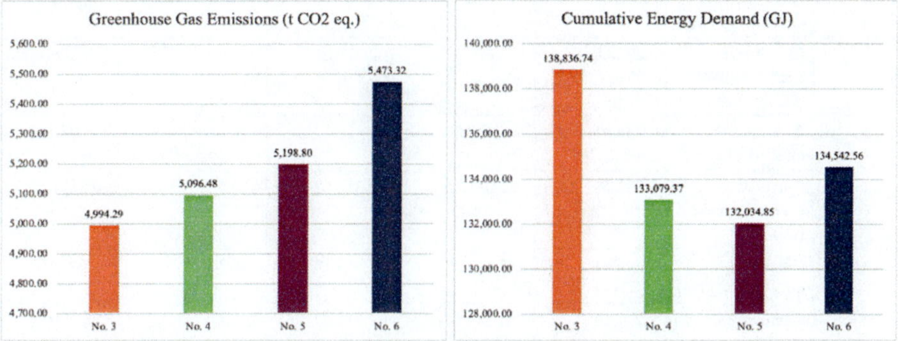

Fig. 7. Greenhouse gas emissions and Cumulative energy demand during Use Phase for Scenarios 3, 4, 5, and 6.

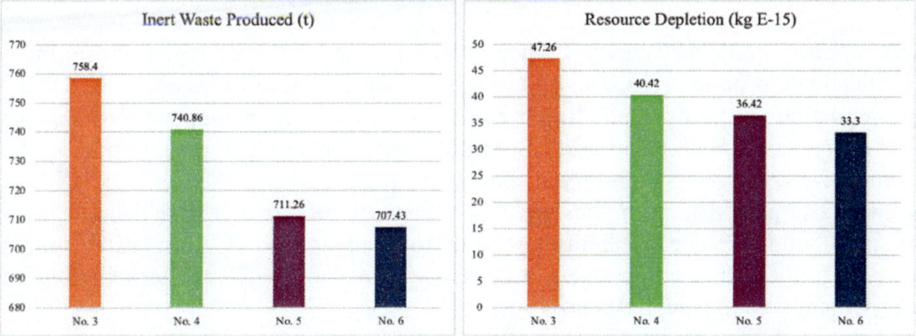

Fig. 8. Inert waste produced and resource depletion during Use Phase for Scenarios 3, 4, 5, and 6

The data obtained from conducting a LCA on Scenarios 3 through 6 (Table 9) demonstrate only the results for the use phase differ from the simulations, as the only varying input originated from the operation conditions.

Scenario No. 6 is the simulation with building operational building conditions most similar to the existing ESTP-ESEO building. In terms of greenhouse effect and cumulative energy demand, the use phase for Scenario No. 6 is responsible for 87.5% and 91.9% of the building's impact throughout its lifecycle. This highlights the elevated impact of the use phase, and how, specifically in this case study, lowering the temperature variation also decreases the environmental impact of the building, as seen in Fig. 3 and Fig. 4.

Table 9. Scenario No. 6 Life Cycle Assessment Results

Scenario	Impact Indicator	Construction	Use	Renovation	Deconstruction	Total
No. 3	Greenhouse effect (t CO_2 eq.)	746.41	4.994.29	19.17	16.98	5,777.04
	Cumulative energy demand (GJ)	9,222.68	138,836.74	2,316.67	277.49	150,653.58
No. 4	Greenhouse effect (t CO_2 eq.)	746.41	5,096.48	19.17	16.98	5,879.24
	Cumulative energy demand (GJ)	9,222.68	133,079.37	2,316.67	277.49	144,896.21
No. 5	Greenhouse effect (t CO_2 eq.)	746.41	5,198.80	19.17	16.98	5,981.56
	Cumulative energy demand (GJ)	9,222.68	132,034.85	2,316.67	277.49	143,851.69
No. 6	Greenhouse effect (t CO_2 eq.)	746.41	5,473.32	19.17	16.98	6,256.08
	Cumulative energy demand (GJ)	9,222.68	134,542.56	2,316.67	277.49	146,359.39

3 Conclusion

Energy consumption analyses in Logiciel Pleiades enables a wide range of user input specifications to be realized, and an even larger scale of data to be obtained. Of the six simulation scenarios applied with emphasis on the indoor temperature input for ideal energy consumption and user comfort, the sixth scenario pertained to possess the highest degree of overall similarity in comparison to the existing ESTP-ESEO building. Idealized energy consumption lies within the range of the fifth and sixth datasets, with the smart building's current ability to live up to the definition of smart sustainability, however further input comparisons a necessity to make complete determinations of energy optimization enhancements.

A thorough life cycle assessment of the ESTP-ESEO smart building with the comparing of the scenario simulations results yielded evidently the use phase is the most consuming life cycle stage. Therefore, the use phase pertains to the greatest impact on the environment. From this case study, possible shortcomings for current building operations were observed and possible impact mitigation strategies discussed. By altering building operation conditions, including the implementation of a lower dynamic temperature variation, the burden buildings pose to the environment can be lowered. LCA serves as a strong decision making support tool because it acts as both a proactive and reactive strategy when it comes to carbon emissions and energy consumption of a building, allowing stakeholders to reach sustainability goals more easily.

4 Perspective

Limitations within Logiciel Pleiades require complete realization in future endeavors, specifically continuations of the current preliminary case study. Weather data, construction data, as well as theoretical input parameters are necessary considerations for success within the intended goal to optimize any areas of energy consumption in the ESEO-ESTP building.

Energy optimization and consumption analyses with a lesser time restriction to engender simulations with individual focuses on parameters (i.e. lighting only example) and combinational to further develop a holistic dataset would be ideal in successfully identifying areas of strengths and weaknesses to develop in the ESEO smart building.

Future opportunities to further the study of life cycle assessment ideally center conducting on smart buildings with a focus on determining the most sustainable use of construction materials, since this case study focused on a preliminary application of LCA and energy optimization only. Moreover, the comparison of results from an LCA application in the design stage of a project, or before construction, and then from that after construction, would be interesting to determine the quantifiable benefits of conducting LCA early on in project creation.

References

1. Author, F., Author, S.: International Organization for Standardization. ISO 14040, environmental management – life cycle assessment – requirements and guidelines. title of a proceedings paper. In: Editor, F., Editor, S. (eds.) CONFERENCE 2016, LNCS, vol. 9999, pp. 1–13. Springer, Heidelberg (2016)
2. Pamu, et al.: Life Cycle Assessment of a building using Open-LCA software, 1968–1978 (2022)
3. Borhani, A., Borhani, A., Dossick, C.S., Jupp, J.: Smart building conceptualization: a comparative analysis of literature and standards. In: Construction Research Congress 2022, pp. 310–318 (2022). https://doi.org/10.1061/9780784483954.032
4. Population, United Nations. https://www.un.org/en/global-issues/population#:~:text=The%20world's%20population%20is%20expected,billion%20in%20the%20mid%2D2080s. Accessed 19 May 2023
5. What is the Sri? Energy. https://energy.ec.europa.eu/topics/energy-efficiency/energy-efficient-buildings/smart-readiness-indicator/what-sri_en. Accessed 25 May 2023
6. Energy efficiency targets, Energy. https://energy.ec.europa.eu/topics/energy-efficiency/energy-efficiency-targets-directive-and-rules/energy-efficiency-targets_en. Accessed 31 May 2023
7. In focus: Energy efficiency in buildings, European Commission. https://commission.europa.eu/news/focus-energy-efficiency-buildings-2020-02-17_en. Accessed 3 June 2023
8. Energy efficiency: Buildings and industry, Energy.gov. https://www.energy.gov/eere/energy-efficiency-buildings-and-industry. Accessed 3 June 2023
9. Hu, S., Yan, D., Azar, E., Guo, F.: A systematic review of occupant behavior in Building Energy Policy. Welcome to IIASA PURE. https://pure.iiasa.ac.at/id/eprint/16353/. Accessed 5 June 2023
10. A review: Application of building information modelling ... – iopscience. https://iopscience.iop.org/article/https://doi.org/10.1088/1755-1315/357/1/012028. Accessed 5 June 2023

11. Krarti, M.: Smart Building Energy Systems. Energy Audit of Building Systems, pp. 361–408, (2020). https://doi.org/10.1201/9781003011613-11
12. Building Energy Simulation and its application for building performance optimization: a review of methods, tools, and case studies. Advances in Applied Energy, https://www.sciencedirect.com/science/article/pii/S2666792423000148. Accessed 5 Jul 2023
13. Campus Métropolitain - Construction de Deux Écoles d'Ingénieurs - Dijon Métropole. SPLAAD, 28 Oct. 2022. www.splaad.com/operations/campus-metropolitain-construction-de-deux-ecoles-d-ingenieurs-dijon-metropole
14. Chevalier, F. : Campus Métropolitain : UN Outil Au Service de l'ingénieur de Demain. Le Journal Du Palais, 11 March 2022. https://journal-du-palais.fr/au-sommaire/collectivites/campus-metropolitain-un-outil-au-service-de-l-ingenieur-de-demain
15. Ingrao, C., Messineo, A., Beltramo, R., Yigitcanlar, T., Ioppolo, G.: How can life cycle thinking support sustainability of buildings? investigating life cycle assessment applications for energy efficiency and environmental performance. J. Cleaner Product. **201**, 556–569 (2018). https://www.sciencedirect.com/science/article/pii/S0959652618324144
16. Optimisation énergétique & environnementale du bâtiment. https://www.izuba.fr/. https://www.izuba.fr
17. Leftheriotis, G., Yianoulis, P.: 3.10 - Glazings and Coatings, ScienceDirect, 01 Jan 2012. https://www.sciencedirect.com/science/article/abs/pii/B9780080878720003103

Beamforming Antenna Array with Circular Polarization for an RF Energy Recovery System an UAV

Salah Ihlou[1](✉), Ahmed EL Abbassi[1], Abdelmajid Bakkali[1], and Hafid Tizyi[2]

[1] Team Electronics, Instrumentation and Measurement, Faculty of Sciences and Techniques, University My Ismail, Meknes, Morocco
ihlou.salah@gmail.com

[2] STRS Laboratory, INPT, Rabat, Morocco

Abstract. Our research proposes a new antenna array for recovering RF energy, which in-corporates pattern and polarization diversity. We conducted experiments to validate the performance of the optimized model, which provides an impedance bandwidth of 5.5 to 6 GHz according to measurements. By switching the input port, the antenna system can generate four circular polarization diversity beams at ±15 and ±45, with two left-hand circular polarization beams and two right-hand circular polarization beams. Additionally, the antenna array achieves low cross-polarization radiation patterns, an excellent axial ratio around the center frequency, and a low envelope correlation coefficient across the working spectrum. The antenna array has recorded gains of 5.3 and 6.6 dB for energy recovery applications. We designed and simulated all circuits, including the 5.8 GHz antenna array, using ADS and CST software. We calculated and optimized all trans-mission line widths and lengths with a 50 Ω characteristic impedance using the LineCalc tool and Tuning function, respectively.

Keywords: Beamforming network · Butler matrix · Antenna array · Circular polarization diversity

1 Introduction

In recent years, there has been a significant increase in the application of smart antenna technology in various research fields, including radar, aerospace, and wireless communication systems such as terrestrial and satellite mobile communication systems, as referenced in sources [1, 2]. A smart antenna is an array that can modify its radiation pattern automatically in real-time based on the propagation conditions, utilizing a combination of electromagnetism and signal processing as explained in [3]. Smart antenna systems can be categorized into two types - Switched Beam Systems that rely on fixed beam forming techniques using analog circuits, and Adaptive Array Systems that use digital technology for channel formation through signal processing [4], both aiming to increase gain in the desired target direction. The level of complexity and effectiveness can vary depending on the specific system being used.

M. Ben Ahmed et al. (Eds.): SCA 2023, LNNS 906, pp. 497–506, 2024.
https://doi.org/10.1007/978-3-031-53824-7_45

The system of antenna that uses beam-switching can be viewed as an expansion of the method of cell sectorization. Typically, a cell site is divided into three or four macro-sectors, each with an angle of 120° or 90°, respectively. This is according to a reference [5]. The switched beam system is made up of four primary blocks, as presented in Fig. 1.

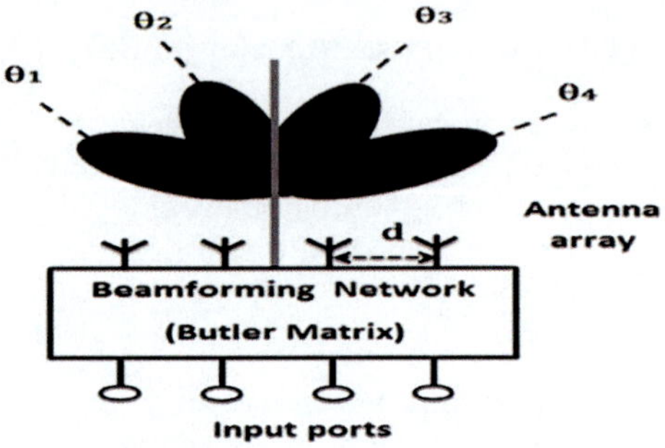

Fig. 1. The general structure of a switched beam system

The adaptive antenna system is designed as an array that can adjust itself continuously to suit the changing radio environment. With the help of active elements like amplifiers and variable phase shifters, it can regulate its radiation pattern, frequency response, and other factors during its operation (as shown in Fig. 2) [6]. To achieve this, it relies on advanced signal processing algorithms that can accurately differentiate between useful and interfering signals. This is made possible by its capability to calculate the directions of arrival (DOA) of the signal sources [5].

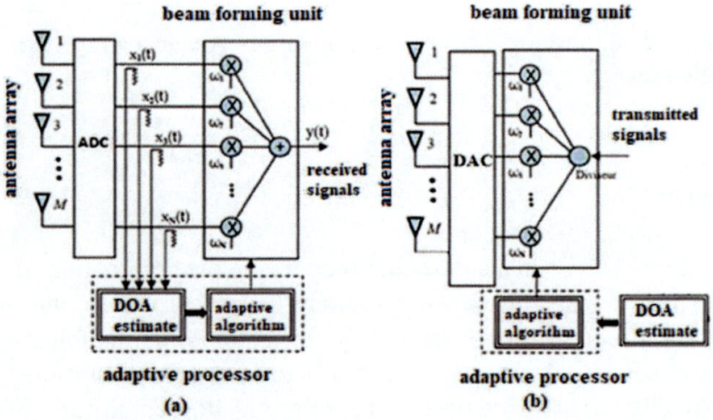

Fig. 2. The general structure of an adaptive system: (a) in receiving, (b) in transmitting [7]

The adaptive antenna system operates by detecting the directions of arrival of sources in its electromagnetic environment using DoA detection algorithms. Then, it separates the desired signal from interfering signals and forms a beam towards the useful signal while creating zeros in the directions of interferences [7]. Compared to the switched beam system, this system provides better performance by optimizing its radiation pattern in real time. However, its implementation is complex and expensive due to the use of active circuits that consume energy. The switched beam system, on the other hand, uses a fixed beam splitter technique, which requires the formation of a beam feed network and results in a fixed antenna radiation pattern. The second technique for fixed beam splitters, which includes the Blass matrix, the Nolen matrix, and the Butler matrix, is commonly used in [7].

We will now provide a brief overview of the underlying concept of Butler's Matrix. The Matrix consists of a symmetrical passive circuit that includes N input and output ports, which are connected to N antennas through equal pathways between the ports. The antenna array is connected to the N outputs, while the beam ports are connected to the N inputs [7].

Comprising directional couplers, fixed phase shifters, and crossovers, the Butler Matrix is primarily utilized for distributing RF signals to the radiating elements of an antenna array, and for creating and directing orthogonal beams [8]. When connected between an RF switch and an antenna array, the Butler Matrix can stimulate two or more ports simultaneously to produce a set of beams. Interconnections between the couplers and the fixed phase shifters allow the Matrix to generate the required phase shift for N orthogonal beams [7]. Waveguide, microstrip, and CMOS technologies can be used to implement the Butler Matrix [9]. In our investigation, we opted for micro-ribbon technology, which is particularly suitable for designing Butler Matrices in the ISM bands (2.45 and 5.8 GHz) [9]. The technology offers various benefits, such as affordability, ease of manufacture, and compactness. Moreover, it allows the array and the antenna array to be implemented on the same substrate.

Upon reviewing the design plans for RF circuits and printed antennas, we determined that the FR4 epoxy substrate would be the best option for constructing the various proposed structures. This substrate has a permittivity of $\varepsilon r = 4.4$, a thickness of $h = 1.6$ mm, and a dielectric loss of $\tan \delta = 0.02$, making it an ideal choice. Additionally, this substrate is readily available in the industry and is cost effective.

2 Design Considerations

2.1 Butler Matrix

This study builds upon our previous investigation detailed in reference [10], in which we introduced an innovative design for a wide bandwidth 4×4 Butler matrix. This matrix was constructed on a substrate with a thickness of 1.6 mm and a dielectric constant of 4.3, known as FR4. Figure 3 illustrates the topology of the proposed Butler matrix, comprising four couplers, one crossover, and two-phase shifters. This particular component achieves exceptional isolation between input ports, thus enhancing bandwidth and preventing mutual coupling. Reference [10] provides a comprehensive account of the analysis and simulation of each constituent element, all of which are integrated to compose the structure simulated through Agilent ADS software.

As depicted in Fig. 4, the transmission coefficients S15, S17, S16, and S18 exhibit deviations from the desired values, registering at approximately −8 dB and −7 dB instead of the intended −6 dB. Consequently, the control of port 1 was distributed among the

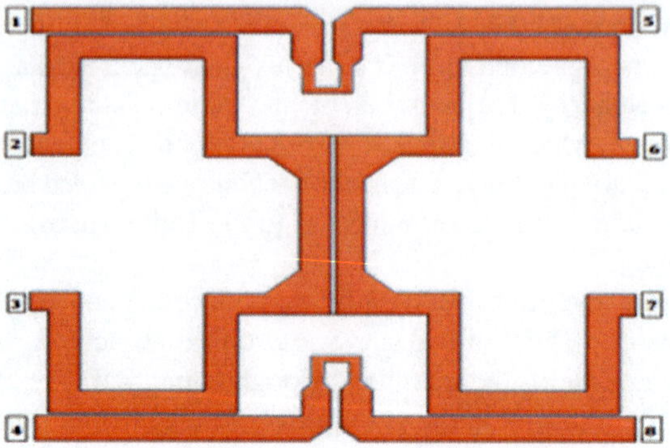

Fig. 3. Proposed 4*4 BM.

output ports. Despite these variations, the authors maintain a positive outlook on the obtained results, deeming them promising.

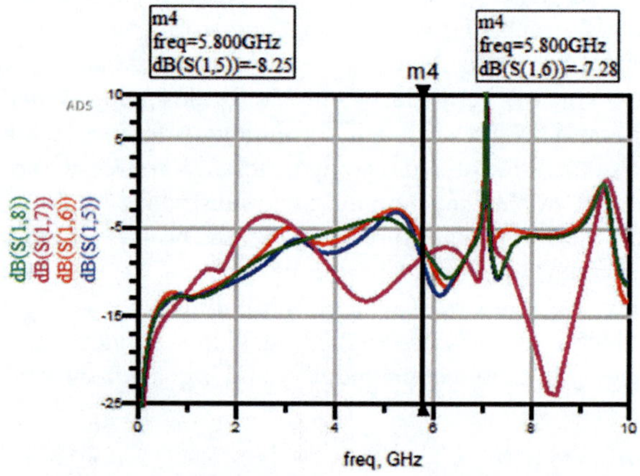

Fig. 4. Simulated insertion losses of the input port 1.

The given text describes the results of a study related to a smart antenna. As shown in Fig. 4, the transmission coefficients S15, S17, S16, and S18 deviate from the ideal values, measuring around −8 dB and −7 dB rather than the target of −6 dB. Consequently, the control of port 1 was distributed among the output ports. Notwithstanding these discrepancies, the authors remain optimistic about the results achieved, considering them promising. To visualize the gain and radiation patterns of the smart antenna, the authors exported its structure to the CST software. Figure 5 displays the radiation patterns of the antenna at 5.8 GHz, and the maximum gain achieved by the multi-beam smart antenna is about 17.31 dB at that frequency.

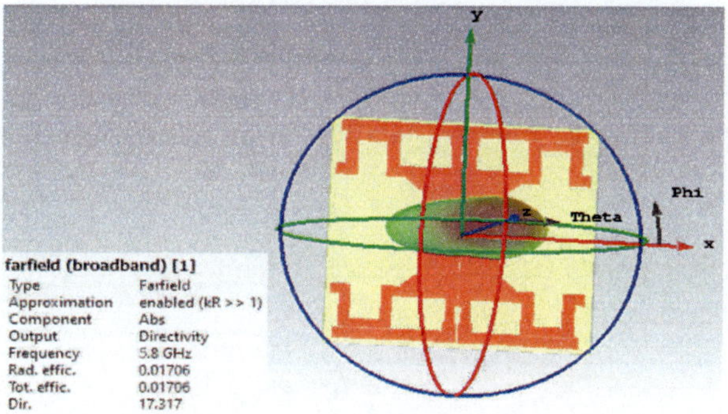

Fig. 5. Simulated (3D) gain diagram of the proposed smart antenna at 5.8 GHz (CST).

2.2 Antenna Element

To validate the performance of the proposed 4x4 Butler Matrix in terms of beamforming, it is essential to connect it to an antenna array comprising four radiating elements. In this regard, the authors will elucidate the design and simulation process of the radiating element, specifically a rectangular patch antenna fed via a microstrip line with a characteristic impedance of 50 Ω. The authors made modifications to the patch feed by introducing a notch to ensure the antenna's suitability for operation at 5.8 GHz. The determination of various patch parameters was accomplished through the utilization of classical Eqs. (1), (2), (3), and (4), followed by optimization using the Tuning function within the ADS software. The geometry of the patch antenna is illustrated in Fig. 6, and the optimized parameter values are presented in Table 1.

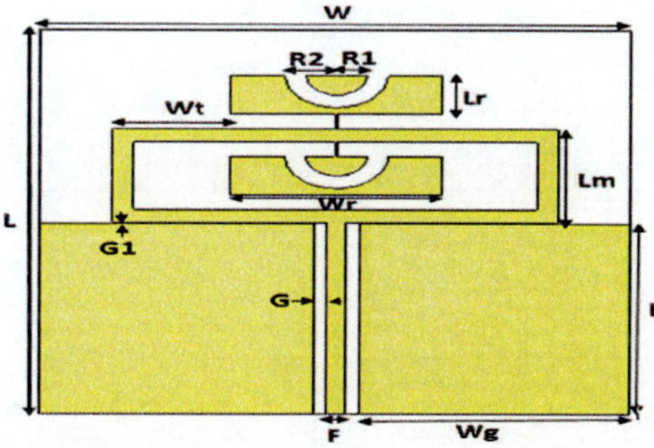

Fig. 6. Geometry of the proposed antenna.

Many optimization processes were conducted through CST Microwave Studio to achieve the optimal performance of the circularly polarized antenna at its operational frequency of 5.8 GHz. The initial design of the proposed antenna was initially formulated based on the classical equations [16].

$$W = \frac{c}{2f_0\sqrt{\frac{(\varepsilon_r+1)}{2}}} \tag{1}$$

$$L = \frac{c}{2f_0\sqrt{\varepsilon_{eff}}} - 2\Delta l \quad (2)$$

With εeff is the effective dielectric constant of the substrate the calculated by the relation:

$$\varepsilon_e = \frac{1}{2}(\varepsilon_r + 1) + \frac{1}{2}(\varepsilon_r - 1)\left(1 + 12\frac{h}{w}\right)^{\frac{-1}{2}} \quad (3)$$

$$\Delta l = 0.412h\frac{(\varepsilon_r + 0.3)(\frac{w}{h} + 0.264)}{(\varepsilon_r - 0.258)(\frac{w}{h} + 0.8)} \quad (4)$$

L = the length of the patch antenna
W = the width of the patch antenna
Δl = fringing field of the antenna
Table 1 shows the various optimized parameters of the proposed design.

Table 1. Optimized Parameters of Proposed Antenna

Parameters	Size (mm)	Parameters	Size (mm)
W	28.4	L	30
Wg	13.45	Lg	14.642
Wr	11	Lr	3
Wt	5.2	Lm	7.3
G	0.358	R1	1.5
G1	0.25	R2	2.5
F	1	M	5.3

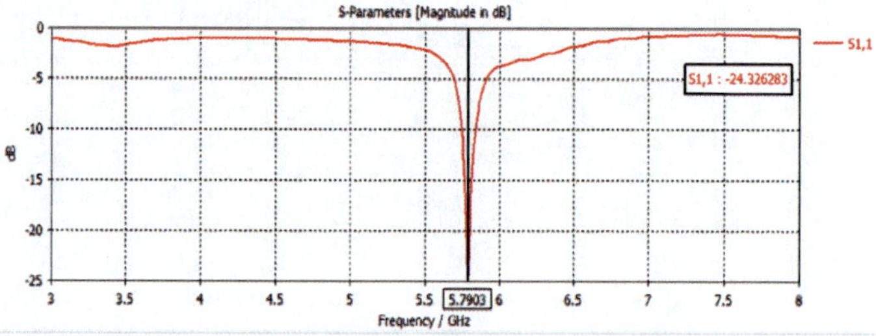

Fig. 7. Simulated S11 versus frequency

The resonance frequency of the antenna is clearly identifiable as 5.8 GHz, with S11 parameter registering –24.32 dB (<−10 dB) at the same frequency, indicating that the antenna has a good input impedance match to the 50 Ω impedance (as shown in Fig. 7). The simulated bandwidth at −10dB is 160 MHz, ranging from 5.72 to 5.88

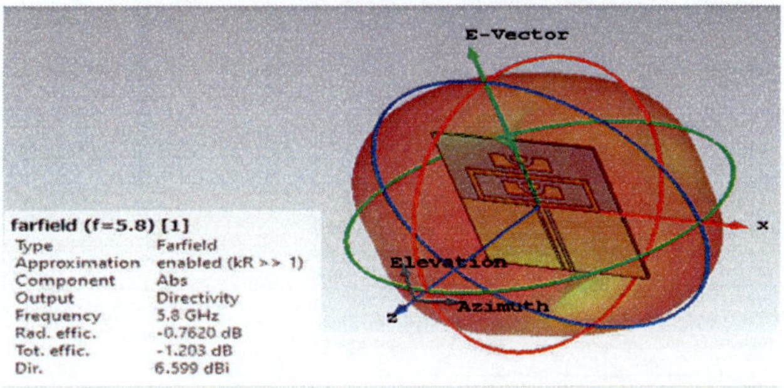

Fig. 8. 3D radiation pattern of the proposed design at 5.8 GHz.

GHz. Figure 8 displays the 3D radiation patterns of the antenna at 5.8 GHz, which demonstrates a radiation directive pattern of 6.6 dB.

Figures 9 and 10 depict the variations in axial ratio and Voltage Standing Wave Ratio (VSWR) concerning frequency for the suggested design. These graphical representations highlight the advantageous characteristics of the antenna design, with an axial ratio that reaches a minimum of 1.9 dB (<3 dB) and a VSWR of 1.05 (<2) at 5.8 GHz. To provide a comprehensive perspective, Table 2 offers a comparative analysis between the performance of the circularly polarized antenna array proposed in the existing literature for rectenna applications at 5.8 GHz and the performance of our proposed design.

Table 2. Comparison of Previous Antenna with the Proposed Antenna at 5.8GH

Ref	S11 (dB)	Bandwidth(MHz)
[10, 11]	−21	130
Our work	−24.30	160

Table analysis reveals that the proposed antenna outperforms other MPT transmitter antennas described in previous studies in terms of quality and overall performance parameters. Moreover, circular polarization, a critical requirement for MAV operations, is a feature of the proposed antenna.

2.3 Smart Antenna System

The smart antenna design suggested for 5.8 GHz (measuring 74.25×67.92 mm^2) integrates a 4×4 Butler Matrix with a group of four microstrip patch antennas, all on the same FR4 substrate. The output ports of the matrix are employed as the feedlines for the array of antennas, with the separation between these output ports mirroring the spacing between the radiating elements. This chosen spacing is approximately half of a wavelength, a selection made to maximize gain, minimize mutual interference among the antennas, and produce orthogonal beams. Figure 11 provides a visual representation of the final configuration of this smart antenna.

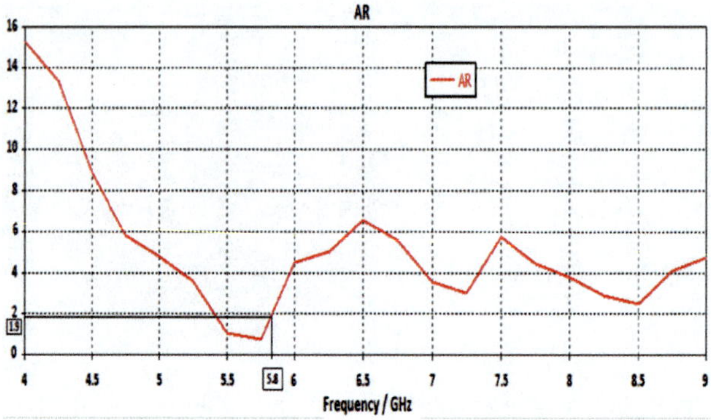

Fig. 9. Axial ratio versus frequency of the proposed antenna

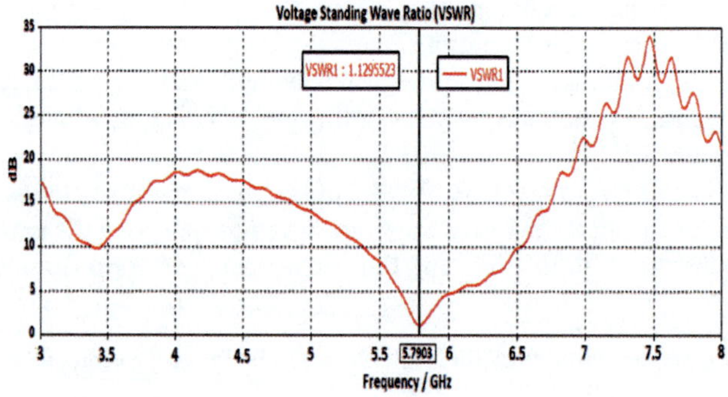

Fig. 10. Simulated VSWR of the proposed CP antenna

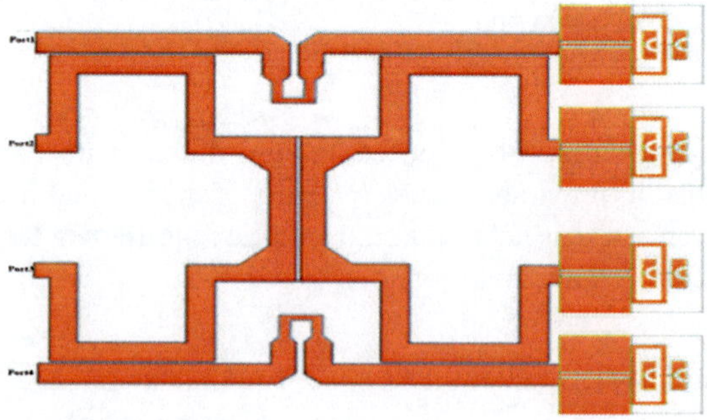

Fig. 11. The geometry of the smart antenna fed by a 4 × 4 Butler Matrix at 5.8 GHz

When observing Fig. 12 at the resonant frequency of 5.8 GHz, it becomes apparent that the reflection coefficients for all four input ports of our intelligent antenna are well below −10 dB. To be precise, the values for parameters S11 and S44 hover around −28.30 dB, whereas the parameters S22 and S33 are approximately −23.46 dB.

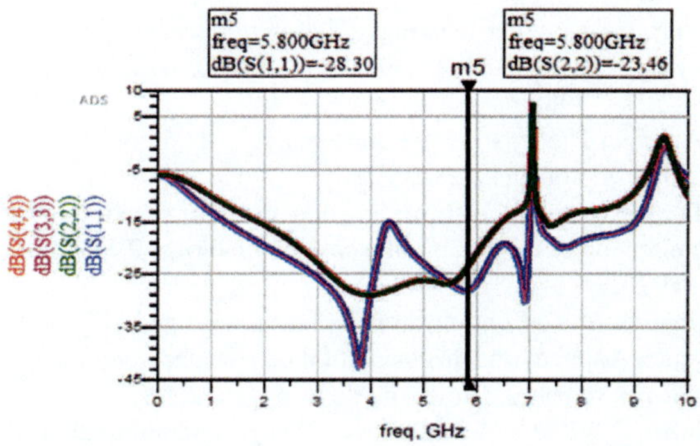

Fig. 12. Simulated S parameters at the inputs of the proposed smart antenna at 5.8 GHz

3 Conclusions

This study presents the simulation results for a patch antenna designed to operate at 5.8 GHz, which outperforms existing references in terms of weight, volume, power, and circular polarization. The antenna has been developed specifically for drones and meets all necessary requirements. To verify the simulation results, a prototype of the antenna will be built and tested.

The proposed antenna boasts dimensions of 30×28.4 mm^2, incorporating a design that simultaneously reduces weight while enhancing performance and user-friendliness. Its activation is achieved through a robust 4×4 patch array coupled with a well-recommended Butler Matrix, which has demonstrated its effectiveness. With a remarkable maximum gain exceeding 17 dB and a return loss of less than -10, this antenna stands out as an optimal choice for integration within the system, thereby enhancing the efficiency of RF-DC power conversion at the Rectenna's input. Given these noteworthy attributes, it is now feasible to expand the research to encompass the production and evaluation of a prototype based on this recommended configuration.

Acknowledgments. The authors are grateful to STRS Laboratory of Institut National des Postes et Telecommunication in RABAT Morocco and members of Electronics, Instrumentation and Measurement Faculty of Sciences and Techniques (FST), University my Ismail in Errachidia, for support with regard to simulation software and facilities.

References

1. Skiani, E.D., Mitilineos, S.A., Thomopoulos, S.C.A.: A study of the performance of wireless sensor networks operating with smart antennas. IEEE Antennas Propag. Mag. **54**(3), 50–67 (2012)
2. Luo, Q., Gao, S., Sumantyo, J.T.S.: Smart antennas for mobile satellite communications. In: IEEE Proceedings of 2014 3rd Asia-Pacific Conference on Antennas and Propagation. IEEE, pp. 104–107 (2014)
3. Alexiou, A., Haardt, M.: Smart antenna technologies for future wireless systems. Trends Challenges IEEE Commun. Mag. **42**(9), 90–97 (2004)

4. International commission on non-ionizing radiation protection: guidelines for limiting exposure to time-varying electric, magnetic, and electromagnetic fields (up to 300 GHz). Health Phys. **74**(4), 494–522 (1998)
5. Balanis, C.A.: Antenna Theory Analysis and Design. 3rd edition, John willey and Son's Inc, ISBN: 0–471–66782-X, vol. 811 (2005)
6. Choi, S., Yun, D.: Design of an adaptive antenna array for tracking the source of maximum power and its application to CDMA mobile communications. IEEE Trans. Antennas Propag. **45**, 1393–1399 (1997)
7. Fadlallah, N.: Contribution à l'optimisation de la synthèse du lobe de rayonnement pour une antenne intelligente, Application à la conception de réseaux à déphasage. Thèse de Doctorat N° 18–2005, UFR des sciences, Université de Limoges (2005)
8. Errifi, H., Baghdad, A., Badri, A., Sahel, A.: Design and simulation of a planar topology butler matrix for 10 Ghz switched multibeam antenna. Mediterranean Telecommun. J. **7**(1) (2017)
9. Bhowmik, W., Srivastava, S.: Optimum design of a 4 × 4 planar butler matrix array for WLAN application. J. Telecommun. **2**(1) (2010)
10. Ihlou, S. Tizyi, H., El Abbassi, A.: Beamforming networks using a broadband 4x4 butler matrix with wideband couplers and crossovers. In: 3RD International Conference on Big Data and Machine Learning (BML'22) (2022)
11. Shimane, E., Diallo, A., Komaru, T., Komurasak, K., Arakaw, Y.: Wireless power transmission system for a micro aerial vehicle. In: URSI (2012)
12. Ishiba, M., Oda, A., Sawahara, H., Komurasaki, K., Arakawa, Y.: Wireless power transmission and telecommunication using a microwave active phased array. In: 9th Annual International Energy Conversion Engineering Conference, p. 5978 (2011)

Enhancing Convergence Speed in Control of Synchronous Motors Using Model Predictive Control–MPC with Reference Model

Said Ziani(✉)

ENSAM, Mohammed V University in Rabat, Rabat, Morocco
s.ziani@um5r.ac.ma

Abstract. This research paper proposes the application of Model Predictive Control (MPC) with a reference model for achieving faster convergence in the control of synchronous motors. Synchronous motors are are utilized in a variety of industrial contexts. Nevertheless, due to the complex dynamics and inherent uncertainties of these motors, attaining rapid convergence in the control of these motors can be a difficult task. The MPC technique is known for its ability to handle multivariable systems with constraints and uncertainties. By incorporating a reference model into the MPC framework, the control algorithm can anticipate the system's behavior and make proactive adjustments, leading to faster convergence. This paper explores the implementation and performance evaluation of the MPC approach with a reference model for controlling synchronous motors.

Keywords: Control of Synchronous Motors · Model Predictive Control · Reference Model

1 Introduction

Because of their high efficiency, precise control capabilities, and sturdy performance, synchronous motors find widespread use in a variety of industrial applications. However, achieving fast convergence in controlling these motors is challenging due to their complex dynamics, non-linearities, and uncertainties. To address this challenge, advanced control techniques, including Model Predictive Control (MPC), have shown promise [1, 2].

MPC is a control strategy that optimizes control inputs over a future prediction horizon by solving an optimization problem at each time step. It considers system dynamics, constraints, and objectives to generate control actions that minimize a predefined cost function. MPC's ability to handle multivariable systems with constraints and uncertainties makes it an attractive choice for controlling synchronous motors.

Traditional MPC methods [3] have a limitation in their reactive nature, where control actions are adjusted based on the current system state. This can result in slower convergence and suboptimal control performance. To overcome this limitation, incorporating a reference model into the MPC framework enables proactive control adjustments by anticipating the system's behavior.

M. Ben Ahmed et al. (Eds.): SCA 2023, LNNS 906, pp. 507–518, 2024.
https://doi.org/10.1007/978-3-031-53824-7_46

The concept of a reference model [4] involves creating a model that represents the desired system response. By comparing the actual system output with the reference model's output, the MPC controller can generate control actions that drive the system towards the desired response. This predictive nature of the reference model enables faster convergence, improved tracking of reference trajectories, and enhanced disturbance rejection.

In this research paper, we investigate the application of MPC with a reference model for controlling synchronous motors to achieve faster convergence. Our objective is to utilize the predictive capabilities of the reference model to proactively adjust control actions, thereby reducing convergence time and improving overall control performance. We evaluate the proposed control strategy through extensive simulations and compare it against conventional control methods.

The remaining sections of this paper are structured as follows: The Sect. 2 describes the approach, which includes the system model and MPC formula. And incorporation of the reference model. Section 3 discusses the simulation setup, presents the results and analysis, and finally, Sect. 4 concludes the paper, highlighting the contributions and potential future directions of this research.

2 Research Methodology

This section discusses this study's research methods and the PMSM's modeling. To simplify and analyze, we shall provide the modeling assumptions [5].

2.1 Model

The stator and rotor winding arrangements determine the electrical equations in the (d-q) reference frame. On the d-axis, the equation shows the dynamics of the direct axis current, id, and its interaction with the stator voltage and magnetic field. In the q-axis, the equation defines the quadrature axis current, iq, and its relationship with the stator voltage and magnetic field. These equations account for back electromotive force (EMF) and winding resistivity and inductance.

However, the magnetic equations in the (d-q) reference frame describe magnetic flux linkage dynamics and torque generation. The equation shows the magnetic flux linkage's rate of change, and its connection with the stator current and rotor magnetic field on the d-axis. The equation represents the magnetic flux linkage's rate of change, q, and its interaction with the stator current and rotor magnetic field on the q-axis. These equations include stator-rotor mutual inductance and permanent magnet properties.

$$\begin{cases} \frac{d_{id}}{dt} = -\frac{R_s}{L_d} i_d + p\frac{L_q}{L_d}\Omega i_q + \frac{v_d}{L_d} \\ \frac{d_{iq}}{dt} = -\frac{R_s}{L_q} i_q - p\frac{L_d}{L_q}\Omega i_d - p\frac{\Omega \varphi_f}{L_q} + \frac{v_q}{L_q} \\ \frac{d\Omega}{dt} = \frac{p}{J}(L_d - L_q) i_d i_q - \frac{f}{J}\Omega + \frac{P}{J}\varphi_f i_q - \frac{C_r}{J} \end{cases} \tag{1}$$

The PMSM model in the (d-q) reference frame includes the following parameters and variables:

- Parameters:

- R_s: Stator resistance (Ω)
- L_d, L_q: Self-inductance in the d-axis and q-axis (H)
- φ_f: Mutual flux due to permanent magnet (Wb)
- Ω: Angular speed (rad/s)
- J: Moment of inertia (kg.m^2)
- f: Damping constant (N/rad/s)
- p: Number of pole pairs
- C_r: Load torque (N.m)

- Variables:
  - i_d, i_q: Currents in the d-axis and q-axis (A)
  - v_d, v_q: Voltages in the d-axis and q-axis (V)

These parameters and variables are essential in formulating the mathematical equations that describe the dynamics of the PMSM in the (d-q) reference frame. They play a crucial role in understanding the motor's behavior, designing control algorithms, and analyzing its performance [6].

2.2 The Model Predictive Control (MPC)

The Model Predictive Control (MPC) method involves formulating an optimization problem to determine the optimal control inputs at each time step based on a prediction model. Here are the mathematical equations used in the MPC formulation [7]:

- State Update Equation:

$$x(k+1) = Ax(k) + Bu(k) + Bd(k) \tag{2}$$

This equation represents the discrete-time state update, where x(k) denotes the state vector at time step k, A is the state matrix, B is the input matrix, u(k) represents the control input vector at time step k, and Bd(k) represents any disturbance affecting the system.

- Output Equation:

$$y(k) = Cx(k) \tag{3}$$

This equation relates the system output, y(k), to the state vector x(k), where C is the output matrix.

- State Constraints:

$$x_{min} <= x(k) <= x_{max} \tag{4}$$

These inequalities represent the state constraints, ensuring that the state variables remain within specified lower (x_{min}) and upper (x_{max}) bounds.

- Input Constraints:

$$\text{u\_min} <= \text{u(k)} <= \text{u\_max} \tag{5}$$

These inequalities denote the input constraints, ensuring that the control inputs stay within specified lower (u_min) and upper (u_max) bounds.

- Cost Function:

$$\text{J} = \sum \left[\text{x(k)}'\text{Qx(k)} + \text{u(k)}'\text{Ru(k)} + (\text{y(k)} - \text{y\_ref(k)})'\text{Wy(k)}'\right] \tag{6}$$

The cost function J is typically a sum of quadratic terms, which penalizes deviations from desired states and control actions. Q and R are positive definite weighting matrices that tune the importance of the state and control penalties, respectively. Wy is a positive definite weighting matrix for the output penalty. y_ref(k) represents the reference output at time step k.

- Prediction Model:

$$\text{x (k} + \text{i} + 1) = \text{A x(k} + \text{i)} + \text{B u(k} + \text{i)} \tag{7}$$

This equation represents the prediction model, which predicts the future states of the system based on the current state, control inputs, and system dynamics. It is used within the optimization problem to predict the system behavior over the prediction horizon.

- Optimization Problem:

  Minimize J subject to system dynamics, constraints, and prediction model.

2.3 Incorporation of the Model Predictive Control (MPC)

To incorporate Model Predictive Control (MPC) [8–10] with a reference model into the PMSM model mentioned above, we can formulate the MPC problem and outline the steps involved in the control algorithm. Here is a general overview of the MPC formulation and the incorporation of the reference model for resolving the PMSM model:

2.3.1 State and Output Variables

Define the state variables that characterize the system's behavior over time. In the case of the PMSM model, the state variables could include the currents (i_d, i_q), the angular speed (Ω), and possibly other relevant variables. Additionally, determine the output variables of interest, such as the torque (C_r) or any other performance measures.

2.3.2 Reference Model

Create a reference model that represents the desired behavior of the PMSM. This model should capture the target trajectories or setpoints for the state and output variables. The reference model can be based on mathematical models, empirical data, or control objectives. It serves as a benchmark for the control algorithm to track and adjust the system's behavior accordingly.

2.3.3 Cost Function

Define a cost function that quantifies the performance objectives and constraints for the MPC algorithm. The cost function typically includes terms that penalize deviations from the reference model and account for control effort or other factors. It captures the trade-offs between tracking the reference trajectories and minimizing control actions.

2.3.4 System Constraints

Specify any constraints that need to be satisfied during the control process. These constraints could include limits on the currents, voltages, torque, or any other relevant variables. Incorporate these constraints into the MPC formulation to ensure safe and feasible operation of the PMSM.

2.3.5 Prediction Horizon and Control Horizon

Determine the prediction horizon, which represents the length of the future time window over which predictions and optimization will be performed. Likewise, establish the control horizon, which indicates the time duration for which control actions will be applied to the system. These horizons influence the trade-off between computational complexity and control performance.

2.3.6 MPC Optimization

Formulate an optimization problem that minimizes the defined cost function while satisfying the system constraints. The optimization problem is solved over the prediction horizon, where the control inputs (voltages v_d, v_q) are adjusted to optimize the system behavior and track the reference model. MPC utilizes online optimization techniques to find the optimal control actions at each time step.

2.3.7 Receding Horizon Control

Implement a receding horizon control strategy, where only the first control actions of the optimized sequence are applied to the system. As time progresses, the optimization problem is solved again at each sampling interval, considering updated measurements and reference trajectories. This allows for adaptability and robustness in controlling the PMSM under changing conditions.

By incorporating the reference model and formulating the MPC problem, the control algorithm adjusts the control inputs (v_d, v_q) based on the prediction of the PMSM model, the desired reference trajectories, and the optimization process. This approach enables proactive control adjustments, faster convergence, and improved tracking of the desired behavior, resulting in enhanced control performance for the PMSM system. These updated equations now incorporate the variables i_d, i_q, v_d, and v_q.

The MPC algorithm as depicted in Fig. 1 follows a predictive control scheme, where at each time step, it predicts the future states of the system, formulates an optimization problem to find the best control actions, and applies the optimal control actions to the actual system. This enables the system to track reference outputs, adhere to constraints, and optimize system performance.

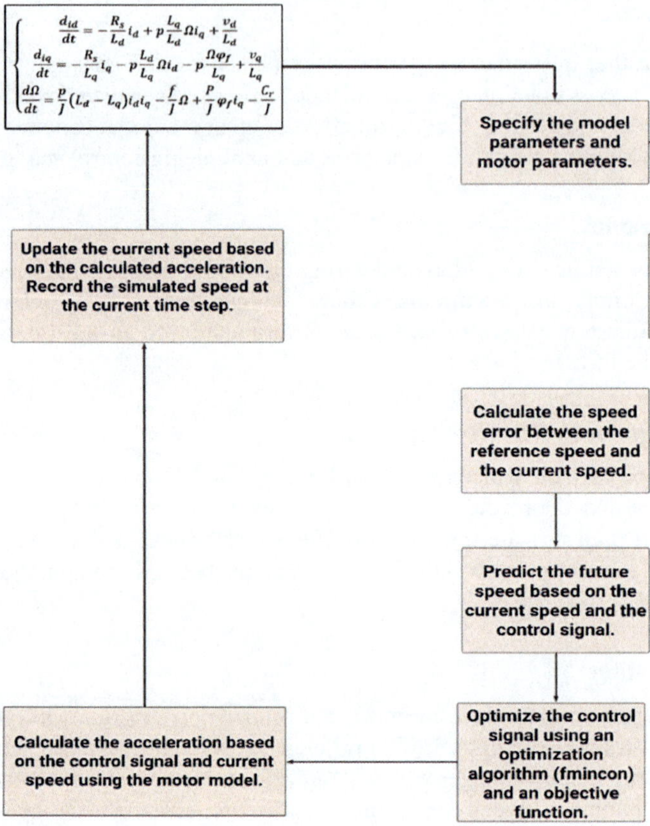

Fig. 1. Algorithm

3 Results and Discussion

The results obtained from the conducted research provide valuable insights into the control of synchronous motors using Model Predictive Control (MPC) with a reference model. This section aims to summarize the key findings and present a comprehensive discussion of the results.

3.1 PMSM Nominal Values

Table 1 lists the PMSM's ratings and actual variables used in simulations:

3.2 Simulation Results and Discussion

The research results of the conducted simulation study using Model Predictive Control (MPC) with a reference model for controlling synchronous motors are presented below:

Table 1. Actual variables

Parameters	Value
P	3
R_s	0.2377 [Ω]
L_d	0.0733 [**H**]
L_q	0.0728 [**H**]
φ_f	0.29562 [Web]
J	0.025942 $\left[N.mS^2/rad\right]$
f	0.02124

Simulation Parameters:

- Time interval (dt): 0.01 s
- Total simulation time (total_time): 10 s
- Time vector (time): Ranging from 0 to 10 s

Motor Parameters Control Parameters are giving in Table 2.

Table 2. Motor Parameters Control Parameters

Motor and Control Parameters	Value
Prediction horizon (N)	5
Weighting factor for speed error (Q)	1
Weighting factor for control signal (R)	1
Moment of inertia (kg.m^2)	0.01
Coupling constant	0.1
Friction constant	0.05

3.2.1 Reference Speed

The reference speed was varied in steps during the simulation to evaluate the motor's response. The following reference speed profile was used:

- Step 1: Reference speed of 20 rpm for 4 s
- Step 2: Reference speed of 8 rpm for 5 s
- Step 3: Reference speed of 20 rpm
- Simulated Speed: The simulated speed of the synchronous motor was recorded throughout the simulation. The algorithm adjusted the control signal at each time step to achieve accurate tracking of the reference speed.

3.2.2 Response Speed

The simulation results for the synchronous motor control system provide valuable insights into its performance. Figure 1 demonstrates the system's ability to accurately track the step-wise reference speed, showcasing its responsiveness to changes in speed requirements. Figure 2 further validates the system's effectiveness by illustrating the alignment between the simulated motor speed and the reference speed. This close correspondence between the two variables confirms the system's capability to maintain the desired speed profile.

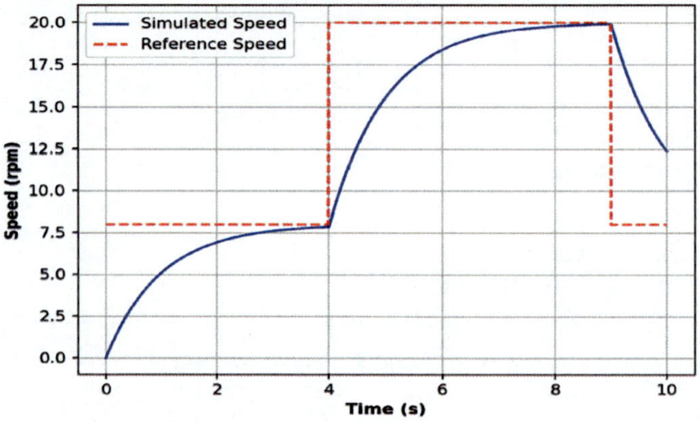

Fig. 2. Reference speed and motor speed.

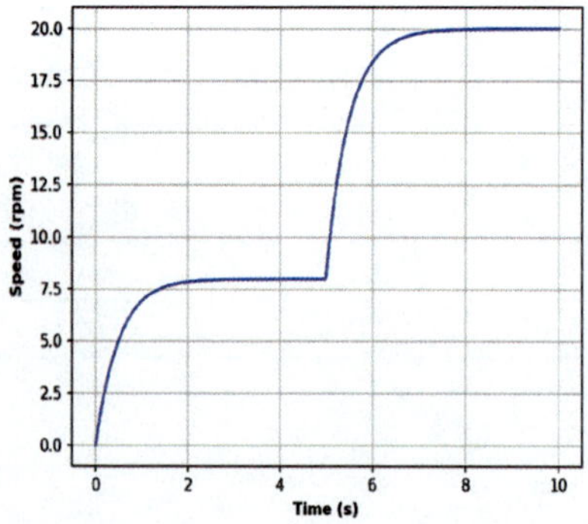

Fig. 3. Motor speed with step-wise changes

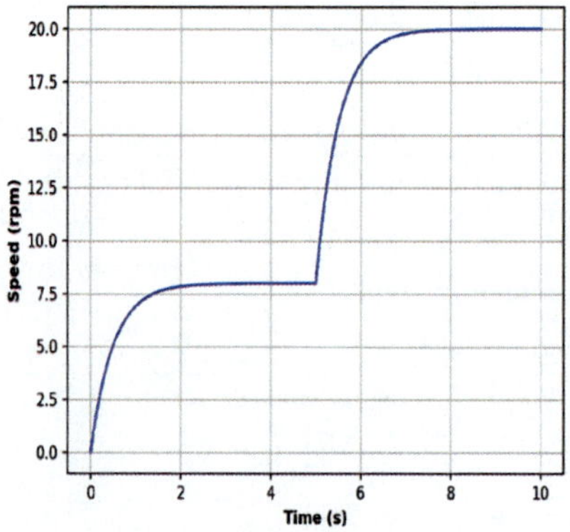

Fig. 4. Rise time improvement

Furthermore, Fig. 3 highlights the motor's response to step-wise changes in the reference speed, revealing its adaptability and ability to smoothly transition between different speed levels. This dynamic behavior is crucial for applications that require rapid adjustments in motor speed. The control system exhibits notable improvements in key performance metrics. Figure 4 shows the reduction in rise time, indicating a faster response and shorter settling time. The system's ability to correct overshoot is demonstrated in Fig. 5, where it effectively minimizes any transient oscillations, ensuring stable

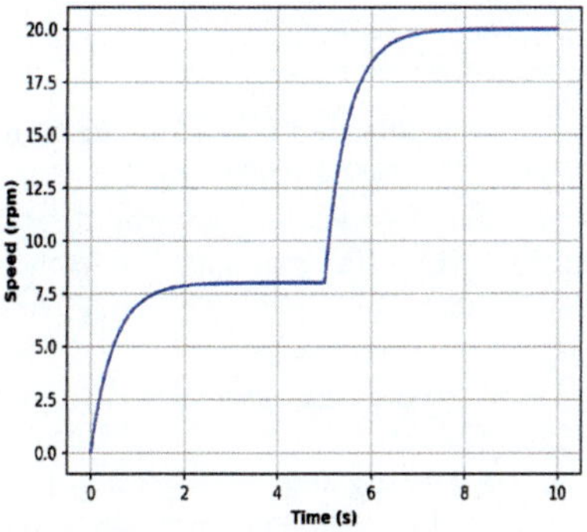

Fig. 5. Overshoot correction

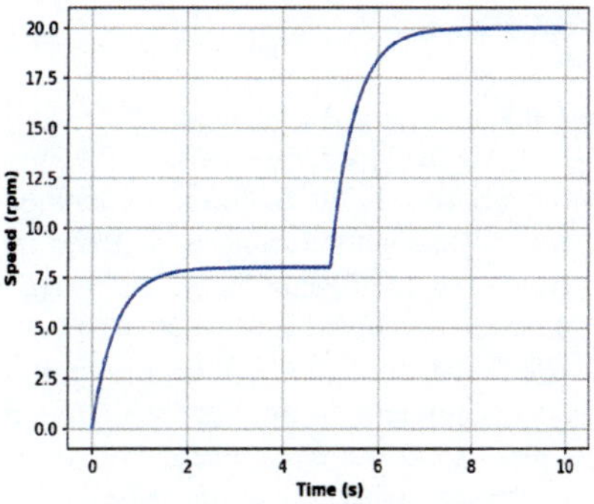

Fig. 6. Speed accuracy

and accurate speed control. These enhancements contribute to overall system stability and reliability. Lastly, Fig. 6 emphasizes the speed accuracy achieved by the control system. The close alignment between the desired reference speed and the actual motor speed confirms the system's precision and its capability to maintain speed consistency.

Overall, these findings demonstrate the effectiveness of the control system in achieving the desired performance goals for the synchronous motor. The system showcases responsive tracking of reference speed, smooth adaptation to speed changes, improved response time, reduced overshoot, and high speed accuracy. These outcomes validate the suitability of the control system for various applications requiring precise and reliable motor speed control. Our research introduces a novel approach in which we propose the utilization of a model predictive control (MPC) strategy with a reference model to expedite convergence and enhance the dynamic performance of synchronous motors. Through extensive simulations, we substantiate the efficacy of our proposed method, demonstrating its ability to deliver the anticipated outcomes. Furthermore, when compared to previous studies [11–18], and [19–24], our methodology surpasses in terms of eliminating static error and achieving the most favorable rising time. The effectiveness of this control can be optimized by combining the methods used in this paper with new recent theories, namely wavelet theory, which has shown very promising results in the biomedical field as illustrated in references [25–31].

4 Conclusion

In conclusion, our research adds to the body of knowledge by introducing an innovative approach known as anticipatory control, accompanied by a reference model. Through our findings, we have successfully showcased the effectiveness of this technique in attaining swifter convergence and superior dynamic performance in synchronous motors. We propose that future investigations delve into the exploration of energy optimization algorithms for mobile applications and the integration of advanced data analysis techniques. These endeavors have the potential to further enhance the results and implications of our research.

References

1. Jiang, J., et al.: Augmented whole-body scanning via magnifying PET. IEEE Trans. Med. Imaging **39**(11), 3268–3277 (2020). https://doi.org/10.1109/TMI.2019.2962623
2. Ito, S., Yoo, H.W., Schitter, G.: Comparison of modeling-free learning control algorithms for galvanometer scanner's periodic motion. In: 2017 IEEE International Conference on Advanced Intelligent Mechatronics (AIM), Munich, Germany, pp. 1357–1362 (2017). https://doi.org/10.1109/AIM.2017.8014207
3. Ghedamsi, K., Aouzellag, D., Berkouk, E.: Control of wind generator associated to a flywheel energy storage system. Renew. Energy **33**, 2145–2156 (2008)
4. Chen, X., Zhou, B., Xie, H., Shi, L., Liu, H., Liu, C.: Investigation of direct and indirect approaches of deep-learning-based attenuation correction for general purpose and dedicated cardiac SPECT scanners. In: 2021 IEEE Nuclear Science Symposium and Medical Imaging Conference (NSS/MIC), Piscataway, NJ, USA, pp. 1–2 (2021). https://doi.org/10.1109/NSS/MIC44867.2021.9875517
5. Ay, M.R., Ahmadian, A., Ghadiri, H., Maleki, A., Ghafarian, P., Zaidi, H.: The influence of X-ray spectra filtration on image quality and patient dose in the GE VCT 64-slice cardiac CT scanner. In: 2009 3rd International Conference on Bioinformatics and Biomedical Engineering, Beijing, China, pp. 1–4 (2009). https://doi.org/10.1109/ICBBE.2009.5163626
6. Miloud, G., Hicham, S., Youcef, B.: Sensorless speed control of synchronous reluctance motors using model predictive control associated with model reference adaptive system. In: 2021 IEEE 1st International Maghreb Meeting of the Conference on Sciences and Techniques of Automatic Control and Computer Engineering MI-STA, Tripoli, Libya, pp. 230–233 (2021). https://doi.org/10.1109/MI-STA52233.2021.9464507
7. Capponi, L.: Direct model predictive control using model based design. In: 2021 IEEE International Conference on Predictive Control of Electrical Drives and Power Electronics (PRECEDE), Jinan, China, pp. 680–685 (2021)
8. Ben Achour, H., et al.: PI controller and quadratic feedback of synchronous machine. In: Farhaoui, Y., Rocha, A., Brahmia, Z., Bhushab, B. (eds.) Artificial Intelligence and Smart Environment. ICAISE 2022. Lecture Notes in Networks and Systems, vol. 635, pp. 668–673. Springer, Cham (2023). https://doi.org/10.1007/978-3-031-26254-8_97
9. Achour, H.B., et al.: Permanent magnet synchronous motor PMSM control by combining vector and PI controller. WSEAS Trans. Syst. Control **17**, 244–249 (2022)
10. Chaou, Y., et al.: Nonlinear control of the permanent magnet synchronous motor PMSM using backstepping method. WSEAS Trans. Syst. Control **17**, 56–61 (2022)
11. Youssef, C., et al.: Electric vehicle backstepping controller using synchronous machine. In: The International Conference on Artificial Intelligence and Smart Environment, pp. 367–373 (2022)
12. Ziani, S., Ghmary, M.E., Youssef, C., Abdelkarim, D., Hafid, B.A., Youssef, E.H.: Developed permanent magnet synchronous motor control using numerical algorithm and backstepping. J. Eng. Sci. Technol. Rev. **16** (2023)
13. Ziani, S., El Ghmary, M., Youssef, A.Z.: Permanent magnet synchronous motor control performed using PI-backstepping with a model of harmonics reduction. Int. J. Power Electron. Drive Syst. **14**(1), 199–208 (2023)

14. Hafid, B.A., et al.: A quadratic observer for sensorless drive system controller. In: Farhaoui, Y., Rocha, A., Brahmia, Z., Bhushab, B. (eds.) Artificial Intelligence and Smart Environment. ICAISE 2022. Lecture Notes in Networks and Systems, vol. 635, pp. 109–113. Springer, Cham (2023). https://doi.org/10.1007/978-3-031-26254-8_16
15. Laabab, I., et al.: A review of the application of artificial intelligence for weather prediction in solar energy: using artificial neural networks. In: Farhaoui, Y., Rocha, A., Brahmia, Z., Bhushab, B. (eds.) Artificial Intelligence and Smart Environment. ICAISE 2022. Lecture Notes in Networks and Systems, vol. 635, pp. 114–119. Springer, Cham (2023). https://doi.org/10.1007/978-3-031-26254-8_17
16. Laabab, I., et al. A literature review of solar cell overheating control. In: Farhaoui, Y., Rocha, A., Brahmia, Z., Bhushab, B. (eds.) Artificial Intelligence and Smart Environment. ICAISE 2022. Lecture Notes in Networks and Systems, vol. 635, pp. 48–53 (2022). Springer, Cham. https://doi.org/10.1007/978-3-031-26254-8_7
17. Youssef, C., et al.: Backstepping control of the permanent magnet synchronous generator (PMSG) used in a wind power system. In: Farhaoui, Y., Rocha, A., Brahmia, Z., Bhushab, B. (eds.) Artificial Intelligence and Smart Environment. ICAISE 2022. Lecture Notes in Networks and Systems, vol. 635, pp. 276–281. Springer, Cham (2023). https://doi.org/10.1007/978-3-031-26254-8_38
18. Echoukairi, A., et al.: Improved methods for automatic facial expression recognition. Int. J. Interact. Mob. Technol. (iJIM) **17**(06), 33–44 (2023). https://doi.org/10.3991/ijim.v17i06.37031
19. Laabab, I., et al.: A solar panels overheating protection: a review. Indonesian J. Electr. Eng. Comput. Sci. **29**(1), 49–55 (2023)
20. Ouhadou, M., et al.: Experimental investigation on thermal performances of SMD LEDs light bar: junction-to-case thermal resistance and junction temperature estimation, Optik **182** (2019)
21. Ouhadou, M., et al.: Experimental modeling of the thermal resistance of the heat sink dedicated to SMD LEDs passive cooling. In: Proceedings of the 3rd International Conference on Smart City Applications (2018)
22. Said, Z., El Hassouani, Y.: A new approach for extracting and characterizing fetal electrocardiogram. Traitement du Signal **37**(3), 379–386 (2020). https://doi.org/10.18280/ts.370304
23. Ziani, S., El Hassouani, Y., Farhaoui, Y.: An NMF based method for detecting RR interval. In: Farhaoui, Y., Moussaid, L. (eds.) ICBDSDE 2018. SBD, vol. 53, pp. 342–346. Springer, Cham (2019). https://doi.org/10.1007/978-3-030-12048-1_35
24. Ziani, S., Jbari, A., Belarbi, L.: Fetal electrocardiogram characterization by using only the continuous wavelet transform CWT. In: 2017 International Conference on Electrical and Information Technologies (ICEIT), Rabat, Morocco, pp. 1–6 (2017). https://doi.org/10.1109/EITech.2017.8255310
25. Ziani, S., Jbari, A., Bellarbi, L.: QRS complex characterization based on non-negative matrix factorization NMF. In: 2018 4th International Conference on Optimization and Applications (ICOA), Mohammedia, Morocco, pp. 1–5 (2018). https://doi.org/10.1109/ICOA.2018.8370548
26. Ziani, S., Jbari, A., Bellarbi, L., Farhaoui, Y.: Blind maternal-fetal ECG separation based on the time-scale image TSI and SVD–ICA methods. Procedia Comput. Sci. **134**, 322–327 (2018). ISSN 1877–0509,
27. Ziani, S., El Hassouani, Y.: Fetal-maternal electrocardiograms mixtures characterization based on time analysis. In: 2019 5th International Conference on Optimization and Applications (ICOA), Kenitra, Morocco, pp. 1–5 (2019). https://doi.org/10.1109/ICOA.2019.8727619
28. Ziani, S.: Contribution to single-channel fetal electrocardiogram identification. Traitement du Signal **39**(6), 2055–2060 (2022)

29. Ziani, S.: Fetal electrocardiogram identification using statistical analysis. In: Farhaoui, Y., Rocha, A., Brahmia, Z., Bhushab, B. (eds.) Artificial Intelligence and Smart Environment. ICAISE 2022. Lecture Notes in Networks and Systems, vol. 635, pp. 447–453. Springer, Cham (2023). https://doi.org/10.1007/978-3-031-26254-8_64
30. Ziani, S., Farhaoui, Y., Moutaib, M.: Extraction of fetal electrocardiogram by combining deep learning and SVD-ICA-NMF methods. Big Data Mining Anal. **6**(3), 301–310 (2023). https://doi.org/10.26599/BDMA.2022.9020035
31. Ziani, S., El Hassouani, Y.: Fetal electrocardiogram analysis based on LMS adaptive filtering and complex continuous wavelet 1-D. In: Farhaoui, Y. (ed.) BDNT 2019. LNNS, vol. 81, pp. 360–366. Springer, Cham (2020). https://doi.org/10.1007/978-3-030-23672-4_26

PV Panel Emulator Based on Arduino and LabVIEW

Catalin Ichim-Burlacu(✉), Cezara-Liliana Rat, Corina Cuntan, Raluca Rob, and Ioan Baciu

Faculty of Engineering, Politehnica University of Timisoara, Hunedoara, Romania
catalin.ichim-burlacu@student.upt.ro, {cezara.rat,corina.cuntan, raluca.rob,ioan.baciu}@fih.upt.ro

Abstract. This paper presents a low-cost PV emulator for testing PV panels in laboratory conditions. The emulator uses a DC power supply and a DC-DC converter. The converter is controlled through a closed-loop control system. The controller is an Arduino UNO board that regulates the output voltage of the converter. The reference value is given by a PV panel model simulated using a LabVIEW program running on a computer. The Arduino controller communicates with this computer via a USB connection, sending measurement values from the sensors (irradiance, temperature, and voltage) and receiving the reference value for the converter. It also displays the reference value on an electronic display. The software component of the emulator is based on the Model-View-Controller (MVC) paradigm. The LabVIEW program houses the PV panel model but also works as an HMI (the view) for the user, displaying the measured and computed values, plotting the current operating point on the PV characteristic curves, etc. The C program running on the Arduino board is the controller.

Keywords: Arduino Uno · LabVIEW · PV panel · Solar panel

1 Introduction

In recent years, the increasing prices of fossil fuels as well as concerns about their effect on the environment have led to an increase in the demand for alternative energy sources. This sparked interest in the development of these sources. Photovoltaic (PV) panels gained interest due to their high usage in daily life. These sources generate electricity from solar energy: PV cells absorb solar energy and convert it into DC electrical energy. Several PV cells mounted on a plate and connected in series and parallel form a PV panel [1]. A PV panel's power output will vary based on a variety of factors including the amount of irradiance received, the temperature of the panel, and the material used to make the PV panel [2].

As opposed to conventional energy sources, PV panels do not cause environmental pollution during electricity generation. However, high costs and the need for infrastructure support have restrained their usage. Solutions for these problems can be found through further research [1].

M. Ben Ahmed et al. (Eds.): SCA 2023, LNNS 906, pp. 519–529, 2024.
https://doi.org/10.1007/978-3-031-53824-7_47

However, the consistent, repeatable conditions necessary for testing and optimizing PV equipment are difficult, if not impossible, to replicate outdoors. Therefore, having an environment with control over the inherent variability of outside temperature and weather conditions that allows repeatable conditions to test PV equipment is required [2].

The purpose of this project is to develop an indoor testing platform to test PV panels under STC (Standard Test Conditions) and other conditions to validate and correct simulation results [3]. In previous research, excellent but relatively complex solutions have been proposed. The best devices for testing PV panels in laboratory conditions are considered to be programmable DC power sources. These can produce the output electrical properties of different PV panels, independent of external conditions [1, 4]. However, they are very expensive and are, therefore, not available in most research laboratories [4–7]. Another option is to use a real PV panel with an artificial light source. The source produces light on the PV panel surface and, by controlling the light intensity, it is possible to emulate irradiance changes. However, the spotlights increase the PV panel temperature dramatically [5, 8]. The third solution is the use of a PV emulator composed of a DC power supply and a controlled DC-DC converter. The DC-DC converter can be controlled by a DSP, a microcontroller, a neural networks-based controller, etc. [5, 9–15]. The emulator provides the same current and voltage characteristics as various types of PV panels under different environmental conditions [2]. However, commercial PV emulators come at a high cost leading to the necessity of finding new solutions for PV emulation [4].

This work presents a low-cost emulator using a DC-DC converter controlled through an Arduino board. Experimental tests that confirm the effectiveness of the developed PV emulator are presented in this study. Papers [16–18] are related to this work.

This paper is organized as follows. Section 1 is the introduction. Section 2 exposes the design of the PV emulator. Section 3 presents the experimental results that verify the performance of the emulator. Finally, the conclusions are given in Sect. 4.

2 Emulator

Emulators are research equipment that include hardware as well as software components. This proposed PV emulator is presented in Fig. 1. It uses a Buck converter controlled by an Arduino UNO board for obtaining the desired PV cell characteristics at different irradiance and temperature conditions. The PV panel characteristics are incorporated into the emulator by using the mathematical model of the panel and its parameters. The emulator dynamics are modeled using LabVIEW and C code.

2.1 Hardware Component

The hardware component of the emulator, shown in Fig. 2, consists of: a DC-DC converter; an Arduino UNO controller; and a computer.

The DC-DC Converter. The DC-DC converter along with a 20V DC power supply form a switching power supply controlled via a PWM signal. This is used to replicate the behavior of the PV panel. This Buck converter is directly connected to the load (a small fan). The electric diagram of the converter can be seen in Fig. 3.

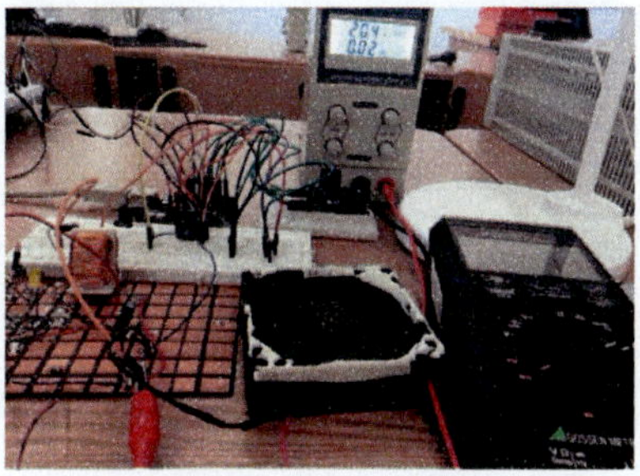

Fig. 1. PV emulator

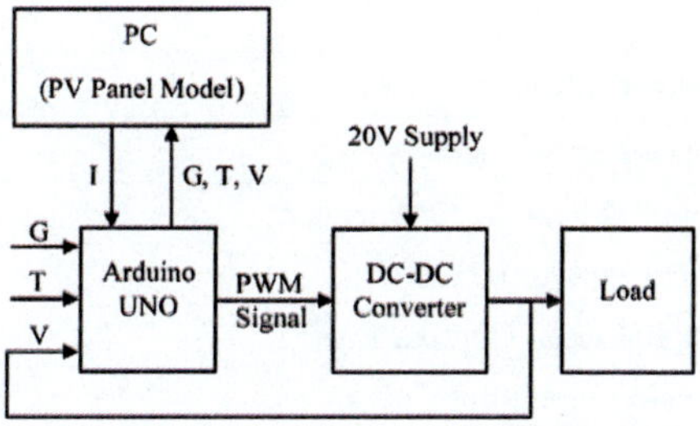

Fig. 2. PV emulator structure

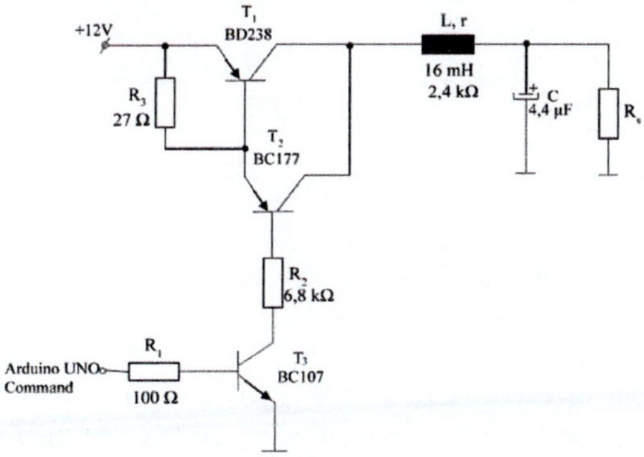

Fig. 3. Buck converter

The Arduino UNO Controller. Arduino is an open-source electronics platform which provides simple inexpensive cross-platform embedded development boards. The board used in this application is a low-cost small-sized embedded board, namely Arduino UNO, selected based on the fact that its performance is comparable to that of an average computer [16]. Its technical specifications are presented in Table 1 [19]. This Arduino board controls the converter. It communicates via a USB connection with the computer which contains the PV panel model, sending measurement values from the sensors [20].

The measurement values are:

- Irradiance (G), measured through a Digilent PmodALS sensor. This sensor communicates with the Arduino board using a 3-wire SPI communication interface.
- Temperature (T), measured through a thermistor.
- The output voltage of the DC-DC converter (V). This is done to form a closed-loop control structure.

The circuit responsible for measuring the voltage is a voltage divider. It is necessary for bringing down the output voltage of the converter (maximum 20V) to a value that can be measured through the analog pin of the Arduino microcontroller (0-5V). This reduces the voltage 5 fold. The 10-bit ADC of the Arduino board will convert the analog

Table 1. Technical specifications of Arduino Uno board

Microcontroller	ATMega328
Operating Voltage	5 V
Input Voltage (recommended)	7 – 12 V
Input Voltage (limits)	6 – 20 V
Digital Input / Output Pins	14 (6 provide PWM output)
Analog Input Pins	6
DC Current per I/O Pin	40 mA
DC Current for 3.3V Pin	50 mA
Flash Memory	32 KB
SRAM	2 KB
EEPROM	1 KB
Clock Speed	16 MHz

value to a digital one between 0 and 1023. This value will then be converted back into volts.through programming [16].

The output of the controller is the duty cycle of the converter determined using the current reference value received from the PV panel model implemented on the PV. The duty cycle is used to provide a PWM switching signal to drive the DC/DC converter.

The Computer. The PC runs a LabVIEW program which contains the PV panel model as well as the HMI. The program uses the measured values from the Arduino board to compute the reference value for the converter.

2.2 Software Component

The emulator software is based on the Model–View–Controller (MVC) paradigm. MVC is a commonly used software design pattern that divides the related program logic into three interconnected elements: the model, the view (or views), and the controller. This separation of the logical components increases flexibility, scalability and extensibility. This paradigm is widely used in web programming [21].

Here, the model and the view are contained within the LabVIEW program running on the PC and the controller is the C program running on the Arduino board.

Model - PV Panel Model. In this study, the used PV panel model was an analytical model, to be more precise an extended version of the single diode model, namely the five-parameter model (I_0, A_i, R_s, R_p, I_{ph}), where I_0 is the reverse saturation current, A_i is the ideality factor of a cell, R_s is the series resistance, R_p is the parallel resistance, I_{ph} is the photocurrent. This model was chosen due to its low complexity and relatively good accuracy. The five-parameter model of a PV panel with N_s cells in series and N_p

cells in parallel is [17]:

$$I = N_P I_{ph} - N_P I_0 \left(e^{\frac{q\left(V+\frac{N_S}{N_P} IR_s\right)}{N_s A_i kT}} - 1 \right) - \frac{V + \frac{N_s}{N_P} IR_s}{\frac{N_s}{N_P} R_p} \tag{2}$$

$$I_{ph} = \frac{G}{G_{ref}} \left[I_{ph_{ref}} + \mu_{sc} \left(T - T_{c_{ref}} \right) \right] \tag{3}$$

$$I_0 = I_{sc_{ref}} \left(\frac{T_c}{T_{c_{ref}}} \right)^3 e^{-\frac{q\varepsilon_G \left(\frac{1}{T_{c_{ref}}} - \frac{1}{T_c} \right) + V_{oc}}{A_i k}} \tag{4}$$

The model inputs are: V – input voltage, G – irradiance, T – cell temperature and the model output is I – output current. The parameters of the PV panel are rendered in Table 2 [17].

Table 2. PV panel parameters and constants

Symbol	Meaning	Value
N_P	Number of cells connected in parallel	1
N_S	Number of cells connected in series	36
V_{OC}	Open-circuit voltage	21.8V
$I_{ph_{ref}}$	Photocurrent to the SRC (298.15 K)	3.11 A
G_{ref}	Irradiance at SRC	1000 W/m^2
μ_{SC}	Relative temperature coefficient of the I_{SC}	0.0013 A/K
R_S	Series resistance	0.18 Ω
R_P	Parallel resistance	310.025Ω
$T_{C_{ref}}$	Cell temperature at SRC	298.15 K
k	Boltzmann constant	$1.381 \cdot 10^{-23}$ J/K
q	Electric charge	$1.602 \cdot 10^{-19}$ °C
A_i	Ideality factor of a cell	1.3
ε_G	Forbidden bandwidth of the cell material	1.12eV

View - LabVIEW Application. The GUI (Graphical User Interface) or the view developed in LabVIEW is presented in Fig. 3. it is used to render the measured data (Solar radiation - G, Voltage - V, Temperature - T).

The data are read from the Arduino board through a serial RS232 connection using a virtual COM port. The LabVIEW application monitors the COM port and reads any data sent through it. The parameters of the serial transmission (baud rate, parity, etc.) can be modified on the front panel of the LabVIEW program in correspondence with

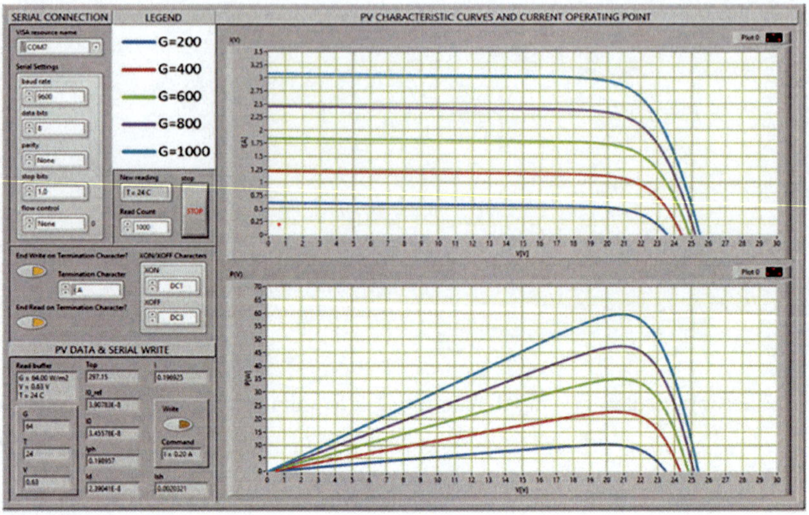

a) Front Panel

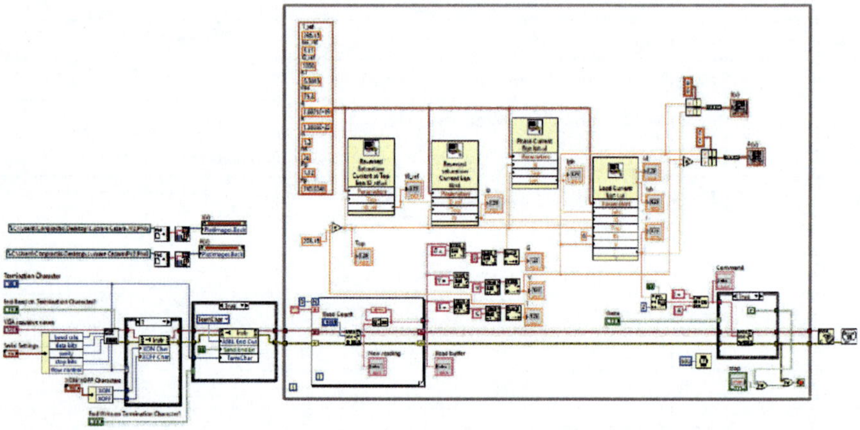

b) Block Diagram

Fig. 4. LabVIEW Application

the transmission parameters of the Arduino board. A single transmission can contain the value of the irradiance, temperature, and voltage. This is read by the LabVIEW program as a string value and, using string functions, the value itself (G, T, and V) is extracted. To determine the output current, the emulator can use either the measurement data or software controls [18]. Either way, the values for irradiance, temperature, and voltage are sent to the PV panel model implemented in the block diagram of the LabVIEW program to compute the current which is sent as a reference value to the Arduino controller in order to regulate the output of the emulator.

The current is also used to calculate the power extracted from the panel:

$$P = I \cdot V \tag{5}$$

The operating point of the panel is plotted on the I-V and P-V curves of the PV panel shown on the front panel of the LabVIEW application.

Controller – C program. The C program written on the Arduino board controls the system. Its functionality is visible in Fig. 4.

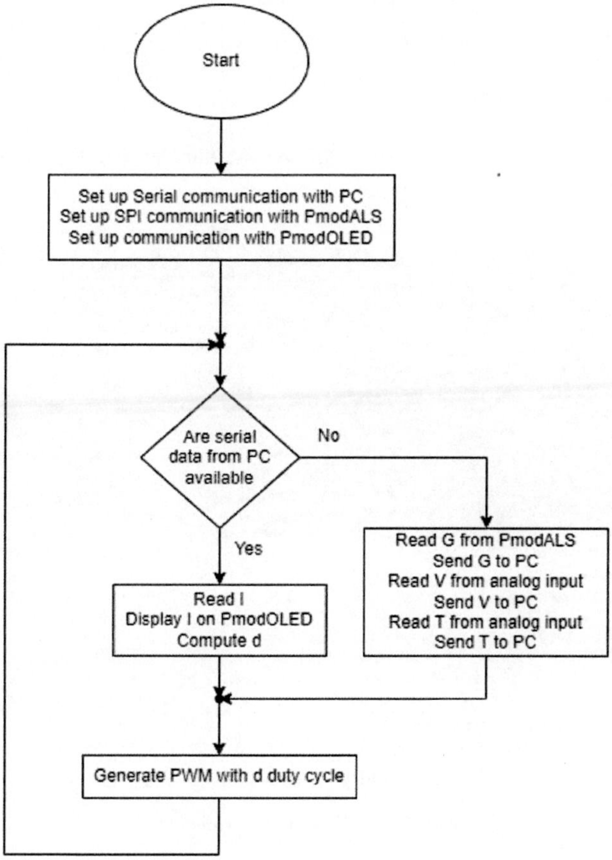

Fig. 5. Flow diagram of the C program

The C program communicates with the sensors: via SPI for the PmodALS light sensor (G) and analog reading for the temperature (T) and voltage (V). This handshake between the Arduino board and the PmodALS sensor is done through the C program.

These measurement values are then sent to the LabVIEW program running on the PC which gives the reference value (I). Upon reception, the reference value is displayed using a PmodOLED digital display and used to compute the duty cycle (d) of the PWM signal used to actuate the converter.

The Pulse-width modulation (PWM) signal is used to control the average power delivered by the converter by switching the supply between 0 and 100% at a rate faster than it takes the load to change significantly. While the switch is closed (on-state), the voltage source is connected, supplying the converter with power. While the switch is opened again (off-state), the voltage source is disconnected and the load is supplied with power stored by the converted. The longer the converter is in on-state, the higher the total power supplied to the load and vice versa. The PWM switching frequency is the standard for DC-DC converters: 20 kHz. The settings for the PWM signal are done through the C program.

3 Experimental Results

The setup was tested using a lamp to replicate the varying irradiance conditions. The results are rendered in Figs. 5, 6, and 7.

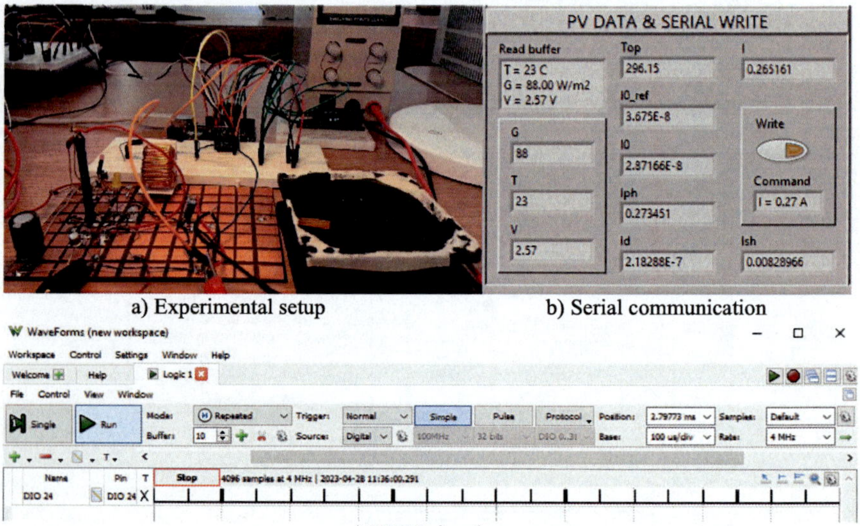

a) Experimental setup b) Serial communication

c) PWM signal

Fig. 6. Low Irradiance (level 1)

Figure 5 shows the behavior of the emulator at low lighting (level 1 – 88 W/m^2). The reference current calculated using the panel model is I = 0.27A. The duty cycle of the actuation signal obtained based on this current is very small and, therefore, does not cause the fan to turn.

Figure 6 shows the behavior of the emulator at medium lighting (level 2 – 360 W/m^2). The reference current calculated using the panel model is I = 1.11A. The duty cycle of the actuation signal obtained based on this current is small but it does cause the fan to turn.

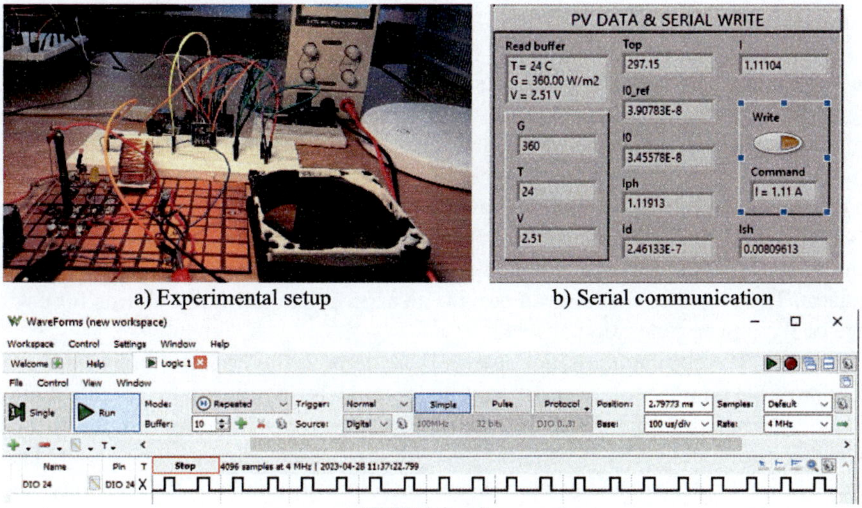

a) Experimental setup b) Serial communication

c) PWM signal

Fig. 7. Medium lighting (level 2

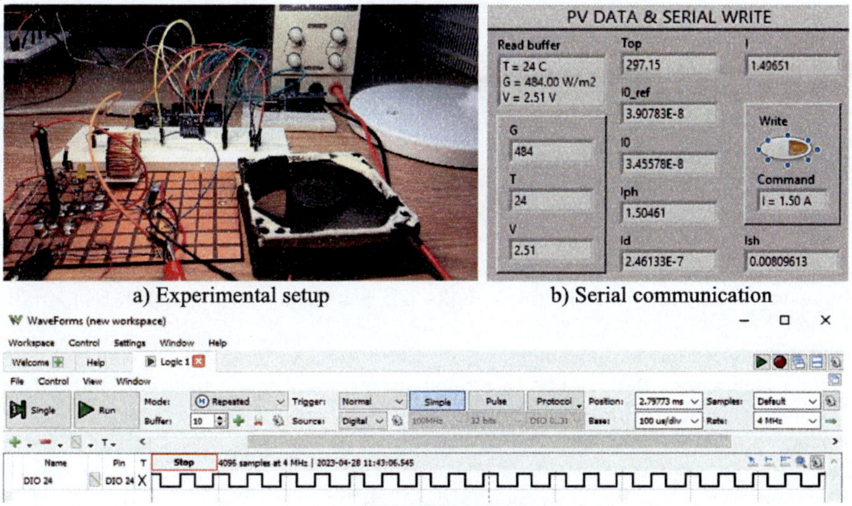

a) Experimental setup b) Serial communication

c) PWM signal

Fig. 8. High lighting (level 3)

Figure 7 shows the behavior of the emulator at high lighting (level 3 – 484 W/m^2). The reference current calculated using the panel model is I = 1.47A. The duty cycle of the actuation signal obtained based on this current causes the fan to turn at a high speed (Fig. 8).

4 Conclusions

The experimental results show that our PV emulator can provide a simple, efficient, and low-cost way for researchers to test PV panels in a laboratory environment. This method decreases dependence on weather conditions. Various types of PV panels can be tested easily by changing the parameters of the software component. The emulator can use data from the sensors or software controls. This PV emulator is simple enough to be easily built in a conventional laboratory environment. The results also show that Arduino UNO is an effective and economical platform for the implementation of the emulator. The proposed method can provide an appropriate low-cost solution for a wide range of PV power generation systems.

However, PV emulators do not always recreate both the transient and steady-state behavior of the panel. Moreover, PV simulation models do not usually contain thermal behavior, degradation processes, or shading behavior, aspects required in the study of modern control techniques. This is the case for the presented emulator as well. These aspects, however, depend on the model used. Therefore the emulator can relatively easily be improved by updating the PV panel model of the LabVIEW program and including the desired aspects.

References

1. Gürkan, S., Karapınar, M., Sorgunlu, H., Öztürk, O., Doğan, S.: Development of a photovoltaic panel emulator and LabVIEW-based application platform. Comput. Appl. Eng. Educ. **28**(5), 1291–1310 (2020)
2. Dolan, D. S., Durago, J.: Development of a photovoltaic panel emulator using Labview. In: 2011 37th IEEE Photovoltaic Specialists Conference, pp. 001795–001800. IEEE (2011)
3. Akiki, T., Eid, L., Abboud, S., Nehme, B.: Developing a test platform for PV panels. In: 2017 29th International Conference on Microelectronics (ICM), pp. 1–5, IEEE (2017)
4. Chalh, A., Motahhir, S., El Hammoumi, A., El Ghzizal, A., Derouich, A.: A low-cost PV Emulator for testing MPPT algorithm. IOP Conf. Ser. Earth Environ. Sci. **161**(1), 012018 (2018)
5. Chalh, A., Motahhir, S., El Hammoumi, A., El Ghzizal, A., Derouich, A.: Study of a low-cost PV emulator for testing MPPT algorithm under fast irradiation and temperature change. Technol. Econ. Smart Grids Sustain. Energy **3**(1), 11 (2018)
6. Abdourraziq, M.A., Abdourraziq, S., Maaroufi, M., Tlemcani, M., Darab, C.: Experimental results of photovoltaic emulator systems. In: 2017 International Conference on Electromechanical and Power Systems (SIELMEN), pp. 261–265. IEEE (2017)
7. Patrascu, C., Rat, C. L., Hulea, D., Vitan, D., Muntean, N.: Mixed PV-wind small power microgrid. In: 2017 International Conference on Optimization of Electrical and Electronic Equipment (OPTIM) & 2017 Intl Aegean Conference on Electrical Machines and Power Electronics (ACEMP), pp. 699–704. IEEE (2017)
8. Nehme, B., Msirdi, N.K., Namaane, A., Akiki, T.: Analysis and characterization of faults in PV panels. Energy Procedia **111**, 1020–1029 (2017)
9. Fărcaş, C., Ciocan, I., Tulbure, A.: Solar emulator for a photovoltaic module. In: 2018 IEEE 24th International Symposium for Design and Technology in Electronic Packaging (SIITME), pp. 314–318. IEEE (2018)

10. Balakishan, C., Sandeep, B.: Development of a microcontroller based PV emulator with current controlled DC/DC buck converter. Inter. J. Renew. Energy Res. **4**(4), 1049–1055 (2014)
11. Nabil, O., Bachir, B., Allag, A.: Implementation of a new MPPT technique for PV systems using a boost converter driven by Arduino MEGA. In: 2018 International Conference on Communications and Electrical Engineering (ICCEE), pp. 1–5. IEEE (2018)
12. Harrison, A., Alombah, N. H.: A new high-performance photovoltaic emulator suitable for simulating and validating maximum power point tracking controllers. Inter. J. Photoenergy (2023)
13. Sharma, S., Joshi, D.: Modeling and control of PV emulator with different controllers and transient load conditions. In: NIScPR-CSIR, pp. 73–79 (2023)
14. Alaoui, M., Maker, H., Mouhsen, A., Hicham, H.: High Power PV array emulator based on state feedback controller under uniform and non-uniform insolation. Trans. Electr. Electron. Mater. **24**, 54–64 (2023)
15. Saraswathi, K.T., Arumugam, P.V., Swaminathan, G., Periasamy, S.: An artificial neural network-based comprehensive solar photovoltaic emulator. Inter. J. Photoenergy (2022)
16. Ichim-Burlacu, C., Rat, C.-L., Panoiu, C., Cuntan, C.: PV Panel monitoring using arduino and LabVIEW. In: 5th Global Power, Energy and Communication Conference (GPECOM), Nevsehir, Turkiye, pp. 270–275. IEEE (2023)
17. Rat, C.L., Ichim-Burlacu, C.I., Bistrian, D.A.: Optimizing the operation of photovoltaic panels using mathematical regression models. In: International Conference on Applied Sciences (ICAS2022), Banja Luka (2022)
18. Rob, R., Rat, C.: Mathematical model for studying the variation of the electrical parameters in functioning of nonlinear loads. In: AIP Conference Proceedings, vol. 1558(1), pp. 1345–1348. American Institute of Physics (2013)
19. El Hammoumi, A., Motahhir, S., Chalh, A., El Ghzizal, A., Derouich, A.: Low-cost virtual instrumentation of PV panel characteristics using Excel and Arduino in comparison with traditional instrumentation. Renewables: Wind, Water Solar **5**(1), 1–16 (2018)
20. Kanth, D., Hariharan, S., Hariharasudhan, S., Saravanan, S.: Design of solar PV emulator using raspberry Pi controller. In 2022 International Conference on Inventive Computation Technologies (ICICT), pp. 581–584. IEEE (July 2022)
21. García, R.F.: MVC: Model–View–Controller. In: iOS Architecture Patterns: MVC, MVP, MVVM, VIPER, and VIP in Swift, pp. 45–106. Apress, Berkeley (2023)

Analysis of Power Consumption After Switching to 5G

Hamza Ben Makhlouf(✉) and Tomader Mazri

National Schools of Applied Sciences, Ibn Tofail University, Kenitra, Morocco
hamza.benmakhlouf@uit.ac.ma

Abstract. A promising new technology, 5G promises a notable improvement in several points, one of these points is the energy part. This study aims to see if this promise is kept in the access network part, and this through the analysis of several radio sites before and after the transition to 5G.

Keywords: F2G · 3G · 4G · 5G · NSA · SA · BBU · RRU · MIMO · Antennas · Radio · Consumption · Energy · Deployment · Frequency

1 Introduction

The 2G, 3G and 4G generations of the mobile network were part of technological continuity, each new generation was an evolution of the previous one. Now 4G networks are reaching a certain stagnation, researchers are moving towards a new generation of wireless networks, 5G.

As shown in Fig. 1; With the introduction of second generation (2G) mobile networks based on Time Division Multiple Access (TDMA) in the 1990s [1], the era of cellular and digital telecommunications officially began. The development of third-generation (3G) networks in the 2000s was based on the wideband code-division multiple access (WCDMA) principle [2]. Despite dominating the market due to an increase in data transmission speed, the third generation never fully supplanted the second generation. Beginning in the early 2010s, fourth generation (4G) networks that use orthogonal frequency-division multiple access (OFDMA) for the downlink and single carrier frequency-division multiple access (SC-FDMA) for the uplink were launched [3].

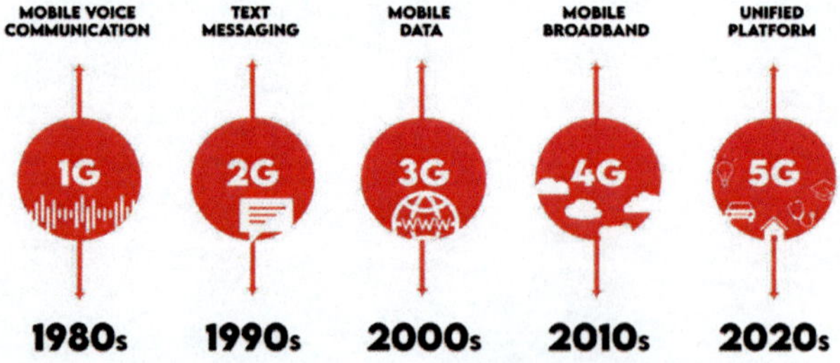

Fig. 1. Evolution of mobile networks

M. Ben Ahmed et al. (Eds.): SCA 2023, LNNS 906, pp. 530–545, 2024.
https://doi.org/10.1007/978-3-031-53824-7_48

Unlike previous generations, 5G will benefit from high frequencies with wide bandwidths and immense connection density [4].

2 Related Works

With 5G, a real technological breakthrough was born, this new mobile network standard. 5G is no longer just high-performance mobile access, but a real integration network comprising a set of technologies making it possible to offer suitable vertical services [5], such as telemedicine, the smart city, the connected car, and access to Very High-Speed Internet. For virtual reality applications, etc. This development is encouraged by the international community via the 3GPP consortium.

Indeed, the requirements of these verticals, such as the qualification of upload and download speeds, acceptable latency thresholds, or the level of system availability are fully taken into account from the initial phase of 5G specifications.

5G will offer increased throughput and very low latency [6], close to that of fiber optics, making it possible to offer high-performance connectivity where fiber is absent, with FWA (Fixed Wireless Access) [7]. On the operator side, network virtualization will make it possible to address all markets with a single infrastructure [8]. The era of dedicated equipment is over, the same hardware architecture will be used for different types of networks and services, thanks to software. But 4G is not living in its last days, it can still provide the basic services of our society, but does not prevent us from beginning to glimpse the limits of this, limits that 5G has as its specifications. Erase:

- Security: the 4th generation was a good student in this area but this is no longer enough today, the new standard of cellular networks brings new coding and modulation technologies that improve security [9].
- Speed: it is true that the speed is sufficient to watch movies in high definition, but does not allow you to take advantage of technology such as VR (virtual reality) or AR (augmented reality), which requires a lot of bandwidth (Table 1), a speed that 5g allows to have [10].
- Connection density: the current network was a hindrance for the IoT because it does not allow a large number of connections per surface, and this need is inevitable with the Internet of Everything. 5G meets this need by providing 1000 times more capacity than 4G networks (Table 1). A large number of connected devices will be served by 5G networks. The vision of an intensely connected network for this new generation [11], it is expected to reach more than 100 billion connected devices by 2030 [12].
- Latency: several services such as online video games have seen a marked improvement with 4G, but if this improvement is sufficient for games, it is not for new services such as autonomous cars or medicine, 5G comes with 1 ms latency (Table 1) [13].
- Mobility: in 2020 we travel using high-speed transport (car, high speed train, etc.), the latest 4G standards only made it possible to have a network at low and medium speed, whereas the 5G network can work even at 350 km/h [14].
- Energy: as 4G is increasingly in demand with an increase in the number of antennas, the energy cost has become enormous, 5G fixes this problem with MIMO (multiple input multiple output) [15] antennas which directs the waves to consume less as

shown in Fig. 2, also cloud servers that centralize and reduce the number of network equipment, all this consumes much less [16].

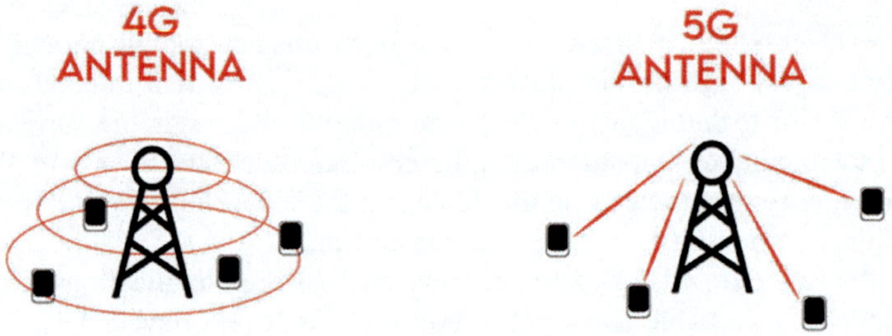

Fig. 2. Difference between normal antenna and MIMO antenna

With the evolution from 4G to 5G, we quickly realized that the problem was much broader, with multiple cross-cutting issues ranging from technological governance to health security, including the environmental impacts of digital and cybersecurity. Faster, more massive connections, and more functionalities, are the main promises that 5G technology intends to hold in a world transformed by the Internet of Things (IoT) [17], particularly in industrial matters (industry 4.0) [18], in the field of health (telecommunication operations) and our daily lives. Let's not forget social networks such as Facebook, Twitter, etc. [19], these have a direct impact on the growing demand for data [20] (Table 2 to Table 15).

Table 1. Comparison between 4G and 5G keys

KPI	4G	5G
Latency	10 ms	< 1 ms
Data traffic	7,2 Eb/m	50 Eb/m
Peak data rates	1 Gb/s	10 Gb/s
Spectrum	3 Ghz	30 Ghz
Capacity	1 K/Km^2	1 M/km^2

These lines show the interest in the latest generation of cellular networks, its importance, and its advantages compared to older generations. They also present the 5G keys by specifying the most important KPIs (Key performance indicators).

The energy promise is very attractive, this same promise that this article aims to study, to see if the current 5G respects it or if the practice deviates from the specifications.

All this development of the network and its capacity came with an increase in energy consumption, an economic as well as an environmental problem.

5G is obliged to remedy this, this article will focus on studying the consumption of some sites after the integration of 5G and see its impact.

3 Material and Method

3.1 Radio Station: Antenna Line

A mobile radio site is a set of transmission equipment used to route information between a UE (user equipment) and the core network. The radio site is part of this channel, it must respect some prerequisites: High Point, a technical area and easy access.

The site's radio chain begins with the BBUs followed by the RRUs, then the latter is connected to the diplexer and the TMA, and the TMA is linked to the radio antennas, the sites can also house FH antennas.

This part will focus on explaining the role of each piece of equipment in order to understand the analysis that follows.

BBU. (Base band Unit) is a unit composed of several slots, it manages the physical layer, performs pre-modulation, allocates resources, and takes care of the coding (Fig. 3).

The BBU is composed of an 8-slot chassis, these slots can accommodate technology cards (2G/3G/4G/5G) and modulation-demodulation cards for each technology, which makes the BBU very modular and its consumption variable electric.

RRU. (Remote Radio Unit) is a radio module that connects directly to the BBU, its role is to convert the baseband signal to a radio frequency signal (Fig. 4).

The RRUs have a specific frequency band, it is for this reason that there are several radio modules used depending on the frequencies wanted on the site.

The use of a large number of RRUs has a strong impact on energy consumption.

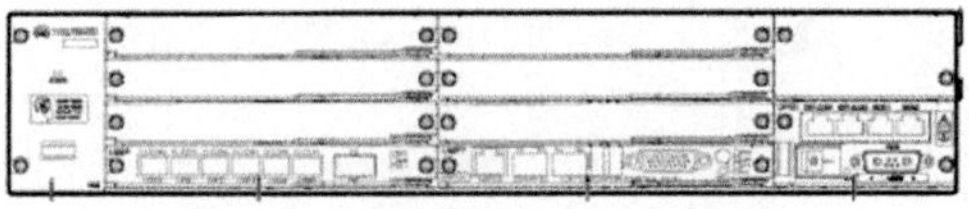

Fig. 3. Representation of BBU[21]

Fig. 4. Representation of RRU[22]

Combiner. Dual Band Combiner is a passive device with four inputs and two outputs, it outputs a signal that combines the two input frequency bands.

This device is used to counter the problem of the ports on the antennas which are generally multi-band.

You can also find trippers that combine frequency bands (Fig. 5).

Fig. 5. Representation of the Dual band combiner [23]

TMA. (Tower Mounted Amplifier) is a device aimed to compensate for the weakening of the signal linked to the length of the coaxial cables, however, the use of TAMs is not essential if the distance between the RRU and the antenna is small (Fig. 6).

Fig. 6. Representation of TMA [24]

Coupler. This is a device that divides a signal channel into two channels according to the frequency range that manages, the coupler can also perform multiplexing and therefore does without diplexers if it is used.

In the antenna line, it is efficient to transit signal to two antennas with the same RRU.

Radio Antenna. An antenna is a transceiver whose mission is to convert electrical signals into electromagnetic waves and vice versa.

Passive antenna. All antennas used for 2G/3G/4G are passive antennas.

Active antenna. For the integration of 5G, and depending on the desired configuration, a 5G antenna can be added to the already existing antenna, the latter is an active antenna because it integrates its radio module.

In another configuration, the existing antenna is removed and replaced by an AIO antenna (all in one) which integrates all the technologies (2G/3G/4G/5G), this also integrates the 3500 MHz radio module (Fig. 7).

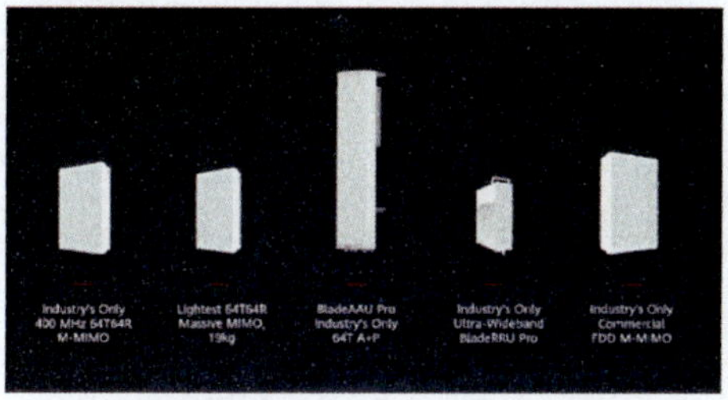

Fig. 7. 5G Huawei equipment [25]

Microwave Link Transmission Antennas. This antenna allows the transmission of information through radio waves from a site A to a site B (Fig. 8).

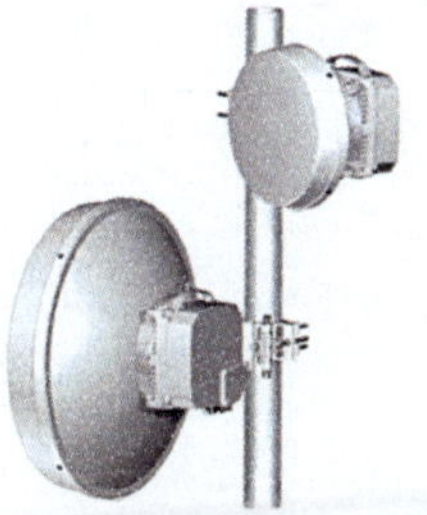

Fig. 8. Microwave link transmission antenna [26]

The frequencies are between 1 GHz and 86 GHz, this technology is disappearing, giving way to fiber optics which is more reliable and less energy consuming.

Sector. The sector of a radio site consists of all the equipment necessary to cover a limited area of the cell by an antenna directed along its azimuth.

Figure 9 represents the synoptic of a sector with all its equipment and explains the recording of this last to have a functional system.

3.2 5G Deployment (Methodology)

In our case, the operator chose to first integrate the 5G NSA without touching the core networks, for this, the engineering team set up several configurations depending on the existing sites.

First, the operator integrates a 5G card into the BBU, then the 2100 MHz or 3500 MHz radio modules, and finally a 5G antenna is added otherwise the existing antenna is replaced by one that manages all frequencies and all technologies (2G, 3G, 4G, and 5G).

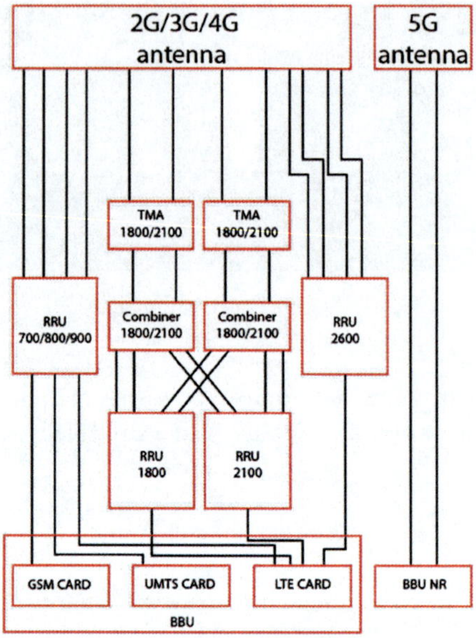

Fig. 9. Synoptic of radio sector

3.3 Method

During this work, the analysis will be performed on a sample of 10 sites belonging to the city of Toulouse in France.

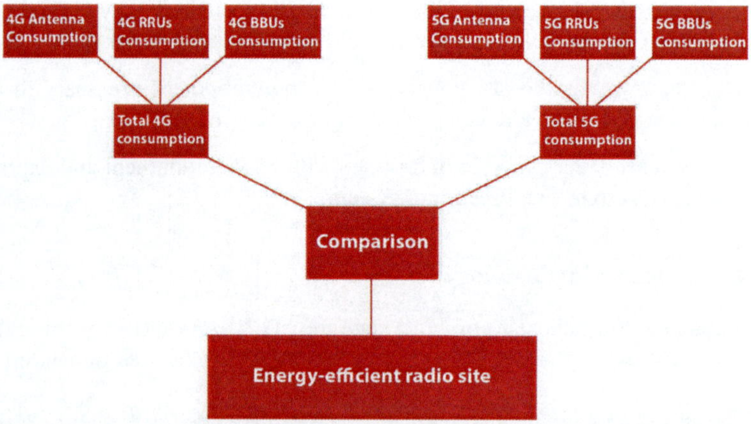

Fig. 10. Flowchart of the methodology

The study will compare the consumption of the antenna's chain equipment in 4G with the consumption of the same sites in 5G.

This study will be based on the energy balances made for the operator, it will consist of comparing the balances before 5G and after 5G.

In this study, the calculation will be made according to the method below (Fig. 10):

Total consumption = consumption of antennas + consumption of RRUs + consumption of BBUs.

The sample of sites was chosen to include all the configurations used by the operator for the 5G project in Toulouse.

3.4 Energy Consumption of 5G Equipment

For this analysis, the data used are communicated by the operator, these data represent the average consumption of each equipment.

Table 2. Consumption of BBU's cards

Equipment	Energy (KW)	Role
UBBPG3B	0,144	5G FDD/TDD card
UMPTG2	0,06	4G/5G FDD card
UMPTG3	0,06	Transmission Card

These numbers are communicated to the design office which calculates the energy balance.

Table 3. Consumption of RRU equipment

Equipment	Energy (KW)
RRU5509t-700M-800M-900M-2x140w(FDR)	0,85
LRFUe-800M 2*60w (FDR)	0,57
MRFUd-900M-2*80 W (FDL)	0,718
MRFUd-1800M-2*80 W (FDR)	0,602
RRU5909-1800M-2*60w (FDL)	0,367
RRU5502w-1800M/2100M-4*80w(FDR)	0,99
MRFUd-2100M-2*80 W (FDR)	0,548
RRU5005-2100M-2*80w estimated (FDL)	0,569
RRU5301-2600M-4*40w(FDL)	0,443
RRU5304w-2600m-4*40w (FDL)	0,577

BBU. For 5G integration, the operator opted for Huawei cards:

RRU. In the RRU part, the list below represents the equipment used:

Table 4. Consumption of AIO antenna

Equipment	Energy (KW)
Blade AAU5835f 3450–3700 MHz	1

Antenna. In our sample of sites, they all integrate passive antennas grouping 2G/3G/4G, to add 5G, the site is enriched with another passive 5G antenna or swap the existing antenna to AIO (All in One) antenna, the latter includes the 3500 MHz radio module directly:

3.5 Energy Balance Calculation

Our work will consist of analyzing and calculating the energy consumption of the antenna chain for each site, to then have a general idea of the difference in consumption between 4G and 5G.

Table 5. Consumption of 4G's BBU

Equipment	Unit energy	Total energy
Huawei BBU GL	0,274	0,274
Huawei BBU U	0,346	0,346

For this, we will take as an example the site 90018 located in Toulouse made up of 2 sectors, we will analyze each equipment of the antenna chain with its consumption in KW.

Table 6. Consumption of 4G's RRU module

Equipment	Qt	Unit energy	Total energy
LRFUe L800 2*60	2	0,395	0,79
MRFUd-900M-2*80 W GU9	2	0,5	1
MRFUd-1800M-2*80 W	4	0,512	2,048
RRU3826 (U2100)-Blade-1*80 W	4	0,308	1,232

4G. The technology cards present in the BUUs are used in all sectors and therefore only one is needed on the radio site.

The site had as a radio module the list below to manage the frequencies 800/900/1800/2100.

This site has no active antenna before the 5G project.

Table 7. Total consumption of 4G radio equipment

Equipment	Energy (KW)
BBU	0,62
RRU	5,07
Antennas	0

Table 8. Consumption of the 5G's BBU

Equipment	Unit energy	Total energy
BBU 1	0,14	0,14
BBU 2	0,66	0,66

Table 9. Consumption of the 5G's RRU module

Equipment	Qt	Unit energy	Total energy
RRU5509t-700M-800M-900M-2x140w(FDR)	2	0,85	1,7
RRU5502w-1800M/2100M-4*80w(FDR)	4	0,99	3,96
RRU5301-2600M-4*40w(FDL)	2	0,443	0,886

Table 10. Consumption of 5G only antenna

Equipment	Qt	Unit energy	Total energy
AAU5636w- 3500M-3700M 64T64R (320w)	2	1	2

Total consumption in 4G:

5G. The 5G project of this site consists of adding the 700/2600/3500 frequencies while keeping the frequencies already mentioned.

The BBU's cabinet will integrate 5G cards.

Sometimes we need to swap the RRU equipment or add someone.

The antenna below directly includes the 3.5 GHz radio module.

Total consumption in 5G:

Table 11. Total consumption of 5G radio equipment

Equipment	Energy (KW)
BBU	0,784
RRU	6,546
Antennas	2
Total	**9,33**

4 Results and Discussion

The same calculation method will be applied on all 10 sites, the results obtained are as follows:

4.1 BBU

Table 12. Comparaison between 4G and 5G BBU's consumption

Site	4G	5G
90002	0,321	0,8
90018	0,62	0,784
190146	0,507	0,612
190477	0,62	0,736
310008	0,62	0,742
310106	1,098	0,791
312264	0,891	0,799
650001	0,593	0,799
650052	0,844	0,799
650355	0,276	0,799
Average	**0,639**	**0,7661**

There is a notable increase in all the sites except for 312264 and 310106 where a drop in consumption is observed and this is due to the change of very old BBUs, in particular the BTS3900L which consumed an average of 140 W by new, less energy-consuming ones. The other 8 sites in this study have seen their energy consumption increase by 120% on average with a peak of 289% for the 650355 which is integrated with 2 UBPg2 cards and 2 UBPg3 cards to activate 2 different systems (L800/L1800 & N2100/N3500) on each card. This drastically increases the total consumption of the site (Fig. 11).

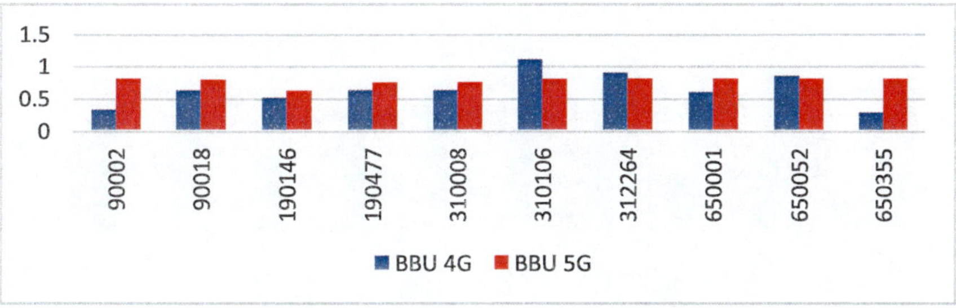

Fig. 11. Charts of BBU's consumption

Table 13. Comparison between 4G and 5G RRU's consumption

Site	4G	5G
90002	2,22	6,546
90018	5,07	6,546
190146	3,711	7,773
190477	5,07	8,33
310008	6,891	12,177
310106	9,771	7,167
312264	8,127	10,779
650001	5,463	9,819
650052	7,761	9,819
650355	7,299	10,221
Average	**6,1383**	**8,9177**

4.2 Wireless Modules

The site 310106 stands out and recorded a drop in consumption of the RRU part, this is explained by the uninstallation of the 1800 frequency on the site and therefore the removal of RRU.

However, the other sites record an increase of 45% on average explained by the addition of RRU of the new frequencies (2100/2600/3500), site 90002 has the consumption of 295% higher than that before 5G, the main reason for this growth is the addition of many radio cards; 700/800/2600 and the pooling of 1800/2100 frequencies with another operator, which forces a card added for the latter too (Fig. 12).

4.3 Antennas

The antenna part consumes much more in 5G on most sites and this is due to the addition of an active antenna that integrates the 3500 MHz radio module, the other sites which

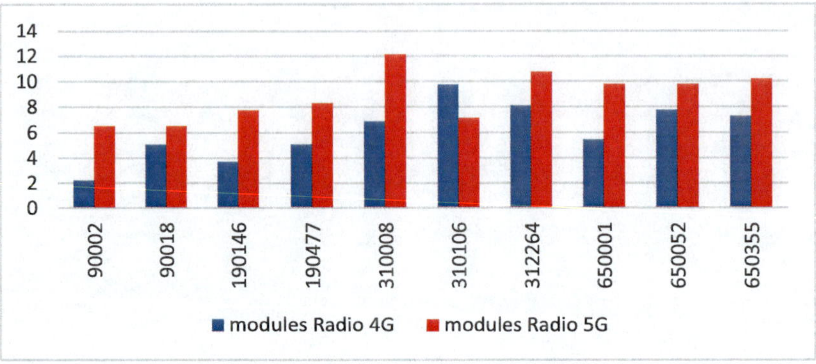

Fig. 12. Charts of RRU's consumption

Table 14. Comparison between 4G and 5G antenna's consumption

Site	4G	5G
90002	0	2
90018	0	2
190146	0	0
190477	0	0
310008	0	0
310106	0	3
312264	0	3
650001	0	3
650052	0	3
650355	0	3
Average	**0**	**2,714285714**

have kept zero consumption, are sites with the fake 5G 2100 MHz which only uses passive antennas (Fig. 13).

4.4 Total Energy

Above, the table showing a comparison of the general consumption of the sites, there is an increase in consumption on all the sites, an average consumption doubled during the transition to 5G, and this is explained by the addition of 5G technology, new frequency band and active antennas (Fig. 14).

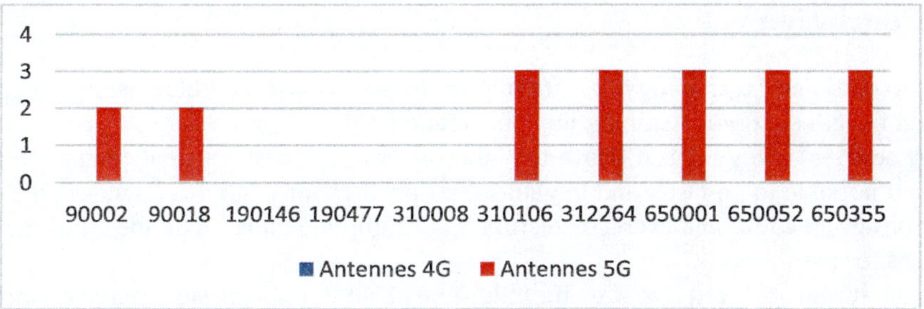

Fig. 13. Charts of the antenna's consumption

Table 15. Comparaison between 4G and 5G sites consumption

Site	4G	5G
90002	2,541	9,346
90018	5,69	9,33
190146	4,218	8,385
190477	5,69	9,066
310008	7,511	12,919
310106	10,869	10,958
312264	9,018	14,578
650001	6,056	13,618
650052	8,605	13,618
650355	7,575	14,02
Average	6,7773	12,39

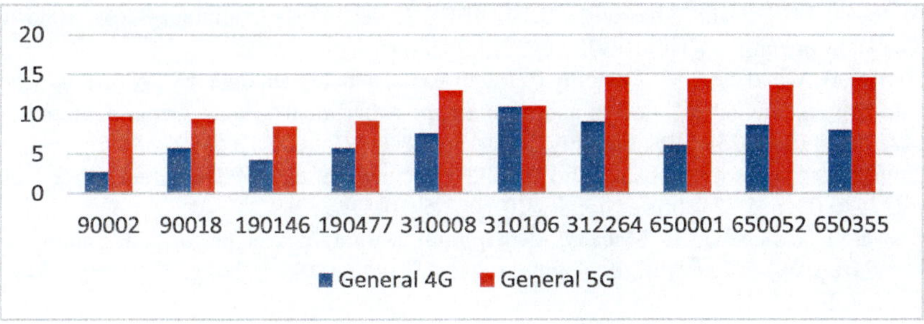

Fig. 14. Charts of site consumption

5 Conclusion

Any revolutionary technology can provide an answer to several problems, on paper 5G should be less energy-consuming than its precursors.

A new technology added above the other generations without fundamental change made it possible to go faster and to gain some specifications, but this counterpart in the field brings an additional average electricity consumption equal to all the consumption of the sites before 5G.

The surplus is partly due to the addition of active all-in-one antennas natively comprising the 3.5 GHz frequency radio module.

In this paper, the purpose of the analysis was to look for radio sites recently updated to 5G, study the addition of equipment to them, and calculate the consumption of the sites before and after, this made it possible to give an estimate of the increase in electricity consumption which is close to 200%. The detailed analysis also made it possible to understand the main causes of this increase.

All these calculations are made on radio sites with an unchanged core and infrastructure, we can hope that 5G SA brings a much less energy-consuming core network which will compensate for this additional energy spent on the radio site.

References

1. Kebede, T., Wondie, Y., Steinbrunn, J., Kassa, H.B., Kornegay, K.T.: Multi-carrier waveforms and multiple access strategies in wireless networks: performance, applications, and challenges. IEEE Access **10**, 21120–21140 (2022). https://doi.org/10.1109/ACCESS.2022.3151360
2. Subotič, I.: 3G, Hspa and 4G Lte Network reliability and achievable internet speeds on airborne aircraft Up To 10.000 Ft above ground level. Informatica **47**(5), 95–110 (2023). https://doi.org/10.31449/INF.V47I5.4513
3. Ahmed, R.T., Jumah Al-Thahab, O.Q.: Comprehensive Review on LTE-BER Reduction with Modern Transform. In: 3rd Information Technology to Enhance e-Learning Other Appl. IT-ELA 2022, pp. 116–122 (2022). https://doi.org/10.1109/IT-ELA57378.2022.10107949
4. Andrews, J.G., et al.: What will 5G be? IEEE J. Sel. Areas Commun. **32**(6), 1065–1082 (2014). https://doi.org/10.1109/JSAC.2014.2328098
5. Martini, B., Gharbaoui, M., Castoldi, P.: Intent-based network slicing for SDN vertical services with assurance: Context, design and preliminary experiments. Futur. Gener. Comput. Syst. **142**, 101–116 (2023). https://doi.org/10.1016/J.FUTURE.2022.12.033
6. Singh, H.: 5G new radio bands and 4G LTE concurrence. In: AIP Conference Proceedings, vol. 2576 (Dec 2022). https://doi.org/10.1063/5.0105769
7. Gheyas, I., Raschella, A., Mackay, M.: Optimal meshing degree performance analysis in a mmWave FWA 5G network deployment. Futur. Internet **15**(6), 218 (2023). https://doi.org/10.3390/FI15060218
8. Bonati, L., Polese, M., D'Oro, S., Basagni, S., Melodia, T.: Open, Programmable, and Virtualized 5G Networks: State-of-the-Art and the Road Ahead. Comput. Netw. **182** (2020). https://doi.org/10.1016/J.COMNET.2020.107516
9. Omerani, H., Mazri, T.: 4G and 5G: Security and privacy analysis. ACM Int. Conf. Proceeding Ser. (2019). https://doi.org/10.1145/3372938.3372967
10. Bansal, H.: Introduction to 5G and Beyond. Appl. 5G Beyond Smart Cities, pp. 1–26 (Jan. 2023). https://doi.org/10.1201/9781003227861-1

11. Gupta, A., Jha, R.K.: A survey of 5G network: architecture and emerging technologies. IEEE Access **3**, 1206–1232 (2015). https://doi.org/10.1109/ACCESS.2015.2461602
12. Challenges, C.: SMARTer2030: ICT solutions for 21st century challenges, p. 9 (December 2015). http://smarter2030.gesi.org/downloads/Chapter_Health.pdf
13. Rohde, U.L., Poddar, A.K., Eisele, I., Rubiola, E.: Next generation 5G radio communication NW. In: 2017 Jt. Conference Eur. Freq. Time Forum IEEE Int. Freq. Control Symp. EFTF/IFC 2017 - Proc., pp. 113–116 (Oct 2017). https://doi.org/10.1109/FCS.2017.8088817
14. Khan, F., Pi, Z., Rajagopal, S.: Millimeter-wave mobile broadband with large scale spatial processing for 5G mobile communication. In: 2012 50th Annual Allerton Conference on Communication, Control, and Computing 2012, pp. 1517–1523 (2012). https://doi.org/10.1109/ALLERTON.2012.6483399
15. Isabona, J., Srivastava, V.M.: Downlink massive MIMO systems: achievable sum rates and energy efficiency perspective for Future 5G systems. Wirel. Pers. Commun. **96**(2), 2779–2796 (2017). https://doi.org/10.1007/S11277-017-4324-Y
16. Joshi, Y., Kabra, G., Tripathi, A.: Energy consumption in smart City projects in the Era of 5G: an analysis of user-generated content. Appl. 5G Beyond Smart Cities 99–110, (2023). https://doi.org/10.1201/9781003227861-6
17. Franchi, F., Marotta, A., Rinaldi, C., Graziosi, F., D'Errico, L:. IoT-based Disaster Management System on 5G uRLLC Network. In: 6th International Conference on Information and Communication Technologies for Disaster Management, ICT-DM 2019 (Dec 2019). https://doi.org/10.1109/ICT-DM47966.2019.9032897
18. Selvam, P.D., Sridhar, J., Ganesan, V., Ravindraiah, R.: The future of Industry 4.0: private 5G networks. Adv. Signal Process. Ind. 4.0, Evol. Commun. Protoc. Appl. Manuf. Syst. **1**, 3.1–3.25 (2023). https://doi.org/10.1088/978-0-7503-5247-5CH3
19. Shurdi, O., Rakipi, A., Biberaj, A.: Performance analysis of Non-Standalone 5G network under different traffic models. In: 2023 33rd International Conference on Radioelektronika, RADIOELEKTRONIKA (2023). https://doi.org/10.1109/RADIOELEKTRONIKA57919.2023.10109072
20. Cisco, Cisco Visual Networking Index: Global Mobile Data Traffic Forecast Update, 2015 -- 2020," Growth Lakel, pp. 2015–2020 (2016). http://www.cisco.com/en/US/solutions/collateral/ns341/ns525/ns537/ns705/ns827/white%7B_%7Dpaper%7B_%7Dc11-520862.html
21. BBU-Base Band Unit in Telecom: Function|Details - Telecomkhabar. https://www.telecomkhabar.com/2019/06/bbu-base-band-unit-function-details-in-telecom/ (Accessed 29 Aug 2022)
22. TD-LTE Distributed Base Station Remote Radio Unit (RRU) from China manufacturer - IFLY. https://www.hk-ifly.com/TD-LTE-Distributed-Base-Station-Remote-Radio-Unit-RRU-pd98411346.html (Accessed 03 Sep 2022)
23. 380–960MHz/1710–2700MHz Dual Band Combiner. http://www.action-telecom.com/380-960MHz-1710-2700MHz-Dual-Band-Combiner.html (Accessed 03 Sep 2022)
24. Kaelus TMA2035 | TWIN TMA 1900 | Kaelus. https://www.kaelus.com/en/rf-conditioning-solutions/tower-mounted-amplifiers/fdd-single-band-tmas/tma2035 (Accessed 03 Sep 2022)
25. Huawei Releases 5G Series Products to Expand Multi-Antenna Technology to All Bands and Scenarios - Mobile World Live. https://www.mobileworldlive.com/huawei-updates/huawei-releases-5g-series-products-to-expand-multi-antenna-technology-to-all-bands-and-scenarios/ (Accessed 03 Sep 2022)
26. Proxim Wireless - Tsunami® GX-824. https://www.proxim.com/fr/products/point-to-point/tsunami-gx-824 (Accessed 03 Sep 2022)

Smart Security Systems

Secure and Efficient Color Image Cryptography Using Two Secret Keys

Mua'ad Abu-Faraj[1(✉)], Abeer Al-Hyari[2], and Ziad Alqadi[3]

[1] Department of Computer Information Systems, The University of Jordan, Aqaba 77110, Jordan
m.abufaraj@ju.edu.jo

[2] Electrical Engineering Department, Al-Balqa Applied University, As Salt 19117, Jordan
abeer.hyari@bau.edu.jo

[3] Computers and Networks Engineering Department, Al-Balqa Applied University, Amman 15008, Jordan
dr.ziad.alqadi@bau.edu.jo

Abstract. The significance of digital color images in various applications necessitates the development of secure and efficient color image encryption techniques. This study presents a simple yet highly secure method for encrypting color images using dual secret keys, enhancing overall image security and safeguarding against hacking attempts. The first key generates index keys for encrypting and decrypting image segments, while the second key determines the starting point in the image_key and divides the unencrypted image into segments. These index keys are then employed to encrypt and decrypt the segments. Encryption and decryption of segments involve a straightforward process, distributing plain data based on the index key and retrieving decrypted data from encrypted data following the index key's indices. The proposed method will be assessed and applied to various color images, with the results analyzed to demonstrate improvements in quality, sensitivity, and speed. The method's speed will be compared with other techniques to highlight the acceleration offered by the proposed approach.

Keywords: Cryptography · Quality · throughput · sensitivity · image_key · Private key · index key

1 Introduction

RGB color model employs the three primary light colors: red, green, and blue. These can be mixed to make all other colors. Conventional RGB digital images will assign a numerical value for red, green, and blue for each pixel; each value is within the range of 0 to 255. Color images are very important data types, and they are used in many vital applications [1,2].

The digital color image is represented by a three-dimensional matrix, one 2D matrix for each color. It is easy to extract each color and combine the three

M. Ben Ahmed et al. (Eds.): SCA 2023, LNNS 906, pp. 549–560, 2024.
https://doi.org/10.1007/978-3-031-53824-7_49

colors to form the color image. Digital color images can be easily reshaped to 1D or to a 2D matrix. Reshaping the color image to a one-row matrix will simplify the process of color image handling [3,4].

Digital color images sometimes require protection from hacking because they may be private or secret or hold valuable secret data. One of the most popular techniques of image protection is image cryptography. Data cryptography means data encryption before sending the data and data decryption after receiving the data [5,6].

The following are the essential requirements for an effective data cryptography method, as stated in [7,8]:

1. Implementation Simplicity: The encryption and decryption process should be straightforward to implement and applicable to various types of images.
2. Encrypted Data Quality: The quality of the encrypted data should be significantly degraded. The measured quality parameters between the original and encrypted data should exhibit the following characteristics (refer to Fig. 1 and Table 1):
   (a) High mean square error (MSE)
   (b) Low peak signal-to-noise ratio (PSNR)
   (c) Low color correlation coefficients (CC)
   (d) High number of pixels change ratio (NPCR)
   (e) Low Structural Similarity Index (SSIM)
3. Decrypted Data Recovery: The quality of the decrypted data should be fully restored, with the measured quality parameters between the original and decrypted data as follows (refer to Fig. 1):
   (a) MSE should be zero
   (b) PSNR should be infinite
   (c) CC should be 1
   (d) NPCR should be zero
   (e) SSIM should be one
4. Security: The method must provide a high level of security by utilizing complex secret keys that create an extensive key space capable of resisting hacking attacks.
5. Consistent Secret Keys: The encryption and decryption phases must employ the same secret keys. Any slight changes in the keys during the decryption phase should be considered a hacking attempt, resulting in a damaged decrypted image.
6. High Cryptography Speed: The method should minimize both the encryption time (ET) and the decryption time (DT) to increase the overall cryptography throughput (number of bytes encrypted/decrypted per second).

Table 1. Quality requirements [2–4]

Quality parameter	MSE	PSNR	Colors CCs	NPCR	SSIM
Measured between encrypted and source messages	High	Low	Low	High	Low
Measured between decrypted and source messages	0	Infinite	1	0	1

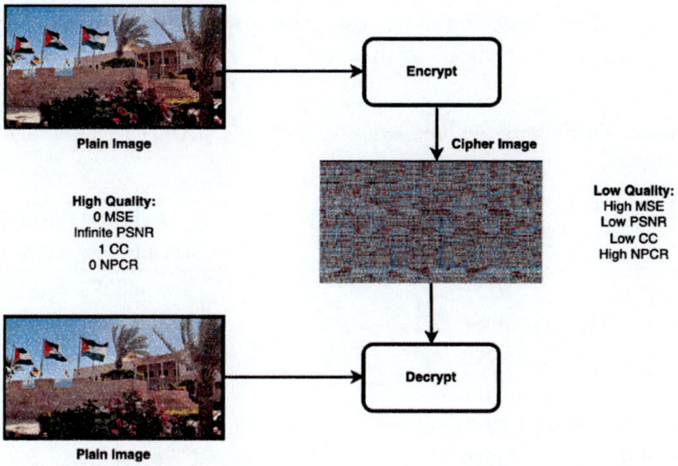

Fig. 1. Cryptography quality measures

Data protection is a very important matter, prompting many researchers to provide various encryption methods with varying efficiency values. Some methods were based on the standards DES, AES, 3DES, and BF; these methods are good for small in-size data; when the data size increased (as long messages), these methods became inefficient, especially if they were to be used for color image cryptography, and Table 2 shows the throughputs of these standard methods [9], the standard method of data cryptography use a fixed length private key and a fixed length of the data block, the processes of encryption and decryption are complicated, the methods discussed here of data cryptography use one **private key** to protect the data, some of these methods are designed to encrypt and decrypt secret messages, and it is difficult to use them for huge messages cryptography [5,6,9].

Some other authors introduced methods that enhanced the throughput; in [10], the authors introduced a method, the throughput was enhanced to reach 169.1 Kbytes per second, while in [11], the introduced method enhanced the throughput to reach 710 Kbytes per second [12,13]. In [14], the authors provided an image encryption scheme based on a mixing technique. In [15], the authors provided a cosine-transform-based chaotic system for image encryption, while in [16], the authors introduced a novel image encryption algorithm based on a polynomial combination of chaotic maps and dynamic function generation. In [17], the authors introduced Multiple-image Encryption Algorithm Based on DNA Encoding and a Chaotic System, while in [18], the authors produced a multiple-image encryption with bit-plane decomposition and chaotic maps; these methods provided good quality and have various speeds. In [19], the authors introduce a color image encryption technique that uses a 128-bit composite key to produce a private key, which is then resized to fit the color matrix size of any image. The speed analysis indicates that the approach enhances performance. Table 2 shows different methods with their encryption time and throughput.

Table 2. Speed parameters of some existing methods [15,17–22] (Average image size = 620.524 Kbytes).

Method	Average encryption time (second)	Average throughput (Kbytes per second)
Kumar et al. [20]	0.9751	636.3379
Yepioda et al. [21]	0.6981	888.8867
Hua et al. [15]	0.9720	638.4082
Asgari-chenaphlu [22]	0.6811	911.0352
Zhang and Wang [17]	1.7217	360.4102
Zhenjun and Sun [18]	1.6119	384.9609
Abu-Faraj et al. [19]	4.3714	302.7968

The aim of this study is to develop a secure and efficient color image encryption technique that utilizes dual secret keys. The proposed method aims to enhance overall image security and protect against hacking attempts. Specifically, the first key is used to generate index keys for encrypting and decrypting image segments, while the second key determines the starting point and segment division in the unencrypted image. The objective is to use these index keys to encrypt and decrypt the image segments through a straightforward process of distributing plain data based on the index key and retrieving decrypted data from encrypted data using the index key's indices. The proposed method will be evaluated by applying it to various color images. The assessment will focus on analyzing the improvements in image quality, sensitivity, and speed achieved by the proposed approach. Additionally, the speed of the proposed method will be compared with other techniques to highlight its accelerated performance.

The paper's structure is outlined as follows: Section 2 presents the proposed method, Sect. 3 discusses the implementation results and provides a discussion, and Sect. 4 presents the conclusions.

2 The Proposed Method

The proposed method uses two secret keys to apply image encryption and decryption. The first key is an image_key; this key must be kept secret between the data sender and receiver. The image_key is used to generate the index keys to be used to encrypt-decrypt image segments as shown in Fig. 2. Figure 3 shows how to use the image_key to generate the required index keys; changing the image_key will change the index keys.

The second secret is a private key containing five parameters: the position parameter, which points to the starting point in the image_key, where to get

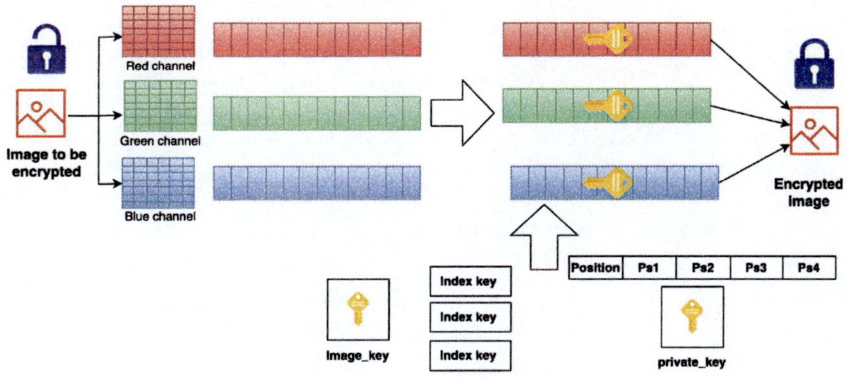

Fig. 2. The proposed method

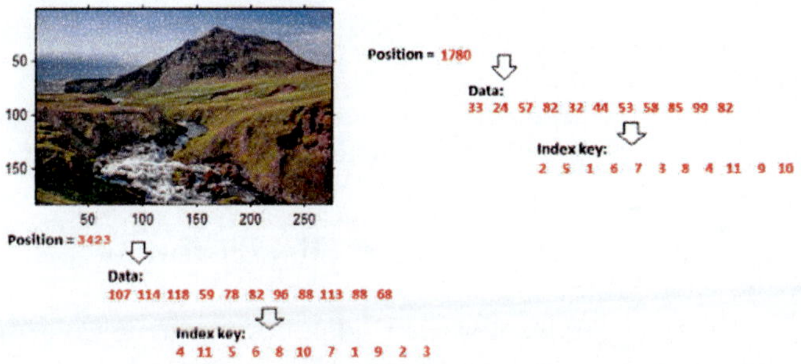

Fig. 3. Generating index keys

the data, the rest four components (Ps) point to the fractions to form the size of each segment. Each component has a double data type; thus, this key will provide a huge keyspace capable of resisting hacking attacks.

The encryption-decryption phases use the index key to apply encryption-decryption. The image segment encryption is performed by distributing the segment values according to the indexes to produce an encrypted segment, while the decryption is applied by retrieving the values from the encrypted segment to form the decrypted segment, as shown in Fig. 4.

The encryption phase can be implemented by applying the following steps:

- Image_key preparation: Get the image, retrieve the image size, and reshape the image into a one-row matrix for each color channel.
- Preparing the image to be encrypted.
- Get the Private key.
- Calculate the lengths of the segments and retrieve each segment.

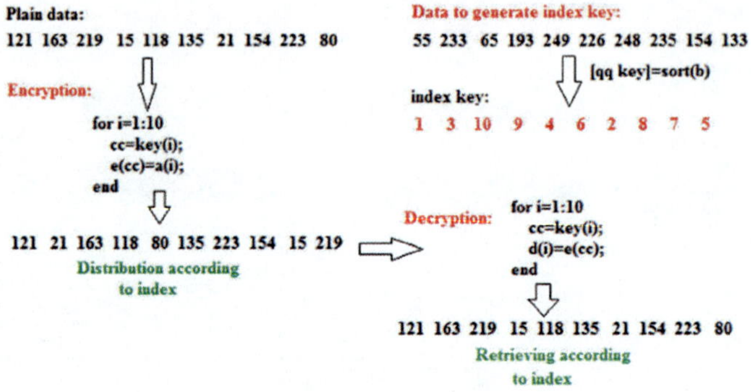

Fig. 4. Encryption-decryption using index key (example)

- From the image_key, get the data and generate the index keys.
- Apply segments encryption.
- Combine the encrypted segments in a one-row matrix and reshape this matrix to a 3D matrix to get the encrypted image.

3 Implementation and Results Discussion

In this article, the utilized dataset is the Kodak Lossless True Color Image Suite, an extensive assortment of top-notch, true-color pictures that are an essential asset for experts and developers working in the fields of image processing, compression, and encryption. This image collection, available at [23], offers a reliable standard for assessing and contrasting the effectiveness of different algorithms and approaches in the realm of image processing. The proposed method was tested for quality; visually, we can prove that it satisfies the quality requirements by looking by eyes at the encrypted and decrypted images and their histograms. Image 1 was encrypted and decrypted using the proposed methods.

Figure 5, 6, and 7 show the obtained images and their histograms. Looking at these images and histograms by eye, we can see that the encrypted image is a damaged image, while the decrypted image is identical to the source image; this proves the quality of the proposed method in both the encryption and decryption phases. The histograms of the encrypted and the source images must be identical because encryption was performed by changing the pixel's positions and keeping this value unchanged. Another way to prove the quality of the proposed method is to test the calculated quality parameters. The selected images were processed using the proposed method. The quality parameters measured between the source and **encrypted** images, all the obtained values satisfied the decryption quality requirements; the results for encryption quality are shown in Table 3 and Table 4, which satisfy the quality requirements.

Table 3. MSE, PSNR, and NPCR results

Image number	MSE	PSNR	NPCR
1	14340	15.1169	99.2894
2	8988.2	19.7886	94.7002
3	11257	17.5380	99.1226
4	7834.9	21.1618	99.4850
5	5076.4	25.5017	99.2742
Remarks (results meet the requirements)	**High**	**Low**	**High**

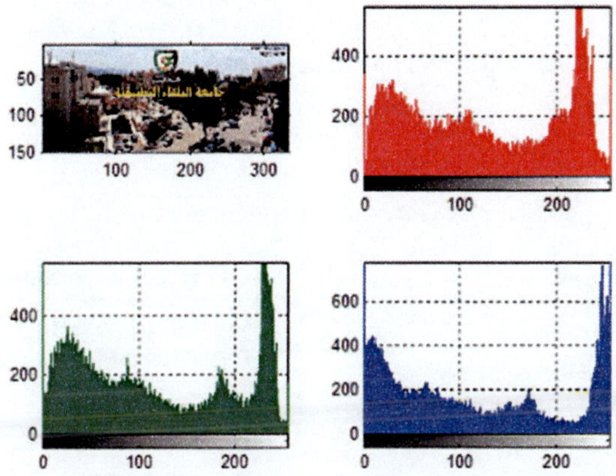

Fig. 5. Source image and its histograms

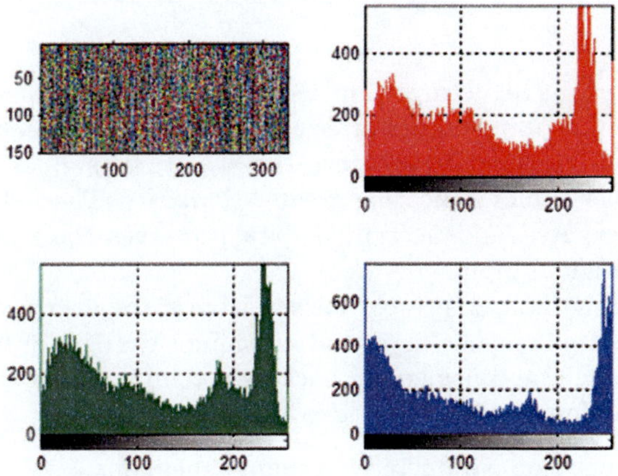

Fig. 6. Encrypted image and its histograms

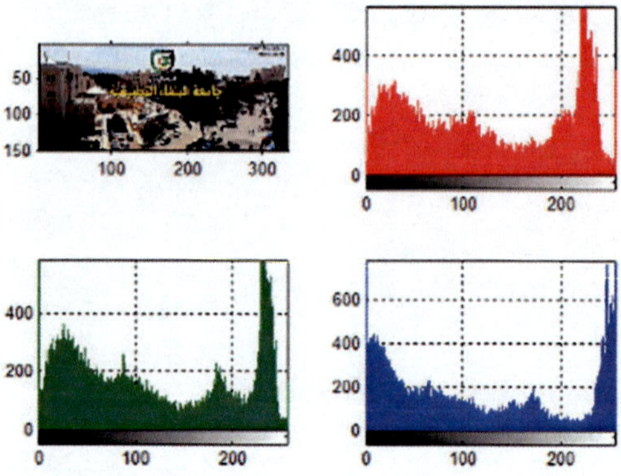

Fig. 7. Decrypted image and its histograms

Table 4. Correlation coefficients results

Image number	CC red	CC green	CC blue
1	0.0193	−0.0001	0.0020
2	0.0061	0.0022	0.0187
3	0.0125	0.0825	0.0303
4	0.0240	0.0092	−0.0104
5	0.0426	0.0053	0.0429
Remarks (results meet the requirements)	**Low**	**Low**	**Low**

Table 5 displays the Structural Similarity Index (SSIM) values for the original input data, as well as for the data after encryption and subsequent decryption. The SSIM is a metric used to measure the structural similarity between two images, with higher values indicating greater similarity. The SSIM values reflect how well the encrypted and decrypted data preserves the visual quality and integrity of the original input.

The proposed method was tested for sensitivity; the encryption and decryption phases must use the same keys, and any minor changes in these keys in the decryption phase will be considered a hacking attempt by producing a damaged decrypted image. Here we have to notice the following:

- Changing the position parameter will change the index keys; thus, the decryption phase will be affected.
- Changing the image key also will change the index keys, and the decryption results will be affected.
- Changing the fraction parameters will change the index keys, affecting the decryption results.

Table 5. SSIM Image Quality Measurement

Image Number	SSIM (Input and Encrypted)	SSIM (Input and Recovered)
1	0.0010	1
2	0.0008	1
3	0.0003	1
4	0.0005	1
5	0.0004	1

Table 6. Speed results

Image	ET (second)	ETP (Kbytes per second)
1	0.0250	5892.5
2	0.0790	6408.2
3	0.4060	4546.1
4	1.2520	3374.5
5	1.2890	3894.7
Average	**0.6102**	**4823.2**

The speed of the proposed method was tested; the selected images were processed using the proposed method, an image was selected as an image_key, the encryption time was measured, and the throughput was calculated. Table 6 shows the obtained results.

From Table 6, we can see that the proposed method provided a good speed; the average throughput was equal to 4823.2 Kbytes per second. The throughput of the proposed method was compared with the throughputs of other existing methods, and the proposed method provided a significant speed up (speedup of the proposed method equals another method throughput divided by proposed method throughput) (see Table 7), the proposed method increased the cryptography throughput and minimized the encryption/decryption time.

Table 7. Method speedup

Method	TPT (Kbytes per second)	Speedup of the proposed method (times)
Kumar et. al [20]	636.3379	7.5796
Yepioda et al. [21]	888.8867	5.4261
Hua et al. [15]	638.4082	7.5550
Asgari-chenaphlu [22]	911.0352	5.2942
Zhang and Wang [17]	360.4102	13.3825
Zhenjun and Sun [18]	384.9609	12.5291
Abu-Faraj et al. [19]	302.7968	15.928
Proposed	4823.2	1.0000

4 Conclusion

A highly secure, efficient, and simple method of color image cryptography was introduced. The method uses a simple data distribution and retrieval to apply data encryption and decryption. The method uses two secret keys; these keys provide a huge keyspace making the hacking process impossible. The first key is an image_key, which generates the required index keys to apply encryption/decryption. The second key is a private key, which is used to divide the image into different segments; each segment is treated alone by encrypting/decryption it using its own index key. The decryption phase results were susceptible to the selected secret keys, the proposed method is tested for sensitivity, and it is shown that making any minor changes in the secret keys during the decryption process is considered a hacking attempt by producing a damaged decrypted image. The quality of the proposed method is tested based on visual testing and the calculated quality parameters, including MSE, PSNR, CCs, NPCR, and SSIM. The results proved the method's encryption and decryption qualities. Different images are selected and treated using the proposed method, and the speed of the proposed method is tested; the results show that the proposed method decreases both the encryption and decryption times, and it is shown that the proposed method provides a significant speedup comparing with other existing methods of data cryptography that can reach up to 15.92.

References

1. Abu-Faraj, M., Zubi, M.: Analysis and implementation of kidney stones detection by applying segmentation techniques on computerized tomography scans. Ital. J. Appl. Math. **43–2020**, 590–602 (2020)
2. Ge, Y., et al.: Enhancing the x-ray differential phase contrast image quality with deep learning technique. IEEE Trans. Biomed. Eng. **68**(6), 1751–1758 (2020)
3. Abu-Faraj, M., Al-Hyari, A., Alqadi, Z.: A complex matrix private key to enhance the security level of image cryptography. Symmetry **14**(4) (2022). https://www.mdpi.com/2073-8994/14/4/664
4. Hua, Z., Zhou, Y., Huang, H.: Cosine-transform-based chaotic system for image encryption. Inf. Sci. **480**, 403–419 (2019)
5. Abu-Faraj, M., Al-Hyari, A., Al-taharwa, I., Alqadi, Z., Ali, B.: Increasing the security of transmitted text messages using chaotic key and image key cryptography. Int. J. Data Netw. Sci. **7**(2), 809–820 (2023)
6. Abu-Faraj, M., Al-Hyari, A., Al-Ahmad, B., Alqadi, Z., Ali, B., Alhaj, A.: Building a secure image cryptography system using parallel processing and complicated dynamic length private key. Appl. Math. Inform. Sci. (AMIS) **16**(6), 1017–1026 (2022)
7. Abu-Faraj, M., Al-Hyari, A., Aldebei, K., Alqadi, Z., Al-Ahmad, B.: Rotation left digits to enhance the security level of message blocks cryptography. IEEE Access **10**, 69388–69397 (2022)
8. Abu-Faraj, M., Al-Hyari, A., Al-Taharwa, I., Al-Ahmad, B., Alqadi, Z.: CASDC: a cryptographically secure data system based on two private key images. IEEE Access **10**, 126304–126314 (2022)

9. Nadeem, A., Javed, M.Y.: A performance comparison of data encryption algorithms. In: 2005 International Conference on Information and Communication Technologies, pp. 84–89. IEEE (2005)
10. Liu, H., Kadir, A., Li, Y.: Audio encryption scheme by confusion and diffusion based on multi-scroll chaotic system and one-time keys. Optik **127**(19), 7431–7438 (2016)
11. Hato, E., Shihab, D.: Lorenz and Rossler chaotic system for speech signal encryption. Int. J. Comput. Appl. **128**, 25–33 (2015)
12. Sathiyamurthi, P., Ramakrishnan, S.: Speech encryption using chaotic shift keying for secured speech communication. EURASIP J. Audio Speech Music Process. **2017**(1), 20 (2017)
13. Tamimi, A., Abdalla, A.: Audio shuffle-encryption algorithm, vol. 1 (2014)
14. Valenza, F., Heucheun Yepdia, L.M., Tiedeu, A., Kom, G.: A robust and fast image encryption scheme based on a mixing technique. Secur. Commun. Netw. **2021**, 6615708 (2021)
15. Hua, Z., Zhou, Y., Huang, H.: Cosine-transform-based chaotic system for image encryption. Inf. Sci. **480**, 403–419 (2019)
16. Asgari-Chenaghlu, M., Balafar, M., Feizi Derakhshi, M.R.: A novel image encryption algorithm based on polynomial combination of chaotic maps and dynamic function generation. Sig. Process. **157**, 11 (2018)
17. Zhang, X., Wang, X.: Multiple-image encryption algorithm based on DNA encoding and chaotic system. Multimedia Tools Appl. **78**(6), 7841–7869 (2019)
18. Tang, Z., Song, J., Zhang, X., Sun, R.: Multiple-image encryption with bit-plane decomposition and chaotic maps. Opt. Lasers Eng. **80**, 1–11 (2016)
19. Abu-Faraj, M., et al.: Protecting digital images using keys enhanced by 2d chaotic logistic maps. Cryptography **7**(2) (2023). https://www.mdpi.com/2410-387X/7/2/20
20. Kumar, B., Karthikka, P., Dhivya, N., Gopalakrishnan, T.: A performance comparison of encryption algorithms for digital images. Int. J. Eng. Res. Technol. **3**(2), 2169–2174 (2014)
21. Heucheun Yepdia, L.M., Tiedeu, A., Kom, G.: A robust and fast image encryption scheme based on a mixing technique. Secur. Commun. Netw. **2021**, 1–17 (2021)

22. Asgari-Chenaghlu, M., Balafar, M.-A., Feizi-Derakhshi, M.-R.: A novel image encryption algorithm based on polynomial combination of chaotic maps and dynamic function generation. Signal Process. **157**, 1–13 (2019)
23. Franzen, R.: Kodak lossless true color image suite (1999). http://r0k.us/graphics/kodak. Accessed 3 Apr 2023

A Comparative Analysis of Deep Learning Approaches for Enhancing Security in Web Applications

Hamza Kadar(✉) and Abdelhamid Zouhair

Faculty Sciences et Techniques de Tanger, University Abdelmalek Essaadi, Tétouan, Morocco

kadar.hamza@etu.uae.ac.ma, a.zouhair@uae.ac.ma

Abstract. This paper provides a comparative analysis of deep learning algorithms for detecting web attacks and code vulnerability, A review of the literature highlights the methodologies, datasets used, achieved accuracies of these models and their limitations. By understanding web attacks and leveraging advanced technologies, it is possible to enhance the security and protection of digital assets. A notable observation is the underutilization of resources to protect JavaScript, a widely used programming language on the internet. To address this gap, our research prioritizes improving JavaScript's security. We aim to develop a system that improve web application protection, ensuring a safer online environment for users.

Keywords: Cyber security · Web attacks · Vulnerability detection · Machine learning · Deep learning · Javascript

1 Introduction

This research paper focuses on two critical aspects: an exploration of web attacks and a comparative analysis of code vulnerability detection models. We conduct a thorough evaluation of these models, with a strong emphasis on their effectiveness in identifying security risks within software applications. Our scrutiny encompasses a detailed examination of their methodologies, datasets, and performance metrics, shedding light on both their strengths and limitations in vulnerability detection.

A significant revelation is the relatively limited attention given to JavaScript, a programming language highly susceptible to vulnerabilities, within the realm of deep learning research. This disparity underscores the necessity of addressing JavaScript's unique vulnerabilities and rectifying its underrepresentation in the deep learning community.

Our research seeks to tackle these issues by highlighting the imperative for innovative approaches. While we acknowledge the prevalent practice of automating data labeling from GitHub commits for vulnerability detection, our perspective arises from the recognition of the complexities involved in distinguishing security fixes within code commits that may also contain unrelated changes. This realization prompts us to explore novel strategies for future endeavors.

M. Ben Ahmed et al. (Eds.): SCA 2023, LNNS 906, pp. 561–570, 2024.
https://doi.org/10.1007/978-3-031-53824-7_50

- By 2025, cybercrime will cost the world $10.5 trillion yearly. [19,20,25]
- The entire cost of cyberattacks in 2022 was $6 trillion. [20,23,24]
- 95% of data breaches are the result of human error. [20,25]
- Globally, 30,000 websites are hacked daily. [20,24]
- 64% of companies worldwide have experienced at least one form of cyber attack. [19,20]
- Every 39 s, there is a new attack somewhere on the web. [25]

This paper aims to provide a comparative analysis of deep learning (DL) approaches for enhancing security in web applications. The paper will review the literature on ML and DL models used for attack detection and vulnerability identification, highlighting the methodologies, datasets used, and achieved accuracies. By analyzing web attacks and leveraging advanced technologies, the paper aims to contribute to the development of robust defense mechanisms for enhancing the security and protection of web-based systems.

1.1 Cyber Attacks

Cyber attacks refer to deliberate and malicious activities aimed at compromising computer systems, networks, or digital infrastructure. These attacks are carried out by individuals, groups, or even nation-states with the intent to disrupt, infiltrate, steal, manipulate, or destroy data, systems, or services. Cyber attacks can take various forms, including malware infections, phishing scams, DDoS (Distributed Denial of Service) attacks, ransomware, social engineering, and insider threats. The motives behind cyber attacks can range from financial gain and espionage to activism and political influence. Understanding and mitigating cyber attacks are crucial to ensure the security and integrity of digital assets and protect against potential harm or loss. [22]

1.2 Web Attacks

Web attacks, also known as cyber attacks, are malicious activities aimed at compromising the security and integrity of websites and web applications. These attacks exploit vulnerabilities in web technologies to gain unauthorized access, steal sensitive information, or disrupt services. Common web attacks include:

- **SQL injection**: Attackers inject malicious SQL code into web forms or URLs to manipulate databases and gain unauthorized access.
- Cross-Site Scripting (XSS): Attackers inject malicious scripts into web pages viewed by users, allowing them to steal sensitive information or per- form unauthorized actions.
- **Cross-Site Request Forgery (CSRF)**: Attackers trick users into performing unintended actions on web applications by exploiting their authenticated session.

- **Distributed Denial of Service (DDoS)**: Attackers overwhelm a website with a flood of traffic or requests, causing it to become unavailable to legitimate users.
- **Phishing**: Attackers create deceptive websites or emails that mimic legitimate ones, tricking users into revealing sensitive information like passwords or credit card details.
- **Brute-force attacks**: Attackers attempt to gain access to web applications by systematically trying all possible combinations of usernames and passwords.
- **Man-in-the-Middle (MitM) attacks**: Attackers intercept and alter communication between users and web applications, enabling them to eavesdrop, modify data, or impersonate both parties.
- Remote File Inclusion (RFI): Attackers exploit insecure file inclusion methods to execute arbitrary code or gain unauthorized access to a server.
- **Clickjacking**: Attackers trick users into clicking on hidden or disguised elements on a web page, leading them to perform unintended actions without their knowledge.
- **Zero-day exploits**: Attackers target previously unknown vulnerabilities in web applications before developers have a chance to patch them, gaining unauthorized access or causing damage.

2 Literature Review

Researchers are actively improving machine learning (ML) and deep learning (DL) systems to make web applications more secure and better protected against web attacks. Their efforts have resulted in significant progress in strengthening defenses and enhancing overall web application security.

2.1 Web Attack Detection

In 2022, J.I. Christy Eunaicy proposed a model [8] for automatic web attack detection using deep learning models (ANN, CNN, and RNN). The study focuses on HTTP-based web attacks and aims to improve detection capabilities while reducing false positives. The proposed model follows a three-phase workflow: data collection, data cleaning, and prediction. The HTTP DATASET CSIC 2010, which contains normal and anomalous requests, including various attacks like SQL injection and XSS, is used for testing. The results show that the RNN classifier achieves a high accuracy of 94% and a low error rate of 6%, outperforming other methods. [13] is deep learning framework for detecting, mitigating, and profiling web attacks in real time conducted by Waleed Bin Shahid , The system combines a deep learning model with a Cookie Analysis Engine to analyze HTTP requests and cookies for malicious activity. The framework was trained on a large dataset using Convolution Neural Network (CNN) and achieved accuracy with 99.94% on a custom dataset and 98.74% on a benchmark dataset. ARTP [16], which is an anomaly-based real-time prevention model for effectively detecting Distributed Denial of Service (DDoS) attacks. With high detection accuracy and linear complexity, this model offers an efficient and reliable approach to address the challenges posed by DDoS attacks in internet networks. PhishNet /citeojewumi is a Google chrome web extension uses the rules extracted from the Random Forest model to alert users when entering a phising website when navigation on their browser, The Random Forest model performed best with an accuracy of 98.35% and a True Positive of 100%, followed by the k-Nearest neighbor model, which had an accuracy of 93.39%, True positive of 98% and a True Negative of 71.43% In 2023, /citeseema DBM-Bi LSTM, introduced by Seema Pillai is an hybrid unsuper- vised system for web attack detection and classification /citeseema using deep learning in order to address the low attack detection ratio found in existing techniques. The model combines De-Noising Autoencoder (DAE) and Stacked Autoencoder (SAE) to enhance feature representation and utilizes a Generative Adversarial Network (GAN) for web attack detection. The evaluation of the model on the CSIC2010v2 dataset showcases high accuracy, achieving 98.6% for attack classi- fication. The model detects various types of web attacks, such as SQL injection, XSS, and parameter tampering. Muralitharan Krishnan's introduces /citemuralitharan an hybrid approach using stacking ensemble techniques to detect and defend against Cross-Site Scripting (XSS) attacks in web applications. By combining machine learning algorithms (NB, SVM, kNN) to prepare data and deep learning frameworks (RNN, CNN, LSTM) for classification, the system achieves 99.5% of accuracy in XSS attack detection.

2.2 Vulnerability Detection

2.2 Vulnerability detection Due to the extensive adoption of deep learning techniques in natural language processing (NLP), researchers have increasingly dedicated their attention to leveraging these methods in the realm of code processing. Wei Tang and Mingwei Tang present CSGVD [18], a new approach for source code vulnerability detection. Their method combines sequence and graph embedding techniques to extract vulnerability features at the function level. CSGVD achieves superior performance compared to existing models, achieving an accuracy of 64.46% on a real-world benchmark dataset from CodeXGLUE [21] The dataset used for evaluation is obtained from CodeXGLUE, which consists of 27,318 samples from two popular C language open-source projects, QEMU and FFmpeg Samuel Ndichus [6] utilizes machine learning to detect XSS and XSRF JavaScript attacks by analyzing Abstract Syntax Tree (AST) features and paragraph vectors. By representing code structure with AST and employing Doc2Vec feature learning, the system achieves accurate and fast classification of malicious JavaScript codes. It outperforms conventional approaches in accuracy and classification speed, even for hard-to-detect malicious codes. The experiments involve a dataset of 2,512 malicious and benign JS codes, using SVM as the classifier and evaluating performance with precision, recall, and F1-score. The proposed system's effectiveness in protecting against JS-based attacks is demonstrated through visualization of results using a ROC graph and calculating the AUC.

JStrong [9] was developed by Yong to detect malicious JavaScript code. It utilizes a graph neural network to analyze the codes syntax and semantic features. By generating an abstract syntax tree and program dependency graph, JStrong achieves near-perfect classification performance. It outperforms existing methods and achieves 99.95% accuracy. The dataset used includes real-world data from top websites and GitHub, demonstrating its effectiveness in detecting JavaScript malware. The limitations include the inability to fully capture code information and the reliance on known malicious samples.

Another system named VUDENC, developed by Dam et al., introduces a novel approach to vulnerability detection through deep learning applied to Python source code. Utilizing a large dataset extracted from GitHub repositories, the system employs LSTM models with word2vec embeddings to predict vulnerabilities in code tokens. The research covers seven major vulnerability types including SQL injections and cross-site scripting. Training data is labeled based on commit contexts. VUDENC achieves recall rates of 78%–87%, precision of 82%–96%, and F1 scores of 80%–90%, successfully identifying vulnerable code segments and providing confidence levels for predictions. However, limitations include potential labeling errors, missed vulnerabilities, and dependency complexities. These findings contribute to automated vulnerability detection in real-world codebases.

Zhou et al.'s VulDeePecker presents a detailed automated vulnerability detection model focused on C/C++ source code, utilizing slicing to capture API and library-related code sections, training a deep neural network. However, its effectiveness is hindered by its dependency on predefined vulnerability libraries. Meanwhile, Xu et al.'s Developer system yields results including a precision of 76.2%, recall of 68%, F1-score of 72.4%, FNR of 31.1%, and FPR of 17.7%. Notably, the system's reliance on existing expert rules for vulnerability data collection impacts data accuracy, revealing inherent limitations.

AutoVAS, an innovative automated vulnerability analysis system that harnesses the power of deep learning. This system transforms intricate source code into comprehensive embedding vectors, utilizing data derived from the National Vulnerability Database (NVD) and Software Assurance Reference Database (SARD). AutoVAS adeptly forecasts and identifies vulnerabilities through advanced techniques such as program slicing and word embedding. Noteworthy outcomes reveal a low 3.62% false negative rate (FNR), a 1.88% false positive rate (FPR), and an impressive 96.11% F1-score. In practical application, AutoVAS successfully pinpoints eleven vulnerabilities, notably three zero-day instances. Acknowledging its scope, the paper recognizes AutoVAS' focus on C/C++ and its current limitation in addressing open-source projects and commits. This pioneering approach marks a significant stride towards automated vulnerability analysis and fortified security practices.

3 Comparison of Existing Web Attack Detection Models

A comprehensive study analyzed the efficiency of current web attack detection models, comparing and evaluating their performance to enhance web security.

The research aimed to identify strengths, weaknesses, and areas for improvement, contributing to the development of more effective detection systems.

Table 1. Comparison of deep learning based systems for web attacks detection

Source	Year	Web-attacks	DL method	Dataset	Acc(%)
[17]	2023	SQL injection, XSS	Generative Adversarial Network	CSIC2010v2	98.6
[10]	2023	XSS	RNN, CNN, LSTM	Github, Kaggle	99.5
[8]	2022	SQL injection, XSS	CNN, RNN	HTTP DATASET CSIC 2010	94
[13]	2022	SQL injection, XSS, others	CNN	HTTP DATASET CSIC 2010	99.94
[16]	2022	DDoS attacks	K-means, ML-IDS	LLDOS	89,23
[11]	2022	Phising attacks	Random Forest , KNN	-	98.35

The comparison of existing web attack detection models (Table 1) reveals several informations. Among the models analyzed, [17] achieved a high accuracy of 98.6% in detecting SQL injection and XSS attacks using a Generative Adversarial Network. [10] demonstrated a remarkable accuracy of 99.5% in detecting XSS attacks by combining RNN, CNN, and LSTM models. [12] focused on a wide range of attacks and achieved an accuracy of 88.4% using a CAPEC classification approach. Notably, [13] achieved an impressive accuracy of 99.94% in detecting SQL injection, XSS, and other attacks using a CNN model

4 Comparison of Existing Code Vulnerability Detection Models

This section provides a thorough comparison of code vulnerability detection models, assessing their effectiveness in identifying security risks in software applications. The analysis examines methodologies, datasets, and performance metrics to reveal strengths, limitations, and potential areas for improvement in code vulnerability detection.

In our evaluation of code vulnerability detection models, we conducted a comprehensive analysis of various approaches. CSGVD [18], utilizing sequence and graph embedding techniques, achieved an accuracy of 64.46% in 2023. Samuel's approach, which employed SVM and Doc2Vec on JavaScript datasets

Table 2. Comparison of deep learning based systems for code vulnerability detection

Name	Data	Label	Methods	Language	Accuracy	FPR	FNR	Precision	F1
CSGVD [18]	from benchmark Code XGLUE	seq and graph embedding	graph neuronal network [21]	C/C++	64.46%	NP	NP	61.8%	60.39%
[6]	D3M [1] [3]	from Dataset	SVM, Doc2Vec	Javascript	NP	NP	NP	NP	NP
JStrong [9]	from GitHub	Pre-existing	Graph neural network	C/C++	99%	NP	NP	NP	NP
[4]	from GitHub	Pre-existing	SVM	Java	63%	NP	NP	67%	65%
VulDee -Pecker [5]	from Github	synth. Patches manual	BLSTM	C/C++	NP	17.7%	31.1%	76.2%	72.4%
AutoVAS [7]	NVD + SARD + CWE + CVE	Pre-existing	RNN, LTSM	C/C++	NP	1.88%	3.62%	95.8%	96.11%
VUDENC [14]	from Github	crawling commit history	RNN, CNN, random forest classifier	Python	82% - 96%	NP	NP	85%	72%
[2]	From open source	Static analysis tool	grid search	Java	87%	NP	NP	85%	85%
DEVELO -PER [15]	from Github + NVD + SARD	crawling commit history	CNN	C / C++ / Java	NP	13.9%	18.1%	84.1%	85.1%

[1,3], yielded results not available (NP) in the provided table. JStrong's graph neural network [9], focused on detecting JavaScript malware, demonstrated an outstanding accuracy of 99%. However, due to limited information in the table, specific values for false positives (FPR), false negatives (FNR), and precision (Pre) are not available (NP).

Other models, such as Pang's SVM-based method for Java [4], showed 63% accuracy with specific precision of 67% and F1 score of 65%. VulDee-Pecker [5] utilized BLSTM for C/C++ code and achieved notable results. AutoVAS [7], which incorporated data from NVD, SARD, CWE, and CVE, exhibited high accuracy (NP) with low FPR and FNR.

Lastly, VUDENC [14] achieved accuracy ranging from 82% to 96% with a precision of 85% and an F1 score of 72% for Python. Hovsepyan's approach [2], using real data and static analysis tools, achieved an accuracy of 87% with precision and F1 scores both at 85%/ and DEVELOPER [15] system achieved a precision of 84.1% and F1 score of 85.1% for C, C++ and JAVA.

These comparative findings shed light on the effectiveness of different machine learning and deep learning models for code vulnerability detection.

5 Conclusion

In conclusion, this research paper provides a comprehensive examination of the current landscape of web attack detection and vulnerability identification. Our study offers insights that can guide future research efforts in enhancing web application security. While we acknowledge significant advancements in addressing specific web attacks like XSS and SQL injections, there remains a noticeable gap when it comes to comprehensive web attack detection, particularly concerning intricate threats like DDoS attacks and all-encompassing vulnerability identification. Our future endeavors will be dedicated to bridging these gaps and advancing web attack detection methodologies.

A crucial observation is the underrepresentation of JavaScript, a highly vulnerable programming language, in deep learning research. This raises pertinent questions about the efficacy of existing security measures in countering JavaScript-based attacks. Our research seeks to redress this imbalance by focusing on innovative approaches to JavaScript vulnerability detection. Another observation of our research is the need for innovative approaches to labeling data for vulnerability detection. While current practices involve automating data labeling from GitHub commits, we acknowledge the challenges in identifying security fixes within code commits that also contain unrelated changes. This insight prompts us to consider new strategies for future work in this area.

To achieve these goals, we plan to explore new strategies and methodologies that consider the evolving nature of web attacks. By adapting and innovating, we can better defend against malicious activities, stay ahead of cyber threats, and ensure the long-term security and integrity of web applications. This research paper serves as a foundation, inviting further investigation and collaboration to create a more secure cyberspace for all.

References

1. Akiyama, M., et al.: Design and implementation of high interaction client honeypot for drive-by-download attacks. In: IEICE Trans. Commun. **93**(5), 1131–1139 (2010)
2. Hovsepyan, A., et al.: Software vulnerability prediction using text analysis techniques. In: Proceedings of the 4th International Workshop on Security Measurements and Metrics, pp. 7–10 (2012)
3. Hatada, M., et al.: Empowering anti-malware research in Japan by sharing the MWS datasets. J. Inform. Process. **23**(5), 579–588 (2015)
4. Pang, Y., Xue, X., Namin, A.S.: Predicting vulnerable software components through n-gram analysis and statistical feature selection. In: 2015 IEEE 14th International Conference on Machine Learning and Applications (ICMLA). IEEE, pp. 543–548 (2015)
5. Li, Z., et al.: Vuldeepecker: A deep learning-based system for vulnerability detection. In: arXiv preprint arXiv:1801.01681 (2018)

6. Ndichu, S., et al.: A machine learning approach to detection of JavaScriptbased attacks using AST features and paragraph vectors. Appl. Soft Comput. **84** 105721 (2019). https://doi.org/10.1016/j.asoc.2019.105721
7. Jeon, S., Kim, H.K.: AutoVAS: An automated vulnerability analysis system with a deep learning approach. Comput. Secur. **106**, 102308 (2021) ISSN: 0167–4048. https://doi.org/10.1016/j.cose.2021.102308, https://www.sciencedirect.com/science/article/pii/S0167404821001322
8. Eunaicy, J.C., Suguna, S.: Web attack detection using deep learning models. Materials Today: Proc. **62**(7), 4806–4813 (2022). https://doi.org/10.1016/j.matpr.2022.03.348
9. Fang, Y., et al.: JStrong: Malicious JavaScript detection based on code semantic representation and graph neural network. Comput. Secur. **118** 102715 (2022). https://doi.org/10.1016/j.cose.2022.102715
10. Krishnan, M., et al.: Detection and defending the XSS attack using novel hybrid stacking ensemble learning-based DNN approach. In: Digital Communications and Networks (2022), pp. 2352–8648. https://doi.org/10.1016/j.dcan.2022.09.024
11. Ojewumi, T.O., et al.: Performance evaluation of machine learning tools for detection of phishing attacks on web pages. Sci. African **16**, 102788 (2022). https://doi.org/10.1016/j.sciaf.2022.e01165
12. Riera, T.S., et al.: A new multi-label dataset for Web attacks CAPEC classification using machine learning techniques. Comput. Secur. **120**, 102788 (2022). https://doi.org/10.1016/j.cose.2022.102788
13. Shahid, W.B., Aslam, B.: An enhanced deep learning based framework for web attacks detection, mitigation and attacker profiling. J. Netw. Comput. Appl. **198**, 103270 (2022). https://doi.org/10.1016/j.jnca.2021.103270
14. Wartschinski, L., et al.: VUDENC: Vulnerability Detection with Deep Learning on a Natural Codebase for Python. Inform. Softw. Technol. **144**, 106809 (2022) 106809. ISSN: 0950–5849. https://doi.org/10.1016/j.infsof.2021.106809, https://www.sciencedirect.com/science/article/pii/S0950584921002421
15. Xu, R., et al.: Detecting code vulnerabilities by learning from largescale open source repositories. J. Inform. Secur. Appl. **69**, 103293 (2022). ISSN: 2214–2126. https://doi.org/10.1016/j.jisa.2022.103293, url: https://www.sciencedirect.com/science/article/pii/S221421262200148X
16. Kishore, P.K., Ramamoorthy, S., Rajavarman, V.N.: ARTP: anomaly based real time prevention of Distributed Denial of Service attacks on the web using machine learning approach. Int. J. Intell. Netw. **4**, 102788 (2023). https://doi.org/10.1016/j.ijin.2022.12.001
17. Pillai, S., Sharma, A.: Hybrid unsupervised web-attack detection and classification - a deep learning approach. Comput. Standards Interfaces **86**, 103738 (2023). https://doi.org/10.1016/j.csi.2023.103738
18. Tang, W., et al.: CSGVD: A deep learning approach combining sequence and graph embedding for source code vulnerability detection. J. Syst. Softw. **199**, 111623 (2023). https://doi.org/10.1016/j.jss.2023.111623
19. comparitech. Cybersecurity crime statistics, facts and trends. https://www.comparitech.com/vpn/cybersecurity-cyber-crime-statistics-factstrends/
20. cybersecurityventures. Cybercrime damages. https://cybersecurityventures.com/cybercrime-damages-6-trillion-by-2021/
21. The United States Department of Justice. Benchmark dataset. https://datasets-benchmarks-proceedings.neurips.cc/paper/2021/hash/c16a5320fa475530d9583c34fd356ef5-Abstract-round1.html

22. The United States Department of Justice. Cyber Attacks. https://www.justice.gov/cyber/cyber-attacks
23. purplesec. Cybersecurity statistics. https://purplesec.us/resources/cyber-security-statistics/
24. techjury. How many cyber attacks per day. https://techjury.net/blog/how-many-cyber-attacks-per-day/
25. zippia. Cybersecurity statistics. https://www.zippia.com/advice/cybersecurity-statistics/

AI-Driven Cyber Risk Management Framework

Yasser Agzayal(✉) and Mohammed Bouhorma

Computer Science, Systems, and Telecommunication Laboratory (LIST),
University Abdelmalek Essaadi, Tangier, Morocco
yasser.agzayal@etu.uae.ac.ma, mbouhorma@uae.ac.ma

Abstract. As technological advancements continue to shape smart cities, the complex and interconnected digital networks within these infrastructures are increasingly susceptible to cyber threats. This PhD thesis explores the application of Artificial Intelligence (AI) as a powerful tool for managing and mitigating cyber risks in these advanced infrastructures and network systems. We focus on developing a proactive approach that combine the early detection, comprehensive assessment, and effective mitigation of impending cyber threats. To achieve this goal, we propose an AI-based framework that harnesses the power of Machine Learning (ML), Reinforcement Learning (RL), Natural Language Processing (NLP), and Graph Theory. By employing these techniques, our framework aims to fortify network and system resilience, providing robust cyber risk management by artificial intelligence and for smart cities. This work paves the way for innovative risk management strategies that not only react to cyber threats but also anticipate and neutralize them effectively.

Keywords: Cybersecurity · Artificial Intelligence · Smart cities · Risk Management · Governance & Data protection

1 Introduction

In the contemporary digital landscape, the backbone of economic, social systems and smart cities are increasingly built on networks and digital systems. These ecosystems have grown exponentially in both scale and complexity. The pervasive interconnectedness of devices, coupled with the transition of data and services to cloud infrastructures, has ushered in a new era of opportunities. However, with these opportunities come new vulnerabilities. Cyber risk management, a multifaceted discipline that aims to identify, assess, and mitigate risks associated with digital assets, has become a cornerstone in protecting the confidentiality, integrity, and availability of these systems. The motivation behind this research lies in the growing urgency to protect critical infrastructures as they continue to be besieged by increasingly sophisticated cyber threats and attacks. The dynamism of the threat landscape necessitates innovative and proactive approaches to cyber risk management. Traditional cyber risk management approaches often suffer from a series of limitations which make them inadequate in addressing the evolving

M. Ben Ahmed et al. (Eds.): SCA 2023, LNNS 906, pp. 571–584, 2024.
https://doi.org/10.1007/978-3-031-53824-7_51

challenges of today's cyber environment. Consequently, the research question is formulated as follows: "How can Artificial Intelligence be applied to improve the identification, assessment, and mitigation of cyber risks in networks and systems? Through this question, our primary objective of this research is to develop an AI-driven cyber risk management framework that can effectively address the challenges posed by the rapidly evolving cyber threat landscape. To achieve this goal, this thesis is organized into several chapters, each focusing on a specific aspect of the research process. The structure of the thesis is as follows, Introduction: This chapter provides an overview of the research problem, background, and contributions for the study. Literature review: This chapter presents a comprehensive review of the existing literature in the areas of cyber risk management, artificial intelligence, and machine learning. AI-driven Cyber Risk Management Framework: This chapter introduces the comprehensive AI-driven cyber risk management framework, it details the methodology, which integrates the risk assessment and decision-making models, as well as other essential components such as data collection, monitoring, and visualization. AI-driven Cyber Risk Assessment Model: This chapter focuses on the development of an AI-driven cyber risk assessment model. AI-based Decision-making Model for Cyber Risk Mitigation: This chapter presents the design and implementation of an AI-based decision-making model for recommending and optimizing cyber risk mitigation strategies. Performance Evaluation: This chapter evaluates the performance and effectiveness of the proposed AI-driven cyber risk management frame- work using both simulated datasets. Results: this chapter provide examples of achievements or model outputs. Conclusion and Future Work: This final chapter summarizes the main findings and contributions of the research.

2 Literature Review

As the study ventures into exploring the potential role of Artificial Intelligence in cyber risk management, it becomes imperative to first comprehend the existing frameworks that have been widely adopted for managing cyber risks. Among these, ISO 27005, EBIOS Risk Manager (RM), and NIST Risk Management Framework (RMF) have been seminal.

2.1 ISO27005

ISO 27005, part of the ISO 27000 series of standards, is an international standard that provides guidelines for risk management processes related to information security. It is characterized by its systematic and structured approach, focusing on risk assessment, treatment, monitoring, and review. It emphasizes the importance of understanding the organizational context and aligning the risk management process with the business objectives.

2.2 EBIOS RM

In contrast, EBIOS (Expression des Besoins et Identification des Objectifs de Sécurité) Risk Manager is a methodology developed by the French Network and Information

Security Agency (ANSSI). EBIOS RM adopts a scenario-based approach and focuses on identifying assets, risks, and proposing relevant security objectives. The methodology is especially known for its ability to facilitate communication between stakeholders by offering a common language and structured method to express and analyze security needs.

2.3 NIST Risk Management Framework (RMF)

On the other hand, the NIST Risk Management Framework (RMF) is a U.S. government standard that encompasses a comprehensive set of guidelines to manage cyber- security risks. The NIST RMF is centered around six steps – categorize, select, implement, assess, authorize, and monitor, and is widely lauded for its emphasis on continuous monitoring and ongoing authorization.

2.4 Related Work on AI-Driven Cyber Risk Management Models

In the realm of cyber risk mitigation, the confluence of Artificial Intelligence (AI) with risk assessment and decision-making models has been the subject of extensive research and development. Here, we review some of the notable literature and related works in this area:

- In paper [1] Deep Learning for Cybersecurity: A Comprehensive and Adaptable Risk Management Model by Rauf, U., et al. (2021): This recent study highlights an adaptable risk management model that incorporates deep learning. The authors focus on how deep learning algorithms can be employed for threat identification, risk assessment, and decision-making in cybersecurity.
- In [2] NISTIR 8286, Integrating Cybersecurity and Enterprise Risk Management (ERM) by NIST (2020): This NIST publication is relevant in the context that it displays how cybersecurity can be integrated with enterprise risk management. While this document does not exclusively address AI risk management, it is indicative of the comprehensive approach that NIST might adopt in integrating AI risks within an enterprise risk management framework.
- In [3] NIST AI 100–1 Artificial Intelligence Risk Management Framework (AI RMF 1.0) by NIST (2023): NIST develop an AI based risk management model in accordance with the mandate set by the National Artificial Intelligence Initiative Act of 2020 (Public Law 116–283), the objective of the AI Risk Management Framework (AI RMF) is to serve as a valuable resource for organizations. They present functions like GOVERN, MAP, MEASURE, and MANAGE which can be applied in AI system-specific contexts and at specific stages of the AI lifecycle. The Framework is designed to be voluntary and to uphold rights.
- In [4] Lanzi et al. proposed a system-centric behavioral model. According to the study, the way malicious software interacts with system resources including directories, files, registries, etc.
- In [5] W. Han proposed a novel malware behavioral-based detection method called APTMalInsight this method identified and cognized Advanced Persistent Threats (APTs) which rely on the system call information and ontology knowledge framework.

- In paper [6] Gibert et al. 2019 Based on the presentation of malware as an image, the following work presents a convolutional neural network (CNN) composed of three convolution layers followed by a fully connected layer used for the classification of malware.
- In this paper [7] Saxe and Berlin propose a system which consists of extracted features from malicious and benign samples, in the second, a deep neural network consisting of an input layer, two hidden layers and output layer is built.
- Ye et al. [8] A heterogeneous deep learning system was suggested for the detection of zero-day malware. The suggested system operates by employing multilayer restricted Boltzmann machines in conjunction with related memory.

3 Proposed AI-Driven Cyber Risk Management Framework

In this chapter, we will delve into the details of the theoretical study of our methodology, we'll also explore the different components of our proposed AI-driven cyber risk management framework.

3.1 Preparation Phase

During this phase we will go through three main activities, the goal is to define the risk management context for a given use case and to load required data by our AI-model for better risk analysis and prediction.

- **Establishing the Context.** Setting Boundaries in an expansive digital realm is essential when it's easy to get lost in vast amounts of data. By defining clear boundaries, the framework is guided to focus only on data relevant to the scenario, ensuring precision and relevancy in its operations. so, the key stages in this step are identifying scope and limits in terms of study area, assets, technologies, and business processes. And define the risk tolerance levels.
- **Data Collection**. Data Collection primarily involves gathering data, and as such, it's not about algorithms but about defining strategies and using tools to collect data like, APIs and Web Scraping for collecting data from external sources like threat intelligence feeds. Network Traffic Capture Tools (such as Wireshark): For collecting network traffic data. Logging Services (such as Elasticsearch, Logstash, and Kibana (ELK) stack): For collecting system logs. Vulnerability scanners or databases such as Nessus, Intruder or else.
- **Data pre-processing**. Data Preprocessing involves cleaning and transforming raw data into a format that can be effectively used by ML algorithms, including techniques like: Managing missing data, imputation, where missing values are filled in based on the values of other data points. Normalization and Standardization: Algorithms like Min-Max Scaler for normalization and Z-score normalization for standardization. Feature Extraction and Engineering: Techniques like Principal Component Analysis (PCA) for dimensionality reduction or using domain knowledge to create new features.

3.2 Assessment Phase

This phase presents our approach for assessing and prioritizing cyber risks using ML algorithms and Graph Theory. The risk score typically considers the likelihood of the threat being exploited and the potential impact if it is. The steps involve identify, assess, and evaluate risk scenarios and they are part of our risk assessment process. I'll explain these steps in detail below.

- **Risk identification**. This step involves recognizing potential threats and vulnerabilities, we used Machine learning algorithms such as Isolation Forest and One-Class SVM to automate threat detection, Pattern Recognition algorithms like Regular Expression Matching for identifying known attack patterns in logs or network traffic, and Natural language processing (NLP) to analyze and extract information from various data sources like threat intelligence reports.
- **Risk analysis.** Risk analysis is about understanding the likelihood and potential impact of the identified risks, we used techniques such as Predictive Modeling like Random Forest, Gradient Boosting to estimate the likelihood of diverse types of threats, and simulation techniques such as Monte Carlo simulations to estimate the potential impact of various risk events.
- **Risk evaluation**. Risk Evaluation involves deciding which risks need to be addressed, we used techniques such as Ranking algorithms like Weighted Sum Model (WSM) to prioritize risks based on their likelihood and impact, and Threshold techniques based on the organization's risk tolerance levels to decide which risks are acceptable and which need mitigation.

3.3 Decision Phase

The decision phase or the AI decision making model implemented in our case uses the output from the risk assessment and the studied environment to take which strategies are most effective, this is where the Reinforcement Learning algorithms comes into play, to find an optimal policy.

- **Define Objectives and Constraints**. Defining objectives and constraints is about establishing a threshold for the level of risk that is acceptable, anything beyond this threshold should trigger mitigation actions. And by determining available resources for mitigation by identifying what resources (financial, human, technological) are available for deploying mitigation strategies.
- **Generate Mitigation Strategies**. Generating mitigation strategies is about analyzing historical data and current network state to suggest potential mitigation strategies. For example, if a certain type of attack is detected, AI can suggest strategies that have worked in the past against this type of attack, our strategy is the application of Game Theory for strategy optimization, by using game theory to model the interaction between the defender (The security system) and the attacker as a game, where each has strategies, they can use.
- **Decision-making with Reinforcement Learning (RL).** We used the Q-learning algorithm to learn the value of actions in states, which means implementing a Q-learning algorithm where each state represents a particular configuration of cyber risk and the actions represent different mitigation strategies, the algorithm learns the

expected future rewards of taking certain actions in different states. For example, in a state where there is a high risk of data breach, one action might be to strengthen firewall rules. Q-learning will learn how effective this is by exploring and exploiting different actions over time. The strategy is to optimize decision-making by maximizing rewards (e.g., reduced risk) and minimizing penalties (e.g., costs), we used the Q-values learned by the Q-learning algorithm to make decisions that maximize the expected future reward, the reward can be defined as a function of the effectiveness of the mitigation strategy in reducing risk, minus any costs associated with the strategy. Continuous Learning and adaptation allow the system to continually learn and adapt, as new data is obtained and the cyber threat environment evolves, the Q-values can be updated, and the system can adapt its strategies.

3.4 Deployment Phase

Mitigating cyber risks effectively hinges on the deployment of countermeasures. As our AI-driven framework has discerned and assessed risks, the critical next phase revolves around crafting and deploying intelligent mitigation strategies. Here's how we've approached this.

- **Approve selected strategies**. We make sure that the actions chosen by the RL agent also adhere to the objectives and constraints defined by stakeholders, even if a certain action has high Q-value, it should not be selected if it violates the risk tolerance levels or consumes resources beyond the available budget.
- **Implementation of selected strategies.** Implement the chosen mitigation strategies within the constraints defined. Rule-based Systems: Implement selected strategies based on rules derived from the decision-making phase. Automated Scripts: Utilize scripts that can be automatically executed based on the chosen strategies.

3.5 Monitoring Phase

The objective of these phase is to continuously monitor the system for new risks and the effectiveness of deployed mitigation strategies, by providing this feedback into the Q-learning algorithm so that it can refine its Q-values and make better decisions in the future.

- **Real time and continuous monitoring.** We used techniques like Time-series Analysis: for monitoring the time-bound data like network traffic and supervising the system status. Also, Anomaly Detection Algorithms (e.g., Isolation Forest, Autoencoders) to detect unusual activities or deviations from the norm, which could indicate an attack or failure of the mitigation strategy.
- **Provide feedback to the model.** We used performance key indicators for continuous improvement, we also used Feedback Loops to update the Q-learning algorithm with new data, so it continuously improves its performance over time.

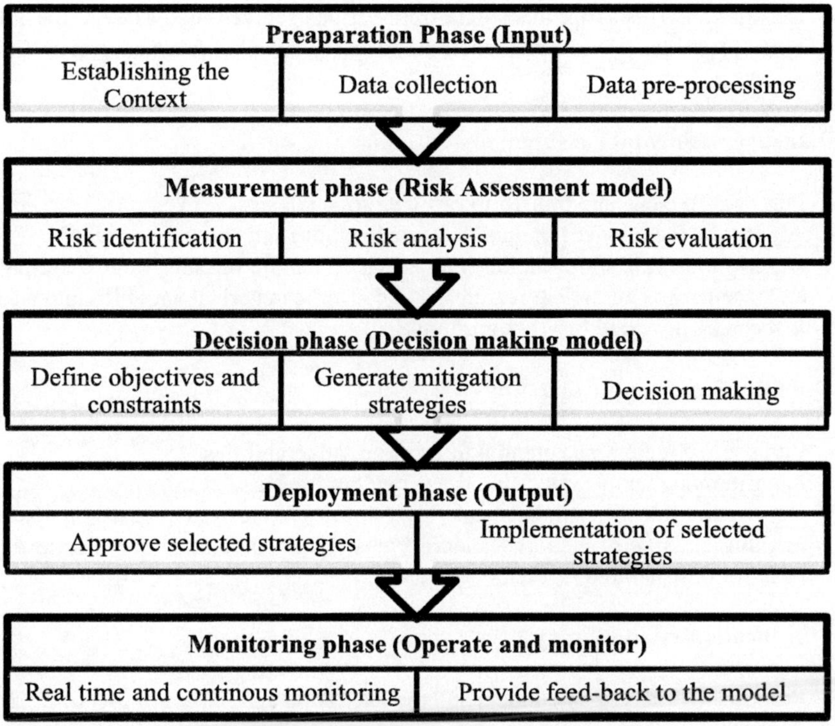

Fig. 1. Graphical illustration of the distinct phases of the proposed AI risk management framework-based on machine learning algorithms.

4 AI-Driven Cyber Risk Management Implementation

The primary objective of this research is to develop an AI-driven cyber risk management framework that can effectively address the challenges posed by the rapidly evolving cyber threat landscape. To achieve this goal, we combine different techniques like data processing, classification, reinforcement learning, game theory, graph analysis and NLP, each one of them are separated on their own module, and each module contains Python scripts for specific functionalities. To ensure that all the components of our AI-based Cyber Risk Management system interact properly, we design the system in a modular way. This involves creating separate classes or modules for each component and using a central orchestrator class to manage the flow of data and control between them. Below is a structure of the project that combines the mentioned components. I'll explain these scripts in detail:

– **Data_processing/data_loader.py**

- Purpose: To load and prepare data for analysis.
- Algorithm/Technique: Pandas for loading CSV, SQL for querying databases.

- Description: This script loads data from various sources such as CSV files, data bases, or APIs. It can have functions to query specific data and prepare it for preprocessing.

– **Data_processing/data_cleaning.py**

- Purpose: To clean and transform raw data for analysis.
- Algorithm/Technique: Imputation, normalization, and one-hot encoding.
- Description: This script includes functions to handle missing values, normalize and standardize numerical features, and convert categorical variables into a form that can be provided to ML algorithms.

– **Classification/random_forest.py**

- Purpose: To recognize potential threats and vulnerabilities.
- Algorithm/Technique: Decision trees, random forests.
- Description: Identifies threats and vulnerabilities in the system based on machine learning classifiers, such as Random Forest. This component can also leverage NLP for unstructured text data.

– **Classification/evaluation.py**

- Purpose: To analyze and understand the likelihood and impact of identified risks.
- Algorithm/Technique: Bayesian networks, decision trees.
- Description: Evaluates the probability of each risk occurring and the potential impact it could have. It could use Bayesian networks to consider conditional probabilities.

– **Game_theory/strategies.py**

- Purpose: To generate possible risk mitigation strategies.
- Algorithm/Technique: Game Theory, Genetic Algorithms, Simulated Annealing.
- Description: This script uses optimization algorithms and game theory to generate a set of possible risk mitigation strategies. It evaluates them based on effectiveness and cost.

– **Reinforcment_learning/environment.py**

- Purpose: To simulate a controlled environment for training the reinforcement learning agent.
- Algorithm/Technique: Custom environment rules and states.
- Description: This script defines the simulated environment where the reinforcement learning agent will be trained. It contains the rules, states, actions, and rewards necessary for the agent to learn how to make decisions. The environment should simulate real-world scenarios that the agent could encounter in cyber risk mitigation.

– **Reinforcement_learning/q_learning.py**

- Purpose: To choose the best mitigation strategies.
- Algorithm/Technique: Reinforcement learning (Q-learning, PPO).
- Description: Uses reinforcement learning to select the best mitigation strategy based on the current state of the system. It learns through Q-learning the value of taking certain actions in given states.

– **Orchestrator/strategy_implementer.py**

- Purpose: To implement the chosen mitigation strategies.
- Description: This script will have functions to execute the chosen strategies, possibly through configuration changes, alerts, or other means.

– **Orchestrator/monitoring.py**

- Purpose: To monitor the system and provide feedback.
- Algorithm/Technique: Time series analysis, anomaly detection.
- Description: Monitors the system after implementing the strategies and provides feedback to ensure that risks are effectively mitigated. It detects anomalies and informs the system if something is not going as planned.

– **Orchestrator/orchestrator.py**

- Purpose: To integrate and manage the workflow among all the components.
- Description: This script contains a class that acts as the central manager for coordinating the flow of data and control between different components. It calls the necessary functions and scripts to carry out the cyber risk management process.

– **Main Script (main.py)**

- Purpose: To integrate and manage the workflow among all the components.
- Description: This script is the main entry point that utilizes all the other scripts. It creates an instance of the orchestrator and uses it to manage the data.

5 Performance Evaluation

This chapter evaluates the performance and effectiveness of the proposed AI-driven cyber risk management framework using simulated datasets. It compares the framework's performance against traditional approaches and assesses its ability to improve the efficiency, effectiveness, and adaptability of the cyber risk management process.

5.1 Confusion Matrix

As the model makes predictions on the dataset, we will need to keep track of:

- True Positives (TP): The cases in which the model correctly predicted the positive class.
- True Positives (TP): 400 means that there were 400 instances that were high-risk, and the model correctly identified them as high-risk.
- True Negatives (TN): 450 means that there were 450 instances that were not high-risk, and the model correctly identified them as not high-risk.
- False Positives (FP): 50 means that there were 50 instances that were not high-risk, but the model incorrectly identified them as high-risk.
- False Negatives (FN): 100 means that there were 100 instances that were high-risk, but the model incorrectly identified them as not high-risk.

If we sum up all these values (400 + 450 + 50 + 100), we get 1000, which is the total number of instances in the test dataset. The confusion matrix would be structured as follow:

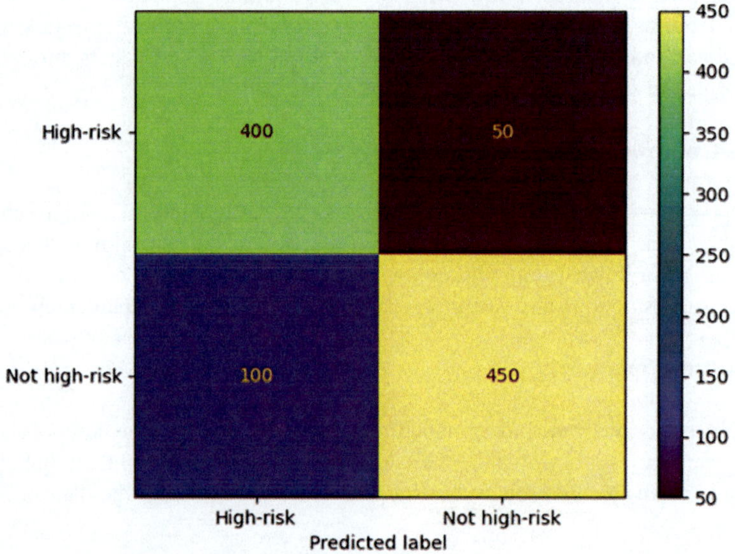

Fig. 2. Graphical illustration of the confusion matrix of the proposed AI risk management framework-based on machine learning algorithms.

5.2 Performance Metrics

Using the values from the confusion matrix, let's calculate the performance metrics:

- Accuracy: It is the ratio of the number of correct predictions to the total number of predictions.

$$\text{Accuracy} = (\text{TP} + \text{TN}) / (\text{TP} + \text{TN} + \text{FP} + \text{FN}) = (400 + 450) / (400 + 450 + 50 + 100) = 0.85 \text{ or } 85\%$$

- Precision: It is the ratio of the number of true positive predictions to the total number of positive predictions made (true positive + false positive).

$$\text{Precision} = \text{TP} / (\text{TP} + \text{FP}) = 400 / (400 + 50) \approx 0.889 \text{ or } 88.9\%$$

- Recall (Sensitivity): It is the ratio of the number of true positive predictions to the total number of actual positive instances.

$$\text{Recall} = \text{TP} / (\text{TP} + \text{FN}) = 400 / (400 + 100) = 0.8 \text{ or } 80\%$$

- F1-Score: It is the harmonic mean of precision and recall and is used when you want to balance these two metrics.

F1-Score = 2 * (Precision * Recall) / (Precision + Recall) ≈ 2 * (0.889 * 0.8) / (0.889 + 0.8) ≈ 0.842 or 84.2%

6 Results

The model has been evaluated on a dataset with 1000 instances, it means that the model is being evaluated on a set of 1000 individual data points or samples. These 1000 instances are a part of the test dataset that the model has not seen during its training phase. The purpose of testing on a separate dataset is to evaluate how well the model can generalize to new, unseen data. These 1000 instances or data points could represent various network activities, system logs, or events that need to be classified as high-risk or not high-risk. There are some examples of what individual instances might look like:

- **Instance 1:** Source IP: 192.168.1.1 - Destination IP: 172.16.0.1 - Source Port: 5060 - Destination Port: 80 - Protocol: TCP - Packet Length: 500 bytes - Timestamp: 2023–06-29 12:00:00 - Number of Failed Login Attempts: 0 - Data Payload: "GET / HTTP/1.1" - Known Malicious Source: No - True Label: Not High-risk - Predicted Label: Not High-risk.
- **Instance 2:** Source IP: 10.0.0.2 - Destination IP: 172.16.0.1 - Source Port: 5061 - Destination Port: 22 - Protocol: TCP Packet Length: 130 bytes - Timestamp: 2023–06-29 12:01:00 - Number of Failed Login Attempts: 5 - Data Payload: "SSH User Authentication" - Known Malicious Source: Yes - True Label: High-risk – Predicted Label: High-risk
- **Instance 3:** Source IP: 203.0.113.1 - Destination IP: 172.16.0.2 - Source Port: 12345 - Destination Port: 22 - Protocol: TCP - Packet Length: 140 bytes - Timestamp: 2023–06-29 12:02:00 - Number of Failed Login Attempts: 1 - Data Payload: "SSH User Authentication" - Known Malicious Source: No - True Label: Not High-risk - Predicted Label: High-risk (False Positive)

And so, on until we have one thousand of these instances. And then our model made the following predictions:

- **Example 1: Risk Scenario Identified:** Risk Scenario: Increase in phishing attempts detected. Potential Impact: Unauthorized access to sensitive data, potential data leakage. Confidence Score (AI Assessment): 87%. AI-based Proposed Mitigations: Implement stricter email filtering rules to reduce the likelihood of phishing emails reaching users. Enforce multi-factor authentication for accessing sensitive resources. Conduct awareness training on phishing for employees.
- **Example 2: Risk Scenario Identified:** Risk Scenario: Outdated server software detected. Potential Impact: Exploitation of known vulnerabilities, potential for unauthorized access or denial of service. Confidence Score (AI Assessment): 92%. AI-based Proposed Mitigations: Schedule and apply security patches and updates to server software. Regularly audit servers for outdated software and vulnerabilities. Implement intrusion detection systems to monitor for unauthorized access attempts.
- **Example 3: Risk Scenario Identified:** Risk Scenario: Unusual data traffic from internal network to an unknown external IP. Potential Impact: Data exfiltration, potential compromise of sensitive information. Confidence Score (AI Assessment): 79%. AI-based Proposed Mitigations: Temporarily block the traffic to the unknown external IP. Investigate to identify the source of the traffic within the internal network. If necessary, update firewall rules to permanently block traffic to suspicious IPs.
- **Example 4: Risk Scenario Identified:** Risk Scenario: Multiple failed logins attempt from a single IP detected. Potential Impact: Brute-force attack, potential unauthorized access to systems. Confidence Score (AI Assessment): 85%. AI-based Proposed Mitigations: Temporarily lock the account after a defined number of unsuccessful login attempts. Implement CAPTCHA challenges to reduce the feasibility of automated brute-force attacks. The confidence score in the examples I provided is a value meant to represent how certain the AI model is of its assessment. In practice, confidence scores are usually derived from the output probabilities of a machine learning model.

7 Comparison with Other Studies

In the realm of AI-Driven cyber risk management, our research introduces a distinctive hybrid approach that harnesses the power of Reinforcement Learning (RL) combined with Graph Theory (GT), Natural Language Processing (NLP) and the principles of continuous improvement. This approach differentiates our framework from other prevalent models in the following ways:

- **Hybrid Approach**: While in [9] Rui Zhang; Quanyan Zhu Department of Electrical and Computer Engineering, New York University, Brooklyn, USA (2019) proposal suggests an A Game-Theoretic Cyber Insurance Framework for Incentive-Compatible Cyber Risk Management of Internet of Things, our methodology diverges by seamlessly integrating RL with continuous improvement. This dual mechanism ensures our framework not only identifies and responds to threats but also constantly refines its strategies, leading to progressively enhanced performance over time.
- **Reinforcement Learning (RL) versus Traditional Machine Learning (ML)** While many studies, including those presented in [10] Diptiban Ghillani, Deep Learning and Artificial Intelligence Framework to Improve the Cyber Security emphasize traditional ML techniques for cyber threat identification, and detection; however,

our model harnesses Reinforcement Learning and Machine Learning techniques to dynamically adapt in changing threat environments.

- **Graph Theory and Natural Language Processing (NLP) Synergy** In [11] Buczak & Guven (2016) elucidate the significance of machine learning in intrusion detection. However, our methodology distinctively blends graph theory and NLP. This dual integration fosters nuanced risk assessments, unveiling intricate threat patterns. Such comprehensive assessments outstrip models that rely on mere categorical and numerical data.
- **Integrated Data Collection**: The foundational strength of our framework lies in its exhaustive data amalgamation, sourcing from network logs, vulnerability databases, and threat intelligence. In [12] J. AL-Saraieh, A. Masarwe, present a novel approach for detecting advanced persistent threats, and our model's data collection paradigm ensures such intricate techniques.
- **Continuous Improvement:** The guiding principle of continuous improvement pervades our model. Unlike in [13] Jun Zhang, and Yang Xiang which highlights Detecting and Preventing Cyber Insider Threats: A Survey, our framework emphasizes self-evolution. It learns, recalibrates, and adapts to threats, both extant and emergent.
- **Holistic Nature**: The combination of Reinforcement Learning and continuous improvement gives our framework a holistic touch. In contrast to singular-focused studies such demonstrated which emphasizes the changing nature of cyber risks, our approach is comprehensive, it not only identifies and reacts to risks but also engages in initiative-taking evolution, ensuring the framework remains relevant in an ever-evolving digital landscape.

While several research have made significant contributions to the domain, our thesis combines techniques such as Reinforcement Learning, Graph Theory, Natural Language Processing with continuous improvement. This hybrid approach, rooted in adaptability and evolution, provides a dynamic solution to the multifaceted challenges posed by cyber threats in the modern era of smart cities and advanced infrastructures.

8 Conclusion

In conclusion, the field of AI-based Cyber Risk Management is ripe for continued innovation and development. This thesis represents a step towards a more secure and intelligent approach that encompasses risk identification, assessment, generation of mitigation strategies, decision-making using reinforcement learning, implementation of strategies, and monitoring, has enabled the system to not only recognize and assess cyber risks but also actively mitigate them in an optimized manner.

9 Future Work

While the framework proposed in this thesis has laid a robust foundation for AI-based Cyber Risk Mitigation, there are several areas where future research and development can bring significant improvements:

- **Real-world Application and Testing**: A critical next step is the application and testing of this framework in real-world scenarios to evaluate its effectiveness and to calibrate the models based on real data.
- **Integration with IoT Devices**: As IoT devices become more pervasive, integrating the framework with IoT networks to safeguard them against cyber threats will be vital.

References

1. Rauf, U., et al.: Deep Learning for Cybersecurity: A Comprehensive and Adaptable Risk Management Model (2021)
2. NISTIR 8286, Integrating Cybersecurity and Enterprise Risk Management (ERM) by NIST (2020)
3. NIST AI 100–1 Artificial Intelligence Risk Management Framework (AI RMF 1.0) by NIST (2023)
4. Lanzi, A., Balzarotti, D., Kruegel, C., Christodorescu, M., Kirda, E.: AccessMiner: using system-centric models for malware protection. In: Proceedings of 17th ACM Conference on Computer Communication Security (CCS), pp. 399–412 (2010)
5. Han, W.: APTMalInsight Fujian Provincial Key Laboratory of Information Processing and Intelligent Control. Minjiang University, Fuzhou, China (September 2020)
6. Gibert, D.: Convolutional neural networks for malware classification (Oct. 2016)
7. Saxe. J., Berlin, K.: Deep neural network-based malware detection using two-dimensional binary program features. In: Proceedings of 10th International Conference on Malicious Unwanted Software, pp. 11–20 (Oct. 2015)
8. Ye, Y., Wang, D., Li, T., Ye, D.: IMDS: Intelligent malware detection system. In: Proceedings of 13th ACM SIGKDD International Conference on Knowledge Discovery Data Mining (KDD), pp. 1043 -1047 (2007)
9. Zhang, R., Zhu, Q.: Department of Electrical and Computer Engineering, New York University, Brooklyn, USA Game-Theoretic Cyber Insurance Framework for Incentive-Compatible Cyber Risk Management of Internet of Things (2019)
10. Ghelni, D.: Deep Learning and Artificial Intelligence Framework to Improve the Cyber Security. American Journal of Artificial Intelligence (2022)
11. Buczak, A., Guven, E.: Computer Science. IEEE Communications Surveys & Tutorials Published 22 January 2016
12. AL-Saraieh, J., Masarwe, A.: A novel approach for detecting advanced persistent threats King Hussein School of Computing Sciences, Princess Sumaya University for Technology, PSUT, Jordan
13. Zhang, J., Xiang, Y.: Detecting and Preventing Cyber Insider Threats: A Survey IEEE Communications Surveys & Tutorials **20**(2) (Second quarter 2018)

Intrusion Detection Using Time-Series Imaging and Transfer Learning in Smart Grid Environments

Firas Abou Naaj(✉), Yassine Himeur, Wathiq Mansoor, and Shadi Atalla

College of Engineering and IT, University of Dubai, Dubai, UAE
S0000002775@ud.ac.ae

Abstract. Intrusion detection systems (IDS) monitor and analyze network traffic and system activity to detect and alert security personnel to potential security breaches or attacks. Although deep learning models have shown great promise in improving the accuracy and efficiency of IDSs, several challenges are associated with their use, including data scarcity and model complexity. Furthermore, to overcome these problems, deep transfer learning is considered in this study. Typically, this article presents a novel intrusion detection (ID) approach using transformed 1D signals into 2D representations and applying pre-trained convolutional neural network (CNN) models. The transformed 2D representations of the signals allow the pre-trained CNN models to effectively learn the features of the signals and accurately classify them as normal or malicious. The performance of the proposed method was evaluated on the CIC-IDS-2018 dataset, and the results showed 92% accuracy in differentiating between normal behavior and malicious activities, which is an improvement compared with other detection methods.

Keywords: Intrusion Detection · Smart Grid · Time-series imaging · transfer learning

1 Introduction

1.1 Background

In recent times, there has been notable progress in interconnected technologies, including smart grids, the Internet of Vehicles, long-term evolution, and 5G communication [21]. A report by Cisco estimates that by 2022, the number of IP-connected devices is projected to be thrice the world's population, creating 4.8 ZB of IP traffic annually [2,13]. This exponential growth raises serious security concerns, as resource-constrained devices exchange vast amounts of sensitive data over the untrusted "Internet" using diverse communication protocols and technologies. According to a recent publication [11], it is crucial to incorporate advanced security measures and resilience analysis at the initial stages of deployment to ensure a secure and sustainable cyberspace [5].

M. Ben Ahmed et al. (Eds.): SCA 2023, LNNS 906, pp. 585–595, 2024.
https://doi.org/10.1007/978-3-031-53824-7_52

Security controls are essential to prevent, detect, and respond to attacks. As reported in a recent study [8], one of the widely used approaches for detecting internal and external intrusions and suspicious activities is the intrusion detection system (IDS). An IDS comprises several tools and mechanisms that monitor computer systems and network traffic, analyze activities, and identify potential threats to the system. There are mainly three types of IDS: signature-based, anomaly-based, and hybrid IDS [24]. Signature-based IDS uses predefined intrusion patterns to identify potential threats, while anomaly-based IDS detects deviations from normal behaviors. Different methodologies, such as machine learning (ML)-based techniques, knowledge-based schemes, and statistical-based approaches, have been used for anomaly detection, with deep learning (DL) methods gaining attention in recent investigations [1].

DL, a complex branch of machine learning, uses interconnected neurons organized in multi-layer networks to represent the mathematical computation of learning processes [6,14]. Its efficacy has been proven in several disciplines, including audio processing, natural language processing, picture and video recognition, robotics, and autonomous systems, to name a few [12]. Deep networks can be divided into three main categories: generative architectures, which use unsupervised learning to learn from unlabeled data, discriminative architectures, which use supervised learning to recognize patterns primarily for prediction tasks and hybrid architectures, which combine generative and discriminative models [4,22].

The advancements in DL techniques have garnered attention from researchers and smart grid actors for their potential applications in cyber-security [3]. DL and ML techniques are commonly used to develop IDSs that utilize traffic data analytics to differentiate between normal network traffic and various cyber-attacks [9,17].

Intrusion detection (ID) is a critical component of cybersecurity that has traditionally relied on rule-based or statistical methods. In recent years, DL models have gained attention due to their ability to learn complex patterns and features in network traffic data [10]. In this literature review, we summarize recent research on ID using DL models [8]. In the realm of network security, ID technology is a highly effective approach. Using DL algorithms for ID has emerged as a trend [16] within the smart grid domain, which proved DL-based ID methods yield promising results.

1.2 Contribution

DL algorithms are commonly used in traditional ID systems, where binary classifiers are employed to classify attacks. However, due to the impact of data dimensions on big data, the training model tends to yield a low attack detection rate, leading to reduced overall detection efficiency. By extracting higher-dimensional features from the original data, DL can generate a more accurate classification model, thereby enhancing the effectiveness of an ID system. This paper proposes an IDS system that first converts one-dimensional (1D) data into 2D space and then applies pre-trained CNN models. Moving on, to train base learners on IDS data, four advanced CNN models, namely Xception, VGG-16, CNN, and our proposed method are employed. The proposed IDS framework's effectiveness and efficiency are evaluated using the CSE-CIC-IDS2018 dataset [20]. The proposed method has shown a 92% accuracy rate in distinguishing between normal behavior and malicious activities, which is an improvement compared to other detection methods.

2 Related Works

This section comprehensively reviews various DL techniques for identifying cyber-attacks in traditional and IoT networks. In machine learning (ML), selecting relevant features from the dataset is crucial for effective model training and classification. However, DL techniques have gained popularity recently due to their ability to extract relevant features, making them more efficient automatically. As a result, DL techniques are widely employed in ID Systems (IDSs) in cybersecurity.

2.1 Deep Learning

Traditional network security methods require improvements to deal with the rising number of network attacks. ID systems have been proposed as a solution but have low detection rates and require significant feature engineering. To address these limitations, Fu et al. [7] propose a DL-based model for network ID (DLNID) that integrates an attention mechanism and the bidirectional long short-term memory (Bi-LSTM) network. The DLNID model employs a convolutional neural network (CNN) for sequence feature extraction and utilizes the adaptive synthetic sampling (ADASYN) method to tackle data imbalance issues. Experimental results indicate that the DLNID model outperforms other comparison methods, achieving an accuracy of 90.73% and an F1 score of 89.65% on the NSL-KDD public benchmark dataset on network ID.

In [28], the author investigates using a deep neural network (DNN) to create a flexible and efficient IDS for detecting and classifying unforeseen and unpredictable cyberattacks. The paper examines various datasets generated through static and dynamic methods to determine the most effective algorithm for detecting future cyberattacks. The DNN model is tested on benchmark malware datasets that are publicly available, and the optimal network parameters and topologies are selected using hyperparameter selection methods. The study shows that DNNs outperform classical machine learning classifiers. Lastly, the author proposes a hybrid and scalable DL scheme, scale-hybrid-IDS-AlertNet, which helps monitor host-level events and network traffic in real-time to notify possible cyberattacks dynamically. In [15], the authors highlight the importance of IDSs in combating network intruders. Still, many existing models struggle with high false alarm rates and detecting User-to-Root (U2R) and Remote-to-Local (R2L) attacks. The paper proposes a bidirectional Long-Short-Term-Memory (BiDLSTM) based IDS to address these challenges. The model is trained and evaluated on the NSL-KDD dataset, where the experimental results have demonstrated that BiDLSTM has outperformed conventional LSTM and existing methods concerning different ML metrics. Additionally, the BiDLSTM approach has a lower false alarm rate and achieves higher detection accuracy for U2R and R2L attacks than conventional LSTM.

In their study, Yang et al. [31] put forth a new approach to IDS for modern vehicle systems using ensemble learning and transfer learning (TL). This IDS utilizes a CNN model and a hyper-parameter optimization scheme to protect against network attacks. The effectiveness of this ID approach was assessed on two public benchmark Internet of Vehicles (IoV) security repositories: the Car-Hacking dataset and the CICIDS2017 dataset, which demonstrated over 99.25% accuracy and F1-scores. Overall, this has shown the effectiveness of TL-based IDS in detecting cyber-attacks in intra-vehicle and external vehicular networks, which is essential due to the vulnerabilities of IoV systems to cyber threats caused by the increasing connectivity of modern vehicles.

2.2 Transfer Learning

The article by [27] discusses the vulnerabilities in the Controller Area Network (CAN) bus used for in-vehicle communications, which can be exploited through network-level attacks. The paper tackles this problem by proposing a new ID method called CANTransfer that utilizes transfer learning (TL). A Convolutional LSTM-based model is trained using known intrusion patterns and then adapted to detect new attacks with limited data using one-shot learning. The experiments demonstrate that the proposed approach outperforms the best baseline model by 26.60% in detecting new intrusions.

The paper [19] suggests a novel IDS model based on deep TL for the security of In-Vehicle Network (IVN) systems. The proposed model involves selecting effective attributes, developing a deep TL-based LeNet model, and testing it with actual data. Based on experimental results, the proposed model performs better than existing models in terms of accuracy and detection rate, making it a viable solution for real-time IVN security. In the paper [29], a CNN-based TL is proposed for network ID, as the available datasets related to network intrusion are often insufficient for effective training. The proposed model comprises two ConvNets concatenated together and utilizes a two-stage learning process. The model is evaluated on the NSLKDD dataset, and the results demonstrate improved detection accuracy for known and novel attacks. This model achieves approximately 22.02% improvement on KDDTest-21 and 2.68% improvement on KDDTest+, in comparison to the traditional ConvNet model. In [25], another IDS scheme is introduced for enhancing the effectiveness of network ID using a wide and deep TL-based stacked GRU (Gated Recurrent Unit) model. This model can memorize a linear regression model with a generalization capacity of a deep GRU model. This method has achieved an evaluation accuracy of 99.92% for multi-class classification on KDDCup 99(10%) dataset and 94.22% on the UNSW-NB15 dataset. The results indicate that the proposed method surpasses most existing ID techniques.

The authors in [26] investigate the use of TL to improve the accuracy of network IDS (NIDS) in a data-scarce environment. Typically, the performance of fine-tuned TL-based NIDS is compared with a NIDS model trained from scratch. It has been found that fine-tuning can enhance the ability of detection models to identify new attacks with a reduced amount of training data. The study suggests that TL can enhance NIDS accuracy while reducing the time and cost of collecting labeled training data. The paper by [18] introduces a new approach for ID in mobile edge computing using a combination of federated learning and TL to enhance the efficiency of training and protect privacy. A reinforcement learning-based client selection scheme is also employed to select reliable and high-quality clients for participation in the training process, leading to better accuracy and communication efficiency. The experimental results show that the proposed framework performs better than traditional federated learning approaches and achieves the highest accuracy within a given budget limit. The article [30] presents a new approach to ID using TL and DL. The method transforms the ID task into an image recognition problem by converting network data into grayscale images and utilizing texture features to detect network intrusion. TL is used to enhance the model's efficiency and adaptability. Experimental results show that this approach outperforms traditional machine learning and DL methods, performs better generalization, and detect novel intrusion techniques effectively.

3 Proposed Method

3.1 Time-Series Imaging for ID

Time-series imaging is a powerful technique that transforms time-series data into images. This conversion allows us to leverage convolutional neural networks (CNNs) and transfer learning (TL), which were previously not feasible due to the absence of graphical representations for such records. Several time-series imaging methods exist, but we utilized the Gramian angular field approach. This technique generates images by calculating the difference or the summation of features compared to other features.

To begin the process, we normalize the features, ensuring their values scale is between -1 and 1. This normalization step is crucial for maintaining consistency and eliminating biases caused by differing scales. Once the features are normalized, the transformation involves turning the data into RGB images that would require to be turned in a grayscale format, allowing us to extract patterns and information from the data. Figure 1 shows how our data looks after applying Gramian Angular Field to our tabular data.

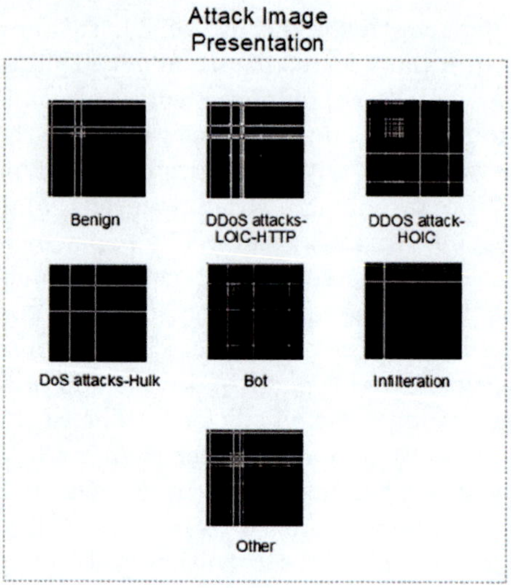

Fig. 1. Image representation of different attacks and normal network traffic behavior.

3.2 Dataset Description

Our method involved using two different datasets, the CIC-IDS2017 as a source domain and CSE-CIC-IDS2018 as a target domain. Both datasets contain 78 features that enable you to identify which class belongs to which.

CSE-CIC-IDS2018 dataset objective [20] was to tackle the shortage of top-notch datasets in the field. The scarcity of datasets is primarily attributed to privacy concerns, resulting in a limited supply of datasets with extensive features. The creators conquered this challenge by ensuring that the data generation process involved establishing an infrastructure comprising two separate entities: attackers and victims. The dataset contains 78 features that allow the differentiation of attack types. The original format of this data is in tabular format, which transforms into a graphical representation. The CSE-CIC-IDS2018 is considered the target domain between both datasets.

CIC-IDS2017 dataset aimed to overcome the limitations of the lack of reliable datasets covering a diverse range of network intrusions. Extensive evaluation of existing datasets revealed their outdated nature, as they no longer aligned with contemporary standards due to the emergence of new intrusion techniques. Recognizing this challenge, the creators of the dataset dedicated their efforts to capturing and reflecting up-to-date network behaviors as of 2017 [23]. The CIC-IDS2017 was used as a source domain in our transfer learning process.

3.3 Transfer Learning

The concept behind transfer learning (TL) is to take one model pre-trained on different features and use that experience to train another model, which improves the performance of the other. There are different types of TL, but in our case, we mainly focused on fine-tuning to train our new model. Figure 2 provides a graphical representation of the intended purpose of the fine-tuning process.

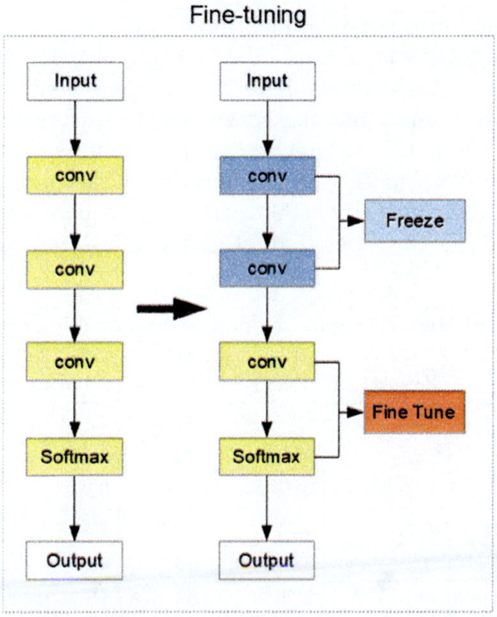

Fig. 2. A graphical representation of the fine-tuning concept, in which layers with features relevant to the current task are frozen.

TL's popularity kept increasing due to its ability to achieve high performance with lower training time compared to training a model from scratch. We use pre-trained models such as Inception, Xception, and VGG-16 for model training purposes. Those Models use ImageNet data for training, which contains over one million records with a thousand different classes. The benefit of using TL is the ability to scale down the number of types and data trained based on our requirements.

4 Experimental Results

Improvements in the proposed method harnessed the capabilities of the Keras API, which built on the robust TensorFlow framework. We employed two widely recognized datasets for conducting the experiments: the CIC-IDS2017 and CSE-CIC-IDS2018. During the training phase, the model underwent 100 epochs using early stoppage. The CIC-IDS2017 (Source Domain) and CSE-CIC-IDS2018 (Target domain) generated over 1.5 million images in the training process.

Before employing transfer learning techniques, we developed a Convolutional Neural Network (CNN) model trained on the source domain. We provided the means to act as a pre-trained model for our target domain. This step aimed to improve the detection rate and duration of training in comparison to existing techniques.

With the creation of our pre-trained model, we began employing it on the CSE-CIC-IDS2018 dataset alongside other techniques such as CNN, VGG-16, and Xception. Our proposed method did not only perform better in all metrics compared to other techniques but also consumed the least amount of time. Looking at Table 1, we can conclude that our proposed method took 301.587% less time to train than a CNN model, 71.43% less time to prepare than VGG16, and 7.9% less time than Xception. Furthermore, our proposed method performed better in all evaluation categories than other techniques.

Table 1. The table represents the performance metrics of the models

Metrics	Xception	VGG-16	CNN	Proposed Method
Precision	92%	92%	91%	92%
Recall	91%	91%	90%	92%
F1-score	91%	91%	90%	92%
Accuracy	92%	92%	90%	92%
Epoch	30	58	69	63
Total Duration	17 h	27 h	63.25 h	15.75 h

The proposed method represents an advantage in contrast to conventional approaches that focus on discerning whether an action constitutes an attack, devoid of insights into its specific type. This deficiency in traditional methods proves inefficient when deployed within an intrusion detection system. Our proposed approach not only identifies the presence of an attack but also analyzes it to classify the type of attack, thus bestowing several pivotal advantages such as tailored incident response, proper resource allocation, and continuous adaptation to new types of intrusions.

4.1 Causal Analysis

Analyzing further the results achieved by our proposed method, we discovered that the model in specific images could not differentiate between two different classes. Even further analysis enabled us to find that certain features are similar, which could cause the images to be nearly identical. Furthermore, this concluded that the Gramian Angular field was limited in creating sufficient differentiable features between different classes.

5 Conclusion

This paper demonstrates a smart grid IDS that uses TL and time-series imaging to detect attacks based on network traffic. We explored grayscale presentation in time-series imaging to evaluate its performance when applying various TL models. After identifying the different models to use for transfer aims to transfer the expertise of a pre-trained model to train a new model. Our research concludes that our proposed method was the optimal model because it is the best-performing and the fastest out of all models. Further developments and research may consider ways to improve the process of 2D representation transformation by enhancing the number of discerning factors visible in the graphical presentation. We could explore different tabular data to graphical representation techniques to improve the proposed method's performance.

References

1. Abd Elaziz, M., Al-Qaness, M.A., Dahou, A., Ibrahim, R.A., Abd El-Latif, A.A.: Intrusion detection approach for cloud and IOT environments using deep learning and Capuchin search algorithm. Adv. Eng. Softw., 103402 (2023)
2. Bousbiat, H., et al.: Crossing roads of federated learning and smart grids: overview, challenges, and perspectives. arXiv:2304.08602 (2023)
3. Diaba, S.Y., Elmusrati, M.: Proposed algorithm for smart grid DDoS detection based on deep learning. Neural Netw. **159**, 175–184 (2023)
4. Dina, A.S., Siddique, A., Manivannan, D.: A deep learning approach for intrusion detection in Internet of Things using focal loss function. Internet Things, 100699 (2023)
5. Elnour, M., et al.: Performance and energy optimization of building automation and management systems: towards smart sustainable carbon-neutral sports facilities. Renew. Sustain. Energy Rev. **162**(112), 401 (2022)
6. Elnour, M., et al.: Neural network-based model predictive control system for optimizing building automation and management systems of sports facilities. Appl. Energy **318**(119), 153 (2022)
7. Fu, Y., Du, Y., Cao, Z., Li, Q., Xiang, W.: A deep learning model for network intrusion detection with imbalanced data. Electronics **11**(6), 898 (2022)
8. Yhaia, H.K.: Deep transfer learning applications in intrusion detection systems: a comprehensive review. Inf. Fusion, 1–32 (2023)
9. Himeur, Y., Ghanem, K., Alsalemi, A., Bensaali, F., Amira, A.: Artificial intelligence based anomaly detection of energy consumption in buildings: a review, current trends and new perspectives. Appl. Energy **287**(116), 601 (2021)
10. Himeur, Y., Alsalemi, A., Bensaali, F., Amira, A., Al-Kababji, A.: Recent trends of smart nonintrusive load monitoring in buildings: a review, open challenges, and future directions. Int. J. Intell. Syst. **37**(10), 7124–7179 (2022)
11. Himeur, Y., Sohail, S.S., Bensaali, F., Amira, A., Alazab, M.: Latest trends of security and privacy in recommender systems: a comprehensive review and future perspectives. Comput. Secur., 102746 (2022)

12. Himeur, Y., Al-Maadeed, S., Varlamis, I., Al-Maadeed, N., Abualsaud, K., Mohamed, A.: Face mask detection in smart cities using deep and transfer learning: Lessons learned from the covid-19 pandemic. Systems **11**(2), 107 (2023)
13. Himeur, Y., et al.: Ai-big data analytics for building automation and management systems: a survey, actual challenges and future perspectives. Artif. Intell. Rev. **56**(6), 4929–5021 (2023)
14. Hnamte, V., Hussain, J.: DCNNBiLSTM: an efficient hybrid deep learning-based intrusion detection system. Telematics Inform. Rep. **10**(100), 053 (2023)
15. Imrana, Y., Xiang, Y., Ali, L., Abdul-Rauf, Z.: A bidirectional LSTM deep learning approach for intrusion detection. Expert Syst. Appl. **185**(115), 524 (2021)
16. Lampe, B., Meng, W.: A survey of deep learning-based intrusion detection in automotive applications. Expert Syst. Appl., 119771 (2023)
17. Liao, P., Yan, J., Sellier, J.M., Zhang, Y.: Divergence-based transferability analysis for self-adaptive smart grid intrusion detection with transfer learning. IEEE Access **10**, 68807–68818 (2022)
18. Masum, M., Shahriar, H.: TL-NID: deep neural network with transfer learning for network intrusion detection. In: 2020 15th International Conference for Internet Technology and Secured Transactions (ICITST), pp. 1–7. IEEE (2020)
19. Mehedi, S.T., Anwar, A., Rahman, Z., Ahmed, K.: Deep transfer learning based intrusion detection system for electric vehicular networks. Sensors **21**(14), 4736 (2021)
20. for Cybersecurity in University of New Brunswick CI (2018) CSE-CIC-IDS2018 on AWS. https://registry.opendata.aws/cse-cic-ids2018/
21. Sayed, A., Himeur, Y., Alsalemi, A., Bensaali, F., Amira, A.: Intelligent edge-based recommender system for internet of energy applications. IEEE Syst. J. **16**(3), 5001–5010 (2021)
22. Sayed, A.N., Himeur, Y., Bensaali, F.: Deep and transfer learning for building occupancy detection: a review and comparative analysis. Eng. Appl. Artif. Intell. **115**(105), 254 (2022)
23. Sharafaldin, I., Habibi Lashkari, A., Ghorbani, A.: Toward generating a new intrusion detection dataset and intrusion traffic characterization, pp. 108–116 (2018). https://doi.org/10.5220/0006639801080116
24. Singh, A., Amutha, J., Nagar, J., Sharma, S.: A deep learning approach to predict the number of k-barriers for intrusion detection over a circular region using wireless sensor networks. Expert Syst. Appl. **211**(118), 588 (2023)
25. Singh, N.B., Singh, M.M., Sarkar, A., Mandal, J.K.: A novel wide & deep transfer learning stacked GRU framework for network intrusion detection. J. Inf. Secur. Appl. **61**(102), 899 (2021)
26. Singla, A., Bertino, E., Verma, D.: Overcoming the lack of labeled data: training intrusion detection models using transfer learning. In: 2019 IEEE International Conference on Smart Computing (SMARTCOMP), pp. 69–74. IEEE (2019)
27. Tariq, S., Lee, S., Woo, S.S.: CANTransfer: transfer learning based intrusion detection on a controller area network using convolutional LSTM network. In: Proceedings of the 35th Annual ACM Symposium on Applied Computing, pp. 1048–1055 (2020)
28. Vinayakumar, R., Alazab, M., Soman, K., Poornachandran, P., Al-Nemrat, A., Venkatraman, S.: Deep learning approach for intelligent intrusion detection system. IEEE Access **7**, 41525–41550 (2019)
29. Wu, P., Guo, H., Buckland, R.: A transfer learning approach for network intrusion detection. In: 2019 IEEE 4th International Conference on Big Data Analytics (ICBDA), pp. 281–285. IEEE (2019)

30. Xu, Y., et al.: Intrusion detection based on fusing deep neural networks and transfer learning. In: Zhai, G., Zhou, J., Yang, H., An, P., Yang, X. (eds.) IFTC 2019. CCIS, vol. 1181, pp. 212–223. Springer, Singapore (2020). https://doi.org/10.1007/978-981-15-3341-9_18
31. Yang, L., Shami, A.: A transfer learning and optimized CNN based intrusion detection system for internet of vehicles. In: ICC 2022-IEEE International Conference on Communications, pp. 2774–2779. IEEE(2022)

Building a Resilient Smart City Ecosystem: A Comprehensive Security and Cybersecurity Management Model

R. G. Guntur Alam[1(✉)], Dedi Abdullah[2], Huda Ibrahim[3], and Ismail Rakip Karas[4]

[1] Fakultas Teknik, Sistem Informasi, Universitas Muhammadiyah Bengkulu, Bengkulu, Indonesia
datuak73@yahoo.com
[2] Teknik Informatika, Fakultas Teknik, Universitas Muhammadiyah Bengkulu, Bengkulu, Indonesia
[3] School of Computing, Universiti Utara Malaysia, 06010 Kedah, Malaysia
[4] Computer Engineering Departemen, Demir Celik Campus 78050, Karabuk, Turkey

Abstract. The development of information and communication technology has spread throughout the world, including Indonesia. There are many benefits, but the risks are unavoidable. Communication is growing massively in cyberspace and thus poses a security threat to smart city services. In Indonesia, cybersecurity is not yet a priority, so policies that regulate all aspects as a standard reference for implementing cybersecurity are still needed. Therefore, the purpose of this research is to propose a cybersecurity model in smart cities in Indonesia. In this study, unstructured interviews were conducted as a method for collecting data. Seven respondents, namely the Director, Head of Cyber and Code Control Section, Head of Operations, and four staff at Jakarta Smart City were selected as respondents. Initial factors for the implementation of cybersecurity in Indonesia are identified and analyzed qualitatively. Actor-Network Theory (ANT) is used as a theoretical framework to explore and validate factors related to cybersecurity. In the end, a new cybersecurity model was proposed. The proposed model focuses on people, technology, and process and follows the flow of government bureaucracy in Indonesia. The proposed cybersecurity model has three main parts, namely: (1) Cybersecurity Stakeholders, (2) Legal Basis for Cybersecurity Management, and (3) Security Management. The new cybersecurity development model resulting from this research can provide input to government agencies to design and implement cybersecurity in smart city projects in Indonesia in the future.

Keywords: Smart City · Smart City Security · Cybersecurity · Cybersecurity Model

1 Introduction

In the era of continuous development in information and communication technology, the construction of smart cities has become a primary focus in improving human life. However, despite the intended progress for civilization and human advancement, modern cities still face various social problems that disrupt social stability. Along with the

M. Ben Ahmed et al. (Eds.): SCA 2023, LNNS 906, pp. 596–607, 2024.
https://doi.org/10.1007/978-3-031-53824-7_53

increasing number of people migrating from rural areas to cities, new problems arise in urban areas. Therefore, the technology becomes an option that must be utilized to provide quality services for urban residents.

The concept of Smart City encompasses the development of economy, human resources, governance, mobility, environment, community life, and modern, intelligent, independent, and integrated telecommunications infrastructure [1]. A smart city also involves monitoring and integrating all city infrastructures, such as tunnels, water networks, bridges, roads, subways, airports, railways, ports, communications, electricity, and government buildings, to optimize resources, urban planning, and enhance public safety [2].

To realize a smart city, cloud computing technology is crucial in providing and storing information and facilitating user connectivity. However, the use of cloud computing technology also brings significant security risks. Security issues can arise in various forms, both non-technical and technical. Many technology providers do not prioritize cybersecurity when providing smart city infrastructure, such as Internet of Things (IoT) devices, because consumers tend to prioritize affordability over the security risks posed by less secure products.

Cybersecurity is a key factor that must be considered in smart city development. In the context of a smart city, the implementation of cybersecurity is influenced by three main actors: individuals involved in smart city development, the process of planning and implementing cybersecurity policies, and the technology of cybersecurity infrastructure, including cloud computing infrastructure, IoT, and various application services provided by the smart city. Some countries have implemented cybersecurity services as part of their smart city initiatives, but other countries face difficulties in implementing cybersecurity policies and strategies.

Indonesia, as a developing country, also faces challenges in developing safe and sustainable smart cities. Cybersecurity has not yet become a top priority in the development of smart cities in Indonesia. The cybersecurity framework for smart cities in Indonesia still refers to existing cybersecurity laws, such as the Information and Electronic Transactions Act No. 11 of 2008 and the Government Regulation on the Implementation of Transactions and Electro-nic Systems No. 82 of 2012. There are still several issues related to cybersecurity development in Indonesia, including a lack of understanding and support from government administrators, inadequate models for handling cyber attacks, and a lack of interagency coordination in addressing cybersecurity issues [3].

Based on this background, this research aims to develop a comprehensive cybersecurity model that can support the development of smart cities in Indonesia. This study will discuss the challenges faced in implementing cybersecurity in smart cities and consider local cultural factors in the development of cybersecurity strategies. By identifying elements that can support or hinder cybercrime prevention, this research will provide a strong foundation for proposing an effective and successful cybersecurity model in the context of smart cities in Indonesia.

2 Smart City and Cybersecurity

The Smart city seen as a city that oversees and consolidates all infrastructure which includes main roads, underground passages, railways, airports, transmission, waterways, electricity, and all government buildings that can be used to optimize resources, urban planning, and supervise programs with maximum security for the convenience of residents [4].

A Smart city implementation relies heavily on computer technology, especially cloud computing, which provides all the information. Cloud computing is not only a source of information but also a location to store data and connect users, simultaneously security risks are unavoidable [5]. Security constraints will appear in all forms and for various reasons, including non-technological and technological causes [6]. User privacy is vulnerable to many security breaches because security systems can penetrate several applications like advertisements, spam emails, and attacks on infrastructure, especially on main internet protocols and network architectures that do not specifically handle [7].

According to [8], almost 40% of all fraudulent transactions start from mobile equipment, where the sectors most affected by these violations are public services, finance, and information. [9] found that some sensys network devices, such as econolite, do not have such encryption, authentication, or security. This device is used to control traffic lights. An attacker can modify signals with fake data causing option errors when setting the configuration and timing of traffic lights, road meters, and traffic signals so that one hundred thousand intersections in Canada and the US can be affected. If an attacker creates a firmware update worm, 200,000 similar sensors used worldwide are vulnerable to being affected [9].

Even though Indonesia is rapidly developing smart cities, this country is still considered weak in cybersecurity. The Ministry of Communication and Information emphasized that Indonesia is after China in cybercrime attacks. In the last three years, there have been 36.6 million cyber attacks [10]. One of them is the hacking of bank customer debit cards. In this case, hackers attempted to penetrate the bank's client card security system in May 2014, indicating how bad the cybersecurity situation was in Indonesia [11].

2.1 Importance of Cybersecurity

In the context of smart cities, the integration of information technology into physical infrastructure has improved the quality of life for city residents but also poses security and privacy challenges due to large amounts of data. Cybersecurity threats are increasing with the increasing number of physical objects connected through the Internet of Things (IoT), which hackers can exploit to take control of devices and threaten the safety of lives. Therefore, cybersecurity becomes crucial in maintaining the critical infrastructure security and protecting personal information.

2.2 Cybersecurity Existing Models

There are several security models implemented in Smart city development. One of the models is the Governance Risk and Compliance (GRC) Model. This model involves three main components: security drivers, management, and stakeholders. This model links legal aspects, organizational objectives, and security threats as security drivers. In addition, there are policy frameworks, process frameworks, and security metrics frameworks to support security implementation at different management levels [12].

Another security model is the Self-Contained Information Security (SCIS) Model, which aims to integrate several security solutions from various urban problems into one independent model. This model includes system-type entities, sensors, actuators, sensitivity levels, and grouping values related to smart city security [13].

The Direct-Control Cycle Model is an information security governance model. This model consists of three levels of management (strategic, tactical, and operational) with three different actions at each level. This model emphasizes the importance of protecting information assets, establishing policies and procedures, and measuring and controlling the implementation of information security policies.The next model, The Information Security Governance Model, operates at three management levels: strategic, tactical, and operational. This model involves directing, implementing, and controlling mutually supporting actions with input/output indicators. This model aims to provide an integrated framework for managing information security, considering executive directives and employee compliance [14].

The IS Security Governance (ISG) Process Model is a model that focuses on the process of implementing information security using the Plan Do Check Act (PDCA) cycle. This model emphasizes a risk-based approach, selecting relevant security frameworks, selecting relevant IT controls, measurement frameworks implementation, and ongoing training to build a security culture. Overall these security models aim to manage risk and

Table 1. Properties and Characteristic of Previous Cybersecurity Implementation Models

Model	**Consideration for aspect of**			**Model Focus**
	People	Tech	Process	
Governance Risk and Compliance (GRC) Model, (Jirasek, 2012)	√	√	√	Focusing on people, technology, and processes with functionality based on laws and regulations, managerial, and stakeholder
Self-Contained Information Security Model (SCIS) (Ferraz and Sampaio, 2013)		√		Focuses on technology and functionality based on available application platforms
Direct-Control Cycle Model (Von Solms and von Solms, 2006)			√	Focus on processes that prioritize strategic, tactical, and operational
Information Security Governance Model (Coertze and von Solms, 2012)			√	Focusing on processes that prioritize governance, namely directing, implementing, and controlling
IS Security Governance Process Model (Nicho, 2018)			√	Focuses on a risk-based approach

integrate information security into Smart city systems and operations. Table 1 summarizes the characteristics of the most common cybersecurity models, their emphasis on human and non-human factors, and the extent to which their focus includes human and non-human factors (people, technology, and processes) in cybersecurity implementation.

3 Research Methodology

This research uses interpretive case studies based on ANT elements to explore the challenges of Cybersecurity in Jakarta Smart city and the factors involved. Interpretive case studies are qualitative research that are descriptive, interpretive, or evaluative in nature to explore certain phenomena, such as programs, events, processes, institutions, or social groups [17]. Case study research conducted when the research focus or the research unit is limited to a particular analysis [18]. This research guided by the philosophy of critical realism, which must go beyond the surface to understand and explain why things happen the way they do and speculate about the structures and mechanisms underlying observed social events [19]. This research is a study that looks back at social phenomena and inquires about their reality to find new solutions. The concern behind this research is to find out the situation, obstacles, challenges, and elements in the cybersecurity aspect of Jakarta Smart City (JSC). Many factors or actors may involved, such as people, technology, processes, policies, and finances. This phenomenon may consist of contradictions that require definite answers to research questions.

A case study approach was use to collect data for this research, and informal and semi-structured interview styles used. This method is adaptable as it allows the researcher to ask additional questions during and after the interview [20, 21]. The main objective of this interview is to identify human and non-human elements in the implementation of cybersecurity in smart cities in Indonesia. The unit of analysis in this study is the network actors implementing cybersecurity at JSC. In analyzing the data, this study uses template concept analysis, which requires the researcher to make list of codes (templates) that represent the themes identified in the text data [22]. Template analysis is a flexible technique that allows researchers to customize the code according to research needs.

4 Instruments and Supporting Factors for Cybersecurity Implementation

Factors that hinder and support the implementation of cybersecurity will be a component of consideration to propose a cybersecurity model explored through four translational moments of the Actor-Network Theory. The first step is to make a basic description of network actors in implementing cybersecurity at JSC. Therefore, this study begins by punctualisation each actor's black box to expand each black box and explaining its contents. The concept of punctualisation allows the conversion of the entire network into a single point [23]. The implementation of a cybersecurity network requires many actors to achieve security in public services at JSC. These network actors can be seen as a combination of various human and non-human actors.

This research follows the main actors in the network of cybersecurity actors, director of the JSC, the DKI Jakarta Provincial Information and Statistics Communication

Agency, and the Operational Implementation Unit. The Director of Jakarta Smart City is responsible for planning, controlling, and managing the JSC system, as stipulated in DKI Jakarta Governor Regulation Number 280 of 2014. Therefore, the Director of JSC has sufficient authority to establish and develop the JSC. In 2020, through the Decree of the Governor of DKI Jakarta Number 17 of 2020, JSC is designated as a technical implementing unit that implements the Regional Public Service Agency (BULD) Financial Management Pattern. With the change in status to BLUD, JSC has more flexibility in managing finances without depending on the Regional Revenue and Expenditure Budget (APBD) and has the authority to establish partnerships to serve interests. Therefore, the Director of JSC can identifie as Main actor in this project.

As the main actor, the JSC director is responsible for coordinating all actors in the JSC. The first responsibility is to lead and coordinate the executors under him. In cybersecurity, the JSC director leads the entire task force to work closely together to ensure data and information technology infrastructure in Jakarta is secure. Second, to coordinate and establish cooperation with other parties from government and private institutions, the smooth running and improvement of JSC management. Third, report and be responsible for all tasks at JSC. The Director of JSC realizes that achieving good cybersecurity management requires compliance with applicable standards, regulations and rules. Standardization, regulation, and rules are non-human actors (processes). Rules and regulations become the basis for focal actors to mobilize other actors (people and technology). In summary, the Main actor must relate to other actors (process, people, and technology) to motivate or coerce the relevant actor to enroll in the network of cybersecurity implementing actors.

In carrying out his work, the director of JSC must comply with the regulations and policies that have been stipulated and is responsible to the DKI Jakarta Province Communication, Informatics, and Statistics Office. In the eyes of the focal actor (director of JSC), the DKI Jakarta Province Communication, Informatics, and Statistics Office is seen as a potential actor in implementing cybersecurity. Therefore it can be registered in the Main actor's network. When this actor is enlarged, it will be clear that this institution has a role as a policy maker, cybersecurity implementation coordinator, and oversight. However, cybersecurity implementation at JSC is entirely the responsibility of the director of JSC. At this point, it can be intepreted that this network of actors consists of (1) the DKI Jakarta Province Communication, Informatics, and Statistics Office, which plays a role in formulating regulations, regulations, and standards for the implementation of promoting cybersecurity, (2) Rules, regulations and standard which is a non-human actor that becomes a reference for the focus actor (JSC directory) in carrying out his work.

The Main actor (JSC director) cannot do all the tasks alone. The JSC Director needs people who can help in carrying out his work. Based on DKI Jakarta Government Regulation Number 280 of 2014, the director of JSC is assisted by the Operational Implementation Unit. In conclusion, the network of cybersecurity actors at JSC consists of:

1. The Operational Implementation Units to assist Main actors in planning, implementing, and reporting for the implementation of cybersecurity
2. Rules, regulations, and standards as non-human actors become the basis of this actor network to do the task

3. Technology as the work object of this actor network.

The details of each punctualization actor are summarized in Table 2.

Table 2. Summary of the Punctualised Actor in the Implementation of Cybersecurity

Puctualised Actor	Actors	Interest	Obstacle
Director of JSC	Process, People, Technology	- Negotiate, motivate, coordinate and then work with other punctulised actors - Initiation, implementation, monitoring and evaluation of cybersecurity	- Flexibility in budgets and collaboration - Freedom of governance for smart city development
The DKI Jakarta Province Communication, Informatics, and Statistics Agency	Process	- Prepare rules, regulations, and standards for the successful implementation of cybersecurity - Leadership in cybersecurity implementation	Must be assisted by other units to carry out the functions assigned by the governor
The Operational Executing Unit	People, Technology	- Implementing cybersecurity for public services in Jakarta Smart City - Improve cybersecurity in public services	Work according to the rules, regulations, and direction of the leadership

In the problematization process, the actors involved agree to be included in the network. In practice, there is no guarantee that they are all consistent within a network of actors, some of whom may consider them unfit to join because of their identity. Therefore interressement moment is needed. The interesement device is a text of rules, conversations, or instructions to convince and lure other actors into a position and follow the requirements of the focus actor [24]. Referring to the network of cybersecurity actors at JSC, the required instruments are standardization, Standard Operating Procedures, Laws and Regulations, policies, and leadership instructions. Standardization is an instrument needed by non-human actors, especially in electronic systems for public services. Standardization acts as a link between the focus actor and other actors and helps the Main actor in coercing and controlling other actors to perform the required activities in the actor-network. Standard Operating Procedures are a necessary tool for human and non-human actors, especially in software and hardware maintenance. If network actors do not have these devices, there is no proper communication between security systems, hardware, and users. In other words, this tool can help non-human actors to have a universal language to interact or communicate in actor networks. Laws and regulations are other necessary instruments needed by actor networks. Legislation is the primary guide for the focal actor (Director of JSC) to lock all human actors in their positions. Like laws and regulations, policies are another necessary instrument needed by a network of actors. Another tool used to carry out the interessement of several actors in the network is the leader's instructions.

This study identifies three sets of interrelated factors that are key to implementing cybersecurity in Jakarta Smart City, namely (1) Cybersecurity Stakeholders, (2) Legal Basis for Cybersecurity Management, and (3) Security Management. From the punctualization actors, two stakeholders related to cybersecurity were identified. These stakeholders are the DKI Jakarta Province Information Communication and Statistics Agency, the Director of JSC, and the Operational Implementation Unit (Smart city management). The DKI Jakarta Province Information Communication and Statistics Agency will prepare rules, regulations and standards for the successful implementation of cybersecurity. Meanwhile, Smart city managers are responsible for negotiating, motivating, coordinating, and collaborating with other stakeholders in analyzing security needs and implementing cybersecurity.

Through registration moments and mobilization moments, other stakeholders are identified. The first is BSSN, in terms of measuring security maturity or external security audits at the national level and the use of electronic certificates. Second, academics from various universities provide technical guidance regarding cybersecurity governance. Third, security professionals in handling certain incidents that have never been handled by the smart city team. Fourth, Cooperation with private companies to realize effective cybersecurity and protect smart cities from various cyber-attacks because most of the technological infrastructure is owned and operated by private companies.

The second factor identified is the Legal Fundamentals for Cybersecurity Management. The implementation of cybersecurity must have a clear legal basis. At the time of interest, two legal instruments were identified as guidelines for policy makers in managing cybersecurity. This instrument is international security law and standards. At the moment of mobilization, the director of the JSC reduces general Laws and regulations to policies at the technical level.In making derivations, the policies made by the JSC director must be aligned with the goals of the JSC. Therefore, the goal of a smart city has also been identified as the basis for cybersecurity management.

The third factor is security management. Security management controls three aspects, security policy control, security implementation control, and security audit and evaluation. Security Policy Controls are developed within a security framework, Includes security policies, security standards, and Security requirements. Meanwhile, Security Implementation Control includes people, technology, and processes, integrated into the implementation framework. Security evaluation and audits at JSC are carried out by internal and external audits. Internal audits are carried out by the JSC team, while external audits invite auditors or consultants to validate the internal audit results.

Based on the analysis of the security implementation model in Sect. 2.2, Cybersecurity Existing Models and the results of the ANT translation process above, factors and recommendations regarding the implementation of cybersecurity at JSC have been identified. The identified human, non-human factors and recommendationsare similar to the Governance risk and compliance (GRC) model proposed by Jirasek, with differences in several sections. Adapting the GRC model by Jirasek with adjustments in several parts and by embedding the identified factors, a new cybersecurity model is proposed. This new cybersecurity model focuses on people, technology, processes and follows the flow of government bureaucracy in Indonesia.

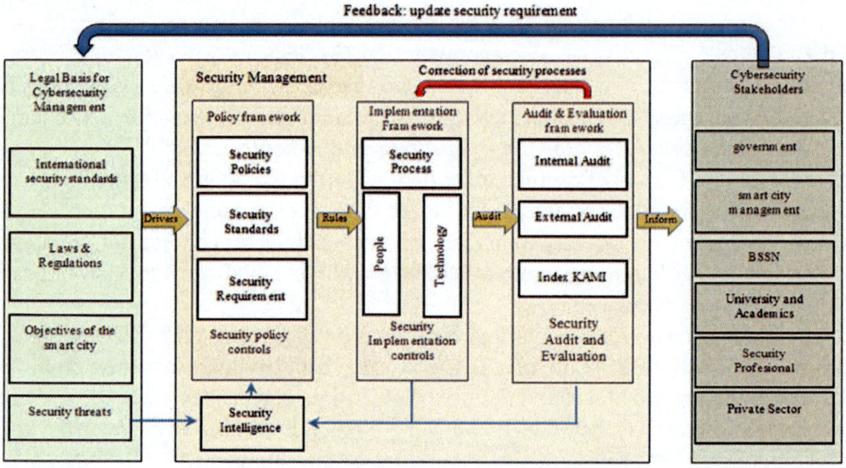

Fig. 1. The proposed cybersecurity model

5 Governance, Risk and Compliance Based Cybersecurity Model

As the main findings of this study, the proposed cybersecurity model consists of three main areas:

- Legal Basis of Cybersecurity Management.
- Security management.
- Cybersecurity Stakeholders.

5.1 Legal Basis of Cybersecurity Management.

The legal basis is the driving force in cybersecurity management. There are four main bases for cybersecurity work in the proposed model:

International Safety Standards. International security standards are international standards in the implementation of information security management systems. This standard model helps Smart city management in protecting and maintaining data and information security.

Laws and Regulations. Laws and regulations must be complied with in cybersecurity, including compliance with the ITE Law and BSSN Regulation Number 8 of 2020 as legal regulations in Indonesia.

Smart Cities Destinations. Smart City integrates information, communication, and Internet of Things technology to improve efficiency and public services in managing city assets. Security is a priority in maintaining the smooth running of Smart city systems by preventing malware attacks and disruptions to IT infrastructure.

Security Threat. Cyber threats can cause severe problems to networks, computer systems, and data. Knowing and understanding security threat trends is an important issue in formulating a responsive cyber attack prevention strategy.

5.2 Security Management

In security management, the proposed bureaucratic-based cybersecurity model has three frameworks that enable smart cities to achieve the goals set on the driver's side.

Policy Framework. Contains a policy includes a set of security policies, standards, and requirements drawn from the area of legal basis. The three scopes of this policy framework simultaneously define security controls in Smart cities.

Security Policy. A security policy is a set of rules that apply to the security activities of a Smart city's computer and communication resources. These rules cover environmental security such as physical, personal, administrative, and network security.

Cybersecurity Standards. Cybersecurity standards are a collection of interconnected features that are often used in smart cities' efforts to protect and regulate cyber environments.

Security Requirements. Security requirements are security features required by system users or qualities that must be owned by a system in a smart city to increase user confidence in the system they use.

Implementation Framework. Security policy implementation involves the role of humans and technology in the control process to achieve smart city goals that enable the tracking of security investments and justification of security budgets by stakeholders.

Audit and Evaluation Framework. Stakeholders need to monitor and evaluate security controls and policy compliance on an ongoing basis. The KAMI Index is used to analyze and assess the implementation of cybersecurity based on governance, risk management, framework, asset management, and technology aspect criteria.

5.3 Cybersecurity Stakeholders

Cybersecurity requires multi-stakeholder collaboration in policymaking. Cybersecurity stakeholders in smart cities, including the government, BSSN, Smart city managers, universities/academicians, and the private sector, face challenges in developing cybersecurity management in smart cities.

6 Conclusion

The results of this study indicate that the development of a cybersecurity model for smart cities in Indonesia is an urgent need. This study identifies and analyzes the key factors that need to be considered in implementing cybersecurity in smart cities.

The proposed cybersecurity model consists of three main areas, the legal basis of cybersecurity management, security management, and cybersecurity stakeholders. The legal basis is a significant foundation in driving the implementation of cybersecurity, including international security standards and compliance with laws and regulations that apply in Indonesia. Security management includes policy frameworks, cybersecurity standards, security requirements, implementation, and audits and evaluations. Security management formulates policies, including security policies, Standards, and Requirements that serve as a reference in managing security controls in Smart cities. The policy framework implementation involves the role of people and technology, while audits and evaluations are significant to monitor and evaluate the effectiveness of implemented security controls. Cybersecurity stakeholders include the government, BSSN, Smart city managers, universities/academicians, and the private sector. Multi-stakeholder collaboration is indispensable in making cybersecurity policies, because they cannot reach individually.

The cybersecurity development model resulting from this research can provide input to government agencies in designing and implementing cybersecurity in smart city projects in Indonesia in the future. By implementing this model, it is hoped that smart cities in Indonesia can strengthen the protection of their infrastructure and public services from cybersecurity threats.

References

1. Campisi, T., Severino, A., Al-Rashid, M.A., Pau, G.: The development of the smart cities in the connected and autonomous vehicles (CAVs) era: From mobility patterns to scaling in cities. Infrastructures **6**(7), 100 (2021)
2. Fakhimi, A.H., Khani, A.H., Sardroud, J.M.: Smart-city infrastructure components. In: In Solving urban infrastructure problems using smart city technologies. Elsevier (2021)
3. Christmartha, M., Gultom, R.A., Aritonang, S.: Strategi Kebijakan Pengembangan Sumber Daya Manusia Siber Nasional Guna Mendukung Pertahanan Negara (Studi Kasus Pada Badan Siber Dan Sandi Negara 2019), J. Pemikir. dan Penelit. Manaj. Pertahanan, vol. 6(2) (2020)
4. Zurita, G., Pino, J.A., Baloian, N.: Supporting Smart Community Decision Making for Self-governance with Multiple Views (2015)
5. Subashini, S., Kavitha, V.: A survey on security issues in service delivery models of cloud computing. J. Netw. Comput. Appl. **34**(1), 1–11 (2011)
6. Allen, N.: Cybersecurity weaknesses threaten to make smart cities more costly and dangerous than their analog predecessors. London Sch. Econ. (2016)
7. Hassija, V., Chamola, V., Saxena, V., Jain, D., Goyal, P., Sikdar, B.: A survey on IoT security: application areas, security threats, and solution architectures. IEEE Access **7**, 82721–82743 (2019)
8. Megdad, M.M., Abu-Naser, S.S.: Fraudulent financial transactions detection using machine learning (2022)

9. Cerrudo, C.: An emerging US (and world) threat: Cities wide open to cyber attacks. IOActive (2015)
10. Danuri, M., Suharnawi, S.: Trend cyber crime dan teknologi informasi di indonesia. INFOKAM **2**, 55–64 (2018)
11. Ardiyanti, H.: Cybersecurity dan tantangan pengembangannya di Indonesia. Politica **5**(1), 95–110 (2018)
12. Jirasek, V.: Practical application of information security models. Inf. Secur. Tech. Rep. **17**(1–2), 1–8 (2012)
13. Ferraz, F.S., Sampaio, C.C.B.: Towards a smart city security Model. In: The Eighth International Conference on Software Engineering Advances, ICSEA, pp. 546–550 (2013)
14. Von Solms, R., Von Solms, S.B.: Information security governance: a model based on the direct–control cycle. Comput. Secur. **25**(6), 408–412 (2006)
15. Coertze, J.: A model for information security governance in developing countries. In: Jonas, K., Rai, I.A., Tchuente, M. (ed.) E-Infrastructure and e-Services for Developing Countries, pp. 279–288. Springer, Berlin, (2013). https://doi.org/10.1007/978-3-642-41178-6_29
16. Nicho, M.: A process model for implementing information systems security governance. Inf. Comput. Secur. **26**(1), 10–38 (2018)
17. Nakamura, Y., Zhang, Y., Sasabe, M., Kasahara, S.: "Capability-based access control for the internet of things: an ethereum. In: 2019 IEEE Global Communications Conference (GLOBECOM), pp. 1–6. IEEE (2019)
18. Arora, A., Nandkumar, A., Telang, R.: Does information security attack frequency increase with vulnerability disclosure? An empirical analysis. Inf. Syst. Front. **8**(5), 350–362 (2006)
19. Buzan, B., Wæver, O., de Wilde, J.: Security: A New Framework for Analysis. Lynne Rienner Publishers (1998)
20. Zhao, J., Zhao, S.: Opportunities and threats: A security assessment of state e-government websites. Gov. Inf. Q. **27**(1), 49–56 (2010)
21. Debruijn, H., Janssen, M.: Building cybersecurity awareness : The need for evidence-based framing strategies. Gov. Inf. Q. **34**(1), 1–7 (2017)
22. Baig, Z., Szewczyk, Z., Valli, C.: Future challenges for smart cities: cyber-security and digital forensics. Digit. Investig. **22**, 3–13 (2017)
23. ITU. Overview of Cybersecurity. Recommendation ITU-T X.1205, Geneva Int Telecommun Union. Publ. online (2008)
24. Ten, C., Liu, C., Manimaran, G.: Vulnerability assessment of cybersecurity for SCADA systems. IEEE Trans Power Syst. **4**, 1836–1846 (2008)

Mobile Applications Security: A Survey About Security Level and Awareness of Moroccan Users

Mouna Sif-Eddine(✉) and Tomader Mazri

Advanced Systems Engineering Laboratory, Ibn Tofail University Morocco, Kenitra, Morocco
{mouna.sif-eddine,tomader.mazri}@uit.ac.ma

Abstract. Today, mobile applications are in widespread use and have become an essential part of their users' daily lives. However, for users and developers alike, the security of mobile applications has become a significant concern. Security flaws in mobile applications can give cybercriminals access to users' personal information, steal sensitive data, and compromise their privacy. Through the creation of a survey consisting of eleven questions, each based on a mobile security practice that users should follow, this study examines the different aspects of mobile application security, including security threats, common attacks, security techniques, and the level of security and awareness of good security practices among users in Moroccan society. The results obtained will then be collected, analyzed, and compared with other research work to deduce the gaps to which we must pay attention. Finally, a list of good security practices will be drawn up to be recommended to these users.

Keywords: Security level · mobile application security · privacy

1 Introduction

Mobile applications are now an integral part of our everyday lives. Online technology lets us buy things, control our money, communicate with people, have fun, and share private details. More and more developers are creating mobile applications, but the biggest challenge is keeping them safe. This violates user privacy. This will be discussed in the following chapters. We will look at how secure mobile phone applications are and what problems they can have. Then, we'll examine how hackers can exploit vulnerabilities to harm phones. We'll discuss types of phone attacks and how they work, find out what they're doing to your phone, and what experts have been doing to prevent it. Finally, we'll look at how people in Morocco protect their phones and give you tips on how to keep your phone safe.

2 The Security Level of Mobile Applications

Mobile operating systems [1], known as smartphone platforms, are designed for portable devices such as smartphones and tablets. With built-in security measures to protect against malware, viruses, data loss, and theft, smartphone platforms provide users with

M. Ben Ahmed et al. (Eds.): SCA 2023, LNNS 906, pp. 608–622, 2024.
https://doi.org/10.1007/978-3-031-53824-7_54

seamless access to messaging services, the Internet, mobile applications, and additional communication features. Depending on factors such as security updates, manufacturer privacy policies, and user security habits, smartphone platforms have varying levels of security. The Android operating system has released new features for its users that aim to improve the security of smartphone platforms. The manufacturers of smartphones are constantly working to achieve this goal, for example through:

- **Google Play Protect**: A malware detection system that uses machine learning to combat emerging threats [2].
- **User account protection**: When you use Google's Password Auto-Fill feature to enter your password in the application, it is checked against a list of compromised passwords. If there is a match, the user is notified so they can take appropriate action [2].
- **Google Phone Antispam Protection**: Filters spam by alerting you when a number is suspicious. This feature is also integrated with Messages. This prevents unwanted messages from reaching your inbox [2].
- **Google Secure Browsing** [2].
- **Application privacy permissions for microphone, camera, and location access** [2].
- **Device encryption when it is locked**: Phone encryption protects your data when it is locked, making it unreadable to potential thieves [2].
- **End-to-end encrypted data is stored in the cloud** [2].
- **Background modification alert system**: All applications are protected to prevent third-party applications from snooping on other people's data. Users can also be alerted if there is an attempt to tamper with the operating system [2].

IOS also incorporates layers of protection to ensure that apps don't contain known malware or have been modified or tampered with. Strict measures such as application sandboxing are also in place to control user access to data.

This section covers some of the security features used on iOS devices, namely iPhone, iPad, and iPod Touch products:

- **Files quarantine**: Defines malicious software or viruses [3].
- **XProtect or YARA signatures**: For virus and malware identification and endpoint protection [3].
- **Gatekeeper**: Used for endpoint protection. Enforces application code signing to ensure only trusted software runs [3].
- **Application Firewall**: Used for endpoint protection [3].
- **Packet filtering**: Firewall solution [3].
- **Eficheck**: Used to protect end nodes and detect hacker control kits [3].

Overall, smartphone platforms have made great strides in security, but it is still important that users are taking extra steps to protect their data and privacy. Users should understand potential risks and follow good security practices like using strong passwords, installing security updates, and limiting access to personal information.

3 Weaknesses in Mobile Applications

An increase in the number of vulnerabilities has accompanied the popularity of mobile applications. Application flaws can enable malicious hackers to take control of your device, steal sensitive information, and compromise your privacy. Therefore, it is essential to understand mobile application vulnerabilities and how to prevent their occurrence. This section examines the principal vulnerabilities of mobile applications and how they affect privacy and security.

Developers can significantly reduce the risks of mobile vulnerabilities by following the OWASP Mobile Top 10 [4] recommendations and applying rigorous security practices. A lesser-known but more preventable vulnerability in mobile applications concerns deep links. This link takes the user to specific content in the mobile application instead of the home page [5]. However, these links can have potential vulnerabilities, including flaws that attackers could exploit to access sensitive data or control your application.

3.1 How Does Deep Link Work?

To create deep links for Android, it is recommended to use the Android URI system (also known as "Android App Links"), which allows you to link a URL directly to an application installed on the phone [5]. Here is how it works:

1. **Add intent filters to the application's manifest file**: The developer should include intent filters in the application's manifest file, which define the URLs or URIs the application should handle.
2. **Register the application with the OS**: The application must be registered with the Android operating system using the manifest file.
3. **Set up the target URL**: Android deep links also require a URL to which the user will be redirected once the application is opened.
4. **Manage the intent in the application**: When the user clicks on the deep link, the Android operating system intercepts it and passes it to the application. The application can then handle the intent and open the specific page for which the deep link was designed.

It is important to note that for deep links to work correctly on Android, the application must be installed on the user's phone and configured to accept deep links. Deep links on iOS work similarly to Android, but use a different mechanism called "universal links" [6]. Here is how it works:

1. **Add the deep link domains to the application**: The developer must add the deep link domains (e.g. myapp.com) by configuring the permissions property file.
2. **Create a landing page**: The application must have a landing page that opens when the user clicks the deep link.
3. **Set up deep links**: Deep links for iOS also require a specific URL format and a "path" that identifies the destination page of the application.
4. **Manage intent in the application**: Once the user clicks on the deep link, the iOS operating system intercepts the intent and passes it to the application. The application can then handle the intent and open the specific page for which the deep link was designed.

3.2 Risks Associated with Their Use

Regarding the transfer of sensitive data via deep links [5], the associated risk lies in the fact that this information can be transmitted to a mobile application. It is crucial to remember that anyone can create a malicious application that can open the same deep links as the legitimate one.

Furthermore, in the case of unsafe use of parameters passed through deep links [7], the problem is not the data passed through the deep links but rather how the passed parameters are used in the application.

The impact of these vulnerabilities depends on how the parameters are used in the application code. It is, therefore, essential to analyze the use of deep links in the application code to avoid malicious exploitation.

4 Cyber-Attacks on Smartphones

Smartphones are ubiquitous in our daily lives and contain much sensitive personal information. However, their increased use has also led to advanced cyber-attacks targeting these devices. How do hackers exploit smartphone vulnerabilities to gain access to sensitive information? What are the different types of cyber-attacks targeting smartphones? How can users protect their data and devices from these threats? In this section, we will explore these questions and examine the impact of cyber-attacks on smartphones.

4.1 Attacks on Smartphones

Malware can cause various attacks, the most common of which are mobile application permission leakage attacks. These attacks have three types: **Confused Deputy**, **Intent Spoofing**, and **Permission Collusion** [9].

- **Confused Deputy**: A confused deputy attack is a form of mobile application permission leakage attack. It involves tricking an application into malicious actions without the user's knowledge. To do this, malware can take advantage of an application with specific permission and use it to access sensitive data or perform unauthorized actions.
- **Intent Spoofing**: Intent spoofing [10] is a mobile application privilege leakage attack. It involves tricking an application into performing a malicious action by impersonating another application. This attack occurs when third-party applications use intent messages to communicate with other applications without proper authentication.
- **Permission Collusion**: Permission collusion is a mobile application permission leak attack technique that allows malicious applications to collect sensitive data by exploiting the cooperation between the permissions of different applications. This technique is difficult to detect because the user authorizes each application individually, but they can perform malicious actions together.

Mobile Spyware, Mobile Phishing, and Mobile Financial Malware are the most common attacks [9]. These attacks will be presented in this section according to their operating algorithm and their impact on the functioning of the mobile phone, without forgetting the security measures proposed by researchers to mitigate their effects, if they exist.

4.2 Mobile Spyware

Mobile spyware is malware that may be unknowingly installed on a mobile device. Spyware can be downloaded through infected applications, malicious websites, or fraudulent text messages.

Once installed, mobile spyware can collect information about the mobile device, such as contacts, call logs, text messages, emails, GPS location, and even photos and videos stored on the device. Some mobile spyware can activate the device's microphone or camera to record conversations or take pictures without user knowledge.

In addition, spyware can access personal and sensitive information stored on the phone. This includes messages, calls, contacts, photos, files, and passwords. This can be a violation of the user's privacy and a threat to their data. It can also allow attackers to access the phone and remotely control its functionality, compromising the security of the phone. Attackers can use the compromised phone to attack other devices or networks, which can cause significant damage. Furthermore, spyware can spread additional malware to the phone or other devices connected to the same network (Fig. 1).

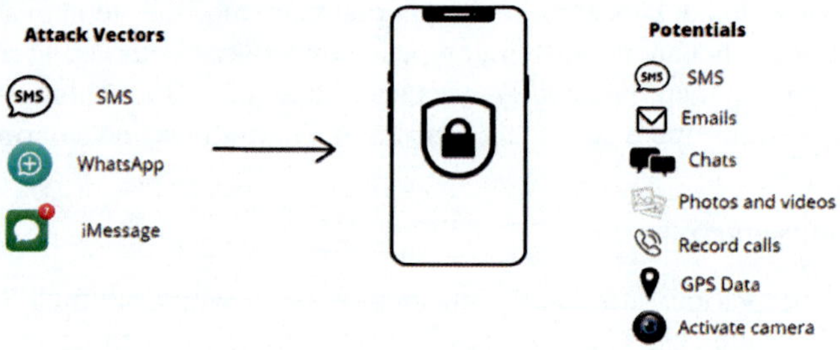

Fig. 1. How does mobile spyware work on smartphones?

4.3 Mobile Phishing

Mobile phishing [8] is an attack method used by criminals to lure mobile phone users into revealing sensitive information such as login credentials, credit card numbers, or passwords. The following is how a mobile phishing attack typically works (Fig. 2); the process of this attack goes through the following steps:

1. Cybercriminals create fake applications or websites that look like trusted ones, such as bank applications, social networks, or popular online stores.
2. Entice users to click on a link that takes them to the fake application or website by sending phishing messages via email, SMS, or instant message.
3. Users are then prompted to enter their login credentials, credit card numbers, or other personal information.
4. Cybercriminals retrieve this information and use it to commit fraud or identity theft.

Mobile phishing can have a significant impact on smartphones, both in terms of security and performance.

- **Data loss**: Mobile phishing attacks can cause loss of sensitive data such as passwords, credit card numbers, bank account information, and other personal information. Cybercriminals can use this information to steal your identity, make fraudulent purchases, or commit various types of fraud [11].
- **Loss of privacy**: Malware can access data stored on your phone, including contacts, messages, emails, and photos. It can also record sensitive information such as keystrokes, passwords, and credit card numbers.
- **System slowdown**: Malware can slow down the operation of your phone by using additional system resources. This can make your phone slower and less responsive, affecting the user experience.

Researchers have proposed solutions to limit the impact of phishing attacks, including content-based filtering, blacklisting, and whitelisting [12]. Content-based filtering involves checking content for suspicious URLs and can be done using rules or statistical differences. Blacklisting explicitly lists sites known to be phishing URLs, while whitelisting allows users to specify which sites they trust so that other sites are scanned. These techniques can help detect smishing, vishing [13], and Wi-Fi phishing attacks.

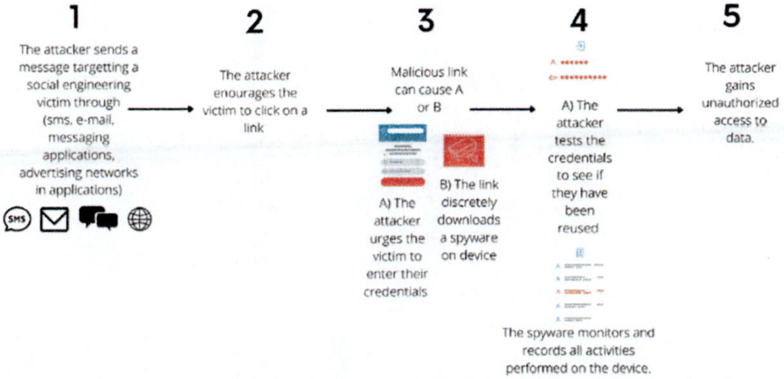

Fig. 2. How mobile phishing is launched on smartphones?

5 Mobile Financial Malware

Mobile financial malware, also known as "banking malware" is a type of malware [9] that targets smartphones and tablets to steal sensitive financial information, such as credit card numbers, login credentials, and banking information.

Mobile financial malware generally works similarly to mobile phishing. Attackers often create malicious apps that mimic legitimate banking or other financial apps, such as mobile payment apps. These applications may be downloaded from unofficial third-party websites or distributed through malicious links via SMS or email.

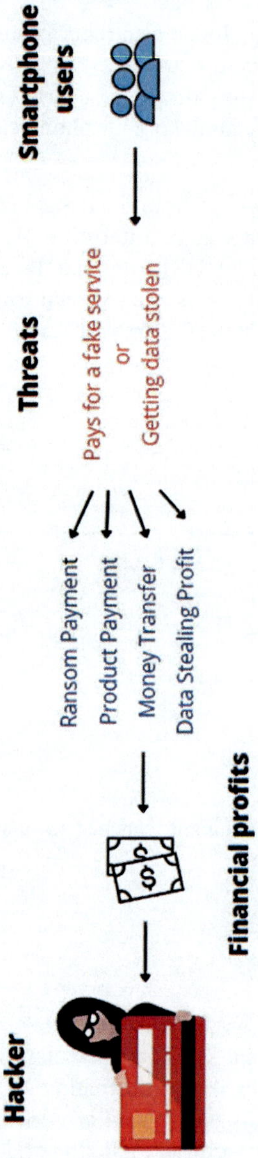

Fig. 3. How does financial malware work on mobiles?

When the user installs the malicious application on the device, the banking malware starts running in the background and attempts to collect sensitive information. The malware can monitor the user's keystrokes, record calls and text messages, access contacts, photos, and files, and even take screenshots.

Financial mobile malware can also use social engineering techniques to trick users into divulging sensitive information. For example, the malware may display fake security notifications or app updates, asking users to enter their passwords or financial information.

Once the sensitive information has been collected, the banking malware can send it to the cyber-criminals' servers, which can then use it to perform fraudulent transactions, online purchases, or bank transfers (Fig. 3).

The previous diagram of the financial malware algorithm describes the various impacts [14] that Financial Malware can have on the user and their mobile. These impacts include the payment of fake products and services, paying a ransom to regain control of the device, stealing online banking login credentials and credit card information, and collecting sensitive information such as banking data or insurance numbers.

Regarding secure attacks on mobile applications, we find brute force attacks [15], which are limited by implementing warning mechanisms in case of password corruption in the application or by incentives to increase password complexity.

We also find spyware attacks limited by a few mechanisms, including device encryption when locked, background modification warning systems, and confidentiality authorization of applications to access the microphone, camera, or location.

6 Methods

In order to study the security level of mobile applications in Moroccan society, it was necessary to elaborate a survey on the security practices of mobile applications among Moroccan users (Fig. 4). This survey is practically dedicated to each Moroccan user with at least one smartphone and of different age groups, levels of education, and professional statuses (stu dents, employees, unemployed…).

Sample size: 80 participants.

Question administration: The questionnaire was posted online and shared on social networks (LinkedIn and WhatsApp).

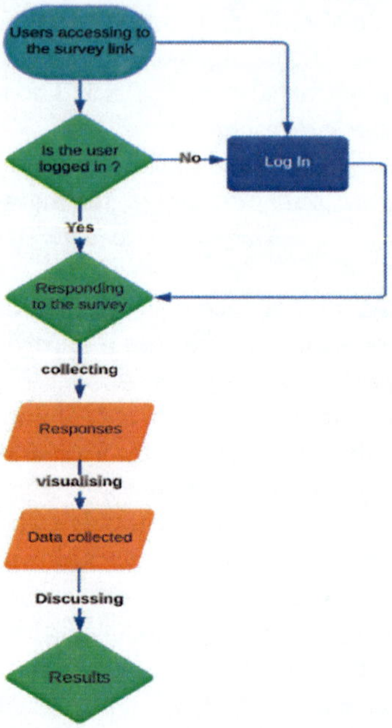

Fig. 4. Methodology used to study the awareness degree among Moroccan users

7 Results and Discussions

The research done in the previous sections shows that the security level of mobile applications is far from being a phenomenon to be taken lightly, as every second smartphone vulnerability multiplies astronomically. However, what is most interesting is that, despite the efforts of cyber security experts, thousands or even millions of users are still victims of cyber-attacks. From this point of view, we would like to conduct a study on Moroccan users' awareness of the dangers associated with using mobile applications. Some hypotheses arise in this context.

We can determine that perhaps it is due to the lack of awareness resources on this subject, the lack of awareness of users, the lack of knowledge of good security practices, the extreme confidence of users in their devices, and the belief that they could not be potential targets, or that the choice of smartphone's technology plays a role in the level of security.

We conducted a field survey to support these hypotheses and understand the causes. A study on mobile application security practices among Moroccan users was shared with 80 people of mixed genders and age ranges.

The survey includes eleven questions and the results were presented as follows:

Regarding the first question: *Have you ever installed an antivirus on your smartphone?*

Of the 80 participants who voted, 53.8% said they had never installed an antivirus on their mobile phone, while 16.3% had installed an antivirus and 30% had never heard of this tool. (see Table 1)

Table 1. Results obtained based on first-question responses

		Frequency	Percent	Valid Percent	Cumulative Percent
Valid	No	43	53,8	53,8	53,8
	No idea	24	30,0	30,0	83,8
	Yes	13	16,3	16,3	100,0
	Total	80	100,0	100,0	

For the second question: *If you received an operating system update message, what do you do?*

Table 2. Results obtained based on second-question responses

	Frequency	Percent	Valid Percent	Cumulative Percent
Valid				
Remind me later	44	55,0	55,0	55,0
Ignore the message	10	12,5	12,5	67,5
Update immediately	26	32,5	32,5	100,0
Total	80	100,0	100,0	

Out of the 80 participants, 32.5% installed the update when they received the message, 12.5% ignored the message completely, and 55% chose the 'Remind me later' option. (see Table 2)

Then, for the third question: *Have you ever received a message that appeared to be from an untrusted source?*

Table 3. Results obtained based on third-question responses

		Frequency	Percent	Valid Percent	Cumulative Percent
Valid	No	26	32,5	32,5	32,5
	Yes	54	67,5	67,5	100,0
	Total	80	100,0	100,0	

Out of 80 participants, 67.5% confirmed receiving fraudulent links through text messages, and emails while 32.5% denied this fact. (see Table 3)

The fourth question was: *Have you ever been a victim of cyber-attack on your smart device?*

Table 4. Results obtained based on fourth-question responses

		Frequency	Percent	Valid Percent	Cumulative Percent
Valid	No	69	86,3	86,3	86,3
	Yes	11	13,8	13,8	100,0
	Total	80	100,0	100,0	

And from the 80 participants, 86.3% have not experienced any attacks on their phones while 13.8% claim to have been victims of cyber-attacks. (see Table 4)

In the fifth question, they were asked to answer the following: *Which operating system is used on your smartphone?*

Table 5. Results obtained based on fifth-question responses

		Frequency	Percent	Valid Percent	Cumulative Percent
Valid	Android	57	71,3	71,3	71,3
	Both	3	3,8	3,8	75,0
	iOS	20	25,0	25,0	100,0
	Total	80	100,0	100,0	

It is concluded that 71.3% use Android smartphones, 25% use iOS phones, and 3.7% use both operating systems. (see Table 5)

Regarding the sixth question: *Do you use the same password for all applications?*

Table 6. Results obtained based on sixth-question responses

		Frequency	Percent	Valid Percent	Cumulative Percent
Valid	No	50	62,5	62,5	62,5
	Yes	30	37,5	37,5	100,0
	Total	80	100,0	100,0	

62.5% answered No and 37.5% with *Yes.* (see Table 6)

The seventh question was: *Have you ever been a victim of money theft through a banking application on your mobile*?

Table 7. Results obtained based on seventh-question responses

		Frequency	Percent	Valid Percent	Cumulative Percent
Valid	No	74	92,5	92,5	92,5
	Yes	6	7,5	7,5	100,0
	Total	80	100,0	100,0	

About the 80 responses collected, 92.5% say they have not experienced money theft through their bank's online application and 7.5% have experienced financial malware attacks. (see Table 7)

The eighth question was: *Do you use public Wi-Fi to get connected to the internet?*

Table 8. Results obtained based on eigth-question responses

		Frequency	Percent	Valid Percent	Cumulative Percent
Valid	Always	15	18,8	18,8	18,8
	Never	12	15,0	15,0	33,8
	Sometimes	53	66,3	66,3	100,0
	Total	80	100,0	100,0	

Of the 80 responses, 66.3% sometimes use public Wi-Fi to connect to the internet, 18.8% always use public Wi-Fi, and 15% never use Wi-Fi in these spaces (see Table 8)

In the ninth question: *Do you leave Bluetooth on without using it?*

Table 9. Results obtained based on ninth-question responses

		Frequency	Percent	Valid Percent	Cumulative Percent
Valid	Always	19	23,8	23,8	23,8
	I deactivate it	61	76,3	76,3	100,0
	Total	80	100,0	100,0	

76.3% turn Bluetooth off when they no longer need to use it, while 23.8% leave Bluetooth on by default on their smartphone. (see Table 9)

Regarding the tenth question: Which authentication mechanism do you use on your smartphone?

Out of 80 responses, 31,3 % use fingerprint as an access control mechanism, 27,5 % use a PIN, 25% use facial recognition, 11,3% use no protection for their smartphone, and the rest (approximately 5.3%) combine more than one security mechanism: Facial

Table 10. Results obtained based on tenth-question responses

	Frequency	Percent	Valid Percent	Cumulative Percent
Valid				
All mechanisms	1	1,3	1,3	1,3
Complex password	1	1,3	1,3	2,5
Face ID	20	25,0	25,0	27,5
Fingerprint	25	31,3	31,3	58,8
Face ID and PIN	1	1,3	1,3	60,0
PIN	22	27,5	27,5	87,5
PIN, Face ID and fingerprint	1	1,3	1,3	88,7
No protection	9	11,3	11,3	100,0
Total	80	100,0	100,0	

recognition and PIN, PIN and facial recognition and fingerprint; some even use mechanisms that were not mentioned among the choices offered such as schematics which can also be considered in the PIN category. (see Table 10)

Finally, for the eleventh question: Are you giving any unnecessary authorizations to mobile applications?

Table 11. Results obtained based on eleventh-question responses

		Frequency	Percent	Valid Percent	Cumulative Percent
Valid					
	Maps	1	1,3	1,3	1,3
	No	64	80,0	80,0	81,3
	Yes	2	2,5	2,5	83,8
	Sometimes	13	16,3	16,3	
	Total	80	100,0	100,0	

Out of 80 responses, 80% never allow applications to use their location data, 3,8% allow access to their location, and 16.3% occasionally allow an application to access their location. (see Table 11).

In the same perspective, Paullet and Al [16] conducted an exploratory study to verify the knowledge and use of security mechanisms to protect smartphones against mobile malware based on an electronic survey of students and alumni of Atlantic Coast University.

In addition, we asked users what operating systems they were using, as well as questions including issues with connecting to open Wi-Fi in public spaces, and knowledge of antivirus software, which was the same for Paullet and Al [16].

However, we have included other issues that the work of Paullet and Al could not address [16] regarding updates installed on smartphones, the likelihood of mobile phishing attacks via SMS or spam as well as the fact of sharing geolocation information with mobile applications installed on the smartphone, without forgetting the problem of using a single password for a number of mobile applications or all mobile applications, which gave an added value to our work.

In terms of the results obtained, our study showed reckless behavior of users in connecting to open Wi-Fi networks, with 66.3% of them indicating that they were able to connect to this type of network occasionally and 15% of them have never been able to connect to public Wi-Fi networks, while for Paullet and Al [16] 73% of users claimed to have used public Wi-Fi networks and 26% for users who rarely or never connect to Wi-Fi hotspots. 76.3% of our users said they did not leave Bluetooth on their phone while for Paullet and Al [16] 69% of users said they did not leave Bluetooth on.

Regarding the use of antivirus software, our results show that 53.8% have never used such solutions and 30% had no idea about this type of security solutions which means a lack of awareness of anti-malware software; while for Paullet and Al [16] 75% have adopted a risky behavior by not using anti-malware software and 9% could not recognize this type of technology.

To conclude, when it comes to security mechanisms, the majority of our surveyed users 89.7% ensure the protection of their smartphones by using a PIN (27,5%), password (1,3%), fingerprint (31,3%), facial recognition (25%), while only 6% do not and other users prefer to combine more than one mechanism. On the other hand, the results provided by Paullet and Al [16] also showed that the majority of participants, 77%, protect their smartphones with a password, PIN, and fingerprint while 23% do not.

8 Conclusion

The small sample size limited this study, as it used only a small sample of 80 individuals. To overcome this limitation, we suggest that future studies be expanded to include individuals from universities and other geographic areas of our country.

Moroccan users are aware of the dangers related to the security of mobile applications as the results of the survey showed that they protect their mobiles with physical security mechanisms to control access to their phones. Apart from that, the participants do still not use some security practices, this is due to their lack of knowledge of these practices or they consider that it might not be necessary for them.Thus, it was necessary to remind the following list of recommendations for the security of mobile applications:

1. Install anti-malware solutions to scan the smartphone in order to detect malicious actions in the background.
2. Install updates on the smartphone to eliminate the risk of being attacked through vulnerabilities detected in mobile operating system versions.
3. Scan and read the messages you receive on your phone before visiting any links.
4. Use strong and different passwords on your applications by combining special characters.
5. Disable Bluetooth when not in use and hide Bluetooth-connected devices on the network.

6. When entering a PIN on your smartphone, try to enter the code discreetly away from the eyes of others.
7. Do not connect to public Wi-Fi to prevent man-in-the-middle and account-hijacking attacks.

References

1. Hewawitharana, C.: Challenges in Mobile OS (Android, iOS, Windows) -A Comparative Analysis (2022)
2. Sécurité et confidentialité sur les appareils mobiles | Centre de sécurité Android Android. https://www.android.com/intl/fr_fr/safety/
3. Aperçu de la sécurité des apps. Apple Support. https://support.apple.com/fr-ca/guide/security/sec35dd877d0/1/web/1
4. OWASP Mobile Top 10 | OWASP Foundation. https://owasp.org/www-project-mobile-top-10/
5. Deep Links: fonctionnement et bonnes pratiques sécurité (2022). https://www.vaadata.com/blog/fr/deep-links-fonctionnement-vulnerabilites-attaques-et-bonnes-pratiques-securite/
6. Universal Links - Apple Developer. https://developer.apple.com/ios/universal-links/
7. Ma, Y., Hu, Z., Liu, Y., Xie, T., Liu, X.: Aladdin: Automating Release of Deep-Link APIs on Android, vol. 1478 (2018). doi:https://doi.org/10.1145/3178876.3186059
8. Goel, D., Jain, A.: Mobile phishing attacks and defence mechanisms: state of art and open research challenges. Comput. Secur. **73** (2017)
9. Muppavaram, K., Rao, M., Rekanar, K., Babu, R.: How Safe Is Your Mobile App? Mobile App Attacks and Defense: ICCII 2017, vol.199–207 (2018). doi:https://doi.org/10.1007/978-981-10-8228-3_19
10. Nirumand, A., Zamani, B., Ladani, B.: VAnDroid: A framework for vulnerability analysis of Android applications using a model-driven reverse engineering technique. Softw. Pract. Exp. **49** (2018)
11. Özdem, A., Gulcu, S.: Causes of Digital Data Loss and Data Recovery Methods(2021)
12. Shahriar, H., Klintic, T., Clincy, V.: Mobile phishing attacks and mitigation techniques. J. Inf. Secur. **06**, 206–212 (2015)
13. Lee, M., Park, E.: Real-time Korean voice phishing detection based on machine learning approaches. J. Ambient Intell. Humaniz. Comput. **14** (2021)
14. Abdul kadir, A.F., Stakhanova, N., Ghorbani, A.: Understanding android financial malware attacks: taxonomy, characterization, and challenges. J. Cyber Secur. Mobil. (2018) doi:https://doi.org/10.13052/2245-1439.732
15. Bošnjak, L., Sres, J., Brumen, B.: Brute-force and dictionary attack on hashed real-world passwords, vol. 1166 (2018). doi:https://doi.org/10.23919/MIPRO.2018.8400211
16. Paullet, K., Pinchot, J.: Mobile malware: Coming to a smartphone near you? **15**, 116–123 (2014)

Author Index

M. Ben Ahmed et al. (Eds.): SCA 2023, LNNS 906, pp. 623–625, 2024.
https://doi.org/10.1007/978-3-031-53824-7